FOURTH EDITION
SATELLITE & TV HANDBOOK

- EVERYTHING YOU SHOULD KNOW ABOUT DISHES
- INSTALLATION ADVICE
- COVERAGE MAPS OF MAJOR SATELLITES
- WORLDWIDE SATELLITE AND TV SURVEY
- NAMES AND ADDRESSES OF ALL MAJOR BROADCASTERS

BILLBOARD BOOKS
an imprint of Watson-Guptill Publications
New York / Amsterdam

Editor-in-Chief: Andrew G. Sennitt
Managing Editor: Bart Kuperus
Production: KSC Holland
Publisher: Glenn Heffernan

EditorialOffice:
P.O. Box 9027
1006 AA Amsterdam
The Netherlands
Fax: +31 (20) 4875120

ISBN: 0-8230-7658-X

Copyright © 1997 by Billboard Books,
an imprint of Watson-Guptill Publications,
a division of BPI Communications, Inc.
1515 Broadway, New York 10036

All rights reserved. No part of this publication may be reproduced,
stored in a retrieval system, or transmitted, in any form or by any
means - electronic, mechanical, photocopying, recording, or
otherwise - without written permission of the publisher.

Manufactured in the United States of America.

MESSAGE FROM THE AUTHOR

Welcome to the Satellite and TV Handbook. The new title was chosen because this fourth edition not only covers satellite radio and TV, but also terrestrial TV channels and transmitter information. In a whole new concept, the Satellite and TV Handbook provides an abundance of information, covering technical details, company details, web sites, complete TV broadcasting information per country and the latest about new digital DBS services around the world.

Satellite radio and TV is just like any conventional terrestrial radio and TV, except that the transmitters are not located here on earth but in space. This book is about how to receive those programs broadcast both via satellite as well as terrestrial transmitters. From the moment that satellite broadcasting became commonplace not only in the Developed World but also in Third World countries, the WRTH editorial team felt that the time was right for a special publication. Despite the fact the number of satellites around the world used for broadcasting purposes has increased to over 200 and with more and more broadcasters switching to satellite, terrestrial transmitters are still widely used. The Satellite and TV Handbook is the only publication that lists both categories.

The Satellite and TV Handbook will be of great use to anybody who plans to buy and install a satellite system, or to anybody who wants to receive more TV channels outside the paved paths of the cable companies. Our main goal is to make complicated technical matter clear and understandable, in such a way that it can be used in practice. The many detailed maps, charts and drawings as well as the extensive programming details make this book an indispensable tool!

good reading,

Bart Kuperus

CONTENTS:

CHAPTER ONE
INTRODUCTION

1.1. Basic Satellite Information 9
1.2. Orbits and Locations 10
1.3. A Variety of Tasks 12
1.4. Some History 14

CHAPTER TWO
TRANSMISSION TECHNIQUES

2.1. Microwaves 15
2.2. Polarization 16

CHAPTER THREE
RECEIVING SATELLITE SIGNALS

3.1. Straight Talking about Satellite Antennas 17
3.2. Straight Talking about Satellite Receivers 25
3.3. How to buy a Satellite TV System 28
3.4. Straight Talking about Co-ax 28

CHAPTER FOUR
BROADCASTING SYSTEMS

4.1. Worldwide Broadcasting Standards 31
   TABLE OF WORLD TV STANDARDS 31
4.2. What is MAC 36
4.3. Global TV Standard vs. Multi-standard Madness 36
4.4. US DBS: Questions and Answers 37

CHAPTER FIVE
INSTALLATION

5.1. Installing the Outdoor Unit 47
5.2. Installing the Indoor Unit 59

CHAPTER SIX
IMPORTANT TABLES

6.1. Various Angles 62
6.2. EIRP/Dish Diameter 65
6.3. American C- and Ku-Band Frequencies 65
6.4. Characteristics of Television Systems and Channels 66

CHAPTER SEVEN
ORGANIZATIONS

7.1. International Telecommunication Union (ITU) 67
7.2. Eutelsat 68

7.3. Intelsat .... .69
7.4. Société Européenne des Satellites (ASTRA) .... .71
7.5. France Télécom .... .73
7.6. Intersputnik .... .74
7.7. NASDA .... .74
7.8. AsiaSat .... .75
7.9. PanAmSat .... .75

CHAPTER EIGHT
SATELLITE COVERAGE ZONES .... .77

CHAPTER NINE
ITU REGION 1 TRANSPONDER LOADING

9.1. Stations in Frequency Order .... .160
9.2. Digital Stations in Frequency Order .... .181

CHAPTER TEN
ITU REGION 2 TRANSPONDER LOADING

10.1. Stations in Frequency Order .... .194
10.2. Digital Stations in Frequency Order .... .213

CHAPTER ELEVEN
ITU REGION 3 TRANSPONDER LOADING .... .219

CHAPTER TWELVE
STATIONS IN ALPHABETICAL ORDER

12.1. ITU Region 1 TV .... .231
12.2. ITU Region 1 RADIO .... .239
12.3. ITU Region 2 TV .... .245
12.4. ITU Region 2 RADIO .... .253
12.5. ITU Region 3 TV .... .256
12.6. ITU Region 3 RADIO .... .262

CHAPTER THIRTEEN
NAMES AND ADDRESSES

13.1. ITU Region 1 .... .263
13.2. ITU Region 2 .... .269
13.3. ITU Region 3 .... .273

CHAPTER FOURTEEN

Useful World Wide Web Addresses .... .275
US TVRO/DSS Dealer Links .... .282

CHAPTER FIFTEEN
WORLD TELEVISION

Europe .... .283
Africa .... .312
Near & Middle East .... .320
Asia .... .326
Pacific .... .336
North America .... .341
Central America & The Caribbean .... .344
South America .... .350

CHAPTER SIXTEEN
WORLD TV BY COUNTRY INDEX 359

CHAPTER SEVENTEEN
GLOSSARY 361

CHAPTER EIGHTEEN

World Time Table 364
Where to obtain the Satellite and TV Handbook 367

CREDITS 368

CHAPTER 1
INTRODUCTION

Satellites in orbit around the earth

1.1. BASIC SATELLITE INFORMATION

As with any specialist subject, there is a lot of satellite-related jargon. The most commonly used terms are printed in italics, but don't let them put you off. If the meaning of a word is still not clear it can also be looked up in the glossary at the back.

According to the Oxford Advanced Learner's Dictionary of Current English the primary meaning of the word satellite is: "A comparatively small body moving in orbit round a planet". The largest satellite orbiting the earth is the moon, which permanently revolves thanks to an interaction of both gravity and centrifugal forces. The secondary meaning - which gradually has become the primary one - is: "Artificial Object, e.g. a spacecraft put in orbit round a celestial body". In this book we discuss these man-made satellites orbiting the "celestial body" we live on, relaying tv and radio signals.

Almost 50 years ago the scientist and science-fiction author Arthur C. Clarke described in an article how satellites can be used for telecommunication and broadcasting. He calculated that if a spacecraft was placed at an altitude of 35,800 km. (22,245 miles) above the *Equator* its orbital velocity (speed) would match the earth's rotation speed. With a revolution time of exactly 24 hours such a spacecraft would appear stationary above the earth. Therefore the narrow plane in which the spacecraft orbits is called a *Geostationary or Geosynchronous Orbit or Arc*. Most people, however, prefer to talk about *Clarke Belt*, named after its discoverer.

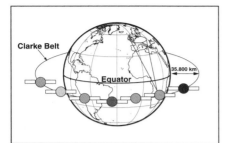

Fig. 1.1-1: Satellites orbit in a narrow plane called the Clarke Belt above the Equator like beads on a string.

Geostationary satellites don't stay in place by themselves, but they have to be kept in place by means of small jet motors. The process of keeping a satellite not only in its orbital slot somewhere above the Equator, but also facing the right way is called *stationkeeping*. This stationkeeping is done from large control centers here on earth and is highly automated. However, manual corrections to the satellite's altitude, attitude and position still have to be carried out regularly.

Modern satellites have a limited lifespan of approximately 10-12 years, which is still considerably longer than older generations of satellites. Solar panels provide the necessary energy to charge the batteries, but the small positioner rockets will eventually run out of fuel. Sometimes the ground station decides to reduce the stationkeeping of older satellites in order to save fuel. As a result the satellite starts to wobble. Also, the plane of the satellite's orbit can deviate from the ideal Equatorial plane, in which case we say that the satellite has gone into Inclined Orbit.(see chapter 1.2. Orbits and Locations)

Before a satellite can relay any signals, the original signal has to be beamed up first. This earth-to-space transmission is called *Uplink*, and can be done from large fixed uplink centers as well as small mobile (even portable) uplink units. Logically, the space-to-earth transmission is referred to as *Downlink*. This downlink beam can be aimed at a certain area, e.g. Europe or the USA, in such a way that it only illuminates the desired area. The coverage area of a satellite beam is called *Footprint* and one satellite can cover various areas from the same orbital location.

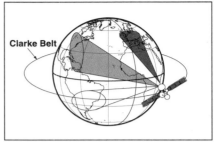

Fig. 1.1-2: Various beams from one satellite. Global Beams cover the entire visible part of the earth. Hemispherical Beams illuminate a certain hemisphere. Zone Beams cover a quarter of the earth's visible service. Spot Beams generally illuminate small areas.

Footprint areas can be large or small, depending on the direction and the shape of the downlink beam. Another frequently used word is *Transponder*. In fact this word was put together out of two words: TRANS(mitter) and (res)PONDER. In general a transponder is considered to be all equipment aboard a satellite necessary to receive and transmit one channel. Nowadays satellites are equipped with up to 20 or 30 transponders, and therefore are capable of receiving and transmitting a large number of channels simultaneously. This capacity, i.e. all transponders together is called the satellite's *Communications Payload*.

1.2. ORBITS AND LOCATIONS

As mentioned earlier, satellites are held in place by two forces; *gravity* and *centrifugal force*.

When a rocket is launched it must have the right speed to escape the earth's gravitational pull. If it goes too slow it will plummet back, if it goes too fast it will escape the earth's sphere of influence and fly towards the sun. Man-made satellites can be placed in all kinds of orbits as they revolve round the earth; it all depends on what speed they were given. All orbits are either *circular* or *elliptical* in shape. The plane of these orbits can be *equatorial, polar* or *inclined*, depending on the specific task of a satellite.

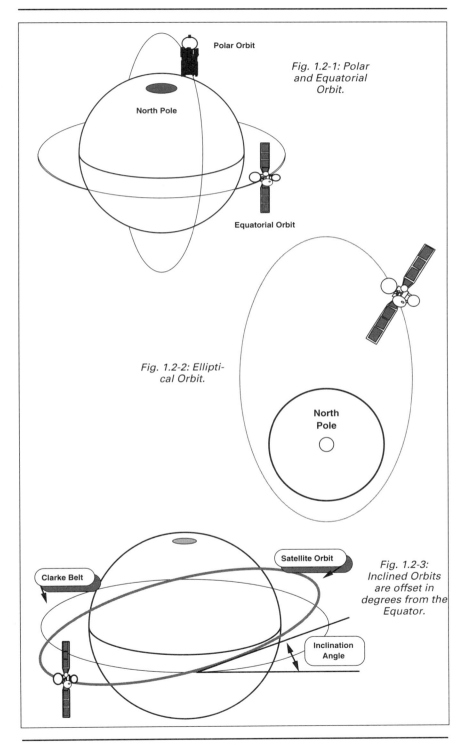

Fig. 1.2-1: Polar and Equatorial Orbit.

Fig. 1.2-2: Elliptical Orbit.

Fig. 1.2-3: Inclined Orbits are offset in degrees from the Equator.

Most geostationary communications satellites are first launched into an elliptical orbit, the *Geostationary Transfer Orbit (GTO)*. This elliptical orbit serves as a first stage in deploying the satellite to a slot in the Clarke Belt. In order to propel the satellite to this position a small rocket motor is fired when the spacecraft has reached a point furthest away from the earth. This point is called the Apogee, and therefore the small jet motor is referred to as *Apogee Kick Motor (AKM)*.

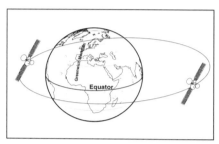

Fig. 1.2-4: The point at which the satellite is furthest away from the earth is referred to as Apogee, whereas the point closest to the earth is called Perigee.

Almost all communications and broadcasting satellites (except for some Russian satellites) are geostationary, i.e. they are "parked" at a certain place in the Clarke Belt. This place is called a *Slot* and all slot coordinates refer to geographical grid coordinates.

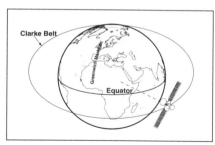

Fig. 1.2-5: The satellite's orbital coordinates refer to geographical grid coordinates.

Sometimes more than one satellite can share a single slot, e.g. the four European ASTRA satellites which are all located at 19.2° East. All ASTRA satellites are *co-located* and this can be best explained if you think of a slot as being a big cube of ± 50 cubic miles in which these satellites are kept. Because these satellites appear to be at the same spot in the sky, viewers can tune in to all available channels without having to move their antenna. There seems to be no limit to the number of satellites that can be kept in one single slot and Société Européenne des Satellites (SES) have already put a seventh ASTRA satellite at the same orbital location.

The seven European ASTRA satellites are co-located at 19.2° East (courtesy SES)

1.3. A VARIETY OF TASKS

Satellites have become tremendously important in our daily life and they serve an array of tasks. During the Gulf War mobile satellite uplink units provided live pictures of what was transpiring in Baghdad and Kuwait. These so called Satellite News Gathering units - or SNG units for short - can be operated from almost any site location. For a number of years, SNG has proved to be an indispensable means of delivering live pictures from a news story location directly to the broadcaster. The Gulf war provided operators with a chance to show the world's broadcasters the benefit of fully transportable, rapid-deployment links facilities. London based SIS-Link, a division of Satellite Information service (SIS), provided three flyaway terminals in Dhahran, Amman and Dubai, two of these units later converging on Kuwait City with the liberating forces. The ability to transmit live pic-

Mobile Satellite News Gathering Unit or SNG Unit.

tures and edited packages from any location, combined with the rapid setup, brought pictures of this conflict direct to the world's television screens in a way which was previously impossible. All around the world, major broadcasting companies operate one or more SNG units such as the Dutch company INTRAX.

Another major task of satellites is of course communications. Some satellites such as the **Inmarsat** satellite are solely equipped to establish direct telephone links with all parts of the world. Both reporters and soldiers made extensive use of the Inmarsat satellite for direct telephone communication links. Small mobile satellite telephone uplink units like the one on the picture were widely operated during the conflict.

At the same time American military "spy satellites" monitored every movement of the Iraqi army and weather satellites provided essential data on cloud

cover above a specific target area. The American navy also used special navigation satellites to determine their exact location.

TWO MAJOR CATEGORIES
In principle, satellites can be sorted into two categories: *Utility satellites* and *Scientific satellites* and they are either stationary or non-stationary. All satellites serve specific purposes or a combination of purposes:
❏ Communication (telephone, fax etc.)
❏ News & Information distribution
❏ Television & Radio transmissions
❏ Navigation
❏ Weather monitoring
❏ Scientific & experimental purposes
❏ Amateur radio

In general, most communications satellites such as Intelsat and Eutelsat also relay radio and tv signals, but there are also special broadcasting satellites. *Direct Broadcasting Satellites (DBS)*, such as the German or TV-Sat, broadcast only four or five tv programs, but with such high power that a 30 cm (i.e. 12 inch) dish is sufficient for reception. Other broadcasting satellites like the European ASTRA and Kopernikus were initially designed for communications purposes, but transformed into de-facto broadcasting satellites later on. These satellites are not as powerful as DBS satellites, but their advantage is that they can broadcast up to 16 tv channels simultaneously. On the other hand, scientific satellites serve as space and earth explorers and are not intended to generate any profit like utility satellites. A famous scientific satellite was the **Giotto** which made invaluable pictures of distant planets and later on had a close encounter with former Halley's Comet, Grigg Skjellerup. Weather satellites are an exception because they are not grouped under scientific satellites but under utility satellites.

1.4. SOME HISTORY

Since the launch of the first artificial satellite, the Russian Sputnik on October 4th 1957, numerous other satellites have been put into orbit. **Early Bird**, the first Intelsat communications satellite, was launched on April 6th 1965. This "bird" was capable of transmitting 240 telephone calls and 1 tv program simultaneously and huge dishes were needed for reception. About the same time the former USSR launched their first TV-satellite called **Molnija**. This Molnija spacecraft was non-geostationary because of its northerly elliptical orbit. This way it could cover areas in the far north of Russia. Nowadays the Russians still have non-geostationary satellites of the Molnija type, but they also operate geostationary satellites called **Gorizont**.

The success of Intelsat's Early Bird was enough reason for the Americans to further develop this system. On October 27th 1966 a second Intelsat was launched under a new name: Intelsat II f-1 (type: Intelsat II, (f)light-1). The first Intelsat III, which was capable of relaying 1500 telephone calls and 1 tv program, was launched in 1968. With another three launches in 1969 Intelsat positioned satellites above the Atlantic Ocean, Indian Ocean and Pacific Ocean. These 3 satellites were the first to provide a global telecommunication and tv network.

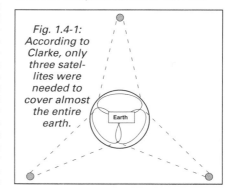

Fig. 1.4-1: According to Clarke, only three satellites were needed to cover almost the entire earth.

CHAPTER 2
TRANSMISSION TECHNIQUES

2.1. MICROWAVES

For technical reasons Television programs are broadcast on high frequencies, i.e. by means of electromagnetic waves with a relatively short wavelength. In many parts of the world, terrestrial tv is already making extensive use of the Very High Frequency (VHF) and Ultra High Frequency (UHF) bands. The problem of overcrowding is the main reason why satellites were seen as the ideal solution. Satellites offer the opportunity to transmit higher frequencies that - because of their propagation characteristics - could thus far not be used by terrestrial relay stations.

As explained earlier, the signal we send to the satellite to be relayed is called uplink, the signal that is transmitted from the satellite down to earth is called downlink. For satellite transmissions downlink frequencies are used in the C- and Ku-Band region. In general *C-Band* (or 4 GHz band) frequencies cover a range from 3700-4200 MHz (or 3.7-4.2 GHz). *Ku-Band* frequencies occupy a range that stretches from 10-17 GHz. However, in practice only the frequencies between 10.7-12.75 GHz are used for broadcasting purposes.

The major advantages of using higher frequencies compared with the regular UHF wavelength are:
1) More frequency space or *bandwidth* is available. Bandwidth is the total range of frequencies over which the signal may vary in the course of its transmission. The wider this range the more information it can store, hence better video and audio quality. (e.g. 1 MHz bandwidth in a 10 MHz spectrum takes proportionately more space than 1 MHz bandwidth in a 1000 MHz = 1 GHz spectrum)
2) Microwaves behave more similarly to light waves than lower frequencies. It is this characteristic which makes it easier to pinpoint a satellite or a certain area on the ground. It is scientifically proven that electromagnetic waves are better concentrated and reflected by an antenna that is considerably larger than the wavelength it directs. Especially for uplinking signals to a satellite without interfering with another one, higher frequencies are used.

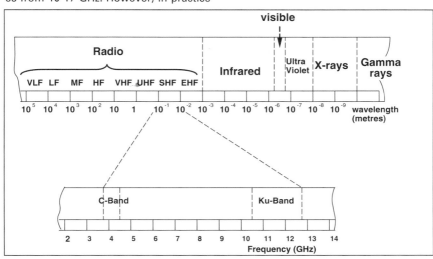

Fig. 2.1-1: The Electromagneticpectrum

3) Microwaves are capable of penetrating the ionosphere without being reflected. The ionosphere is influenced by solar activity and therefore it is sometimes more and sometimes less permeable to electromagnetic signals. In general, signals on a frequency lower than 30 MHz are reflected by this layer.
4) Microwave frequencies are less susceptible to atmospheric distortion. We all know that from time to time radio and tv signals are impaired and distorted because of certain weather conditions or the effects of solar activity. Many international broadcasters even avoid using certain frequencies over a period of time because of their specific susceptibility to atmospheric interference and solar activity. Of course satellite signals are also less affected because of the relatively short distance they have to travel trough the earth's atmosphere.

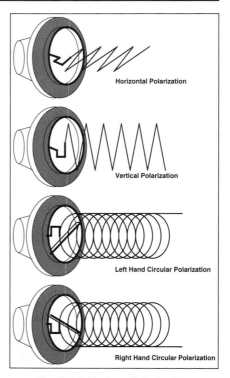

FREQUENCY TABLE

P-Band	200-400 MHz
L-Band	1530-2700 MHz
S-Band	2700-3500 MHz
C-Band	3700-4200 MHz
	4400-4700 MHz
	5725-6425 MHz
X-Band	7900-8400 MHz
Ku1-Band (FSS-Band)	10.7-11.75 GHz
Ku2-Band (DBS-Band)	11.75-12.5 GHz
Ku3-Band (Télécom)	12.5-12.75 GHz
Ka-Band	17.7-21.2 GHz
K-Band	27.5-31.0 GHz

Fig. 2.2-1: Various Polarizations

2.2. POLARIZATION

The propagation of all electromagnetic waves is a combination of an electric and a magnetic field. If the vibration of the electrical field of a signal radiated from a transmitting antenna is parallel to the ground it is called *Horizontal Polarization*. If the electrical field vibrates perpendicular to the ground it is defined as *Vertical Polarization*. Vertically and horizontally polarized signals and signals vibrating in any plane in between are called *Linear Polarization*. When the electrical field of a signal is not vibrating but rotating we call this *Circular Polarization*. Circularly polarized signals can be either *Left* or *Right Handed*.

The reason for using either V/H polarization or RH/LH circular polarization on the same satellite is to create more usable frequency space. If the polarization of adjacent transponder channels is alternately horizontal and vertical, they can overlap halfway without causing any interference. The same goes for circularly polarized channels. This way more channels can be squeezed into the available space, e.g. DBS channels of 27 MHz bandwidth can be placed only 18 MHz apart.

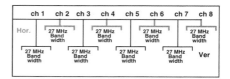

Fig. 2.2-2: The reason for using linear or circular polarizations is known as frequency re-use. This way almost twice the number of channels can be transmitted within the same bandwidth.

CHAPTER 3
RECEIVING SATELLITE SIGNALS

3.1. STRAIGHT TALKING ABOUT SATELLITE ANTENNAS

Satellite antennas come in many varieties and designs. They are either made of metal or fiberglass or a combination of the two. The most common type of satellite antenna is the parabolic reflector antenna. The reflecting surface has to be smooth and the shape kept to a close tolerance in order to achieve a high gain at Ku-Band frequencies. Most antennas are circular and have a parabolic dish-shape, but they can also be rectangular, horn shaped or even square and totally flat. Depending on their specific function (uplinking, downlinking or receiving signals) microwave antennas can be as large as 30 meters or as small as 30 cm. But whatever their shape or size, all microwave reflector antennas have one thing in common; they reflect and concentrate (magnify) signals to a focal point or reflect a transmitted signal from a focal point.

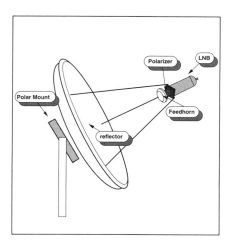

Fig. 3.1-1: This is the most commonly used satellite dish configuration

THE ANTENNA AND ITS COMPONENTS
1) REFLECTOR: Because of its shape this is the most characteristic part of the outdoor unit.
2) FEEDHORN/WAVE GUIDE: This is the front part or receptive part of the small unit in front of a satellite dish.
3) POLARIZER: Electromagnetic waves can be either linear (H/V) or circular (LHCP/RHCP) polarized. Technically, a polarizer is a device positioned in the wave guide in between the feedhorn and the LNB which converts between linear and circular polarizations. Commonly, however, a polarizer is considered to be the part in a satellite's wave guide system which allows receivers to pick up only one polarity. Usually the polarizer consists of a probe, activated by a little servo motor. In this way we can mechanically rotate the probe to lie in the plane of the required signal *(skew)*. Another method widely used is the magnetic or ferrite polarizer. This polarizer has no moving parts but uses a magnetic field around a coil to influence the magnetic plane of an incoming electromagnetic signal. Now the signal is twisted to lie in the probe's plane. Satellite Master Antenna TV systems *(SMATV)* use a different method ie. the Orthogonal-Mode Transducer *(OMT)* or Ortho-Coupler. With this method a polarizer is not needed because two LNBs receive both polarizations simultaneously.
4) LNB/LNA/LNC: These are 3 names for the same device. Low-Noise Block *(LNB)*, Low-Noise Amplifier *(LNA)* or Low-Noise Converter *(LNC)*. Generally we talk about LNB. This Block consists of an integrated amplifier and converter. The amplified satellite signal has to be downconverted to a lower frequency range that is more easily transported by means of a regular coaxial cable without too much signal loss. Usually Ku-Band signals are downconverted to a frequency between 900 and 2000 MHz. The performance of an LNB is based upon 2 factors. Firstly, its

capability to amplify the incoming satellite signal and secondly the amount of noise the LNB itself adds to that signal. The noise temperature is measured in decibels (dB). Thanks to the development of the High Electron Mobility Transistor *(HEMT)* modern Ku-Band LNBs have a noise figure of between 1.6 and 0.6 dB. C-Band LNBs generally have even lower noise figures.

5)MOUNT: There are various ways of mounting a satellite dish. Fixed dishes usually are equipped with a ridged mount which allows you to aim at one specific satellite. Motorized dish installations have a pivoting construction to allow multi satellite reception.

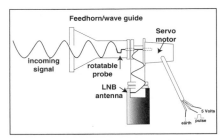

Fig. 3.1-2: This is an example of a mechanical polarizer. The rotatable probe can be adjusted to lie in the plane of the incoming signal

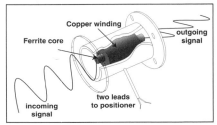

Fig. 3.1-3: This is an example of a magnetic polarizer. An electric current in the copper winding influences the magnetic plane of the incoming signal

The Orthogonal-Mode Transducer (OMT) or Ortho Coupler

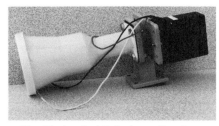

The Mechanical Polarizer + Feedhorn

The Magnetic or Ferrite Polarizer

ANTENNA PERFORMANCE
The quality of a satellite antenna is determined by a large number of interrelated factors:

1) ANTENNA GAIN: This is the overall capability of an antenna to reflect and concentrate satellite signals. The gain of a reflector antenna is inherent in its size and surface accuracy. In general one could say that the bigger the reflective area *(aperture)* the higher the gain. However, the antenna gain is also proportional to the wavelength of the incoming signal. The higher the frequency, the higher the gain. Surface inaccuracies, though, are better tolerated when receiving C-Band frequencies than when receiving the much higher Ku-Band frequencies.

2) ANTENNA EFFICIENCY: In practice the antenna efficiency is a combination of antenna gain and the performance of the feedhorn unit. Even if two antennas have identical reflectors there can be a difference in the actual percentage of signal intercepted by the feedhorn. If, for instance, the feedhorn is positioned outside the antenna's focal point, or if the

feedhorn was designed for another type of antenna with a different Focal length-to-antenna Diameter Ratio *(F/D Ratio)* this would certainly detract from the total efficiency. (see Fig. 3.1-4)

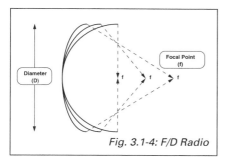

Fig. 3.1-4: F/D Radio

3) BEAMWIDTH & SIDE LOBE PATTERN: The beamwidth or acceptance angle of a reflector is measured in degrees. This is done at the point where the main lobe signal is half its power (-3dB). The larger the reflector, the narrower the beamwidth. Also, deeper dishes usually have a smaller beamwidth angle than a more shallow configuration (e.g. offset fed antennas). With a lot of satellites spaced closely together it becomes more and more important to have an antenna with a narrow beamwidth. An ideal antenna should amplify signals coming directly from the center of the reflector *(main lobe)* while attenuating or ignoring signals coming from an off-axis angle *(side lobes)*. The latter is extremely difficult and therefore the engineers strive for a reflector design that produces a very

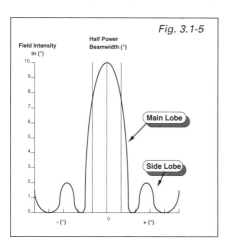

Fig. 3.1-5

powerful and narrow main lobe. (see Fig. 3.1-5)

4) ANTENNA NOISE: Although antenna gain is one of the most important factors a high noise temperature could spoil the total performance. The antenna noise temperature is a measure of how much noise the reflector detects from the surrounding environment.

The Gain/(noise)Temperature Ratio *(G/T Ratio)* is also a combination of factors:

a) Side lobe attenuation characteristics: lower side lobe levels mean less cosmic noise (noise from outer space) will be detected.

b) Elevation angle: a lower elevation angle of the antenna means more ground noise will be detected.

c) F/D Ratio: The farther away the focal point, i.e. a higher F/D Ratio (e.g. shallow dishes), also means more ground noise detection.

d) Other sources: e.g. fluorescent lights, high voltage power cables, terrestrial radio and television transmitters, are also noise generators.

A good quality antenna should collect and concentrate signals coming from any chosen satellite and reject cosmic noise and signals from any adjacent satellite. To accomplish this the beamwidth or acceptance angle of the antenna should be as narrow as possible. A larger antenna not only performs better (higher antenna gain) but generally has better beamwidth characteristics than a smaller one. Also, a center focus antenna usually has a smaller beamwidth angle than a similarly sized offset fed antenna. The size of a satellite antenna is absolutely crucial for good video and audio quality - the bigger the better. The LNB in front of the satellite dish not only detects satellite signals, but also cosmic and ground noise. The LNB amplifies both satellite signals and noise. The intensity of cosmic noise is so high that it totally obliterates unconcentrated satellite signals. However, unlike cosmic noise, satellite signals are highly directional and parallel beams which can easily be reflected and concentrated by the antenna's surface. The larger the dish the more it magnifies the satellite signal. Therefore we say that a larger dish has a better overall Carrier to Noise Ratio *(C/N Ratio)* than a smaller type.

Thus, you cannot compensate for the disadvantages of a small dish by using a better LNB because:

❑ A small dish reflects less signals to the LNB.
❑ A small dish is far more susceptible to the effects of rain (signal attenuation).
❑ A small dish has a less hyperbolic shape, hence a wider beamwidth.
❑ A small dish has a higher noise temperature.

CONVENTIONAL CONFIGURATIONS
Basically, parabolic satellite antennas are grouped in 2 categories: *Center Focus antennas* and *Offset Fed antennas*.

1) CENTER FOCUS ANTENNA
By far the most commonly used antenna configuration is the Center Focus Para-

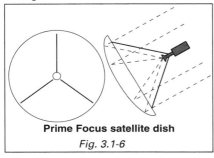

Prime Focus satellite dish
Fig. 3.1-6

bolic Reflector, also known as *Prime Focus Antenna*. A *parabolic* reflector provides the best compromise between high effectiveness (antenna gain) and low production costs. The Center Focus antenna always has a perfectly round circumference, but the parabola can be either deep or shallow.
PROS: 1) very small beamwidth angle 2) simple construction 3) high dish surface accuracy (mainly because of simple fabrication process) 4) low production costs 5) feedhorn facing downwards, meaning no water ingress.
CONS: 1) masking of the dish surface by the LNB (aperture blockage) 2) dish surface collects lot of dirt 3) poor side lobe attenuation characteristics.

2) OFFSET FED ANTENNA
Whereas the center focus antenna is always round, an offset fed antenna is usually slightly egg shaped. The offset configuration is particularly useful in smaller units because the LNB doesn't project a shadow onto the main reflector

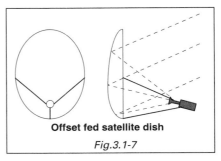

Offset fed satellite dish
Fig.3.1-7

(aperture blockage). This is a major advantage over a center focus antenna. It is not so hard to understand the offset principle if you imagine the offset reflector as being a cut out section of a much larger center focus reflector.
However, the feedhorn is still positioned at the focal point of this larger dish and therefore it looks offset. This also explains the almost upright position of the antenna when it receives signals coming in from a ± 30° elevation angle. The offset fed antenna even appears to be looking downwards when aiming at satellites barely visible over the horizon (e.g. satellites with an extreme eastern or western azimuth). It is difficult to calculate the gain of an offset fed antenna accurately because in order to do so we need to know the total effective area (i.e. the larger center focus "parent" reflector) facing directly towards the satellite.

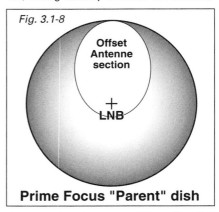

Prime Focus "Parent" dish

Another big advantage of the offset fed antenna lies in the position and construction of the feedhorn/LNB assembly. The feedhorn in a typical center focus dish is facing the warm earth and thus detects a considerable amount of ground noise. The feedhorn in the offset fed configura-

tion faces upwards aiming towards outer space. Hence, noise figures are generally lower for offset fed antennas than for similarly sized center focus versions.
PROS: 1) no aperture blockage, hence high dish area efficiency (the reflective surface area doesn't suffer from any masking by the LNB) 2) very good side lobe attenuation characteristics 3) better noise temperatures than center focus antennas 4) no water, snow, ice or dirt problems on the dish's surface 5) very easy to put up (upright position, flat to the wall)
CONS: 1) compared with a similarly sized center focus dish the offset version has a larger beamwidth angle 2) less effective intercept area due to a 30° angle to the satellite's line of sight 3) because of the upright position this type of antenna is more susceptible to high winds 4) feedhorn facing slightly upwards and therefore it could suffer a degree of signal attenuation due to the adherence of water droplets to the feedhorn cap. (also more susceptible to water ingress.) 5) dish surface inaccuracies have more negative effects on the overall performance compared with center focus antennas.

OTHER CONFIGURATIONS:
A) GREATER WIDTH THAN HEIGHT OFFSET ANTENNA.
This type of offset fed antenna was specially designed to compensate for the typical offset dish disadvantage of a wider beamwidth.

Currently the Clarke Belt is pretty congested at some places and some smaller dishes have suffered interference problems caused by adjacent satellites already. Therefore, in the near future when there will be even more Ku-Band satellites spaced closely together (± 3°), this type of antenna could solve potential

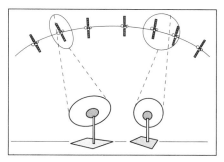

Fig. 3.1-9: Greater Width than Heigth Offset Dish. The larger the Dish the narrower the Beamwidth.

interference problems. Fig. 3.1-9 illustrates the view at the Clarke Belt by a regular offset fed antenna (right) and the compressed beamwidth of a greater width than height offset fed antenna (left). In fact, the principle is very simple; the larger the dish the narrower the beamwidth. When we give the dish a greater width than height we thus artificially create a narrower beamwidth at the horizontal plane.

B) DUAL REFLECTOR ANTENNAS
1) Cassegrain = Dual reflector antenna using a convex hyperbolic subreflector and a parabolic main reflector
2) Gregory = Dual reflector antenna using a concave hyperbolic subreflector and a parabolic main reflector

These types of antennas are named after two scientists in optics who lived way before there were any artificial satellites. As you might have noticed, the jargon used in this article contains a lot of optics related phraseology. The Cassegrain and Gregory antenna are also often referred to as Back-Fire antennas and they can be either center focus or offset fed. This back-fire technique in a center focus configuration is very useful in tropical areas because the the LNB is positioned behind the main reflector and thus stays relatively cool. Because of their complexity these antennas should meet certain very accurately determined parameters in order to perform satisfactorily. If the subreflector is too small (less than approx. 5 x the wavelength) a lot of signal will overflow. If it is too big it will cast a bigger shadow onto the main reflector and thus reduce the antenna gain (this only refers to the center focus version). If the feedhorn in an offset fed

A modern offset fed Gregory antenna (courtesy DeltaStar)

PROS: 1) very high surface efficiency 2) no direct effects of solar heat on the LNB. CONS: 1) center focus version below 2 meters suffers from disproportionate masking of the main reflector by the sub-reflector. 2) limited applications 3) complex geometry and construction 4) very costly 5) high noise temperature (because the subreflector also amplifies side lobes and noise)

C) MULTI FOCUS ANTENNA

configuration is too big or placed too closely to the subreflector it will opaque part of it. In general dual reflector antennas are mainly used in larger systems, although there are some smaller offset fed Ku-Band Gregory and Cassegrain antennas available in Europe.

Also referred to as Spherical Antenna or Multi-Beam Torus Antenna (so called because of the toroidal reflector).

These types of antennas are capable of reflecting signals from a cluster of satellites with different orbital positions to multiple feed points. So without any tracking 2 or more satellites can be watched simultaneously. However, these satellites should not be spaced too widely across the arc and therefore this system is better suited for the American market. The major disadvantage of the multi focus system is that the antenna gain is relatively low because of signals being reflected to an off-axis focal point. In practice a multi focus antenna should also be considerably larger to compensate for the high noise figure. Furthermore, multi focus antennas are less suitable for receiving Ku-Band signals.

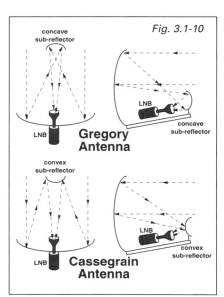

Fig. 3.1-10

PROS: 1) no need for tracking 2) the possibility to receive signals from different satellites simultaneously. 3) very suitable for SMATV (Satellite Master Antenna TV) systems.
CONS: 1) extremely costly because the LNB is the most expensive part of the outdoor unit 2) high noise temperature (especially at Ku-Band) 3) less suitable for the consumer electronics market, however, Ku-Band satellite dishes are available in Europe that can receive both ASTRA and some Eutelsat satellites.

D) FLAT PLATE ANTENNA

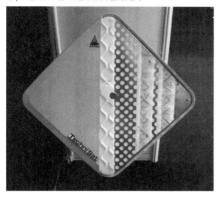

The *Flat Plate Antenna*, Microstrip Antenna (so called because of its meticulously designed circuitry) or Squarial® is the newest generation of satellite antennas. The Flat Plate Antenna (FPA) principle is totally different from the reflector antenna. In this case the satellite signal is not concentrated and reflected to a focal point, but detected by an array of elements concentrated on the antenna's flat surface. All the separate elements are interconnected to a waveguide system with an equal phase length, leading to the LNB in the middle. The FPA comprises very enhanced technology in order to achieve good side lobe attenuation qualities and a relatively small beamwidth angle. The same technology makes this small unit (40x40 cm.) even more efficient than a conventional reflector antenna of the same size. However, the FPA technique is still under development and larger units are not yet available.

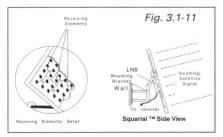

Fig. 3.1-11

PROS: 1) high efficiency 2) slim structure (environmentally friendly) 3) relatively favorable side lobe attenuation and beamwidth characteristics 4) excellent cross-polar performance (-25dB)
CONS: 1) limited applications 2) complicated production process

E) ZONE PLATE ANTENNA (ZPA)

The Flat Antenna Company's Zone Plate Antenna is a new approach to the Fresnel related antenna design. Fresnel himself was a scientist in optics like Gregory and Cassegrain. A conventional ZPA is comprised of a number of concentric rings, placed onto a transparent plate in a way that they create a lens effect. By mounting a transparent zone plate onto a metallized reflective sheet effectively a second antenna is created. This way the antenna's gain is drastically increased. Contact: Flat Antenna Company, 68 Bentfield Rd., Stanstead, Essex CM24 8HS, UK

PROS: 1) high efficiency 2) environmentally friendly 3) very good beamwidth characteristics.
CONS: 1) larger antennas are relatively expensive 2) less suitable for multi-focus purposes.

MOUNTS
1) AZ/EL MOUNT

This is the ideal mount for a fixed dish installation. The Az(imuth)/El(evation) mount enables you to accurately aim your dish at any particular satellite, by allowing you to adjust the elevation and azimuth angles individually. After aligning, the whole setup is fixed with nuts and bolts, so you can only pinpoint one particular satellite at the time. Most dishes in Europe are equipped with some

making an outdoor unit steerable. Some mount/jack combinations however are not capable of tracking the entire visible Clarke Belt, so in this case make sure that you can watch the most interesting satellites at your location. Another important aspect when buying an outdoor unit is the quality of the material. Most mounts have a ball bearing construction and are made of galvanized steel, but there are still some jacks of very inferior quality. Because of the possibility of water intrusion and the risk of rust some manufacturers have decided to produce stainless steel jacks, so whenever you buy a motorized satellite system: *Ask for a Stainless steel jack.*

3) HORIZON-TO-HORIZON MOUNT or H-H Mount

kind of Az/El mount, that can either be mounted on a pole or fixed to the wall.

2) POLAR MOUNT
Polar mounts are designed to enable the satellite antenna to track the visible part of the geosynchronous arc from almost any site location. Initially polar mounts were designed by astronomers, who discovered that it was much easier to track a celestial body by creating an axis that would compensate for the earth's rotation. This axis is parallel to the earth's North/South pole axis, hence the name polar mount.

The device that actually moves the antenna west- and eastwards is called *Actuator* or *Jack* (because of its shape). The jack is activated by an indoor unit called *Positioner*. This positioner can be separate or part of a more expensive multi-satellite receiver. The mount/jack combination is the cheapest solution for

This is practically the same principle as the polar mount/jack combination, but here the separate mount and actuator are integrated in a single unit. The major advantage of a H-H mount is its simple and robust construction. It also allows the dish to make a full 180° swing from east to west. Its compact design and its totally enclosed gear box ensure sufficient protection against all outdoor elements. The H-H mount is totally compatible with all indoor units, but the antenna mounting plate needs some slight modifications to fit some antennas, especially offset fed. Contact: Gentact International Inc., 14F-3, No. 75, Sec. 1, Hsin-Tai 5 Road, Hsichih Town 221, Taipei Hsien, Taiwan, ROC. Fax: +886 (2) 698 2411

SATELLITE ALIGNMENT EQUIPMENT
It is absolutely essential to fine adjust the total antenna/mount assembly in order to receive any satellite signal at all or to improve picture and audio quality. An experienced rigger uses a sensitive measuring device only, but it is always advisable to use a combination of a tv set and a test instrument. In this way you rule out the possibility of mistaking one satellite for another. A very helpful device for aligning the outdoor unit is the SAMM 3® Satellite Dish Alignment Monitor Meter. This very small and handy meter also produces a high pitched audio tone, that increases as soon as a satellite comes into alignment. The SAMM 3® is

fitted with a belt clip, making it ideal for installing dishes at awkward places. Contact: Oakbury Components Ltd., Oakbury House, Mill Lane, Lambourn, Berkshire RG16 7YP, Great Britain. tel: +44 (488) 71458, fax: +44 (488) 73172.

3.2. STRAIGHT TALKING ABOUT SATELLITE RECEIVERS

Satellite receivers are designed to detect and select the available audio and video information and to translate this information in a way that it can be reproduced by a regular tv set/monitor or Hi-Fi amplifier. Modern satellite receivers look like regular modern home entertainment components and serve as a satellite channel control module. The average receiver consists of a Power Supply Unit (PSU) to feed the whole system and the outdoor polarizer/LNB unit, a *downconverter* or *front end*, an *Intermediate Frequency (IF)* station, picture and audio processors and an *Radio Frequency (RF)* modulator to produce the UHF signal to your tv set. Some receivers come with a built-in positioner to power the outdoor antenna actuator unit.

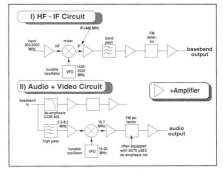

Fig. 3.2-1: Simplified Block Diagram of a regular Satellite Receiver

Power Supply
The PSU's used in satellite receivers are similar to regular ones used in all kinds of consumer electronics. Basically the incoming 110 or 220 AC mains voltage is downconverted by a step-down transformer and fed into a rectifier circuit. This way the AC (Alternating Current) is transformed into a lower DC (Direct Current). A rectifier circuit is nothing more than a series of diodes which conduct the current only one way, and series of *Low-Pass Filters (LPF)*, which get rid of the 50 or 60 Hz vibration. Finally this filtered output is fed to a number of voltage regulators which supply the various parts of the receiver with the necessary voltages. Voltage regulators come in a 24, 18, 12 and 5 Volt direct current (Vdc) version. Most receivers use 12 Vdc regulators, because most transistors and *Integrated Circuits (IC's)* are powered by this voltage. LNB's need between 12-24 Vdc of 100-200 milliAmps (mA), whereas polarizers only need 5-6 Vdc, but with 500 mA. The outdoor actuator needs between 24-36 Vdc of 6 A. So the PSU of modern receiver/decoder/positioner combinations has to be extremely well designed to avoid overheating or loss of voltage in other - usually vulnerable - parts of the receiver.

Downconverter/Tuner
The purpose of this amplifier is to select one particular channel from the total range of frequencies relayed from the LNB. This channel then has to be converted down to the receiver's final Intermediate Frequency. The heart of the tuner comprises a *Voltage Tuned Oscillator (VTO)* that mixes its output with the signal coming from the LNB. Then this combined signal is passed on to the receiver's final IF stage.

The most commonly used IF is 70 MHz, although some receivers downconvert to 480 MHz. Some new receivers intended for the Japanese DBS market have an IF of 403 MHz. The 70 MHz IF was widely adopted by the industry because a circuit named a *Phase Locked Loop (PLL)* detector - required for conventional tv sets - was already available.

Selecting channels
Selecting Ku-Band channels is far more complicated than selecting the North American C-Band channels because the center frequencies and bandwidths are not standardized. If we wanted to receive just one Ku-Band channel on a certain receiver, a fixed VTO frequency would suffice. However, in practice we would like to receive various channels, so some improvising is needed. In the old days potentiometers were used to alter the voltage to the VTO. These rotary tuning knob devices are still very popular amongst satellite enthusiasts. Modern satellite receivers have pre-programmed channel settings as well as the option to tune in channels electronically. Most receivers allow you to store up to 200 preference channels, but some cheaper models offer only 100 settings or less. The majority of receivers are equipped with some kind of *Automatic Frequency Control (AFC)* to eliminate any frequency drifting by the receiver. Another method to eliminate drifting is to use a PLL circuit or a quartz stabilized configuration.

Final IF Stage
This part consists of a *bandpass filter* and an amplifier. The filter sets the channel bandwidth to a typical 28 MHz or less by eliminating all other signals. The amplifier compensates for any signal loss incurred during the downconversion process and amplifies the signal passed through to the detector/*demodulator*. The latter processes the FM modulated satellite signal and converts it to a combined *composite video* and *baseband* signal. This baseband signal is used as an input to decoders and digital audio processors. Basically there are two types of demodulators which both have their own pros and cons. The Phase Locked Loop (PLL) circuit is the most commonly used, because it is stable and better capable of discriminating between weak signals and noise. However, PLL circuits tend to produce a somewhat unstable picture.

The Video Processor
This part of the satellite receiver extracts the composite video signal from the baseband signal by relaying it through a *low pass filter*. Then this signal is fed through a *de-emphasis* network, which attenuates the upper frequencies in order to restore the original video wave pattern. Now the video signal is almost back in the state it was before it was uplinked, but not quite. The filtered and de-emphasized video signal is still mixed with the wave form on top of which it was broadcast. When looking at the picture at this stage we would see an annoying flicker caused by this dispersal wave form. A clamping circuit finally smoothens the picture by removing the dispersal wave form.

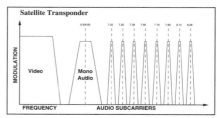

Fig. 3.2-2: General Transponder Layout

The Audio Processor
The audio processor demodulates the audio information on any chosen *subcarrier*. Normally audio is broadcast on frequencies between 5.0 and 8.5 MHz. Many European Ku-Band satellites broadcast audio on 6.65, 6.60 or 6.50 MHz. Extra narrowband frequencies are available on 7.02/7.20, 7.38/7.56, 7.74/7.92 and 8.10/8.28 MHz. The end result of an audio processor is called mixed audio baseband signal, which is fed to a variable fre-

quency demodulator to allow the various audio subcarriers to be tuned in separately. Then this signal also has to be fed through a de-emphasis network to match the audio *pre-emphasis* transmitted on various satellites. The most commonly used standards are 50, 75 or 150 microseconds and many receivers allow you to select these settings. Finally, most broadcasters use some kind of noise reduction system on their stereo subcarriers and there are many variations in use worldwide. Most European satellite broadcasters use the *Wegener Panda 1®* system, so receivers in this part of the world should be equipped with a Wegener noise reduction circuitry, or "sound-alike" variant.

The RF Modulator

In general it is possible to feed both audio and composite video signals directly to the tv set, but some older sets only accept UHF of VHF antenna input. The satellite receiver's RF modulator converts baseband audio and composite video signals in such a way that it can be relayed directly to the antenna socket of a conventional tv set. Usually a UHF channel between 30-40 is chosen, but in the USA VHF channels 3 and 4 are also widely used. Most satellite receivers have a little trimmer at the back, allowing you to tune to a clean UHF channel.

Important receiver specifications

The quality of a satellite receiver is determined by the clarity of both its picture and sound. All of these are subjective criteria, but sound and picture quality do have a concrete technical basis, which can be measured. Although a watchable picture can be broadcast with only 10 MHz of bandwidth, most satellite tv stations broadcast with a video bandwidth of 36 or 28 MHz. Most Intelsat and Eutelsat satellites use 36 MHz, whereas the European ASTRA satellites all broadcast with 28 MHz of bandwidth. The overall picture quality will deteriorate when the receiver's bandwidth is reduced below the bandwidth of the original signal. When receiving a 36 MHz signal with a receiver bandwidth that is considerable narrower (e.g. 15 MHz), the picture starts to show all kinds of abnormalities. Sharp vertical edges will show the effects of truncation (tearing effect) and sparklies will appear in saturated colors. Most receivers have a 28 MHz bandwidth as standard, because this way the detection of impulse noise (sparklies) caused by weak signals is reduced, without suffering any negative visible effects. However, receivers with tunable bandwidth - e.g. between 36-15 MHz - are to be preferred.

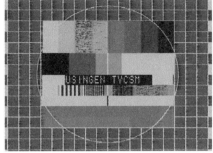

Heavy Truncation in this USINGEN test pattern caused by a narrow receiver bandwidth

Sparklies in this test pattern are caused by a weak signal

Another important criterion is the sensitivity of a satellite receiver. White and black streaks - generally called sparklies or spikes - are caused when an input signal (C/N Ratio) is weaker than the receiver's *threshold*. In other words: The higher the threshold, the less sensitive the receiver. For a given C/N Ratio, or signal coming from the satellite dish, the receiver will generate its own *Signal-to-Noise Ratio (S/N Ratio)*. Just imagine the receiver's noise as being a classroom full of children. When they all would be chatting, someone would have to shout to be heard. The quieter the classroom (low receiver noise level) the lower the signal level (C/N Ratio) would have to be. The receiver's threshold is the point at which it cannot longer clearly distinguish between satellite signal and noise. At this point we will see sparklies appearing, first in saturated colors and than throughout the whole picture. The receiver's threshold is measured in dB and most modern receivers have a threshold at a C/N Ratio of 7dB or lower.

3.3. HOW TO BUY A SATELLITE TV SYSTEM

It is not easy to give general advice for any part of the world on buying a satellite system. It is very important to know beforehand what type of system and which model to choose, as this will affect which programs you will be able to watch. In general there are 3 groups of satellite systems.
❑ The fixed dish, single satellite system
❑ The fixed dish, multi satellite system
❑ The motorized dish, multi satellite system.

Then of course you have to find out what satellite and frequencies the programs you want to watch are being broadcast on. Most satellites broadcast both in the C- and Ku-Band region simultaneously, but there are also a lot of Ku-Band only satellites around. However, even a Ku-Band only satellite such as the German Kopernikus satellite can broadcast in various Ku-Band frequency ranges, i.e. Ku1- and Ku3-Band. In order to receive all of the programs available on this satellite, one would need two separate LNBs or a Dual-Band LNB. Fortunately this satellite is an exception and in most cases one system with either a C- or a Ku-Band LNB will do. Another important issue is encoding. Most commercial tv stations around the world use some form of encryption. There are so many encryption systems around that it is absolutely essential to first ascertain what method your favorite stations are using. In the worst case scenario you will have to install multiple decoders, but a lot of satellite operators and program providers have standardized on one particular system. In some countries in Europe (e.g. UK and Scandinavia) and in the US it is possible to buy receivers with built-in decoders. Most receivers in the US have built-in VideoCipher II decoders whereas receivers in the UK can have built-in *VideoCrypt* decoders. Receivers in Scandinavia usually come with built-in D- or D2-MAC *EuroCrypt* decoders.

3.4. STRAIGHT TALKING ABOUT CO-AX

The importance of a good connection between the LNB and the receiver is often greatly underestimated. Yet in many cases the overall picture and sound quality is impaired because of the use of inferior coaxial cable. The cable should be regarded as being a link in a chain and if this link is weak, it affects the strength of the entire chain. Coaxial cable, or short co-ax, comes in all kinds and thicknesses, produced by dozens of manufacturers. The characteristics and quality of a cable determine the overall signal quality offered to the receiver. Any signal loss between the LNB and the receiver will result in sparklies or other visible distortions of the picture.

The following criteria are of great importance for maintaining a good quality signal:
❑ The impedance of the coaxial cable
❑ The length of the coaxial cable
❑ The frequency of the signal
❑ The strength of the coaxial cable

Looking closer at Co-ax
Co-ax is composed of two concentric conductors, separated by an insulating material called *dielectric*. This dielectric

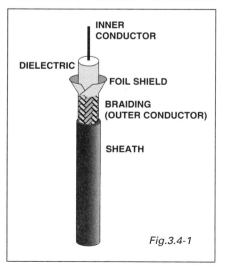

Fig.3.4-1

serves as a spacer to keep the distance between the center conductor and the shield to a close tolerance, ensuring an accurate impedance value. The impedance of a cable is determined by the ratio between the center conductor and the shield.

The dielectric material used also has an influence on the impedance or attenuation characteristics of co-ax. The ideal dielectric material would be air, allowing the inner conductor to be thicker without having to increase the outer cable's diameter, whilst maintaining the same impedance. For practical reasons it is not possible to construct a totally air-spaced co-ax. Therefore dielectric materials such as teflon, polypropylene (a hard, translucent substance) and polyethylene foam (a soft, white material), or a combination of these materials with air, are used in modern co-ax. All manufacturers strive to put in as much air in their dielectric materials as possible. The advantages are clear. The impedance is constant, whereas the attenuation is kept to a minimum. Another advantage is that the cable becomes less rigid, hence easier to work with.

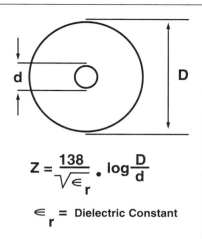

$$Z = \frac{138}{\sqrt{\epsilon_r}} \cdot \log \frac{D}{d}$$

ϵ_r = Dielectric Constant

Fig.3.4-2: Coaxial Impedance Formula

Signal Losses

Another part of the coaxial cable which influences signal losses, is the outer shield. It is most important that this shield is as "high-frequency proof" as possible, to avoid any interference from outside. A solid shield such as copper foil is better than copper braiding, but in most cases (see Fig. 3.4-1) both materials are used together. Signal losses also increase as frequency increases for any given type of cable. For this reason, co-ax is not used for frequencies above 2.5 GHz. For frequencies higher than 2.5 GHz a metal pipe of a certain diameter (e.g. a Ku-Band feedhorn) is used as a kind of wave guide. With higher frequencies air becomes a better conductor than copper and therefore cannot be used as dielectric. Another good example is the ferrite polarizer, which actually is a piece of wave guide. A wave guide can be used for frequencies up to 70 GHz.

Matching Impedance

If we short-circuit one end of the co-ax while connecting a high-frequency energy source to the other (open) side, the entire signal will be fully reflected. In this case the standing wave ratio is indefinite. Maybe this requires a short explanation. The high-frequency signal produces electrical charges in the inner and outer conductors, resulting in a maximum voltage level at certain places in the coaxial cable. This maximum voltage level occurs after every half wavelength/co-ax length and together with the center conductor's resistance is referred to as Characteristic Impedance. The wavelength is determined by the frequency of the high-frequency signal, i.e. a signal with a frequency of 100 MHz has a wavelength of 300 cm, whereas a 1000 MHz frequency has a wavelength of 30 cm. If the impedance of a coaxial cable matches the impedance of the high-frequency transmitting antenna, only a very small part will be reflected. In this case the standing wave ratio will be approx. 1. The more the impedance of a coaxial cable differs from the transmitting antenna or energy source, the more the standing wave ratio will increase. This criterion is also appropriate for the manufacturing of the coaxial cables. If the diameter of the center conductor is not kept constant, this mismatching will result in signal losses.

Cables used in the distribution and reception of all tv signals are typically rated at 75 Ohms. All components and equipment are developed to match this impedance value. There is a wide variety of cables available, intended for various specific tasks. Roughly there are four categories which use 75 Ohm cable.

co-ax stripper

1) CATV, Community Antenna TeleVision
CATV's are big cable networks that relay radio and tv signals over large areas, sometimes several cities. The cables used for these large networks have to meet certain standards, because high power signals are relayed over vast areas. These mechanical and electrical standards are usually set by the local PTT.

2) (S)MATV, (Satellite) Master Antenna TeleVision
SMATV networks are usually a lot smaller than CATV's (approx. 1000 connections), designed to relay signals within housing communities, bungalow parks or hotels. A number of (satellite) antennas on top of the roof receive the desired programs, which are then relayed to the various apartments.

3) SATV, Satellite Antenna TeleVision

SATV's are direct-to-home satellite systems, with a direct connection between the LNB and the satellite tuner in the living room. It is essential to use good quality cables, preferably with a foam dielectric for this purpose. *F-connectors* are the industry standard used for attaching all kinds of coaxial cables to the LNB, satellite tuner, or tv set. F-connectors can be of the crimp type, screw-on type or watertight click-on type. When using a regular crimp type F-connector outdoors, it should be made waterproof with self amalgamating tape to avoid any water ingress.

4) Leads between components
Pre-fabricated coaxial leads are also used to interconnect tuners, VCR's, laserdisc players and satellite tuners. The connection between the antenna wall socket and your radio and tv set is also done by means of this type of lead. However, these cables are usually of very poor quality and should not be used in between the satellite tuner and the LNB!!

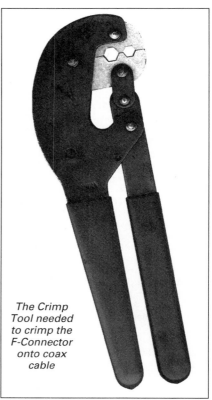

The Crimp Tool needed to crimp the F-Connector onto coax cable

CHAPTER 4
BROADCASTING SYSTEMS

4.1. WORLDWIDE BROADCASTING STANDARDS

Basically there are three tv broadcasting systems worldwide: NTSC, PAL and SECAM. These three formats were developed in the early days of television. In mid-1940 a group of American electronics engineers formed the National Television Standards Committee (NTSC), to establish a standard for b/w tv broadcasts in the US. In the early Fifties, when color transmissions were introduced, the NTSC standard was modified to allow color transmissions to be viewed as b/w broadcasts on a b/w tv set. The SECAM system is based on a modified version of the original NTSC system, but works with 625 scanning lines instead of 525. It was developed by French engineers who were not satisfied with some technical aspects of the NTSC system. The poor color reproduction of the American system led to the NTSC standard being nicknamed "Never Twice the Same Color". However, the SECAM (Séquence Couleur à Mémoire) system itself has some serious limitations. Freely translated SECAM means sequential color or with memory, which describes the technique used to eliminate color inaccuracies. Unfortunately it cannot accomodate simple, but frequently used special effects like fades and wipes. Therefore modern studios in France first produce their programs in PAL and then convert them to SECAM for transmission. The PAL system was developed by Telefunken in Germany in the Sixties as a counterpart of both SECAM and NTSC systems, because neither of the two existing systems reproduced color satisfactorily. The PAL (Phase Alternating Lines) format uses 625 scanning lines like SECAM, but eliminates color distortion by inverting the phase of the color signal every alternate line. Actually the PAL system combines all the advantages of the NTSC and SECAM systems that were developed earlier.

All tv broadcasting countries in the world use one of five variations on the three basic standards: PAL-B, SECAM, NTSC, PAL-N and PAL-M. PAL-N is used in Argentina (625 lines) and PAL-M (525 lines) is the Brazilian standard. In Europe, except for France which uses SECAM, the standard is PAL-B. Former Eastern Bloc countries and the CIS still use SECAM, although many countries are changing over to PAL. In 1993 Czechoslovakia split into the Czech Republic and Slovakia. Soon afterwards Slovakia changed to PAL whereas the Czech Republic still uses SECAM. Most tv sets sold in Europe are dual-standard, i.e. they can detect and reproduce both PAL and SECAM color systems.

WORLD TELEVISION BROADCAST STANDARDS BY COUNTRY

Country	VHF	UHF	Color	Sat Standard	Voltage	a/c
Afghanistan	B	None	PAL/SECAM		220	50 Hz
Alaska	M	None	NTSC		120/240	60 Hz
Albania	B	G	PAL		220	50 Hz
Algeria	B	None	PAL	PAL	127/220/380	50 Hz
Andorra	B	G	PAL	PAL	220	50 Hz
Angola	I	None	PAL		220	50 Hz
Antigua & Barbuda	M	None	NTSC		220	60 Hz
Argentina	N	None	PAL	PAL-N	220	50 Hz
Armenia	D	K	SECAM		127/220	50 Hz
Aruba	M	None	NTSC		127/220	50 Hz 60 Hz
Australia	B	G	PAL		240/415	50 Hz

Country	VHF	UHF	Color	Sat Standard	Voltage	a/c
Austria	B	G	PAL	PAL/D2-MAC	220/380	50 Hz
Azerbaidzhan	D	K	SECAM		127/220	50 Hz
Azores	B	None	PAL		220/380	50 Hz
Bahamas	M	None	NTSC		120/220	60 Hz
Bahrain	B	G	PAL		230	50 Hz
Bangladesh	B	None	PAL		220/240	50 Hz
Barbados	M	None	NTSC		110/220	50 Hz
Belarus	D	K	SECAM	SECAM	127/220	50 Hz
Belgium	B	H	PAL	PAL/D2-MAC	220	50 Hz
Belize	M	None	NTSC		110/220	60 Hz
Benin	K	None	SECAM		220/380	50 Hz
Bermuda Island	M	None	NTSC		115/230	60 Hz
Bolivia	M	None	NTSC		110/120/230/380	50 Hz
Bophuthatswana	I	None	PAL	PAL	220/380	50 Hz
Bosnia/Hercegovina	B	H	PAL		220	50 Hz
Botswana	K	None	SECAM		220	50 Hz
Brazil	M	None	PAL	PAL-M	110-127/220	50 Hz 60 Hz
British Indian Ocean Territory	M	None	NTSC		110/220	60 Hz
Brunei Darussalam	B	None	PAL		110-127/220	50 Hz
Bulgaria	D	K	SECAM	SECAM	220	50 Hz
Burkina Faso	K	None	SECAM		220/380	50 Hz
Burma	M	None	NTSC		115	60 Hz
Burundi (Rep.)	K	None	SECAM		220/380	50 Hz
Cambodia	B	G	PAL		120/208/220	50 Hz
Cameroon	B	None	PAL		220/380	50 Hz
Canada	M	None	NTSC	NTSC	115/230	60 Hz
Canary Islands	B	G	PAL		110/220	50 Hz
Cayman Island	M	None	NTSC		110	60 Hz
Central African Republic	K	None	SECAM		220	50 Hz
Chad	D	None	SECAM		220	50 Hz
Chile	M	None	NTSC		230/380	50 Hz
China (Peoples Rep. of)	D	None	PAL	PAL	220	50 Hz
China (Rep. of) Taiwan	M	None	NTSC		110/220	60 Hz
Colombia	M	None	NTSC		110/120	60 Hz
Congo (Peoples Rep. of)	D	None	SECAM		220	50 Hz
Cook Islands	B	None	PAL		220	50 Hz
Costa Rica	M	None	NTSC		120/220	60 Hz
Cote D'Ivoire	K	None	SECAM		220/380	50 Hz
Croatia	B	H	PAL		220	50 Hz
Cuba	M	None	NTSC	NTSC	115/120	60 Hz
Cyprus	B	G	PAL		240	50 Hz
Czech Republic	D	K	SECAM	PAL/SECAM	220	50 Hz
Denmark	B	None	PAL	D2-MAC	220	50 Hz
Djibouti	K	None	SECAM		220	50 Hz
Dominica (Commonwealth)	M	None	NTSC		240	50 Hz
Dom Republic	M	None	NTSC		110	60 Hz
Easter Island	B	None	PAL		220/380	50 Hz
Ecuador	M	None	NTSC		110/220	60 Hz
Egypt	B	None	SECAM		220	50 Hz

Country	VHF	UHF	Color	Sat Standard	Voltage	a/c
El Salvador	M	None	NTSC		110/220	60 Hz
Equatorial Guinea	B	None	SECAM		220/380	50 Hz
Estonia	D	K	SECAM		220	50 Hz
Ethiopia	B	None	PAL		127/220	50 Hz
Falkland Islands	I	None	PAL		220	50 Hz
Faroe Islands	B	G	PAL		220	50 Hz
Fiji	M	None	NTSC		240	50 Hz
Finland	B	G	PAL		220/380	50 Hz
France	L	None	SECAM		220/240	50 Hz
Gabon	K	None	SECAM		220	50 Hz
Galapagos Islands	M	None	NTSC		110/220	60 Hz
Gambia	B	None	PAL		230	50 Hz
Georgia	D	K	SECAM		127/220	50 Hz
Germany	B	G	PAL	PAL/D2-MAC	220	50 Hz
Ghana	B	None	PAL		220/240	50 Hz
Gibraltar	B	G	PAL		240	50 Hz
Greece	B	G	SECAM	SECAM	220	50 Hz
Greenland	B	None	PAL		110/220	60 Hz
Grenada					220	50 Hz
Guadeloupe	K	None	SECAM		220	50 Hz
Guam (US Terr.)	M	None	NTSC		110/220	60 Hz
Guatemala	M	None	NTSC		110/120/ 127/220	60 Hz
Guiana (French)	K	None	SECAM		127/220	50 Hz
Guinea (Rep.)	K	None	PAL		220/380	50 Hz
Guyana	M	None	NTSC		120/240	50 Hz 60 Hz
Haiti	M	None	NTSC		110/220	50 Hz 60 Hz
Hawaii (US State)	M	none	NTSC		115	60 Hz
Honduras (Rep.)	M	None	NTSC		110	60 Hz
Hong Kong	I	None	PAL		200	50 Hz
Hungary	D	K	PAL	PAL	220	50 Hz
Iceland	B	G	PAL		220	50 Hz
India	B	None	PAL	PAL	230/440	50 Hz
Indonesia	B	None	PAL	PAL	127/220	50 Hz
Iran	B	None	SECAM	SECAM	220	50 Hz
Iraq	B	None	SECAM	SECAM	220	50 Hz
Ireland	I	I	PAL		220/380	50 Hz
Israel	B	G	PAL	PAL	230/400	50 Hz
Italy	B	G	PAL	PAL	220/380	50 Hz
Jamaica	M	None	NTSC		110/220	50 Hz
Japan	M	None	NTSC	NTSC	110/220	50 Hz 60 Hz
Johnston Istan	M	None	NTSC		110	60 Hz
Jordan	B	None	PAL	PAL	220	50 Hz
Kazakhstan	D	K	SECAM		127/220	50 Hz
Kenya	B	None	PAL		240	50 Hz
Korea (Dem. Peoples Rep. of)	D	K	PAL		220	50 Hz
Korea (Rep.)	M	None	NTSC		100/200	60 Hz
Kuwait	B	G	PAL	PAL	240	50 Hz
Kyrgyzstan	D	K	SECAM		127/220	50 Hz
Laos	M	None	PAL		127/220	50 Hz
Latvia	D	K	SECAM		220	50 Hz
Lebanon	B	G	SECAM		110/190	50 Hz
Lesotho	I	None	PAL		240	50 Hz
Liberia	B	None	PAL		120	60 Hz
Libya	B	None	PAL	PAL	127/130	50 Hz
Lithuania	D	K	SECAM		220	50 Hz
Luxembourg	B	G & L	PAL/SECAM	PAL/D2-MAC	110/220	50 Hz

Country	VHF	UHF	Color	Sat Standard	Voltage	a/c
Macau	I	None	PAL		220	50 Hz
Macedonia	B	H	PAL		220	50 Hz
Madagascar	B	None	PAL	PAL	110/220	50 Hz
Madeira	B	None	PAL		220/380	50 Hz
Malawi	B	G	PAL		220	50 Hz
Malaysia	B	None	PAL	PAL	240	50 Hz
(Maldives (Rep. of)	B	None	PAL		220	50 Hz
Mali (Rep. of)	K	None	SECAM		220	50 Hz
Malta (State of)	B	None	PAL		240	50 Hz
Marshall Islands	M	None	NTSC		110/220	60 Hz
Martinique	K	None	SECAM		220	50 Hz
Mauritania (Islamic Republic of)	B	None	SECAM	SECAM	220/380	50 Hz
Mauritius	B	None	SECAM		240	50 Hz
Mayotte	K	None	SECAM		220	50 Hz
Mexico	M	None	NTSC	NTSC	110/220	60 Hz
Micronesia	M	None	NTSC		110/220	60 Hz
Midway Island (Fed. State of)	M	None	NTSC		110	60 Hz
Moldova (Moldavia)	D	K	SECAM		127/220	50 Hz
Monaco	L	G	PAL	PAL/D2-MAC	127/220	50 Hz
Mongolia	D	None	SECAM		220	50 Hz
Montserrat	M	None	NTSC		220	60 Hz
Morocco	B	None	SECAM	SECAM	127/220	50 Hz
Mozambique	B	None	PAL		220	50 Hz
Myanmar	M	None	NTSC		230	50 Hz
Namibia	I	None	PAL		220	50 Hz
Nepal	B	None	PAL		220	50 Hz
Netherlands	B	G	PAL	PAL/D2-MAC	220	50 Hz
Netherlands Antilles	M	None	NTSC		127/220	50 Hz 60 Hz
New Caladonia	K	None	SECAM		220	50 Hz
New Zealand	B	None	PAL		230	50 Hz
Nicaragua	M	None	NTSC		120	60 Hz
Niger	K	None	SECAM	SECAM	220/380	50 Hz
Nigeria	B	None	PAL		230	50 Hz
Norfolk Island	B	None	PAL		240	50 Hz
Northern Mariana Islands (US Comm.)	M	None	NTSC		220	50 Hz
Norway	B	G	PAL	PAL/D2-MAC/ D-MAC	230	50 Hz
Okinawa	M	None	NTSC		110	60 Hz
Oman (Sultanat of)	B	G	PAL		220	50 Hz
Pakistan	B	None	PAL		230/400	50 Hz
Palau	M	None	NTSC		220	60 Hz
Panama	M	None	NTSC		110/115/ 120/126	60 Hz
Papua N. Guinea	B	G	PAL		240	50 Hz
Paraguay	N	None	PAL		220	50 Hz
Peru	M	None	NTSC		220	60 Hz
Philippines	M	None	NTSC		110/220	60 Hz
Poland	B	G	PAL	PAL	220	50 Hz
Polynesia (French)	K	None	SECAM		220	60 Hz
Portugal	B	G	PAL	PAL	220	50 Hz
Puerto Rico	M	None	NTSC		120	60 Hz
Qatar	B	None	PAL		240	50 Hz
Reunion	K	None	SECAM		220	50 Hz

Country	VHF	UHF	Color	Sat Standard	Voltage	a/c
Romania	D	K	PAL		220	50 Hz
Russia	D	K	SECAM	SECAM	127/220	50 Hz
Rwanda	D	K	PAL		220	50 Hz
Samao (American)	M	None	NTSC		120	60 Hz
Saudi Arabia	B	G	None	PAL/SECAM	127/220	50 Hz
Senegal	K	None	SECAM		220	50 Hz
Seychelles	B	None	PAL		240	50 Hz
Sierra Leone	B	None	PAL		230	50 Hz
Singapore	B	None	PAL		230	50 Hz
Slovakia	D	K	PAL	PAL	220	50 Hz
Slovenia	B	H	PAL		220	50 Hz
Somalia (Rep.)	B	None	PAL		230	50 Hz
South Africa	I	None	PAL	PAL	220/380	50 Hz
Spain	B	G	PAL	PAL	127/220	50 Hz
Sri Lanka	B	None	PAL		230	50 Hz
St. Grenadines	M	None	NTSC		110	60 Hz
St. Kitts & Nevis	M	None	NTSC		220	60 Hz
St. Lucia	M	None	NTSC		220	50 Hz
St. Marino	B	G	PAL		220	50 Hz
St. Pierre & Miquelon	K	None	SECAM		220	50 Hz
St. Vincent	M	None	NTSC		230	50 Hz
Sudan	B	None	PAL		240	50 Hz
Suriname	M	None	NTSC		110/115/127/220	60 Hz
Swaziland	B	G	PAL		220	50 Hz
Sweden	B	G	PAL	PAL/D2-MAC/D-MAC	220	50 Hz
Switserland	B	G	PAL	PAL	220	50 Hz
Syrian Arab Rep.	B	G	PAL/SECAM		115/200	50 Hz
Tadzhikistan	D	K	SECAM		127/220	50 Hz
Tanzania	B	None	PAL		230	50 Hz
Thailand	B	M	PAL	PAL	220/380	50 Hz
Togo	K	None	SECAM		127/220	50 Hz
Trinidad & Tobago	M	None	NTSC		115	60 Hz
Tunisia	B	G	SECAM		115/220/380	50 Hz
Turkey	B	None	PAL	PAL	220/380	50 Hz
Turkmenistan	D	K	SECAM		127/220	50 Hz
Uganda	B	None	PAL		240/415	50 Hz
Ukraine	D	K	SECAM		127/220	50 Hz
United Arab Emirates	B	G	PAL	PAL	20	50 Hz
United Kingdom	I	I	PAL	PAL/B-MAC	240	50 Hz
Uruguay	N	None	PAL		220	50 Hz
USA	M	M	NTSC	NTSC/B-MAC	110/220	60 Hz
Uzbekistan	D	K	SECAM		127/220	50 Hz
Venezuela	M	None	NTSC	NTSC	120/240	50 Hz 60 Hz
Vietnam	M	None	NTSC/SECAM		120/127/230	50 Hz
Virgin Islands (American)	M	None	NTSC		110	60 Hz
Virgin Islands (British)	M	None	NTSC		220	50 Hz
Wallis & Futuna	K	None	SECAM		220	50 Hz
Yemen	B	None	PAL/NTSC		230	50 Hz
Yugoslavia	B	G	PAL	PAL	220	50 Hz
Zaïre (Rep.)	K	None	SECAM	SECAM	220	50 Hz
Zambia	B	None	PAL		220	50 Hz
Zimbabwe	B	None	PAL		220	50 Hz

4.2. WHAT IS MAC

Because the three broadcasting standards PAL, SECAM and NTSC are not compatible with each other, the need arose to develop at least one universal European standard. The *MAC (Multiplexed Analogue Components)* standard is a very controversial system amongst experts, because it still uses analogue picture information (albeit separated) together with digital sound. MAC comes in five varieties: A-,B-,C-,D-, D2-MAC. B-MAC (525 lines) is widely used for satellite broadcasts in North America, Australia and South Africa, D-MAC in parts of Scandinavia and D2-MAC in the rest of Europe.

At the World Administrative Radio Conference (WARC) in 1977 so called DBS satellite frequencies and orbital positions were allocated to various European countries. At this conference it was more or less decided to use a MAC version as a DBS broadcasting standard. In 1986 this "advice" was adopted by the various European governments and it became obligatory for broadcasters to use the MAC system for broadcasts on DBS satellites. An EC Directive set D2-MAC - conceived as an intermediate step to *HDTV* - as the sole television transmission standard for all high-power satellites over Europe. However, it left the way open for new services to appear on rival low and medium powered satellites using cheaper, less sophisticated PAL technology.

In practice only the D2-MAC variant proved viable in Europe and is still widely used as a satellite broadcasting standard. It has become clear that D2-MAC as the only standard for satellite tv in Europe has flopped.

In 1985 the German and French Ministers of Communications agreed to adopt the D2-MAC version for satellite broadcasting and to feed pictures to and through cable networks. D2-MAC combines analogue (video) and digital (audio) signal components (D2 = dual binary coding). The two main advantages of D2-MAC are the brilliant digital audio quality and the separation of luminance (brightness) and chrominance (color) signals. The latter eliminates the so called *moire effect* - the phenomenon that changes a newsreader's gray striped jacket into a brightly colored clown's suit. Another advantage is that digital sound only occupies a fraction of the bandwidth, so various sound channels can be broadcast along with the picture.

Since the early days of television the screen proportions have always been 4:3. For compatibility reasons the D2-MAC screen proportions were initially kept the same. Later on the idea of changing the screen proportions to a cinemascopic format of 16:9 gained ground. The 16:9 format has certain advantages, e.g. feature films can be shown in full without having to cut off the edges or to show black bars above and below. Also the vision angle resembles the natural human angle of sight, so watching a 16:9 screen is more pleasant and natural.

The next step of course is High Definition Television (HDTV). With HDTV, or the European HD-MAC variant, the number of scanning lines is doubled from 625 to 1250. Of course all these developments require new end user products like HDTV TV-sets and VCR's. It is this requirement, unfortunately, that has triggered competition amongst the big consumer electronics manufacturers. At the moment there are various HDTV versions under development, which are - again - not compatible!

4.3. GLOBAL TV STANDARD VS. MULTI-STANDARD MADNESS

There is still a glimmer of hope for one universal global tv standard. Currently the Moving Picture Experts Group (MPEG) is developing a new digital video and tv technique called *MPEG-2*. Their main goal is to spell an end to the multi-standard madness of PAL, SECAM, NTSC and even MAC by ensuring true worldwide compatibility of tv hardware and software. MPEG-2 has already been accepted by manufacturers represented in the European Launch Group, which is ascertaining an alternative, non-MAC future for European broadcasting. The new system covers broadcast and video standards as well as high-definition and widescreen formats by encompassing high speed digital compression rates of up to 15 megaBits per second. It is fully compatible with existing digital video standards used for CD-i and computer applications. Currently MPEG-2 is already used for various DVB video and audio transmissions on a number of satellites in Europe and in the USA

4.4. US DBS: QUESTIONS AND ANSWERS

What is DBS?
A new class of television services have recently become available to viewers in the continental United States. These services allow households to receive television programming directly from satellites on small (18 inch to 3 foot diameter) satellite dishes which are not movable but instead are aimed at one position in the sky.

The signals are digitally compressed, allowing several programs to be broadcast from a single satellite transponder thereby allowing up to 200 channels receivable with a dish pointed at one orbital position in the sky. Programming on the various services includes most major cable services, sports, Pay Per View (PPV) movies, audio services, and specialized "niche" programming aimed at smaller audiences. These services are often referred to as Direct To Home (DTH) services but the term Direct Broadcast Satellite (DBS) services is more generally used.

There are currently six DBS services in operation with others expected to begin in the coming months. Although they still have far fewer subscribers than the Cable TV industry, DBS services are rapidly adding subscribers and the industry has very strong growth potential. As a result, many companies are interested in getting into the DBS business.

What services are available today?
Six companies currently offer all-digital DBS services in the U.S. with their services called Primestar, DIRECTV, USSB, DISH Network, Sky Angel, and AlphaStar. Primestar is offered by a group of Cable TV companies and operates from conventional satellites using 27-36 inch dishes. Primestar has been very successful logging over 1.75 million subscribers and capturing over 30% of the DBS market in its first two years of digital operation.

DIRECTV, Inc. which is a subsidiary of Hughes Communications offers a service called DIRECTV which operates from specially designed High Powered DBS satellites receivable with 18 inch dishes. DIRECTV is considered the premier DBS service in the U.S. today. They have the most channel capacity available today and have signed up over 2.5 million subscribers since starting in 1994 capturing over half of the DBS market.

The United States Satellite Broadcasting Company (USSB) has partnered with DIRECTV to deliver a complimentary 25 channel service which use the same satellites and reception systems and a merged program guide making their services appear as a single service. USSB has logged over 1.5 million subscribers to date.

The DISH Network is provided by EchoStar Communications, Inc. They operate from specially designed High Power satellites receivable with 18 inch dishes. DISH started in spring of 1996 so they have a relatively small market share at this time but their service has proven to be very popular adding subscribers at a rate comparable to the other services. Their success is primarily because they have entered the market with very inexpensive hardware and programming which has proven to be very popular with price-sensitive subscribers and which has caused other providers to lower hardware and programming prices.

Dominion has partnered with EchoStar to deliver a 6 to 10 channel Christian religious service using a merged program guide making their services appear as a single service. It is too early to tell how successful this service will be but it is clearly targeted toward a niche not served by the other services.

Tee-Comm electronics offers a service called AlphaStar which uses a 36" dish and operates from conventional satellites. Their service started in mid 1996 and has added subscribers at the slowest rate giving them far less subscribers than the other services. They say their service is the only one available to U.S. residents of Alaska and Hawaii and it appears to be most popular outside the continental U.S.

Because some services share a common reception system and operate from the same orbital location, viewers may want to think of the four possible DBS services as 1) PrimeStar, 2) DIRECTV/USSB, 3) EchoStar/Sky Angel, and 4) AlphaStar.

What equipment is needed?
Each service provider sells or leases reception hardware that includes a dish, a decoder, and a remote control. A single decoder can decode a single channel

which can then be routed to several TV sets and VCRs throughout a household.

A separate decoder is needed for each TV or VCR that subscribers want to be able to view a different channel on simultaneously. Therefore to watch two different channels simultaneously or to tape one channel while watching another a household must have two decoders.

DIRECTV, USSB, and EchoStar use systems that allow subscribers to self-install their equipment, although many choose professional installation. PrimeStar and AlphaStar require professional installation.

Why is a decoder necessary?
DBS is now a viable alternative to cable TV for one important reason: recent advances in real-time digital video compression technology allow a very large number of channels to be carried on the same frequency range where only a few could in the past. Each of the DBS services uses a recently developed real-time lossy compression system which allows an average of about six channels to be broadcast from a single satellite transponder where only one channel was possible before. This results in the DBS services' ability to broadcast approximately 200 channels from a single orbital location in the sky. Without digital compression only 32 channels would be allowed making the DBS services much less desirable.

What determines the number of channels on each transponder?
The number of channels which can be compressed into a single transponder depends on a lot of things such as desired image quality (i.e. resolution), frame rate of the source material, amount of movement in the source material, degree of allowable visible artifacts, and other factors.

Programming containing frames with many fast-moving small objects such as a basketball game can be compressed about 3 or 4 to a transponder before significant digital artifacts appear. Programming containing mostly large still images can be compressed at a higher rate, perhaps 5 or 6 to 1 transponder. Movies are filmed at 24 frames per second rather than 30 for video so they contain less source material. In addition, film is not interlaced and is in general fairly constant from frame to frame. As a result, film can be compressed more, perhaps 7 or 8 to 1 transponder for near laser disc quality.

What compression systems do the DBS services use?
Each of the DBS services transmits a bitstream which contains compressed audio, compressed video, authorization information, program guide information, and other information. The decoders in each subscriber's homes decode the digital bitstream converting it into video and audio which can be displayed by conventional TV sets. Each service requires a decoder designed to work with its system.

Primestar uses a proprietary video compression system developed by General Instruments called DigiCipher-1. The format used by all the other services is based on the MPEG-2 compression standard but also uses some proprietary components. The EchoStar/Sky Angel and AlphaStar services use a transmission system based on the DVB standard being pushed by some companies as a world broadcast standard. DVB uses the standard MPEG-2 and also attempts to standardize more of the systems.

Does this mean the decoders are interchangeable between services?
No. There is absolutely no decoder standardization in the DBS world.

DIRECTV and USSB use a common decoder as do EchoStar and Sky Angel. Therefore there are four possible decoder types from which potential subscribers can choose today. These four are all different and not interchangeable. While many parts of the systems are common, each of the four broadcast bitstreams contain some proprietary information which only their decoder can understand. In order to change to another service subscribers must either sell or otherwise exchange their decoders for one designed for the new service. This includes services which use DVB compliant decoders.

It doesn't really make much difference what transmission system is used by the DBS providers since they all can create similar quality audio and video and none are interchangeable with other services.

Can you tell me more about PrimeStar?
A group of major Cable Multi System Operators (MSOs) have joined together to form Primestar Partners, Limited which offers a digital 90 channel Direct-To-Home service to North America called Primestar. This will soon be expanded to about 160 channels. They were the first DBS service and started with 30 analog channels several years ago. They converted to digital in 1994 and claim to be the first digital service beating DIRECTV and USSB by a few weeks.

Primestar is now in over 1.75 million homes. Their success is primarily because they do not require customers to buy the decoder or dish. Instead, they lease it and include the lease cost in the monthly subscription fees. They operate from a conventional medium power satellite so they use a 3 feet diameter dish which must be professionally installed. They is using a smaller dish measuring about 27" which should make them more popular with some consumers.

Primestar uses the medium power Satcom K1 satellite using General Instruments' DigiCipher 1 digital broadcasting system. They plan to migrate to the GE-2 medium-powered satellite scheduled to launch into the 85 degree west orbital position early in 1997. They will then be able to increase to about 140 channels.

Primestar decoders are manufactured by General Instruments. Their decoders are hardware upgradable meaning hardware update modules can be attached which allow parts of the system to be updated without replacing any components. They have hinted at plans to migrate to a new MPEG-2 system called DigiCipher 2, but it is now questionable whether this conversion will ever be made.

Primestar services can be purchased at Radio Shack stores nationwide.

What is the DIRECTV/USSB service?
The DIRECTV/USSB service is the premier DBS service available today offering up to 200 video channels. This service uses three specially designed High Power Ku-band satellites which operate from one fixed position in the sky. The first (DBS1) uses sixteen 120-watt transponders while the other two (DBS2 and DBS3) are configured to each use eight 240-watt transponders. This results in a total of 32 broadcast transponders.

DIRECTV has sold five of the 120-watt transponders to USSB. (Actually USSB owns 5/16 of one of the entire satellites since federal regulations require DBS broadcasters to own their broadcast facilities.) The two competing companies both offer programming receivable with a common dish and decoder.

The hardware used by DIRECTV and USSB is called Digital Satellite System or DSS(TM). Sony, Thomson Consumer Electronics (owner of the Proscan, RCA, and GE names), Hughes Network Systems, Toshiba, Matsushita (Panasonic) and possibly others now build the DSS receiving equipment. Additional manufacturers have been licensed to sell the units including Uniden, Samsung, Sanyo, and Daewoo. The decoders are sold through both satellite dealers and consumer electronics retailers.

DIRECTV and USSB customers must purchase their decoders. Prices range from about $300 to $600 depending on the models. Rebates and other promotions have dropped the net price of some models to under $200 when pre-paid programming subscriptions are purchased.

Each manufacturer of DSS equipment differentiates their product by providing a unique user interface including their own on-line program guide with a different look and feel and different remote controls. The program guides contain programming information such as descriptions of upcoming episodes and scheduled talk-show guests. Each manufacturer also can choose whether or not to include certain features such as a Favorite Channel list or a Universal remote control.

AT&T also sells DSS hardware and DIRECTV and USSB programming directly to their long-distance customers. They have invested significantly in DIRECTV.

What DIRECTV and USSB programming is available?
DIRECTV and USSB share the rights to all 32 broadcast frequencies at the 101 degree west orbital position. The channels carried by each service are unique and do not appear on each other's services. There is little or no free programming on either service.

The DIRECTV/USSB programing is approximately as follows:

70 Channels of major cable services
50 Channels of subscription sports
20 Channels of special interest/niche scs.
50 Channels of Pay Per View (PPV) movies
10 Promotional Channels
200 Total Channels

The FCC has issued USSB five of the frequencies so USSB broadcasts from five transponders on one of the 120-watt satellites giving them about 25 channels. Their service is made up primarily of Premium Movie channels including HBO and Showtime.

DIRECTV has been issued 27 frequencies and programs about 175 channels of programming which can be broken down into five areas: cable programming, subscription sports, music services, Pay Per View (PPV) movies, and special interest/niche services. DIRECTV's cable programming is the basis of their service.

DIRECTV offers the most television sports coverage available anywhere. They have subscriptions to most major professional and some college sporting events. They use the addressable nature of the decoders to allow reception only in certain geographic locations such as outside the local broadcast coverage areas. Significant local black-out rules apply meaning many games viewers may want to watch are not available in their area.

DIRECTV offers Pay Per View (PPV) movies time-shifted on about 50 channels with many starting at intervals of at most 30 minutes. Prices for PPV movies are usually $3 when ordered through the DSS remote control and $5 when ordered over the phone. Occasionally movies are offered at a slightly lower or higher price.

The DSS system has built in copy protection technology which can control whether or not a PPV movie can be recorded. The degree to which this is used is unclear. Some viewers say a few things they have tried to record have been protected, but others say nothing they have tried to record has been protected. USSB says they have no plans to ever copy protect any of their programming, but some of their PPV events may have been protected.

Customers with more than one DSS decoder in a household pay an additional $1 to $5 fee per decoder.

Can you tell me more about the EchoStar/Sky Angel Service?
EchoStar and Sky Angel provide the second High Power DBS service in the U.S. EchoStar started operation in early March 1996 and have already signed up nearly half a million subscribers thus far. They are currently offering about 100 channels from their 119 degree orbital position where they control 21 broadcast frequencies. Their High Power satellites are called EchoStar-1 and EchoStar-2.

EchoStar's service, called the DISH™ Network, uses a DVB compliant system which sells for $200 for the basic system and $300 for a step-up when a prepaid programming subscription is purchased. Their decoder has a high-speed data port for future use. EchoStar says their system can be self-installed so they sell an installation kit as well.

EchoStar leases a single transponder to Dominion Satellite from which Dominion provides eight channels of religious programming called Sky Angel. The Sky Angel service uses EchoStar's reception hardware and are complementary to EchoStar's service in the same way DIRECTV and USSB's services are today

EchoStar programming is currently priced lower than the other DBS services. EchoStar is the only DBS service to offer any of the Star Trek series programming.

Can you tell me more about Alpha-Star?
Canadian based Tee-Comm has recently launched a medium-power service they call AlphaStar to the U.S. They are broadcasting 100 channels of video and audio services to 24-inch dishes from AT&T's Telstar 402R satellite in the 89 degree orbital position. They use a DVB compliant system like EchoStar manufactured by Tee-Comm Electronics. They hope to have up to 200 channels by sometime in 1997.

A key feature of AlphaStar's service is its ability to serve Alaska, Hawaii, and Puerto Rico. Also, they say they are the only service to carry X-rated adult programming.

Amway recently backed out of plans to sell AlphaStar equipment through their marketing system which was a major blow to AlphaStar.

Does the compression used by the DBS systems really work ?
Yes, but the resulting quality seems to be open to debate. There are occasional dig-

ital artifacts resulting from the heavy compression used on most of the services. There is a significant trade-off each service provider needs to make regarding quality vs number of channels.

Many customers report that the video and audio quality are excellent and the systems work extremely well. Others report noticeable digital artifacts on at least some channels.

The quality seems to vary significantly across channels in part due to variances in the source material. At times the video and audio quality on all of the services is stunning.

What about High Definition TV?
DBS is expected to be the first means in which most Americans have access to broadcast High Definition Television programming. Terrestrial broadcasters are concerned over the investment necessary to move to terrestrial HDTV which will still leave them with one channel when multiple lower-resolution channels are possible. DBS will no doubt prove to be the most cost-effective means of delivering HDTV to homes in the U.S. for years to come.

Most DBS decoder models contain a very high speed data port which can be directly connected to an HDTV decoder. Whether or not the DBS companies plan to use it remains to be seen. Significant HDTV broadcasts will likely be many years away although a few experimental broadcasts could occur in 1997 on at least some of the systems.

How do I connect a DBS decoder to my home audio/video system?
There are plenty of options for hooking the decoders into a home A/V system. TV sets with S-video inputs can use the S-video output jack on the decoder allowing the display of pure component (Y/C) video as it was uplinked to the satellite. This appears to be most advantageous on those channels which are broadcast using digital tape or fiber optic cable as the source. On those channels, use of the Y/C port can avoid the conversion from the digital component signal to NTSC making very high quality images possible.

Viewers who choose to use the RF output to connect their TV sets do not get stereo or surround sound audio to their TV speakers. Stereo sound is available only through the direct audio output jacks from a DBS decoder.

How is the DBS equipment installed?
The DSS and DISH hardware were both designed to be easy to install with no professional equipment required. The dishes can be installed anywhere there is a direct line of sight to the satellite with no trees or buildings in the way. Each service broadcasts all its channels from one position in the sky so the dish does not need to move. The dishes typically have a built-in audible signal meter or blinking LED to indicate signal strength to help position them during installation.

Homeowners can install the hardware, but professional installation is recommended. Thomson says the suggested retail price of a DSS basic installation is $200 but some installers charge lower fees. More complex installations cost more.

Self-installation kits are also available for both the DSS and DISH systems for about $70. These typically contain cables, a compass, a grounding block, a telephone T connector, and all hardware necessary to mount the dish and connect it up. They also typically include a videotape which demonstrates the installation process. All the necessary cables can be purchased at Radio Shack or other similar retailers, but those who don't have easy access to supplies may want to consider the self-installation kit.

What about watching the broadcast networks and local channels?
Local channels are not carried on any DBS services so local news and other local programming must be received over the air (or perhaps via cable).

A network programming package is available on most if not all services, but it can legally be received by only a fraction of the DBS customers. Network programming over the satellite is available only to those outside the terrestrial coverage areas of network affiliates as specified in the Satellite Home Viewer Act recently renewed in 1994. Those who can receive network affiliates will not be able to purchase this package and therefore must get network programming over the air (or by some other means).

According to the SHVA, if you can

receive the networks using a roof-top antenna if you are not eligible to receive them over the satellite. You must also not have subscribed to cable for 90 days. Note that this is Congressional legislation, not an FCC regulation. The full text of the law is available on the Web at: http://www.law.cornell.edu/uscode/17/119.html

This legislation may change in the future, however. EchoStar has plans to uplink local channels from several major cities and spot-beam them back for reception over their DISH system. This could make reception of local channels possible which many consider a necessity for widespread acceptance of DBS.

I live in the city where there are tall buildings. Can I receive a DBS service?
You must have a direct view of the satellite to get any DBS service with no trees or buildings in the way.

Where in the sky are the DBS satellites positioned?
The DBS satellites operate from 22,300 miles above the equator at various positions across the U.S. Dishes are pointed toward the south with the angle above the horizon dependent on the distance north of the equator. Those in the northern part of the U.S. (such as Minnesota) see the satellites about 25-35 degrees above the horizon. Those in the southern part see it higher in the sky.

The DIRECTV/USSB satellites are at the 101 degree West orbital location which is above a North/South line running through western Nebraska. Viewers in the central portion of the U.S. (such as Texas or the Dakotas) see the satellite about straight to the south. On the East coast it is slightly west of south and on the West coast it is slightly east of south.

The EchoStar/Sky Angel satellites are at the 119 degree orbital position. This is above a north/south line running through western Nevada. Those on the west coast will see the satellite about straight to the south. All others will see it to the southwest.

Can I connect more than one TV to a single dish?
The DBS dishes connect to the decoders with coaxial cable. The dish electronics have either one or two coaxial connections depending on the model so at most two decoders can be connected to one dish. It is important to purchase a package which can allow more than one decoder to connect to a dish if viewers ever want to hook up more than one decoder in a household.

Channel Master and perhaps other companies sell a MultiSwitch which takes both coaxial outputs from the dish and allows up to four decoders to be connected to it. Note that the base units from most manufacturers can only be connected to one decoder, so the Deluxe unit is necessary in this configuration.

DIRECTV and USSB say in order to authorize more than one decoder at a location, there must be a telephone connection at each decoder.

Is the telephone connection really necessary for the DIRECTV/USSB service?
Yes if you want to take advantage of all services available from DIRECTV. The phone line is used to verify the location of the DSS unit and manage the blackout restrictions imposed by the professional sports leagues.

You must be connected to a phone line to be authorized to receive regional sports networks or pro sports packages. It is also required to purchase impulse PPV movies and other special events.

Many viewers have never connected their systems to a phone line and don't feel they need it. Some recreational vehicle owners take the DSS system with them when they travel and others take systems to cabins or other remote locations occasionally where no phone line is available.

How do the conditional access systems work?
On most systems, the DBS decoders accept a credit-card sized processor board called a SmartCard which plugs into the front and allows the decoder to receive authorized programming. The authorization stream is sent on each transponder along with the video and audio information. The SmartCard can be inexpensively and easily replaced by the owner if necessary to help curb piracy.

Pirate DSS SmartCards have been developed and are being sold now, mostly in Canada. Other systems have not yet been broken. DIRECTV has issued several Electronic Counter Measures (ECMs) which have temporarily shut down the pirate cards and are also distributing smart-card replacements which are expected to make the existing pirate cards unusable. They have pushed for several indictments of pirate card manufacturers and have helped successfully prosecute some. They say they will vigorously fight against piracy to protect their programmers.

What other DBS services might be available?

The potential for data services is perhaps the most exciting aspect of the DBS services. Because the signals are sent as digital packets, the systems can send video, audio, and computer data in any combination to the decoders. Most if not all of the decoders contain a high-speed data port which can be connected to a computer or another external decoder. The 24 MHz bandwidth of each transponder can send an enormous amount of information (at least 23 MBits of data per second.

DIRECTV seems to be the first company who will take advantage of data service capabilities. They haved partnered with MicroSoft Corp. to produce a Windows 95 based PC system which can receive DIRECTV programming in combination with data services. Hardware will be manufactured by Adaptec and perhaps others. This is expected to be available late in 1997 or early in 1998. DIRECTV says data services will be a very important part of their business in the future.

We may discover that data services are available only through the use of a PC-card based decoders and those with standard decoders will not be able to receive data services though the data port(s).

What are some of the disadvantages of the DBS systems?

Network affiliates provided on the DBS services cannot be received by the majority of U.S. households.

Because of the broadcast frequencies used by the DBS providers, outages can occur as a result of severe thunderstorms in all DBS systems. The satellites are focused to send more power to rainier areas to help minimize this problem, but it does exist.

There are occasional visible digital artifacts which some viewers find objectionable. Some claim this is very distracting while others hardly notice it. It appears to be quite subjective. Nevertheless digital artifacts are a part of the DBS services.

Some cable TV customers with cable-ready VCRs and TVs are used to being able to watch one channel and record another or set their VCR to record two different cable channels while they are out. The DBS systems, like any system which requires a decoder, can only decode one channel at a time so a separate decoder must be purchased for each TV or VCR which are to be used at the same time. Also, some models don't have a program timer to use with a VCR's timer to record programs on more than one channel at a time.

Many on the west coast are disappointed that programs appear very early since the services use east coast feeds for most of their programming.

Although most if not all decoders contain a parental lockout feature, occasionally violent and sexually explicit programs are broadcast with no rating so they are available to all viewers who set the lockout limit at a typical setting. This makes the parental lockout ineffective.

Where are the DBS uplinks located?

DIRECTV uses a state-of-the-art all digital facility in Castle Rock, Colorado to uplink all programming to the DIRECTV satellite. The center includes several receiving stations and four 13-meter uplink dishes. Programming is provided to the uplink facility via satellite, over fiber optic cable, and through the use of digital tape.

Equipment in DIRECTV's broadcast center includes more than 300 Sony digital Betacam video recorders, a digital routing system that includes more than 800 inputs and outputs, and 50 automated playback and recording systems.

USSB uses a 20,000+ square foot all-digital uplink facility in Oakdale, Minnesota which is near Saint Paul. They are using two 9-meter Ka-band uplink dishes which are inside a specially constructed microwave-transparent atrium which shields them from exposure to the weather.

EchoStar uses a $40 million all-digital uplink facility in Cheyenne, Wyoming.

AlphaStar uses a $40 million all-digital uplink facility in Oxford, Connecticut which was recently purchased from GTE Spacenet. They use two 13-meter uplinks and one Simulsat 7-meter downlink.

Can you tell me more about the DSS system?
The three satellites are called DBS-1, DBS-2, and DBS-3. Each has 16 transponders powered by 120-watt traveling-wave tube amplifiers (TWTAs) suitable for both digital and analog transmissions.

The satellites operate in the Broadcast Satellite Services (BSS) portion of the Ku-band spectrum (12.2-12.7 GHz) and use circular polarization. They can deliver 58 to 53 dBW radiated power over the contiguous U.S. and southern Canada.

Each spacecraft weighs 3800 pounds and measures 7.1 meters across and 26 meters long with antennas and solar panels deployed. The solar panels generate 4300 watts of electrical power.

The DSS system uses Quadrature Phase Shift Key (QPSK) modulation to encode digital data on the RF carriers. The audio is MPEG-1 Layer II encoded. Surround sound can be achieved by encoding the audio with Dolby Pro-Logic before MPEG encoding. The video is encoded using MPEG-2 syntax with up to CCIR 601-1 sampling rates which is capable of up to 720 x 480 images although lower resolutions are currently being used.

The system uses a statistical multi-program encoder called a StatMux that dynamically varies the bit rate according to video content taking into consideration other programs multiplexed on the same transponder.

Each of the DBS satellites can be configured for either sixteen 120 Watt transmissions or eight 240 Watt. This is based on the DC power generating capability of their solar panels.

The DSS architecture can broadcast 40 Mbits/sec per transponder in either of two error control modes. In High mode, 30 Mbps is allocated to information and 10 Mbps to error control. In Low mode, 23 Mbps is allocated to information and 17 Mbps is allocated to error control.

High mode requires about 3dB more signal power to achieve an end-to-end availability equivalent to Low mode.

DBS-1 is running in Low mode while DBS-2 and DBS-3 are running in High mode. Therefore DIRECTV and USSB have 16 transponders at 240 Watts in High mode and 16 at 120 Watts in Low mode. A fourth satellite could be added to bring them all to 240 Watts, but DIRECTV says there are no plans for a fourth satellite at this time.

How many of these systems have sold and how many do they expect to sell?
DIRECTV and USSB claim over 2.5 million authorized decoders to date with that number climbing by thousands every day. Primestar claims about 1.75 million subscribers to date and they expect a comparable growth rate to DIRECTV and USSB. EchoStar now has nearly 0.5 million subscribers and AlphaStar has about 50,000.

DIRECTV has forecasted 10 to 12 million systems sold within six years of the start of their operation (which started in 1994). USSB has said they expect to have 40 million subscribers within 10 years of the start of operation. EchoStar expects to have 3 million customers by the year 2000. Most of these seem wildly optimistic.

Industry experts say they expect to see about 12-16 million DBS subscribers by the year 2000. The DBS companies continue to fall short of their very optimistic estimated subscriber numbers and are selling at a slower rate than anticipated.

DIRECTV believes it will break even in mid 1997 when they expect to have 3 million subscribers. USSB says their break-even point is closer to 1.5 million subscribers which they expect to hit in mid-1997.

Will DBS hardware prices drop in the near future?
DBS decoder prices have dropped drastically in the last few months mostly in response to very low prices introduced by competition from EchoStar. DISH equipment is now priced at $200 to $300 when purchased with a pre-paid programming subscription which has severely undercut other providers' prices. This promotion has proven to be

very popular and is rapidly expanding EchoStar's customer base. DIRECTV and USSB have now matched the prices using rebates and other promotions so a $200 DSS hardware system is also available with a pre-paid programming subscription. This represents a drastic drop from the $700 to $900 price of first generation decoders two years ago.

Can you straighten out some of the acronyms?
The term DBS should be used when referring to all of the available Direct To Home services so PrimeStar, DIRECTV, AlphaStar, EchoStar/Sky Angel and USSB are all DBS services. DSS refers to the equipment used only by the DIRECTV/USSB service and should not be used to describe any other system.

DISH refers to the equipment used only by EchoStar/Sky Angel. TVRO usually refers to traditional large dish systems although technically all dishes could be considered TVRO systems.

What DBS system should I get?
That depends on your desired programming, your tolerance for dish size, what you can afford to spend, what orbital slots you can see from your location, and several other issues.

DIRECTV and USSB have the highest channel capacity and are probably the best for subscribers who want the most possible choice and can afford a number of services. NFL football and other sports enthusiasts will also want DIRECTV as well as those who like a lot of Premium services and Pay Per View movies and events. They will likely be the first with significant data services as well.

PrimeStar requires the lowest initial investment and does not require the subscriber to be responsible for equipment repairs. Although they use a larger dish, they have the lowest cost of entry and are very popular with rural customers.

EchoStar is the low-price leader with the most inexpensive hardware and programming costs. Although they have less channel capacity than DIRECTV, they have more superstation type programming and are the only source for Star Trek programming. Their Sky Angel partner will be the first choice for those looking for Christian religious programming as well.

AlphaStar appears to be the choice for those outside the continental U.S. as well as those who want X-Rated adult programming.

Each of the service providers are severely in debt and are continuing to operate at a loss. When choosing a DBS service provider, viewers should note whether or not it has enough cash to continue to operate so they are not left holding unusable hardware if a service were to cease operation. It appears AlphaStar is having some difficulty at this point, but all other providers seem as though they will be around for the foreseeable future.

What is High Power DBS and how does it differ from DBS?
Several years ago the FCC reserved a portion of broadcast spectrum and reserved several U.S. satellite orbital positions for a class of television service they called Direct Broadcast Satellite (DBS). The satellite locations are spaced nine degrees apart from others broadcasting in the same frequency range (rather than two for conventional satellites) and these satellites are allowed to broadcast at a higher power providing interference-free reception on very small satellite dishes. This is the FCC's definition of DBS and they have specifically licensed several companies to provide DBS services including DIRECTV, USSB, EchoStar, and MCI Communications.

It is also possible for companies who are not licensed DBS broadcasters to offer Direct To Home services from conventional satellites. To the consumer these Direct To Home services look identical to licensed DBS services, except that they generally require a somewhat larger dish (although still much smaller than conventional dishes) and they also require professional installation. As a result, the definition of DBS is now generally used for any Direct To Home service using small satellite dishes from a fixed satellite position but the term High Power DBS is used for FCC defined DBS services.

The FCC has set aside eight orbital positions at the equator for U.S.-owned High Power DBS services of which four are to provide service over the east coast and four over the west. At each of these slots the FCC is permitting a maximum of 32 broadcast frequencies (transponders). The FCC assigns DBS frequencies to applicants in a way that gives them an equal number of orbital positions from east coast satellites and west coast satellites. The idea is that each company can provide service to the entire continental U.S. by broadcasting from both their east and west satellites.

However, with today's technology, three of the four eastern positions (101 degrees west longitude, 110 degrees w, and 119 degrees w) are at longitudes which can actually provide coverage to the entire continental U.S. These are the most desirable slots and they are in very high demand by the DBS companies.

Are any other DBS services planned?

Communications giant MCI has partnered with Rupert Murdoch's News Corporation and together they have purchased rights for 28 transponders at the 110 degree orbital slot. They have recently announced plans to merge with EchoStar to provide a single 500 channel service called simply "Sky" which will operate from several orbital positions including EchoStar's current 119 degree location and 110, 61.5, and 148 degree spots.

There are several regulatory hurdles to overcome, but if allowed to occur the Sky service will likely become the premier DBS service available and a very signifcant competitor to Cable TV. They plan to uplink local affiliates from several major cities and spot-beam them thereby allowing DBS viewers to receive their local affiliates in many parts of the country. Local affilate broadcast is considered by many to be the most important impediment to real competition for cable TV. It will be interesting to see how this plays out in the coming months.

USSB owns the rights to 3 transponders at the 110 degree orbital slot. It appears they are preparing to launch some kind of service at that location as well.

Cable giant TCI has failed in their attempt to get approval to use orbital slots assigned to Canada and/or Mexico for their own high-power DBS service. They currently own 11 transponders at 119 degrees but they were trying to get more since that would likely restrict them to a service with 100 or less channels. They have announced plans to launch satellites into the 119 location and offer some kind of service from there but details are scarce at this point.

Cable giant TCI has failed in their attempt to get approval to use orbital slots assigned to Canada and/or Mexico for their own high-power DBS service. They currently own 11 transponders at 119 degrees but they were trying to get more since that would likely restrict them to a service with 80 or less channels. There is still some chance they will reach an agreement with the other DBS provider at 119 (EchoStar), but other deals are also still possible.

Where can I get more information?

Richard R. Peterson
The DBS Connection
1480 Lark Avenue
Maplewood, MN 55109
USA

You can get more information using the following phone numbers:

DIRECTV Consumer Information
1-800-DIRECTV

DIRECTV Dealer Information
1-800-323-1994

USSB Consumer Hotline
1-800-BETTERTV

USSB Dealer Hotline
1-800-898-USSB

Sony Information
1-800-838-7669

Primestar Information
1-800-PRIMESTAR

EchoStar Information
1-800-333-DISH

AlphaStar Information
1-888-ALPHASTAR

This document is updated and submitted every few months to the rec.video.satellite.dbs Internet news group.

It is also available on the World Wide Web from John Hodgson's DBS Home Page at:

http://www.dbsdish.com/index.html-
http://www.dbs-online.com/DBS/

CHAPTER 5
INSTALLATION

BEFORE MAKING ANY SETTINGS WHATSOEVER, FAMILIARIZE YOURSELF WITH ALL SPECIFIC FUNCTIONS OF YOUR SATELLITE SYSTEM AND READ THE INSTALLATION INSTRUCTIONS THOROUGHLY

5.1. INSTALLING THE OUTDOOR UNIT

Trying to find a satellite is easier said than done. In the old days, finding any satellite at all was quite difficult because satellite receivers didn't have a digital frequency readout. Nowadays, receivers come with pre-set channels, which allow you to tune accurately to a station on a desired satellite.

INSTALLING A FIXED SATELLITE SYSTEM
What Satellite Where?
To understand the following graphics, which will help you to find any satellite at any site location, some essential jargon needs to be learned. For aligning a satellite dish we use two terms: *Azimuth and Elevation*. For determining the elevation angle we use an inclinometer or a plumb line and a protractor and for the azimuth we use a compass. The first figure explains the correlation between azimuth and elevation, showing that due south has a 180 degrees azimuth.

Figure 5.1-2 shows a graph with azimuth and elevation details for any given satellite. At the intersections of the receiving site's latitude grids (vertical)

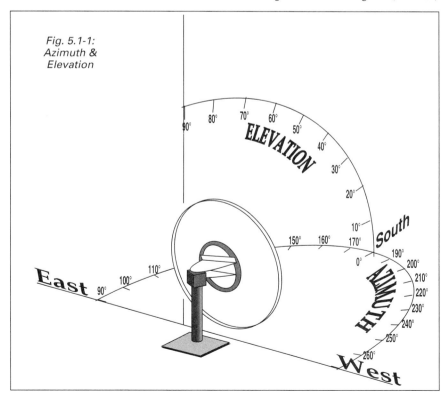

Fig. 5.1-1: Azimuth & Elevation

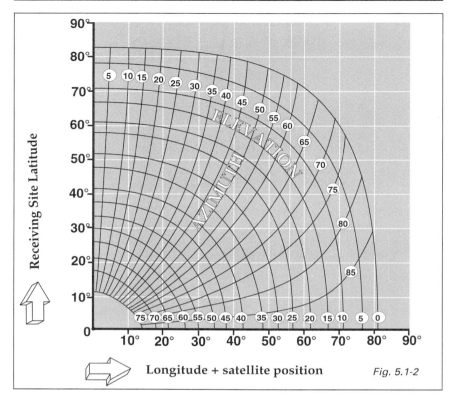

Fig. 5.1-2

and the longitude + satellite position grids (horizontal) the elevation/azimuth values for a certain satellite can be found.

EXAMPLE 1:
Intelsat 603 (11.003 GHz V, Muslim TV), location 34.5° West. Receiving site Amsterdam, longitude 5.5° East. The difference is 40°. With a receiving site latitude of 52.5° North we will find an elevation value of 20° and an azimuth value of 47.5° relative to true south. So the angle between the dish and the satellite is 7.5 degrees wider than we would have anticipated based on the information of receiving site + satellite location (5.5 + 34.5 = 40). The actual angle is 180° + 47.5 = 227.5°.

EXAMPLE 2:
For Amsterdam (latitude 52.5° North) the elevation is zero with a longitude + satellite position value of 75°. So with a receiving site longitude coordinate of 5.5° East, the most westerly located satellite (with zero elevation) is at 69.5° West.

Choosing the Receiving Site
Although sometimes planning regulations require a satellite dish to be placed in an obscured position, it is essential that no objects block a clear view to the satellite. But when does a tree, a roof ridge, a chimney or an entire housing block become a problem?

The following graphic and calculation method provide the necessary information which allows you to choose the best place to put up your satellite dish.

The European ASTRA Satellites (19.2° East) have an elevation of ± 30° when seen from the Netherlands. Suppose your neighbour's house appears to block a clear view to the satellites. Using figure 5.1-3, you can accurately calculate the distance you have to keep by measuring the exact height of that house. Let's say the house is 10 meters high. With an elevation angle of 30° you would need a rising of 57.74 cm per meter for a clear view. Knowing that, you can easily calculate that you have to install your dish at least 20 meters away to avoid blockage.

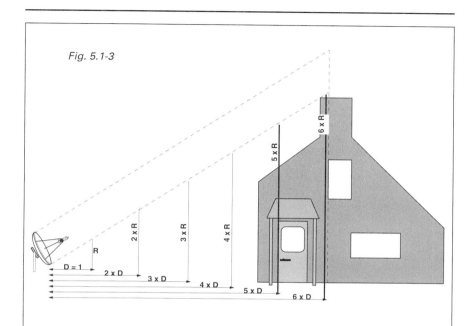

Fig. 5.1-3

ELEVATION ANGLE "A" IN DEGREES

Elevation Angle "A"	Rising "R"				
1°	1.75 cm	20°	36.40 cm	46°	103.55 cm
2°	3.50 cm	22°	40.40 cm	48°	111.06 cm
4°	7.00 cm	24°	44.52 cm	50°	119.17 cm
6°	10.51 cm	26°	48.77 cm	52°	127.99 cm
8°	14.05 cm	28°	53.17 cm	54°	137.64 cm
10°	17.64 cm	30°	57.74 cm	56°	148.26 cm
12°	21.26 cm	32°	62.49 cm	58°	160.03 cm
14°	24.93 cm	34°	67.45 cm	60°	173.20 cm
16°	28.67 cm	36°	72.65 cm	62°	188.07 cm
18°	32.49 cm	38°	78.13 cm	64°	205.03 cm
		40°	83.91 cm	66°	224.60 cm
		42°	90.03 cm	68°	247.51 cm
		44°	96.57 cm	70°	274.75 cm

Distance "D" = 1 meter, rising "R" in centimeters. $R = D \times \tan A$

INSTALLING A MOTORIZED SATELLITE SYSTEM

Needed: inclinometer, water level, square, signal strength meter, small tv set (LCD TV), F-connectors, crimp tool, self amalgamating tape

Installing a motorized system is much more complicated than installing a fixed satellite dish. Professional satellite riggers need the best part of a day to accurately install and align a motorized satellite system, but enthusiastic do-it-yourselfers are more likely to need a whole weekend to get the job done. If you add a few more evenings to carry out some fine adjusting you seriously have to consider whether you are prepared for such a delay before watchable pictures are available. On the other hand, you can safe yourself a lot of money adding up to a few hundred dollars. Before attempting any installation at all first try to understand some of the installation theory.

BEFORE YOU START, MAKE SURE THAT THE CENTRAL MAST IS PERFECTLY PLUMB! A VERY SLIGHT DEVIATION AFFECTS THE TRACKING CURVE OF THE SATELLITE ANTENNA AND WILL SERIOUSLY IMPAIR THE RECEPTION QUALITY

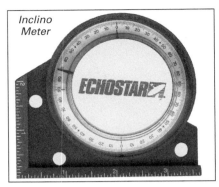

Inclino Meter

Polar Mount Principle. Polar mounts are designed to enable the dish to track the visible part of the Clarke Belt from almost any site location. The basic idea of the polar mount is an invention of astronomers, who require a mount that enables telescopes to follow the stars as they appear to rotate round the earth. However, the satellite system's polar mount is not a real polar mount, but a modified version. The secret of the modified polar mount is that by tilting the pivoting axis slightly, just one motorized axis is needed. This is done by giving the polar axis the same angle as the site latitude. The problem with a polar mount is that the polar axis angle varies depending on the location of the dish site. Near the Equator, where the latitude is 0 the polar axis angle will be 0° too, meaning that it stands perpendicular to the mounting pole, but all other sites north or south require different settings. When looking at the polar mount's view of the sky it is clear that the shape of the dish track curve is similar to the shape of the satellite arc, but not quite.

Declination. For proper tracking of the satellite arc both the dish track curve and the satellite arc have to coincide. In order to alter the shape of the dish track curve we have to create another angle in between the dish and the polar axis. This angle is usually referred to as the *declination angle*, or *declination offset*. By altering this declination offset the swing of the dish can take account of the apparent flattening of the visible satellite arc at more northerly receiving site latitudes.

True South. For proper tracking of the satellite arc (in the northern hemisphere) it is absolutely essential that the total base plate assembly is facing True South. There are various ways of finding out True South. Two of the best ways are: a) using a compass, b) using the sun.

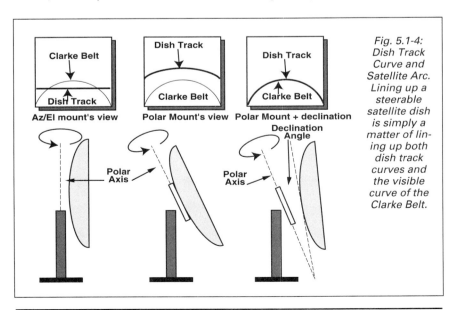

Fig. 5.1-4: Dish Track Curve and Satellite Arc. Lining up a steerable satellite dish is simply a matter of lining up both dish track curves and the visible curve of the Clarke Belt.

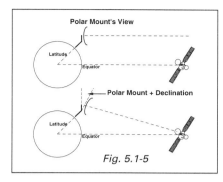

Fig. 5.1-5

FINDING TRUE SOUTH

using a compass: Any compass points along a magnetic line to the *Magnetic North Pole* - which is located near the north coast of Canada - and not to the *Geographical North Pole* (True North). At almost any site location in the world there is a difference in the reading on the compass and the actual North Pole, except when this site is located at an *agonic line*, an imaginary line with zero magnetic variation. Agonic lines can be found in all areas of the globe as the magnetic variation maps will show. In the northern hemisphere (north of the Equator), for all sites East of the agonic line the geographical North Pole is located East of the compass reading, whereas for sites West of the agonic line true north is located West of the compass reading. In the southern hemisphere, the whole situation is reversed. South of the Equator, from all sites East of an agonic line the geographical North Pole is located West of the compass reading, from all sites West of an agonic line true north is located East of the compass readout. If this is a little bit confusing, just have a look at Fig. 5.1-6 in which you will find the compass readings for Los Angeles and Boston. North of the Equator magnetic variation is negative (-) when West of the agonic line, positive (+) when East. South of the Equator settings are again reversed.

Magnetic variation at most locations in the world can be found from the maps shown. However, you could also ascertain the exact value for your area by phoning your local airport information service. Because the magnetic North Pole is never stationary, but drifting slightly northwards from year to year, the current magnetic variation figures for a certain area are likely to change slightly as well. You should also take into account that a compass readout is always inaccurate near iron objects, e.g. the total mount structure.

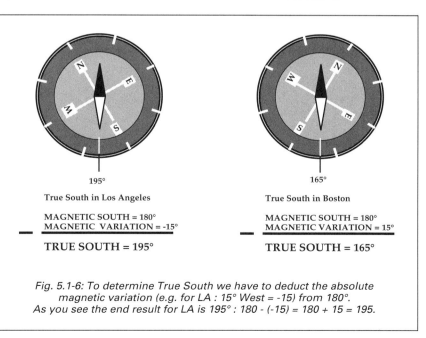

Fig. 5.1-6: To determine True South we have to deduct the absolute magnetic variation (e.g. for LA : 15° West = -15) from 180°. As you see the end result for LA is 195° : 180 - (-15) = 180 + 15 = 195.

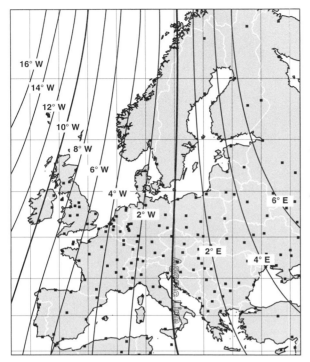

Fig. 5.1-7: EUROPE.
The Isogonic Lines connect points with equal magnetic variation, the Agonic Lines are lines with zero magnetic variation.

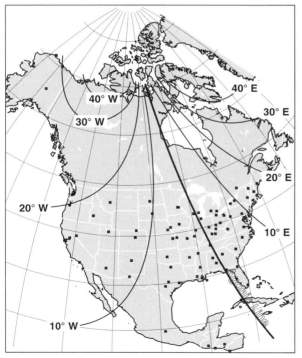

Fig. 5.1-8: NORTH AMERICA.
The Isogonic Lines connect points with equal magnetic variation, the Agonic Lines are lines with zero magnetic variation.

Fig. 5.1-9: ORIENT & NEW GUINEA. The Isogonic Lines connect points with equal magnetic variation, the Agonic Lines are lines with zero magnetic variation.

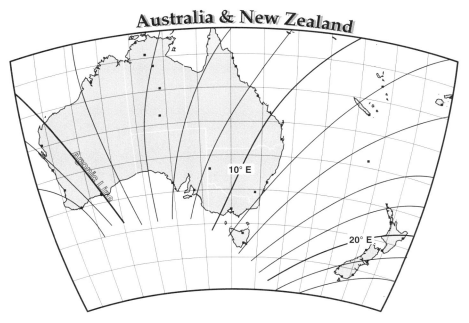

Fig. 5.1-10: AUSTRALIA & NEW ZEALAND. The Isogonic Lines connect points with equal magnetic variation, the Agonic Lines are lines with zero magnetic variation.

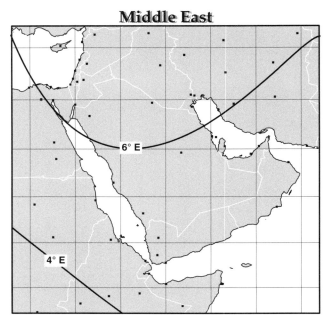

Fig. 5.1-11: MIDDLE EAST. The Isogonic Lines connect points with equal magnetic variation, the Agonic Lines are lines with zero magnetic variation.

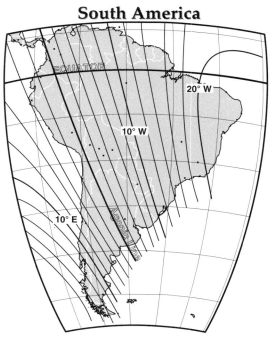

Fig. 5.1-12: SOUTH AMERICA. The Isogonic Lines connect points with equal magnetic variation, the Agonic Lines are lines with zero magnetic variation

using the sun: In the northern hemisphere the sun is due south when the shadow it casts is shortest. In other words, when seen from earth the sun reaches its highest point above a certain location exactly in the south. Unfortunately, for most locations and on most days of the year this is not exactly at noon. You can determine where and when the sun is at its highest point with a few simple calculations. At Greenwich (UK) at 1200 o'clock Greenwich Mean Time (GMT) the sun is at its highest point. Now, the earth makes a full revolution every 24 hours (= 360°), therefore 1° difference in longitude is 4 minutes: (24x60)/360 = **4**. Amsterdam is located at a longitude of 5° East, which means that in Amsterdam the sun reaches its highest point at 5 x 4 = 20 minutes earlier, i.e. **1140 GMT**. In local Dutch time this will be one hour later and even two hours when summer time is in force. Another means to ascertain the sun's culmination is by subtracting the sunrise time from the sunset time. These times vary from location to location, but most local papers print this information daily. For the same day and location I made the following calculation:
  sun up 05.50
  sun down 21.30
  difference 15.40 hours
  divided by 2 = 7.50
  sunrise time 5.50 + 7.50 = 12.100 = 13.40 (Dutch Summer Time)

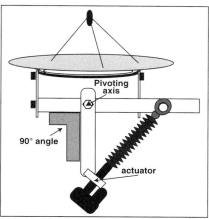

Fig. 5.1-13: Apex Position

GET TRACKING
The figures show the basic layout of all polar mounts. The exact mechanical details concerning adjusting the necessary angles vary widely from system to system. However, basic principles stay the same, except when installing an offset dish!! In general, offset dishes are somewhat harder to install yourself, but with specific manufacturer's information it should be feasible.
(**NB:** for exact angle figures see chapter 6.1.)

1) Set the total construction in its apex position, i.e. the mounting plate of the antenna should be at a 90° angle to the

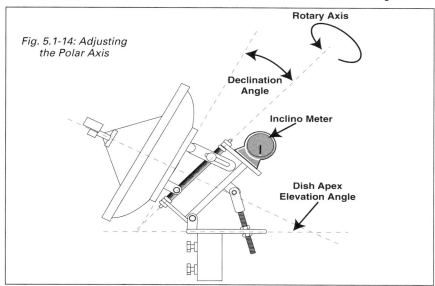

Fig. 5.1-14: Adjusting the Polar Axis

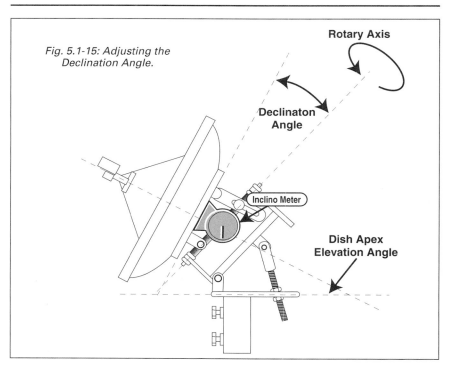

Fig. 5.1-15: Adjusting the Declination Angle.

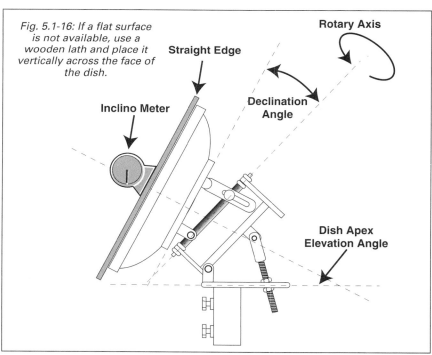

Fig. 5.1-16: If a flat surface is not available, use a wooden lath and place it vertically across the face of the dish.

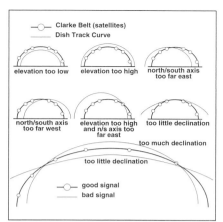

Fig. 5.1-17: Dish Tracking Errors. Most tracking problems are caused by an incorrect North/South orientation of the entire mount. However, if the polar axis or declination angles have not been set correctly, tracking will also be inaccurate.

pivot assembly. Use a carpenter's square to ensure a 90° angle at any time. The actuator (jack) can be used to secure this position by sliding the jack in its clamp until the right orientation is acquired. Then loosen the bolts of the mount's base slightly and aim the total assembly roughly to the south. (make sure that the antenna plate stays at a 90° angle to the pivot assembly.)

2) First adjust the polar elevation angle. (roughly the same number of degrees as the receiving site latitude, e.g. Amsterdam 52.5° / polar elevation 52.5°) Place the inclinometer on the rear of the polar axis and adjust this axis for the correct polar elevation.Then tighten the polar axis bolts.

3) Now we have to adjust the declination offset angle. Determine the correct declination offset angle for your receiving site

EXCEPTION: DECLINATION ANGLE FIGURES IN CHAPTER 6.1 ONLY APPLY TO CENTER FOCUS DISHES. FOR OFFSET DISHES USE THE MANUFACTURER'S RECOMMENDED MEASURING POSITION AND FIGURES.

latitude from the table in chapter 6.1. Make sure the total assembly is still in its apex position. Place the inclinometer on the mounting plate at the back of the antenna (If a flat surface cannot be found at the back of the antenna use a wooden lath and place it vertically across the face of the dish).

4) Now we have to determine whether the resultant of the polar angle minus the offset angle is equal to the elevation figure listed in the elevation/declination table (chapter 7.1). When measuring with a lath across the dish's surface you should read the inclination angle value (also listed) for your site location. In case the elevation/inclination angle is not equal to the given values, all angles should be rechecked. Just keep in mind that the total mount assemble must be measured in its apex position!!

5) At this stage the jack should be disconnected to ensure free movement of the total assembly. If all these steps have been carried out correcly, the mount should be roughly in its correct position. For proper fine adjustment we now have to start scanning the skies for any picture, even the weakest. A small tv set is a tremendous help, because a weak picture or even just a sync bar shows up much earlier than any reading on a signal strength meter. Choose a widely used frequency and polarity to start your scanning with. Then, when pictures start to show up try to determine the satellite that they are broadcast on. The following figure and schedule will help you to fine adjust the total mount assembly.

POLAR MOUNT CORRECTIONS

East	South	West	Correction
up	up	up	increase axis inclination
down	down	down	decrease axis inclination
up	down	up	decrease declation
down	up	down	increase declination
up	-	down	rotate east
down	-	up	rotate west

*Note: the directions listed above only north of Equator.

MAINTENANCE OF THE OUTDOOR UNIT

The biggest enemy of the outdoor unit is water. Therefore it is absolutely essential to protect all parts sufficiently from the effects of water. Let's start with the most expensive part, the feedhorn/LNB assembly. Usually this unit is constructed in 3 separate elements, the feedhorn, the polarizer and the LNB. The rubber rings supplied should seal off every joint, but to make sure they really do you could add some Vaseline® (petroleum jelly) in between the various parts (not too much because the Vaseline could get into the waveguide, spoiling the exact shape). The LNB itself is totally water resistant, but the connection to the cable (the F-

Picture above: This professional Kathrein offset dish is equipped with covers that protect the entire LNB/polarizer unit against water ingress (courtesy Kathrein)

socket) usually isn't. Some manufacturers supply water resistant F-connectors, but generally the F-connector is of the conventional "crimp" type. Any water intrusion into the connection between the LNB and the antenna cable could seriously impair reception or even totally ruin the LNB or receiver. The best way to protect both LNB and the cable connection is by shielding them off with a protective cover, like this professional Kathrein satellite dish.

Most C-Band antennas in the USA are equipped with such a cover as standard. The next best solution is to use self-amalgamating tape to seal off the cable connection. This is not real tape but some sort of sticky compound that you wrap around the connector and squeeze. The only disadvantage is that when you want to readjust your dish and therefore need to link up an alignment meter, you have to get rid of the sticky stuff first before you can disconnect the cable. Whenever you do this, be careful and make sure that the center conductor stays clean. The third solution is a more permanent one. Some shrink sleeves have a solid compound on the inside which becomes fluid once heated. This shrink sleeve together with the compound is usually a sufficient (but permanent) solution. However, the big danger with shrinking is overheating the delicate LNB circuitry, so be very careful to avoid this.

The antenna reflector itself needs very little maintenance. In general, it is sufficient to clean the surface regularly for this will not only guarantee proper reception but also enhance the life of the reflector. To avoid water droplets build-

Picture left: This is Swedish MicroWave's solution to water ingress. This cover protects both the magnetic polarizer and the LNB.

ing up on the reflector's surface it might be advisable to wax the reflector with ordinary car wax. This will also make it easy to remove dirt and dust.

The antenna mount does need some regular attention. Immediately after installing the total outdoor unit and having tightened all nuts and bolts it is wise to cover nuts, bolts and other parts with a thick layer of Vaseline. Do not use regular car rust preventives like Tectyl® because they might obstruct the moving mechanism inside the jack. Also make sure the jack is mounted the right way up. Types with drip holes should be mounted with the motor up whereas with some other types without drip holes the motor should be down. Anyway, read the instructions on the jack carefully, then you are safe. Make sure all the cable entry points are water tight and don't forget to put drip loops into the cables. H-H mounts are relatively maintenance free, but they also do tend to rust.

5.2. INSTALLING THE INDOOR UNIT

After installing and aligning the outdoor unit, the indoor unit has to be fitted in with the rest of the tv and video equipment. Before connecting any cables and leads it is important to familiarize yourself not only with the new satellite system, but also with your VCR and tv set. The various operating manuals and especially the chapters on connection possibilities should be read carefully. Inferior pictures with odd patterning, e.g. the so called "fish-bone" distortion, are often a result of not fully exploiting each component's possibilities.

INTERCONNECTING ALL UNITS
First of all make sure that your new satellite system is working satisfactorily by attaching it directly to your TV. Most modern TVs allow you to input a signal in two ways. One is by means of the regular antenna socket and the other is a form of direct video connection. In general satellite receivers are pre-tuned to one or more of the most popular satellite channels. In case there is no satellite signal a switchable test bar signal can be generated on the antenna output socket. In most cases this UHF signal is tunable between ch 30-39 by means of a little

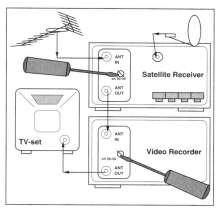

Fig. 5.2-1: Interconnecting Units with a Coaxial Cable.

screw at the back of the receiver. The next step is similar to tuning in a VCR. Put the tv in scanning mode and allocate a channel number as soon as the test bar has been found. Now you can check if all the functions of the new satellite system are working fine.

The coaxial connection is only one way of linking various video sources. However, the process of modulating audio and video signal on to a UHF carrier surely detracts from the overall picture and sound quality. Most satellite receivers and VCR's are not even capable of reproducing stereo audio over the UHF RF-socket. Fortunately most video and satellite equipment offers the possibility of direct video and audio connecting. In case of a complicated setup with a VCR, a satellite receiver and probably an amplifier, we need to create a kind of dialogue between the various components. There are various ways of achieving this dialogue, but there is an even larger variety of connectors and sockets available. A few examples of video and audio connectors and sockets are explained below.

The SCART Connector Virtually all modern European video equipment is fitted with one or more *SCART* connectors. SCART - like DIN (Deutsche Industrie Norm) - is an acronym for Syndicat des Constructeurs d'Appareils Radio recepteurs et Televiseurs (syndicate of radio and tv manufacturers). The SCART connector, also known as Euro connector or Peritel connector, was developed by French engineers in 1980 to spell an end to the incompatibility of the Japanese

PHONO and European DIN standards. Although the SCART pin assignment is uniform among manufacturers, not all pins are used in many cases.

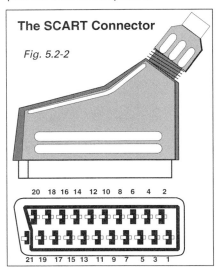

The SCART Connector

Fig. 5.2-2

Regular SCART pin assignment:
1) Right Audio **Out**
2) Right Audio **In**
3) Left Audio **Out**
4) Audio Ground
5) Blue Ground
6) Left Audio **In**
7) Blue **In**
8) Switching Volt.
9) Green Ground
10) Intercomm. Line
11) Green **In**
12) Intercomm. Line
13) Red Ground
14) Intercomm. Line Ground
15) Red **In**
16) Fast RGB Blanking
17) Comp. Video **Out** Ground
18) Comp. Video **In** Ground
19) Comp. Video **Out**
20) Comp. Video **In**

Decoder SCART pin assignment:
1) Right Audio **Out**
2) Right Audio **In**
3) Left Audio **Out**
4) Audio Ground
5) -
6) Left Audio **In**
7) -
8) Switching Voltage
9) -
10) -
11) -
12) -
13) -
14) -
15) -
16) -
17) Baseband **Out** Ground
18) Comp. Video **In** Ground
19) Baseband **Out**
20) Comp. Video **In**

6 Pin DIN-AV Connector. This German standard connector is still used in some tv sets in Europe, but most sets nowadays are equipped with one or more SCART sockets. Connections for stereo audio, video signal (also FBAS, German acronym for Farb-, Bild-, Austast-, and Synchronsignal) and a switching voltage pin are available. When connecting a satellite receiver or VCR by means of the 6 in DIN-AV connector, this socket is automatically choosen by the tv set. All signals coming from the TV's own tuner will be interrupted.

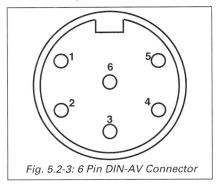

Fig. 5.2-3: 6 Pin DIN-AV Connector

6 Pin DIN-AV Connector pin assignment
1) Switching Voltage
4) Main (Left/Mono audio)
2) FBAS (Comp. video)
5) 12 Volts
3) Common Ground
6) Sub (Right audio)

D-15 Sub Connector. This connector is widely in use in Europe to connect decoders and satellite equipment.

D-15 connector

Fig. 5.2-4: D-15 Sub Connector

D-15 Sub Connector
1) Audio Left **In**
2) Video **In**
3) Video Switch Signal
4) Baseband **Out**
5) Clamped Video **Out**
6) Audio Right **In**
7) Audio Switch Signal
8) -
9) -
10) -
11) -
12) Audio Left **Out**
13) Audio Right **Out**
14) -
15) -

CHAPTER 6
IMPORTANT TABLES

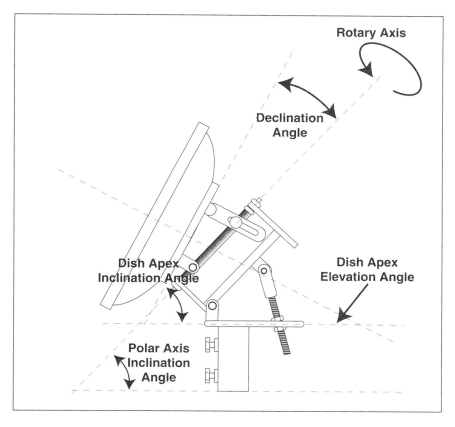

Fig. 6.1-1: Various Angles

6.1. VARIOUS ANGLES

MODIFIED POLAR MOUNT ANGLES
(center focus dishes only)

Latitude	Polar Axis	Declination	Inclination	Elevation
0.00°	0.00°	0.00°	0.00°	90.00°
0.50°	0.51°	0.08°	0.59°	89.41°
1.00°	1.02°	0.16°	1.18°	88.82°
1.50°	1.53°	0.23°	1.76°	88.24°
2.00°	2.05°	0.31°	2.36°	87.64°
2.50°	2.56°	0.39°	2.95°	87.05°
3.00°	3.07°	0.47°	3.54°	86.47°
3.50°	3.58°	0.54°	4.12°	85.88°
4.00°	4.09°	0.62°	4.71°	85.29°
4.50°	4.60°	0.70°	5.30°	84.70°
5.00°	5.11°	0.77°	5.88°	84.11°
5.50°	5.63°	0.85°	6.48°	83.52°
6.00°	6.14°	0.93°	7.07°	82.93°
6.50°	6.65°	1.01°	7.66°	82.35°
7.00°	7.16°	1.08°	8.24°	81.76°
7.50°	7.67°	1.16°	8.83°	81.17°
8.00°	8.18°	1.24°	9.42°	80.58°
8.50°	8.69°	1.31°	10.00°	79.99°
9.00°	9.20°	1.39°	10.59°	79.41°
9.50°	9.71°	1.47°	11.18°	78.82°
10.00°	10.23°	1.54°	11.77°	78.23°
10.50°	10.74°	1.62°	12.36°	77.65°
11.00°	11.25°	1.69°	12.94°	77.06°
11.50°	11.76°	1.77°	13.53°	76.47°
12.00°	12.27°	1.84°	14.11°	75.89°
12.50°	12.78°	1.92°	14.70°	75.30°
13.00°	13.29°	2.00°	15.29°	74.72°
13.50°	13.80°	2.07°	15.87°	74.13°
14.00°	14.31°	2.14°	16.45°	73.55°
14.50°	14.82°	2.22°	17.04°	72.96°
15.00°	15.33°	2.29°	17.62°	72.38°
15.50°	15.84°	2.37°	18.21°	71.79°
16.00°	16.35°	2.44°	18.79°	71.21°
16.50°	16.86°	2.52°	19.38°	70.62°
17.00°	17.37°	2.59°	19.96°	70.04°
17.50°	17.88°	2.66°	20.54°	69.46°
18.00°	18.39°	2.74°	21.13°	68.88°
18.50°	18.90°	2.81°	21.71°	68.29°
19.00°	19.41°	2.88°	22.29°	67.71°
19.50°	19.92°	2.95°	22.87°	67.13°
20.00°	20.42°	3.03°	23.45°	66.55°
20.50°	20.93°	3.10°	24.03°	65.97°
21.00°	21.44°	3.17°	24.61°	65.39°
21.50°	21.95°	3.24°	25.19°	64.81°
22.00°	22.46°	3.31°	25.77°	64.23°
22.50°	22.97°	3.38°	26.35°	63.65°
23.00°	23.47°	3.45°	26.92°	63.07°
23.50°	23.98°	3.52°	27.50°	62.50°

Latitude	Polar Axis	Declination	Inclination	Elevation
24.00°	24.49°	3.59°	28.08°	61.92°
24.50°	25.00°	3.66°	28.66°	61.34°
25.00°	25.51°	3.73°	29.24°	60.76°
25.50°	26.01°	3.80°	29.81°	60.19°
26.00°	26.52°	3.87°	30.39°	59.61°
26.50°	27.03°	3.93°	30.96°	59.04°
27.00°	27.53°	4.00°	31.53°	58.46°
27.50°	28.04°	4.07°	32.11°	57.89°
28.00°	28.55°	4.14°	32.69°	57.32°
28.50°	29.05°	4.20°	33.25°	56.74°
29.00°	29.56°	4.27°	33.83°	56.17°
29.50°	30.07°	4.33°	34.40°	55.60°
30.00°	30.57°	4.40°	34.97°	55.03°
30.50°	31.08°	4.47°	35.55°	54.46°
31.00°	31.58°	4.53°	36.11°	53.89°
31.50°	32.09°	4.59°	36.68°	53.32°
32.00°	32.59°	4.66°	37.25°	52.75°
32.50°	33.10°	4.72°	37.82°	52.18°
33.00°	33.60°	4.78°	38.38°	51.61°
33.50°	34.11°	4.85°	38.96°	51.04°
34.00°	34.61°	4.91°	39.52°	50.48°
34.50°	35.12°	4.97°	40.09°	49.91°
35.00°	35.62°	5.03°	40.65°	49.35°
35.50°	36.12°	5.09°	41.21°	48.78°
36.00°	36.63°	5.15°	41.78°	48.22°
36.50°	37.14°	5.21°	42.35°	47.65°
37.00°	37.63°	5.27°	42.90°	47.09°
37.50°	38.14°	5.33°	43.47°	46.53°
38.00°	38.64°	5.39°	44.03°	45.97°
38.50°	39.14°	5.45°	44.59°	45.41°
39.00°	39.65°	5.51°	45.16°	44.84°
39.50°	40.15°	5.57°	45.72°	44.28°
40.00°	40.60°	5.62°	46.22°	43.78°
40.50°	41.15°	5.68°	46.83°	43.17°
41.00°	41.65°	5.74°	47.39°	42.61°
41.50°	42.16°	5.79°	47.95°	42.05°
42.00°	42.66°	5.85°	48.51°	41.49°
42.50°	43.16°	5.90°	49.06°	40.94°
43.00°	43.66°	5.96°	49.62°	40.38°
43.50°	44.16°	6.01°	50.17°	39.83°
44.00°	44.66°	6.06°	50.72°	39.28°
44.50°	45.16°	6.11°	51.27°	38.73°
45.00°	45.66°	6.17°	51.83°	38.17°
45.50°	46.16°	6.22°	52.38°	37.62°
46.00°	46.66°	6.27°	52.93°	37.07°
46.50°	47.16°	6.32°	53.48°	36.52°
47.00°	47.66°	6.37°	54.03°	35.97°
47.50°	48.16°	6.42°	54.58°	35.42°
48.00°	48.66°	6.47°	55.13°	34.87°
48.50°	49.16°	6.52°	55.68°	34.32°
49.00°	49.65°	6.57°	56.22°	33.78°
49.50°	50.15°	6.61°	56.76°	33.24°
50.00°	50.65°	6.66°	57.31°	32.69°
50.50°	51.15°	6.71°	57.86°	32.14°
51.00°	51.65°	6.75°	58.40°	31.60°
51.50°	52.14°	6.80°	58.94°	31.06°
52.00°	52.64°	6.84°	59.48°	30.52°
52.50°	53.14°	6.89°	60.03°	29.97°

Latitude	Polar Axis	Declination	Inclination	Elevation
53.00°	53.63°	6.93°	60.56°	29.44°
53.50°	54.13°	6.98°	61.11°	28.89°
54.00°	54.63°	7.02°	61.65°	28.35°
54.50°	55.12°	7.06°	62.18°	27.82°
55.00°	55.62°	7.10°	62.72°	27.28°
55.50°	56.12°	7.14°	63.26°	26.74°
56.00°	56.61°	7.19°	63.80°	26.20°
56.50°	57.11°	7.23°	64.34°	25.67°
57.00°	57.60°	7.27°	64.87°	25.13°
57.50°	58.10°	7.30°	65.40°	24.60°
58.00°	58.59°	7.34°	65.93°	24.07°
58.50°	59.09°	7.38°	66.47°	23.53°
59.00°	59.58°	7.42°	67.00°	23.00°
59.50°	60.08°	7.45°	67.53°	22.47°
60.00°	60.57°	7.49°	68.06°	21.94°
60.50°	61.07°	7.53°	68.60°	21.41°
61.00°	61.56°	7.56°	69.12°	20.88°
61.50°	62.05°	7.60°	69.65°	20.35°
62.00°	62.55°	7.63°	70.18°	19.82°
62.50°	63.04°	7.66°	70.70°	19.30°
63.00°	63.53°	7.70°	71.23°	18.77°
63.50°	64.03°	7.73°	71.76°	18.24°
64.00°	64.52°	7.76°	72.28°	17.72°
64.50°	65.01°	7.79°	72.80°	17.20°
65.00°	65.51°	7.82°	73.33°	16.67°
65.50°	66.00°	7.85°	73.85°	16.15°
66.00°	66.49°	7.88°	74.37°	15.63°
66.50°	66.98°	7.91°	74.89°	15.11°
67.00°	67.47°	7.94°	75.41°	14.59°
67.50°	67.97°	7.97°	75.94°	14.06°
68.00°	68.46°	7.99°	76.45°	13.55°
68.50°	68.95°	8.02°	76.97°	13.03°
69.00°	69.44°	8.05°	77.49°	12.51°
69.50°	69.93°	8.07°	78.00°	12.00°
70.00°	70.42°	8.10°	78.52°	11.48°
70.50°	70.92°	8.12°	79.04°	10.96°
71.00°	71.41°	8.15°	79.56°	10.44°
71.50°	71.90°	8.17°	80.07°	9.93°
72.00°	72.39°	8.19°	80.58°	9.42°
72.50°	72.88°	8.21°	81.09°	8.91°
73.00°	73.37°	8.23°	81.60°	8.40°
73.50°	73.86°	8.25°	82.11°	7.89°
74.00°	74.35°	8.27°	82.62°	7.38°
74.50°	74.84°	8.29°	83.13°	6.87°
75.00°	75.33°	8.31°	83.64°	6.36°
75.50°	75.82°	8.33°	84.15°	5.85°
76.00°	76.31°	8.35°	84.66°	5.34°
76.50°	76.80°	8.37°	85.17°	4.83°
77.00°	77.29°	8.38°	85.67°	4.33°
77.50°	77.78°	8.40°	86.18°	3.82°
78.00°	78.27°	8.41°	86.68°	3.32°
78.50°	78.76°	8.43°	87.19°	2.81°
79.00°	79.25°	8.44°	87.69°	2.31°
79.50°	79.74°	8.46°	88.20°	1.80°
80.00°	80.23°	8.47°	88.70°	1.30°
80.50°	80.71°	8.48°	89.19°	0.81°
81.00°	81.20°	8.49°	89.69°	0.31°

6.2. EIRP/DISH DIAMETER

ADVISED DISH DIAMETER TABLE
(prime focus dish with a max. cable length of 25 meters)

EIRP	LNB typ. 0.9dB max. 1.1dB	LNB typ. 1.1dB max. 1.3dB	LNB typ. 1.4dB max. 1.6dB
38dBW	1.80 m	1.80-2.40 m	2.40 m
39dBW	1.50 m	1.80 m	1.80-2.40 m
40dBW	1.35 m	1.50 m	1.80 m
41dBW	1.20 m	1.50 m	1.80 m
42dBW	1.20 m	1.35 m	1.50 m
43dBW	1.20 m	1.20 m	1.50 m
44dBW	0.99 m	1.20 m	1.35 m
45dBW	0.99 m	0.99 m	1.20 m
46dBW	0.90 m	0.99 m	1.20 m
47dBW	0.90 m	0.90 m	0.99 m
48dBW	0.75 m	0.75 m	0.90 m
49dBW	0.60 m	0.65 m	0.75 m
50dBW	0.60 m	0.65 m	0.75 m
51dBW	0.60 m	0.60 m	0.65 m
52dBW	0.55 m	0.55 m	0.60 m
53dBW	0.50 m	0.55 m	0.60 m
54dBW	0.50 m	0.55 m	0.60 m
55dBW	0.45 m	0.50 m	0.55 m
56dBW	0.40 m	0.44 m	0.48 m
57dBW	0.38 m	0.41 m	0.44 m
58dBW	0.36 m	0.38 m	0.42 m
59dBW	0.34 m	0.36 m	0.40 m
60dBW	0.32 m	0.34 m	0.36 m
61dBW	0.30 m	0.32 m	0.34 m
62dBW	0.28 m	0.30 m	0.32 m
63dBW	0.26 m	0.28 m	0.30 m
64dBW	0.23 m	0.25 m	0.27 m

6.3. AMERICAN C- AND KU-BAND FREQUENCIES

AMERICAN C- AND KU-BAND CHANNEL FREQ. CHART (Standardized)
Ku-Band frequency chart based on a 16 channel subdivision. For 32 channel format use half-spacing. No univeral standard is set.

Channel 1	3720/11730 MHz	Channel 15	4000/12157 MHz
Channel 2	3740/11743 MHz	Channel 16	4020/12170 MHz
Channel 3	3760/11791 MHz	Channel 17	4040
Channel 4	3780/11804 MHz	Channel 18	4060
Channel 5	3800/11852 MHz	Channel 19	4080
Channel 6	3820/11865 MHz	Channel 20	4100
Channel 7	3840/11913 MHz	Channel 21	4120
Channel 8	3860/11926 MHz	Channel 22	4140
Channel 9	3880/11974 MHz	Channel 23	4160
Channel 10	3900/11987 MHz	Channel 34	4180
Channel 11	3920/12035 MHz		
Channel 12	3940/12048 MHz		
Channel 13	3960/12096 MHz	**NB:** All TV in 525-lines NTSC, unless stated otherwise	
Channel 14	3980/12109 MHz		

6.4. CHARACTERISTICS OF TELEVISION SYSTEMS AND CHANNELS

CHARACTERISTICS OF TELEVISION SYSTEMS
(as indicated in CCIR Report 624-3, XVIth Plenary Assembly, Dubrovnik, 1986)

System	Number of lines	Channel width MHz.	Vision band-width MHz.	Vision/Sound separation MHz.	Vestigial side-band MHz.	Vision mod.	Sound mod.
B	625	7	5	+5.5	0.75	Neg.	FM
D	625	8	6	+6.5	0.75	Neg.	FM
G	625	8	5	+5.5	0.75	Neg.	FM
H	625	8	5	+5.5	1.25	Neg.	FM
I	625	8	5.5	+5.996	1.25	Neg.	FM
K	625	8	6	+6.5	0.75	Neg.	FM
L	625	8	6	+6.5	1.25	Pos.	AM
M	525	6	4.2	+4.5	0.75	Neg.	FM
N	625	6	4.2	+4.5	0.75	Neg.	FM

N.B: Channels L2, L3, L4, Vision/Sound separation is -6.5 MHz. (France)

CHANNEL INFORMATION
(frequencies in MHz.)

VHF Channels:

West European "E" Channels
2 = 48.25
2A = 49.75
3 = 55.25
4 = 62.25
5 = 175.25
6 = 182.25
7 = 189.25
8 = 196.25
9 = 203.25
10 = 210.25
11 = 217.25
12 = 224.25

Italy
A = 53.75
B = 59.75
C = 82.75
D = 175.25
E = 183.75
F = 192.25
G = 201.25
H = 210.25
H1 = 217.25

Ireland
A = 45.75
B = 53.75
C = 61.75
D = 175.25
E = 183.75
F = 191.25
G = 199.25
H = 207.25
I = 215.25
J = 223.25

France
2 = 55.75
3 = 60.50
4 = 63.75
5 = 176
6 = 184
7 = 192
8 = 200
9 = 208
10 = 216

East European "R" Channels
1 = 49.75
2 = 59.25
3 = 77.25
4 = 85.25
5 = 93.25
6 = 175.25
7 = 183.25
8 = 191.25
9 = 199.25
10 = 207.25
11 = 215.25
12 = 223.25

North/South America
2 = 55.25
3 = 61.75
4 = 67.25
5 = 77.25
6 = 83.25
7 = 175.25
8 = 181.25
9 = 187.25
10 = 193.25
11 = 199.25
12 = 204.25
13 = 211.25

Japan
1 = 91.25
2 = 97.25
3 = 103.25
4 = 171.25
5 = 177.25
6 = 183.25
7 = 189.25
8 = 193.25
9 = 199.25
10 = 205.25
11 = 211.25
12 = 217.25

Australia
0 = 46.25
1 = 57.25
2 = 64.25
3 = 86.25
4 = 95.25
5 = 102.25
5A = 138.25
6 = 175.25
7 = 182.25
8 = 189.25
9 = 196.25
10 = 209.25
11 = 216.25

New Zealand
1 = 45.25
2 = 55.25
3 = 62.25
4 = 175.25
5 = 182.25
6 = 189.25
7 = 196.25
8 = 203.25
9 = 210.25

China (P.R.)
1 = 49.75
2 = 57.75
3 = 65.75
4 = 77.25
5 = 85.25
6 = 168.25
7 = 176.25
8 = 184.25
9 = 192.25
10 = 200.25
11 = 208.25
12 = 216.25

South Africa
4 = 175.25
5 = 183.25
6 = 191.25
7 = 199.25
8 = 207.25
9 = 215.25
10 = 223.25
11 = 231.25
13 = 247.43

Morocco
4 = 163.25
5 = 171.25
6 = 179.25
7 = 187.25
8 = 195.25
9 = 203.25
10 = 211.25

French Overseas Territories
4 = 175.25
5 = 183.25
6 = 191.25
7 = 199.25
8 = 207.25
9 = 215.25

UHF Channels:

North/South America
14 = 471.25
15 = 477.25
16 = 483.25
17 = 489.25
18 = 495.25
19 = 501.25
20 = 507.25
21 = 513.25
22 = 519.25
23 = 525.25
24 = 531.25
25 = 537.25
26 = 543.25
27 = 549.25
28 = 555.25
29 = 561.25
30 = 567.25
31 = 573.25
32 = 579.25
33 = 585.25
34 = 591.25
35 = 597.25
36 = 603.25
37 = 609.25
38 = 615.25
39 = 621.25
40 = 627.25
41 = 633.25
42 = 639.25
43 = 645.25
44 = 651.25
45 = 657.25
46 = 663.25
47 = 669.25
48 = 675.25
49 = 681.25
50 = 687.25
51 = 693.25
52 = 699.25
53 = 705.25
54 = 711.25
55 = 717.25
56 = 723.25
57 = 729.25
58 = 735.25
59 = 741.25
60 = 747.25
61 = 753.25
62 = 759.25
63 = 765.25
64 = 771.25
65 = 777.25
66 = 783.25
67 = 789.25
68 = 795.25
69 = 801.25

UHF TV channels 70-83 were discontinued. (to be used by radio)

Europe/Africa
21 = 471.25
22 = 479.25
23 = 487.25
24 = 495.25
25 = 503.25
26 = 511.25
27 = 519.25
28 = 527.25
29 = 535.25
30 = 543.25
31 = 551.25
32 = 559.25
33 = 567.25
34 = 575.25
35 = 583.25
36 = 591.25
37 = 599.25
38 = 607.25
39 = 615.25
40 = 623.25
41 = 631.25
42 = 639.25
43 = 647.25
44 = 655.25
45 = 663.25
46 = 671.25
47 = 679.25
48 = 687.25
49 = 695.25
50 = 703.25
51 = 711.25
52 = 719.25
53 = 727.25
54 = 735.25
55 = 743.25
56 = 751.25
57 = 759.25
58 = 767.25
59 = 775.25
60 = 783.25
61 = 791.25
62 = 799.25
63 = 807.25
64 = 815.25
65 = 823.25
66 = 831.25
67 = 839.25
68 = 847.25
69 = 855.25

Australia
28 = 527.25
29 = 534.25
30 = 541.25
31 = 548.25
32 = 555.25
33 = 562.25
34 = 569.25
35 = 576.25
36 = 583.25
37 = 590.25
38 = 597.25
39 = 604.25
40 = 611.25
41 = 618.25
42 = 625.25
43 = 632.25
44 = 639.25
45 = 646.25
46 = 653.25
47 = 660.25
48 = 667.25
49 = 674.25
50 = 681.25
51 = 688.25
52 = 695.25
53 = 702.25
54 = 709.25
55 = 716.25
56 = 723.25
57 = 730.25
58 = 737.25
59 = 744.25
60 = 751.25
61 = 758.25
62 = 765.25
63 = 772.25
64 = 779.25
65 = 786.25
66 = 793.25
67 = 800.25
68 = 807.25
69 = 814.25

China (P.R.)
13 = 471.25
14 = 479.25
15 = 487.25
16 = 495.25
17 = 503.25
18 = 511.25
19 = 519.25
20 = 527.25
21 = 534.25
22 = 543.25
23 = 551.25
24 = 559.25
25 = 605.25
26 = 613.25
27 = 621.25
28 = 629.25
29 = 637.25
30 = 645.25
31 = 653.25
32 = 661.25
33 = 669.25
34 = 677.25
36 = 693.25

N.B: Japan Channel 13-62 = No./So. America Channel 14-63

CHAPTER 7
ORGANIZATIONS

7.1. INTERNATIONAL TELECOMMUNICATION UNION (ITU)

INTERNATIONAL TELECOMMUNICATION UNION (ITU)
Addr: Place des Nations, 1211 Geneva 20, Switzerland
Tel: +41 (22) 730 51 11
Fax: +41 (22) 733 72 56
WWW: http://www.itu.ch

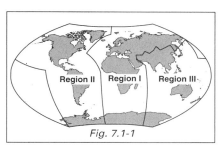

Fig. 7.1-1

The International Telecommunication Union was founded in 1865 and as such is the oldest intergovernmental organization. Since 1947 it is a subdivision of the United Nations and now has a membership of 172 countries. It is the international organization responsible for the regulation and planning of telecommunications worldwide, for the establishment of equipment and systems operating standards, for the coordination and dissemination of information required for the planning and operating of telecommunications networks and services and for the promotion of and contribution to the development of telecommunications and the related infrastructure. It is comprised of four permanent organs:
1) The International Frequency Registration Board (IFRB)
2) The International Radio Consultative

Committee (CCIR)
3) The General Secretariat
4) The International Telegraph & Telephone Consultative Committee (CCITT)
It is the IFRB that allocates orbital slots and frequencies to member countries of the ITU for domestic, regional and international communications satellites as well coordinating and allocating terrestrial frequencies.

The International Telecommunication Union has divided the world into three separate regions:
1) Region I (includes Europe, Africa, Near & Middle East)
2) Region II (includes the Americas)
3) Region III (includes Asia, the Indian subcontinent and the South Pacific)

7.2. EUTELSAT

EUROPEAN TELECOMMUNICATIONS SATELLITE ORGANIZATION (Eutelsat)
Addr: Tour Maine-Montparnasse 33, Avenue de Maine, 75755 Paris Cedex 15, France
Tel: +33 (1) 45 38 4757
Fax: +33 (1) 45 38 3700
WWW: http://www.eutelsat.com
Eutelsat is the largest operator of satellites in Europe. It controls eight communications satellites, used throughout the European continent for telephone, telex, fax messages and data transmission and to distribute tv and radio programs. In 1977, Europe's Conference of Postal and Telecommunications Administrations (CEPT) founded Eutelsat. Today shares in the organization are split up amongst 45 member countries: Andorra, Armenia, Austria, Azerbaijan, Belgium, Rep. of Bulgaria, Croatia, Cyprus, Czech Republic, Denmark, Estonia, Finland, France, Germany, Greece, Hungary, Iceland, Ireland, Italy, Latvia, Liechtenstein, Lithuania, Luxembourg, Malta, Moldova, Monaco, the Netherlands, Norway, Poland, Portugal, Romania, Russian Fed., San Marino, Serbia, Slovakia, Slovenia, Spain, Sweden, Switzerland, Turkey, the UK, Vatican State. Belarus and the Ukraine also indicated that they will be applying for Eutelsat membership status in the near future. Eutelsat is best known for carrying tv and radio programs to cable headends and *TVRO (TV receive only)* owners. The European Broadcasting Union uses six

transponders on the ECS II f4 satellite for its daily Eurovision service. The ECS II f5 satellite was destroyed in early 1994 after the third stage of an ARIANE rocket failed and will be replaced by the ECS II f5r.The newest Eutelsat satellite, called ECS II f6 (Hot Bird 1) was launched in early 1995. Hot Bird 2 was successfully launched in 1996. These two satellites are the most powerful of all ECS satellites and are co-located together with the ECS II f1 at 13° East.

7.3. INTELSAT

INTERNATIONAL TELECOMMUNICATIONS SATELLITEORGANIZATION (Intelsat)
Addr: 3400 International Drive, N.W., Box 63, Washington D.C. 20008, USA
Tel: +1 (202) 944 6872
Fax: +1 (202) 944 7925
WWW: http://www.intelsat.int

Intelsat began on August 20th 1964, when representatives of eleven nations signed agreements for a global commercial communications satellite system. When Early Bird (Intelsat I) was launched, 46 nations had become Intelsat members. In actual size, Early Bird was a very small satellite in comparison to today's spacecraft. It weighed about 85 pounds and stood only two feet tall. In terms of its impact on how the world communicates, however, it was one of the most powerful forces for change that the world has ever witnessed.

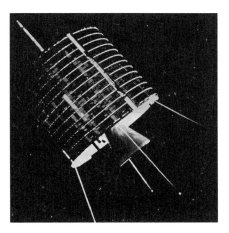

Early Bird set in motion a new way of communicating. Four years after its launch, Intelsat established the world's first global communications system, with satellites above all three ocean regions, just in time to provide coverage to an audience of half a billion people that watched the first man on the moon.

Since 1973, Intelsat operates with an organizational structure comprising of four ranks: the Assembly of Parties, the Meeting of Signatories, the Board of Governors and the Executive Organ.

Currently, INTELSAT has more satellites in operation than any other commercial organization, a fleet of over 20 high powered, technically advanced spacecraft in geostationary orbit: the INTELSAT V/V-A series; the INTELSAT VI series; and the INTELSAT VII/VII-A series. In addition, INTELSAT has a single all Ku-band satellite in service, known as INTELSAT K. The next generation of INTELSAT spacecraft, the INTELSAT VIII/VIII-A series is under construction. The six VIII/VIII-A satellites will all be launched in 1997.

The first in the new generation of satellites, the Intelsat 801, was launched on Friday, Feb. 28, 1997. The Intelsat 801 satellite is based on the first in a new series of platforms designed to provide more powerful and higher capacity C-band communications services. Built by Lockheed Martin Telecommunications, it offers a number of technical innovations, including two steerable C-band spotbeams, interconnection between the C and Ku bands, satellite news gathering (SNG) service, and compatibility with series VII and VIIA satellites.

Intelsat 801 weighed 3,420 kg at liftoff. It provides a capacity of 3 TV channels, 22,500 telephone circuits, and up to 112,500 digital telephone circuits.

Intelsat member countries are: Afghanistan, Algeria, Angola, Argentina, Armenia, Australia, Austria, Bahrain, Bahamas, Bangladesh, Barbados, Belgium, Benin, Bolivia, Brazil, Brunei Darussalam, Burkina Faso, Cameroon, Canada, Cape Verde, Central African Rep., Chad, Chile, China (People's Rep. of), CIS, Colombia, Congo, Costa Rica, Cote d'Ivoire, Cyprus, Denmark, Dominican Rep., Ecuador, Egypt, El Salvador, Ethiopia, Fiji, Finland, France, Gabon, Germany, Ghana, Greece, Guatemala, Guinea, Haiti, Honduras, Hungary, Iceland, India, Indonesia, Iran, Iraq, Ireland,

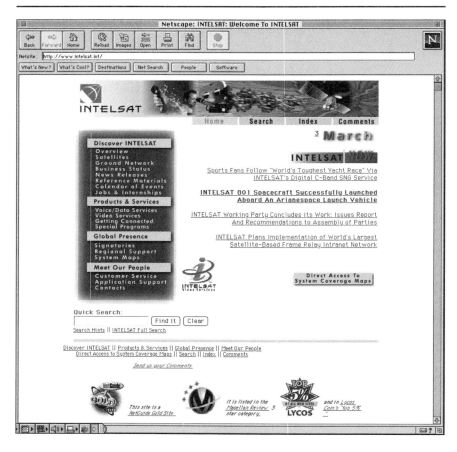

Israel, Italy, Jamaica, Japan, Jordan, Kazakhstan, Kenya, Korea, Kuwait, Kyrgyz Rep., Lebanon, Libya, Liechtenstein, Luxembourg, Madagascar, Malawi, Malaysia, Mali, Malta, Mauritania, Mauritius, Mexico, Monaco, Morocco, Mozambique, Namibia, Nepal, the Netherlands, New Zealand, Nicaragua, Niger, Nigeria, Norway, Oman, Pakistan, Panama, Papua New Guinea, Paraguay, Peru, Philippines, Portugal, Qatar, Romania, Rwanda, Saudi Arabia, Senegal, Serbia, Singapore, Somalia, South Africa, Spain, Sri Lanka, Sudan, Swaziland, Sweden, Switzerland, Syria, Tanzania, Thailand, Togo, Trinidad & Tobago, Tunisia, Turkey, Uganda, UAE, UK, USA, Uruguay, Vatican State, Venezuela, Vietnam, Yemen, Zaire, Zambia, Zimbabwe.

Intelsat on the world wide web
If you are interested in more information about Intelsat and the Intelsat satellites, you can now find all the details on Internet.

A World Wide Web Home Page at:
http://www.intelsat.int
Electronic document distribution at:
ftp.intelsat.int/pub/
E-mail to any staff member at:
first_name.last_name@intelsat.int

All of the information on the World Wide Web is cross-referenced by Hypertext links. You will find annual reports, Signatory information, INTELSAT satellite fact sheets, newsletters, press releases, satellite coverage maps, contact information, job opportunities, a calendar of events, and more. Technical documentation can also be accessed and downloaded via anonymous File Transfer Protocal at ftp.intelsat.int.

7.4. SOCIÉTÉ EUROPÉENNE DES SATELLITES

SOCIÉTÉ EUROPÉENNE DES SATELLITES (SES)
Addr: L-6815 Château de Betzdorf, Luxembourg
Tel: +352 71 72 51
Fax: +352 71 72 53 24/227
Telex: 60625 ASTRA LU
WWW: http://www.astra.lu

SES is a private European organization, currently operating the four ASTRA satellites at 19.2° East. In focusing on the needs of the European television viewers and satellite TV programmers, SES introduced a new approach to satellite television: the design of a market- and service-oriented company which offers entertaining and high quality packages of television programs for the various European language markets. Today the company employs over 130 staff from more than ten nations and of different cultural backgrounds, who devote their talents to the highly specialised jobs at the forefront of satellite technology and the commercialization of satellite technology. SES operates the ASTRA satellite system under a franchise agreement with the grand Duchy of Luxembourg.

Recently expanded, the SES franchise runs until the year 2010, with a possible extension, and covers audio-visual services as well as possible new business applications. The revenue from the ASTRA satellite operation is generated by leasing channels on the satellites to television and radio broadcasters. They generate their revenue by selling advertising space on their channels, or, in the case of premium programming, by collection subscription fees from their audiences.

The thought behind the pan-European ASTRA concept was to broadcast as many radio and tv stations as possible, in a way that they can be received on a 60 cm. diameter dish in most parts of Europe. SES decided to use the FSS frequency range, the same range that was already used by medium-power satellites such as Eutelsat and Intelsat. Also the idea of co-locating various satellites at one orbital location was part of the main strategy. As a result of all these requirements a 16-channel medium-power satellite, broadcasting with a bandwidth of 28 MHz, was developed. Using the FSS frequency band (10.9-11.7 GHz) it was possible to create 50 channels with linear polarization on three satellites, to be used solely for broadcasting.

The first ASTRA satellite was launched in December 1988 atop a ARIANE-4 rocket and on February 5th 1989 the first broadcasts from 19.2° East were witnessed. Within a few months all transponders were occupied and because of this success the second ASTRA satellite was launched a little more than a year later. The third ASTRA satellite was launched in May 1993 and the fourth late 1994. A fifth satellite, the ASTRA 1E, was launched in late 1995. The ASTRA 1D satellite broadcasts in a lower Ku-Band frequency range (10.7-10.9 GHz) and the ASTRA 1E/F/G in a higher (11.7-12.75 GHz) frequency range. Currently a total of seven satellites are co-positioned at the same orbital location. A further location at 28° East is under option

ADR, DMX, MUSICAM.
DMX has launched a multi-channel music service to European direct to home satellite dish owners in Europe. The new service offers subscribers up to 90 channels of digital, non-stop, commercial free music. The service, already available via cable in Europe and via satellite and cable in the USA, has been launched on Astra using a subscription based service built on the Astra Digital Radio (ADR) standard.

Astra subscribers will receive 90 distinctly programmed channels, including classical, rock, jazz and country, along with many specialty and international channels such as Swiss Folk, five German-language channels, Norwegian, Flemish and Hebrew.

DMX is implementing a roll-out strategy for its European launch with Germany, Austria and Switzerland having the first retail access to the DMX direct-to-home receivers. TechniSat and Kathrein-Werke KG are the first to bring DMX receivers to market in Europe through consumer electronics retailers.

The addition of DMX to Astra was made possible by the development of several proprietary technologies. DMX, working with Societe Europeenne des Satellites (SES) developed a joint specification to be used by all receiver manufacturers using the MPEG II digital audio compression system known as MUSICAM. MUSICAM will also be used for future digital audio broadcasting (DAB) applications, and has been selected Europe-wide as the audio standard for digital video broadcasting for future cable and DTH compressed digital transmission.

In addition, DMX and Wegener developed the enabling technology for a new system that reconfigures existing analog audio subcarriers of Astra transponders into digital subcarriers. This will allow DMX to be the first digital signal carried on Astra.

The DMX signal on Astra will be combined with digitized signals from several European digital broadcasters and delivered to subscribers as ASTRA Digital Radio (ADR). The signal will then be received in subscribers' homes by a proprietary DMX-ADR receiver. These receivers are already being produced by two manufacturers, Kathrein and TechniSat, and discussions are continuing with various other manufacturers, including Pace, in Europe and Asia. The DMX MUSICAM signal, which will be distributed in encrypted form, will be decoded by a smart card developed by News Datacom (NDC) for use in Astra DTH receivers.

DMX will be distributed through three subscription management partnerships, including Selco in Germany, Austria and parts of Switzerland; Nethold/Multichoice in the Netherlands, Belgium, Scandinavia, Italy, Russia, Hungary, the Czech Republic and several other European countries; and BSkyB in the United Kingdom and Ireland.

7.5. FRANCE TÉLÉCOM

FRANCE TÉLÉCOM
Addr: 6, Place d'Allenery, F-75740, Paris Cedex 15, France
Tel: +33 (1) 4444 2222
Fax: +33 (1) 45 31 5211
WWW: http://www.francetelecom.com

France Télécom test pattern

France Télécom, the operator of the French Télécom satellites, was founded in 1984. This national French satellite network is not only used for broadcasting and telecommunications purposes, but also for military assignments. Furthermore, the Télécom satellite system was designed in a way that it can cover both France and a larger part of Europe with its Ku-Band transponders, as well as the French overseas territories with strong C-Band transponders. The strong X-Band transponders are used for military purposes.

The Télécom 2A satellite (8° West) was launched on December 16, 1991, while the 2B (5° West) was launched on April 15, 1992. The stationkeeping of the three Télécom satellites is done from the control room at the French space center in Toulouse. All Télécom satellites broadcast to France in the 12 GHz or Ku3-Band range, but the new Télécom 2 satellites are much more powerful than their predecessors. In comparison with the first generation, the 2 series has 26 transponders (used to be 12) with a power of 10-55 Watt. Each Télécom 2 satellite has an advanced antenna system aboard cosisting of semi-global, shaped beam and steerable spotbeams. The antenna pointing accuracy is better than 0.15°. The C-Band antenna, which provides both spot- and semi-global beam coverage, consists of a large reflector and a 10-element power source with single and dual polarization possibilities. The Ku3-Band antenna is powered by two primary sources, one for each linear polarity. The X-Band military antenna, aimed at central Europe, provides both LHCP and RHCP and can be steered by ground command to provide coverage of almost any location in the satellite's line of sight.

7.6. INTERSPUTNIK

INTERSPUTNIK
Addr: 2 Smolensky Lane 1/4, Moscow 1210 99, Russia.
Tel: +7 (095) 22 40 333
Fax: +7 (095) 25 39 906

Intersputnik is an intergovernmental organization that can be joined by the government of any state which shares the organization's principles. The Intersputnik system was established to provide cooperation and co-ordination of efforts in designing, establishing, operating and developing a communications system based on satellites with a view to meeting requirements for international radio and TV exchange, telephony, telegraphy, data transmissions and other services. Its board is the main governing body composed of one representative from each membering country with one vote regardless of the investment share in the company. Day-today management is done by the Operations Committee of the Board.

The history of Russian satellites started on October 4th 1957, with the Sputnik 1 satellite. In fact, the Sputnik 1 was the first satellite ever. Its main task was to orbit round the earth producing a bleeping signal, but nevertheless it became the biggest sensation of the 20th century. The Sputnik 1 also carried measuring equipment to determine the temperature and density of the atmosphere and ionosphere and to measure the propagation of electromagnetic waves. The frequency on which the data was transmitted was 20.000 kHz, so right in the shortwave frequency spectrum. Another frequency of 40.000 kHz was also used for data transmission. The Sputnik 1 had a diameter of only 58 cm (i.e. 22.8 inch) and weighed 84 kg.

The first communications satellite operated by Intersputnik was launched in 1965. This Molnija satellite weighed over 100 kg. and had a diameter of 1.58 meter . The satellite described a highly elliptical orbit with an inclination of 63.4°, a perigee of 500 km. and an apogee of 40,000 km. The Molnija spacecraft had a revolution time of twelve hours of which only eight could be used to broadcast television pictures. The elliptical orbit was necessary to provide the northerly areas, which cannot be reached with a geostationary satellite, with television.

Intersputnik is an international satellite organization, which was founded by the former USSR and some allied countries back in 1971. There now are 22 member nations: Afghanistan, Belarus, Bulgaria, Cuba, Czech Republic, Gabon, Georia, Germany, Hungary, Kazakhstan, Kirghizstan, Korea (Dem. Peoples Rep.), Laos, Mongolia, Nicaragua, Poland, Romania, Russia, Slovakia, Syria, Tajikistan, Turkmenistan, Vietnam, Yemen. In addition, other countries such as Algeria, Iraq, Cambodia and Libya sometimes use the Intersputnik network for communication and video exchanges.

7.7. NASDA

NATIONAL SPACE DEVELOPMENT AGENCY OF JAPAN (NASDA)
Addr: World Trade Center Building 2-4-1, Hamamatsu-cho, Minato-ku Tokyo 105, Japan.
Tel: +81 (3) 5470 4111
Fax: +81 (3) 3433 0796
Telex: J28424 (AAB:NASDA J28424).
WWW: http://www.goin.nasda.go.jp/

NASDA was established in October 1969, for the purpose of advancing space exploration in the interest of peace. Besides numerous other space related tasks NASDA operates a large number of satellites such as the Yuri (BS-2A/B) series. The first Yuri, launched in April, 1978, saw action in a wide range of experiments by Japanese television and government agencies. The Yuri 2-series - 2A and 2B - were respectively sent into orbit in January 1984 and February 1986 with Japanese N-II rockets. The BS-3 series (Yuri 3A and Yuri 3B) were launched in August 1990 and August 1991 respectively with Japanese H-I rockets. The Telecommunications Satellite Corporation of Japan (TSCJ) controls the satellites on geostationary orbit and NHK (Japan Broadcasting Corporation) and JSB (Japan Satellite Broadcasting Inc.) conduct broadcasting services on these satellites. TSCJ also provides one of the channels of the BS-3B, the backup satellite for the BS-3A, to be used for HDTV transmissions in the near future.

The BS-3 satellite is a box type, three-axis stabilized satellite. The broadcast

antenna is always directed towards earth and two large solar panels supply the necessary power. All BS-2 series satellites will be replaced in due course by the BS-3 series satellites. The BS-3 will be used for DBS purposes covering the entire Japanese archipelago including outlying islands such as Okinawa and Ogasawara.

7.8. ASIASAT

ASIA SATELLITE TELECOMMUNICATIONS Co. Ltd. (ASIASAT)
Addr: 23-24/F East Exchange Tower, 38-40 Leighton Road, Causeway Bay, Hong Kong.
Tel: (852) 805 6666
Fax: (852) 576 4111
Telex: 68345 ASAT HX

The Asiasat 1 satellite - launched on April 7, 1990 atop a Chinese Long March rocket - is the first commercial Asian communications satellite. In a way this satellite is unique because it had been launched before as the American Westar VI in 1985. However, the spacecraft failed to reach its orbital slot and later on was retrieved by a special Space Shuttle mission. It then was sold to the Asiasat company, which redeployed the satellite at an 105.5° East. A second Asiasat satellite - Asiasat 2 - was stationed at the same orbital location.

Most of the programming on the Asiasat 1 satellite is provided by STAR TV (acronym for Satellite Television Asian Region). STAR TV is a Direct-To-Home (DTH) satellite television service comprised of a sports channel (Prime Sports), a music channel (MTV), a news/information channel (BBC World Service Television), a Chinese language channel (Chinese Channel) and a general entertainment channel (Star Plus). Other Asiasat channels include an Indian entertainment channel in Hindi (Zee TV) and three domestic stations: Pakistan TV, CCTV (Chinese) and Myanmar TV.

7.9. PANAMSAT (ALPHA LYRACOM)

PANAMSAT
Addr: One Pickwick Plaza, Greenwich, CT 06830, USA
Tel: +1 (203) 622 6664
Fax: +1 (203) 622 9163.
WWW: http://www.panamsat.com

Alpha Lyracom, an international commercial satellite consortium, was founded in 1984 and operates under the name Pan American Satellite. In the same year the highest American body for broadcasting, the FCC (Federal Communications Commission), granted PanAmSat all the rights to launch and exploit an independent international satellite system. This made PanAmSat the world's first private international satellite operator and a direct competitor to Intelsat. PanAmSat's main goal was to offer cheap satellite communication and broadcasting capacity in the USA, Latin America and Europe. In the meantime PanAmSat has plans to launch another three satellites to expand their services worldwide.

The first PanAmSat satellite *PAS 1* (initially called Simon Bolivar) was manufactured by RCA Electronics and launched on June 15, 1988 atop an ARIANE rocket. The spacecraft has a communications payload of 24 transponders, 18 C-Band and 6 Ku-Band. The relatively wide bandwidth of the various transponders makes it possible to split them up easily, but with *half-transponder* mode the radiated power (EIRP) is reduced too. The PAS 1 has six target areas or footprint areas. The C-Band areas are called North, Central, South and Latin American Beams, The Ku-Band Areas are referred to as European and Conus Beam. (see chapter 8).

The second PAS satellite, the Hughes built PAS 2, was launched in July 1994 and positioned over the Asia/Pacific region at 168° The PAS-4 was launched in July 1995 and positioned at 68.5° East. The PAS-3r was launched on January 12, 1996.

PanAmSat now provides coverage to 97 percent of the world's population. The PAS-3 Atlantic Ocean Region satellite completes the company's global expansion effort. PAS-3 commenced service on February 19, 1996. The company has two additional satellites under construction that will serve the Atlantic Ocean Region by early 1997.

Over these six satellites, PanAmSat has service agreements going forward worth approximately $3.2 billion. PanAmSat is investing over $1 billion to build and operate its ever-expanding global satellite system.

Beyond the six-satellite system, PanAmSat already plans to launch an additional Indian Ocean Region satellite and an additional Pacific Ocean Region satellite. By early 1998, it will operate at least two satellites in each ocean region.

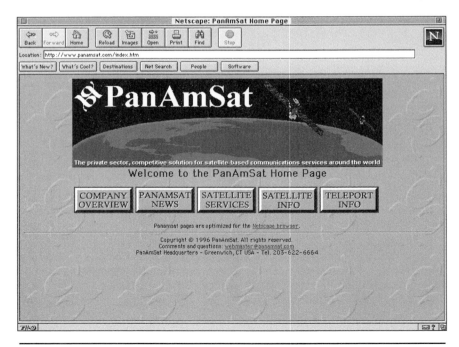

CHAPTER 8
SATELLITE COVERAGE ZONES

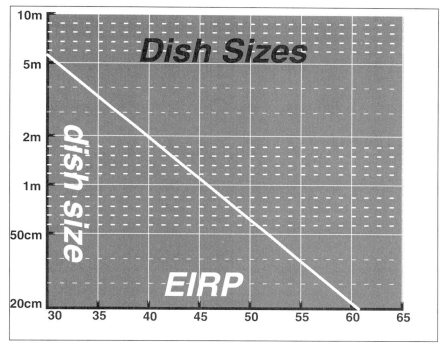

Fig.8-1: Graph with Dish Diameter and EIRP Level Figures.

All footprint contour maps in this chapter show an Effective Isotropic Radiated Power (EIRP) value in decibel Watts (dBW). The above graph gives a rough translation of these values into required dish sizes for a certain area. Exact dish diameter figures can be found in chapter 7.2. Smaller footprints generally have higher EIRP levels than larger ones. The outer edge of a footprint indicates the lowest receivable EIRP level. The radiated power is highest near the footprint's center.

NB: Ku-Band signal strength deteriorates with heavy rain or snow fall.

ANIK

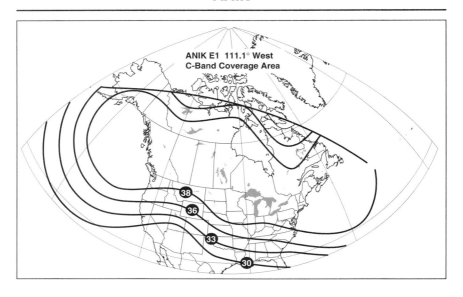

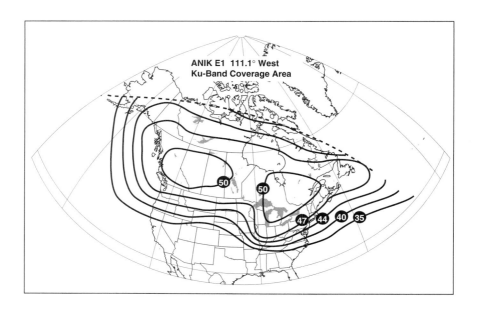

APSTAR

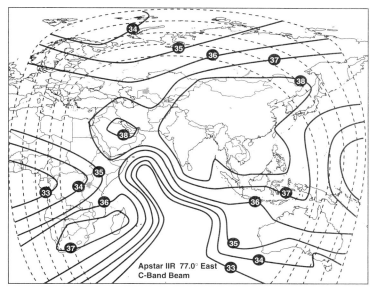

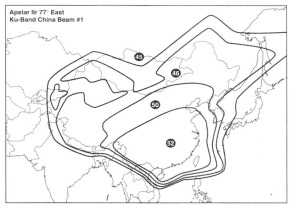

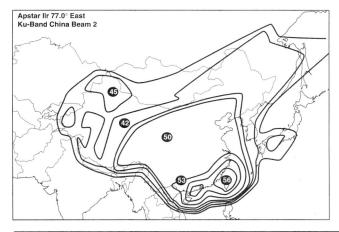

ARABSAT

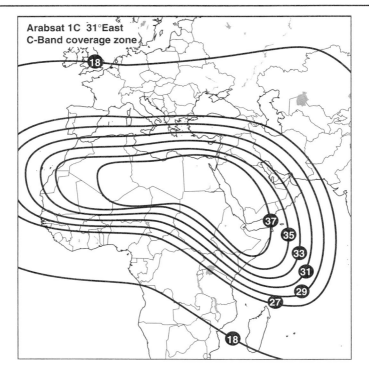

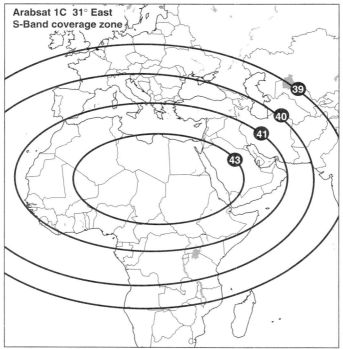

ARABSAT

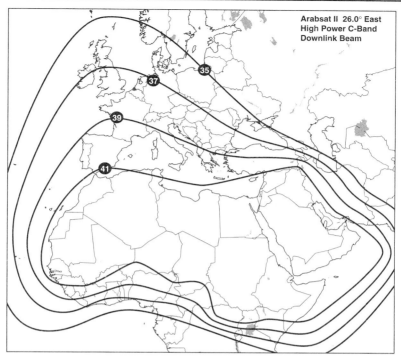

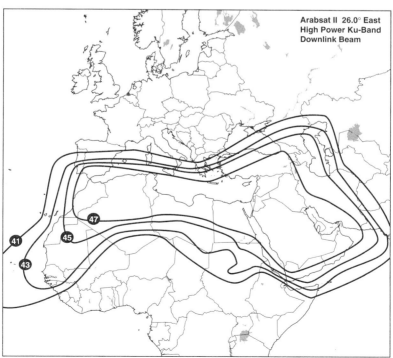

ASIASAT

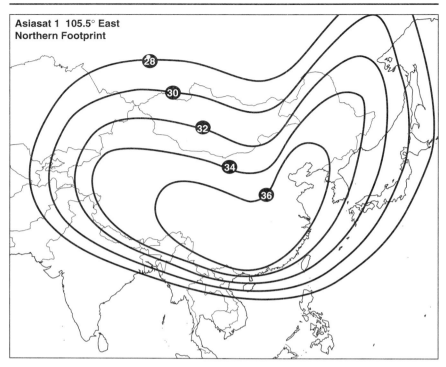

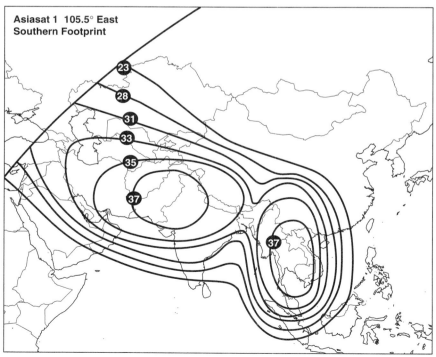

ASIASAT

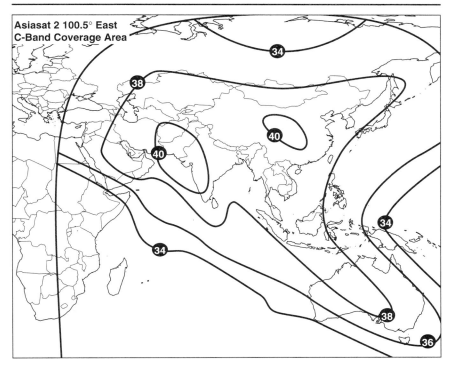

Asiasat 2 100.5° East
C-Band Coverage Area

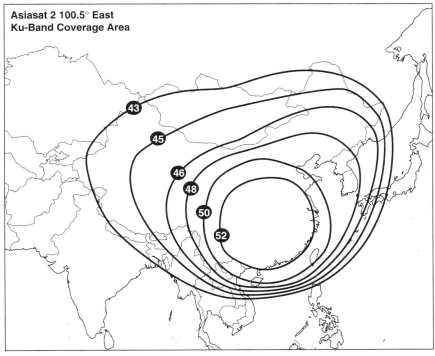

Asiasat 2 100.5° East
Ku-Band Coverage Area

ASTRA

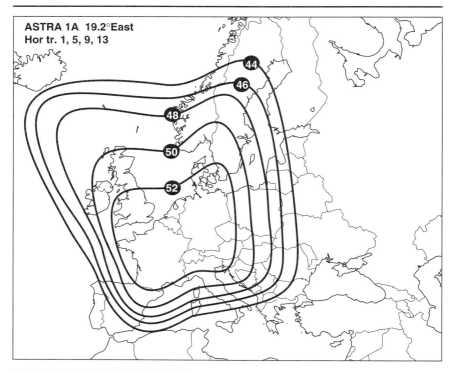

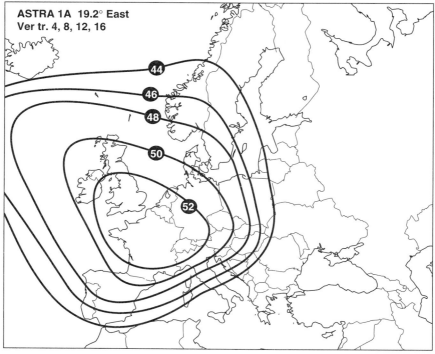

ASTRA

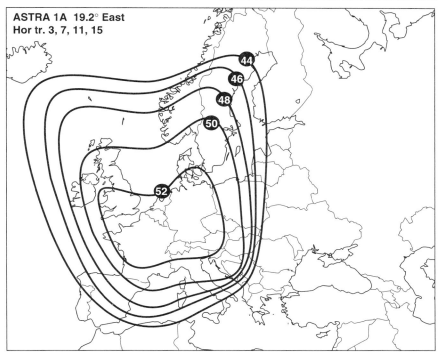

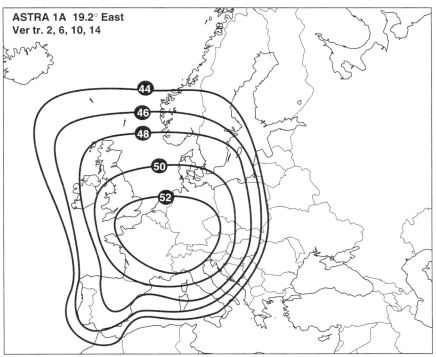

ASTRA

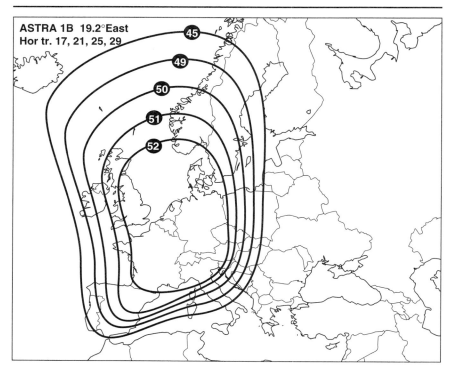

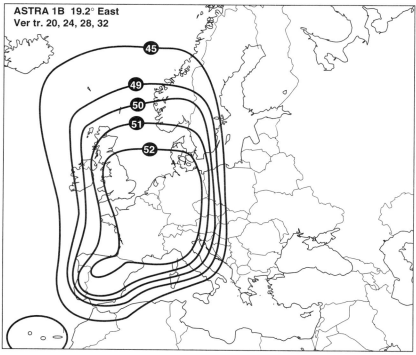

ASTRA

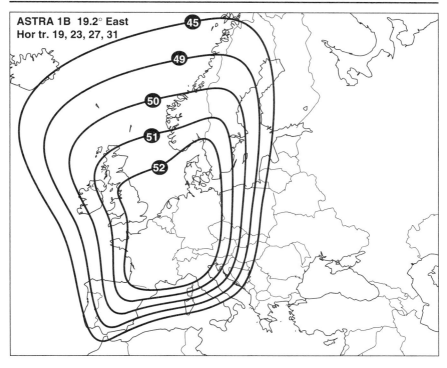

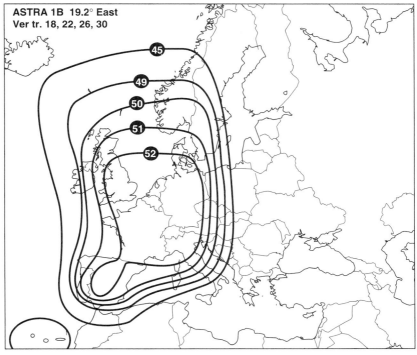

ASTRA

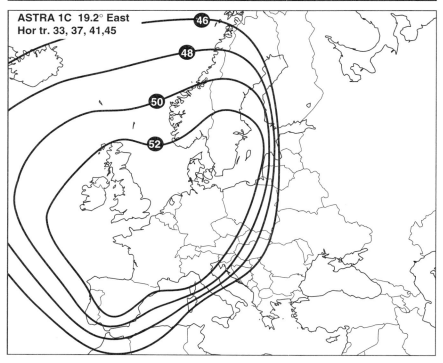

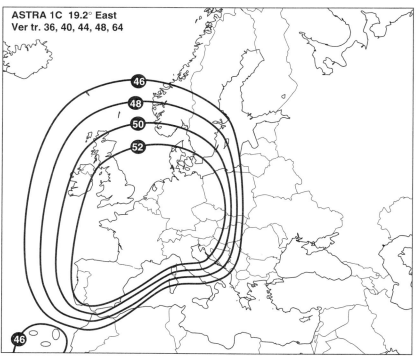

ASTRA

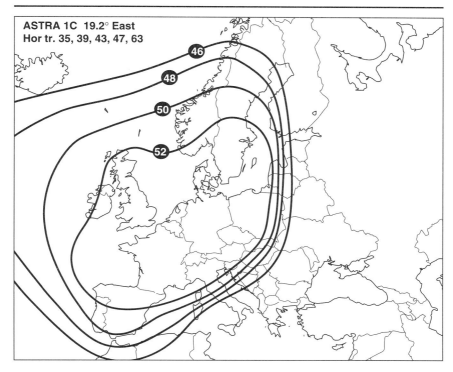

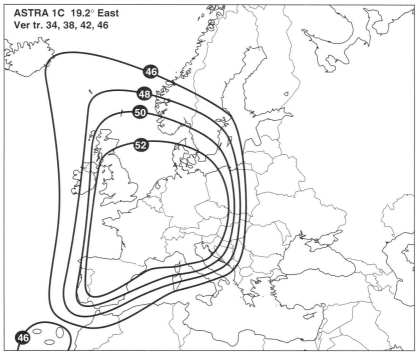

ASTRA

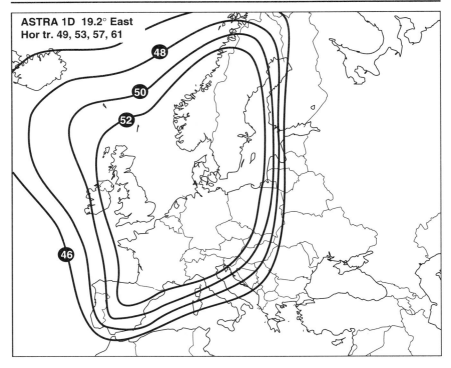

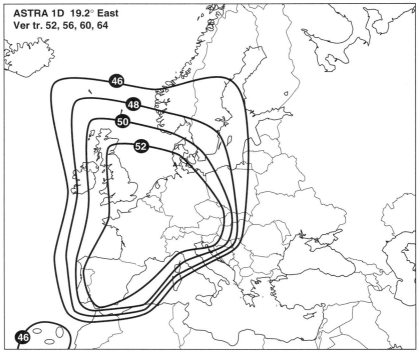

ASTRA

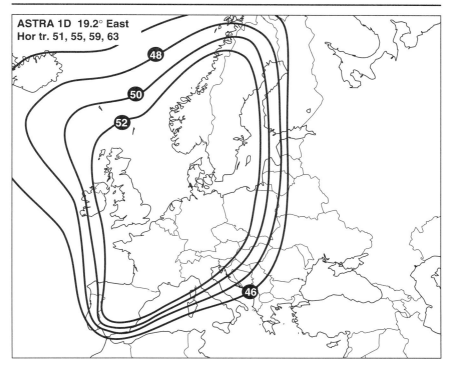

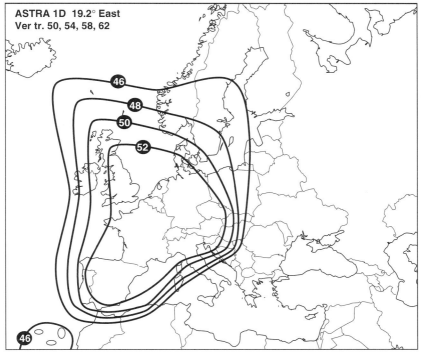

ASTRA

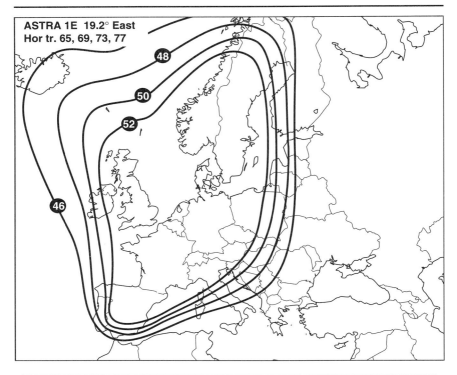

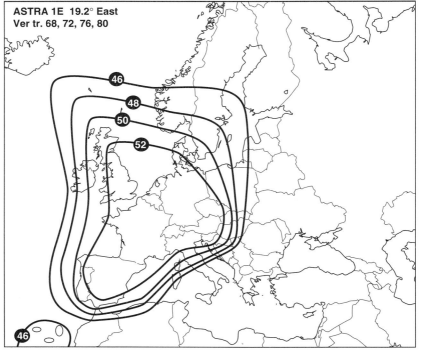

ASTRA

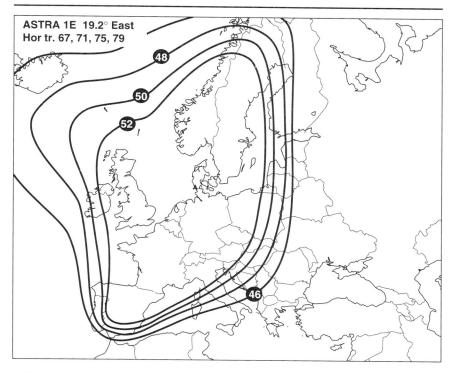

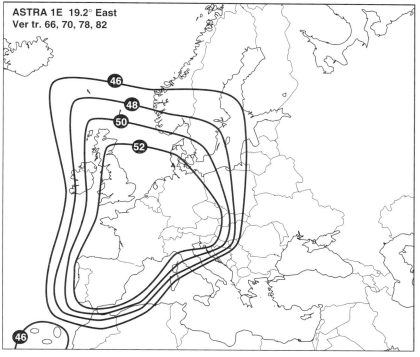

ASTRA

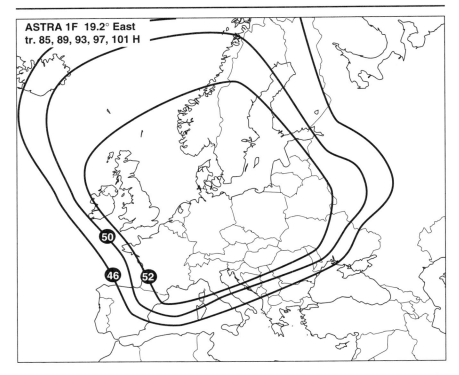

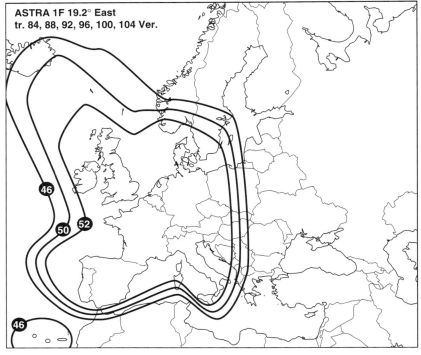

ASTRA

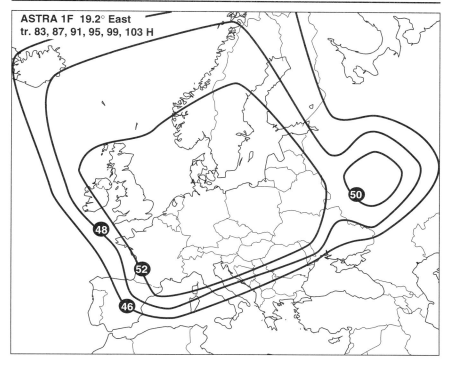

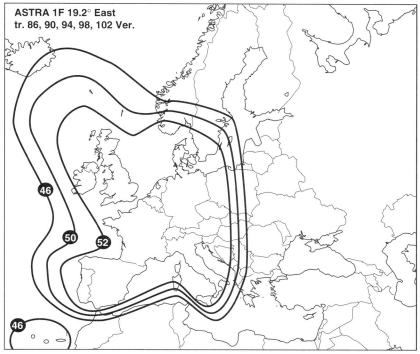

BRASILSAT

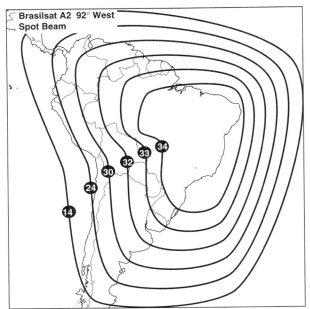

Brasilsat A2 92° West Spot Beam

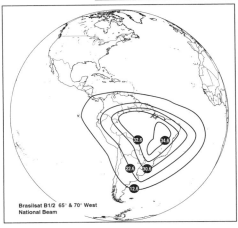

Brasilsat B1/2 65° & 70° West National Beam

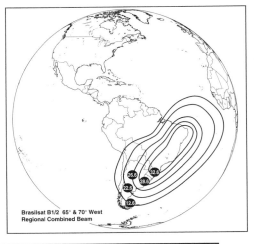

Brasilsat B1/2 65° & 70° West Regional Combined Beam

BS-3A-B/DFS KOPERNIKUS▼

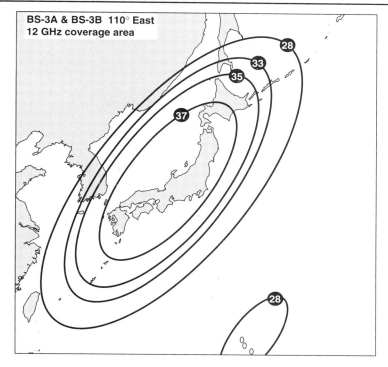

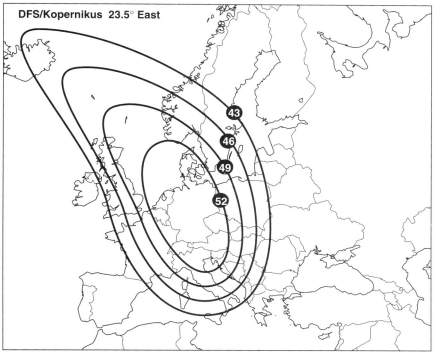

EUTELSAT

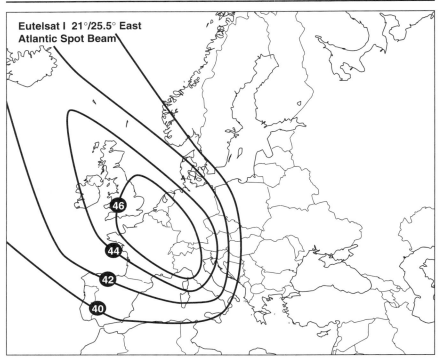

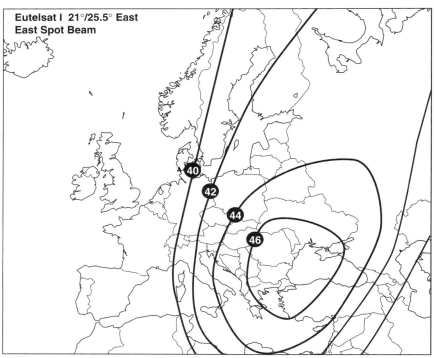

EUTELSAT

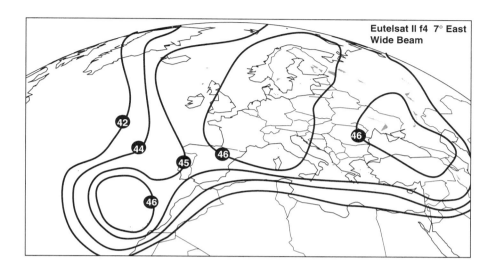

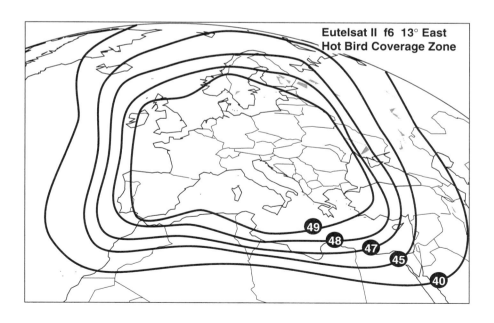

EUTELSAT

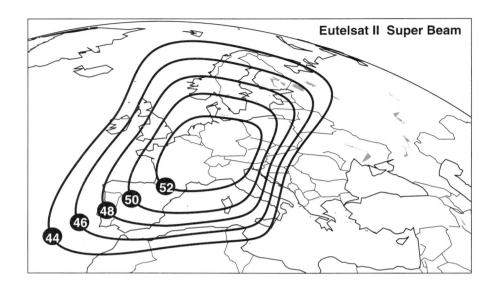

Eutelsat II Super Beam

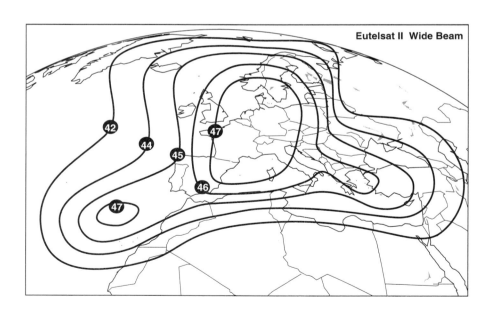

Eutelsat II Wide Beam

EXPRESS

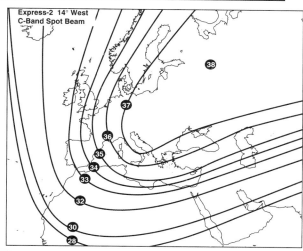
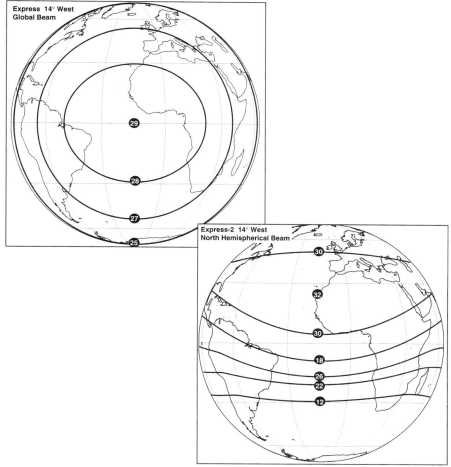

GALAXY

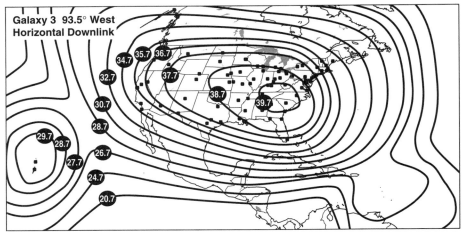

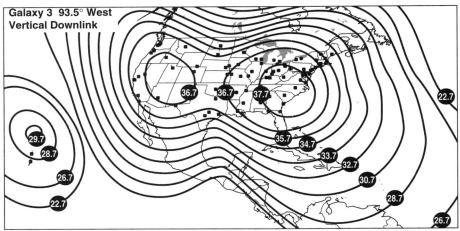

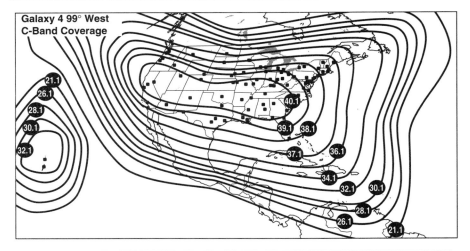

GALAXY

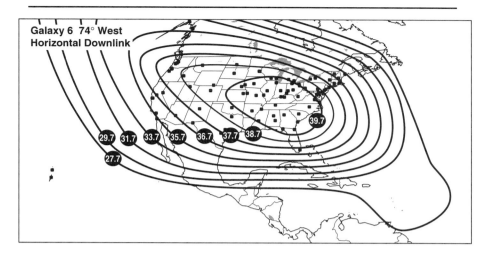

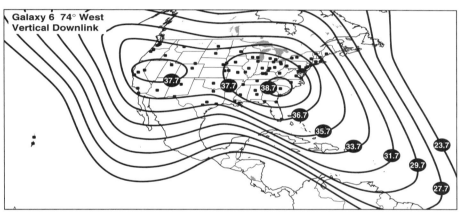

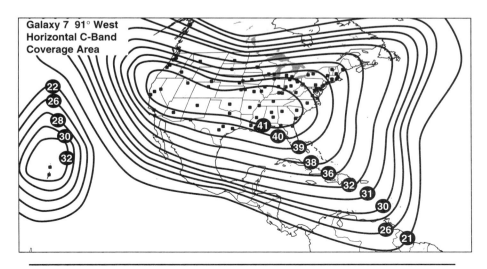

GALAXY

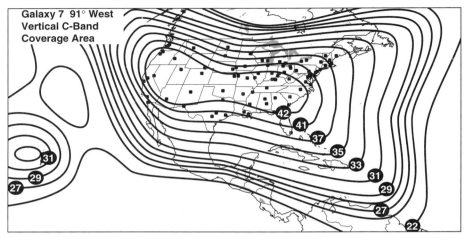

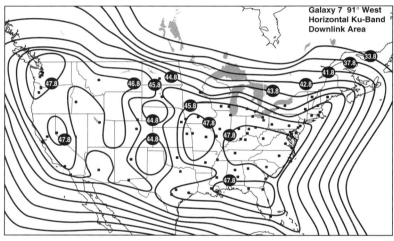

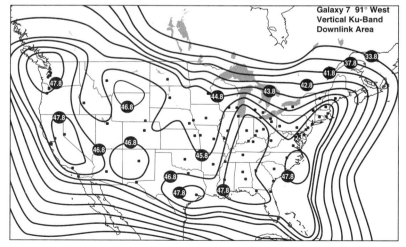

GORIZONT

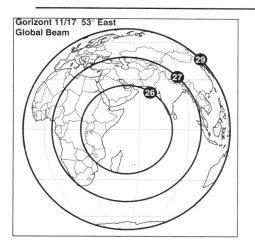
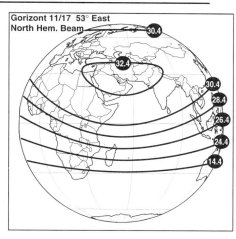
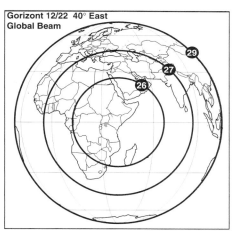
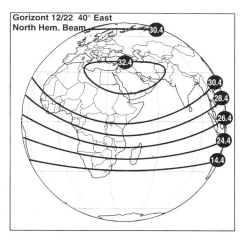
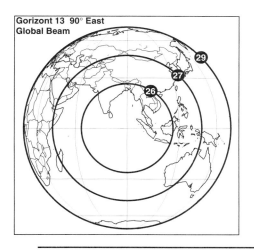
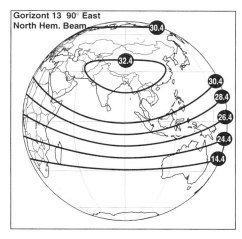

GORIZONT

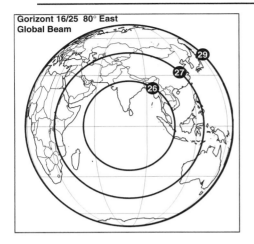
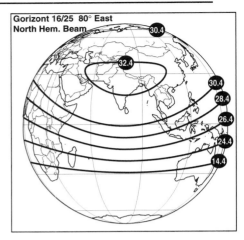
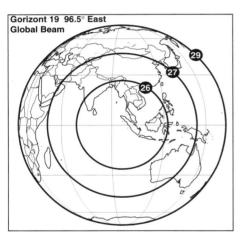
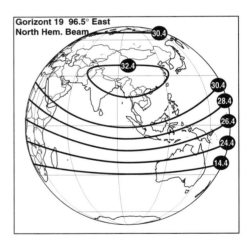
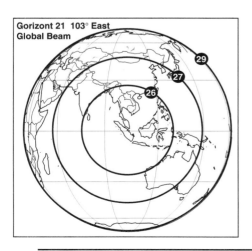
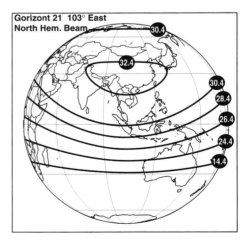

GSTAR

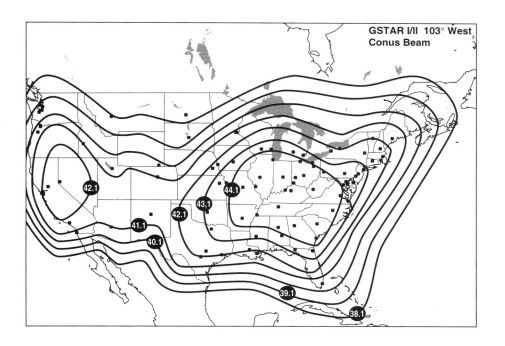

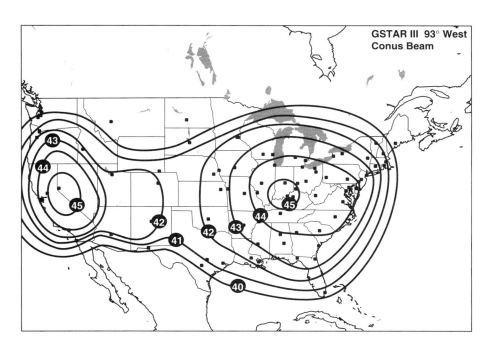

HISPASAT

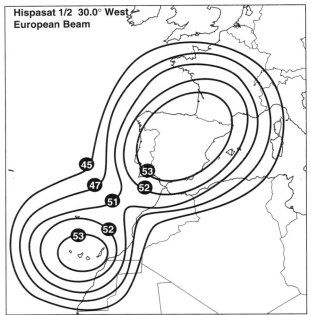

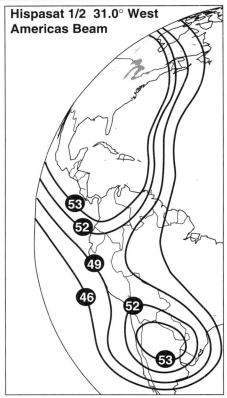

INSAT

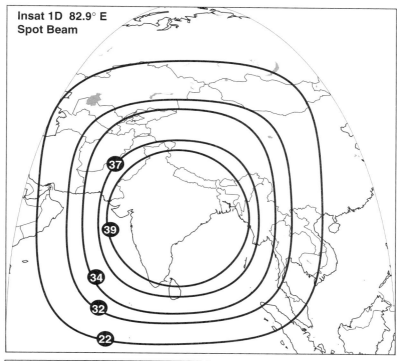

Insat 1D 82.9° E Spot Beam

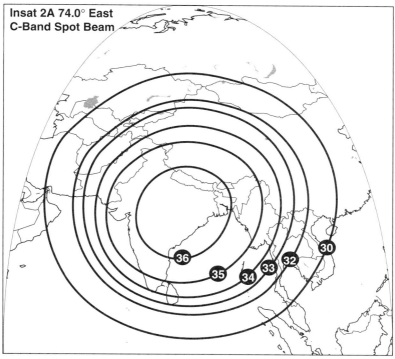

Insat 2A 74.0° East C-Band Spot Beam

INSAT/INTELSAT▼

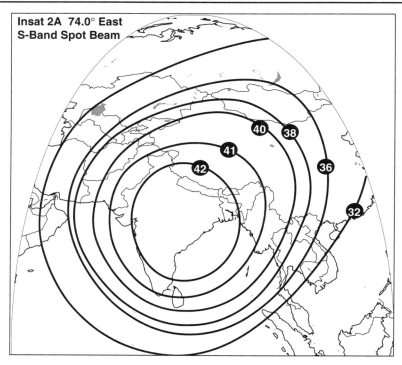

Insat 2A 74.0° East
S-Band Spot Beam

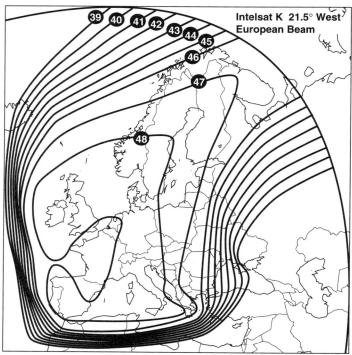

Intelsat K 21.5° West
European Beam

INTELSAT

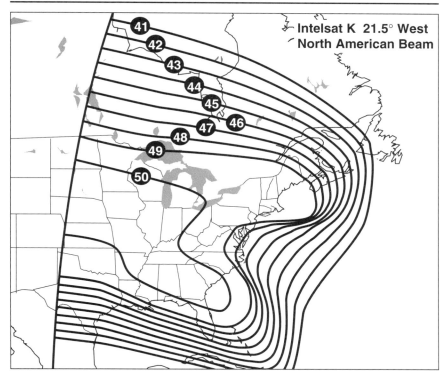

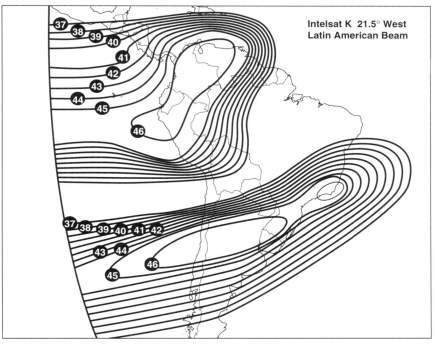

INTELSAT

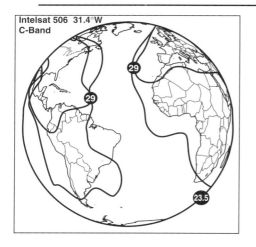
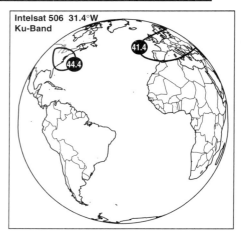
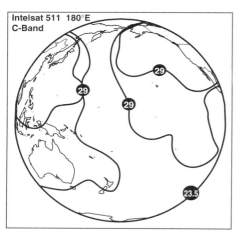
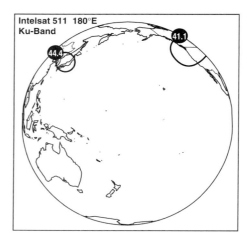
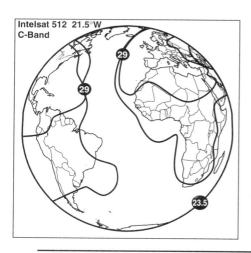
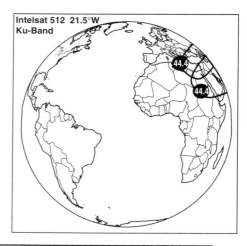

INTELSAT

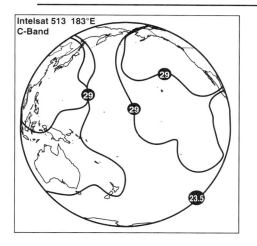
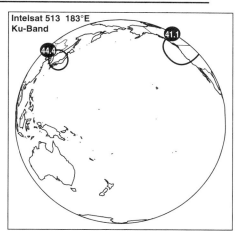
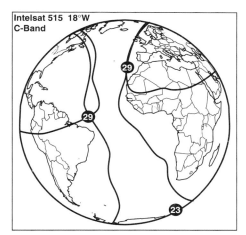
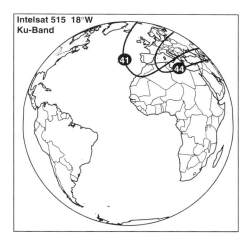
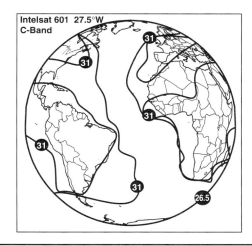

INTELSAT

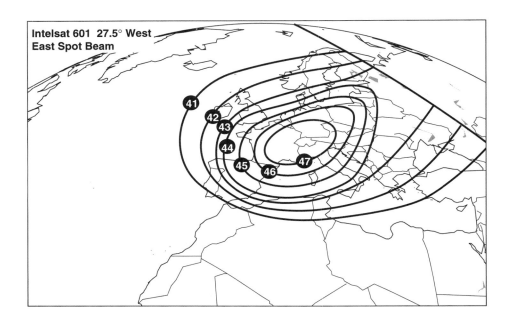

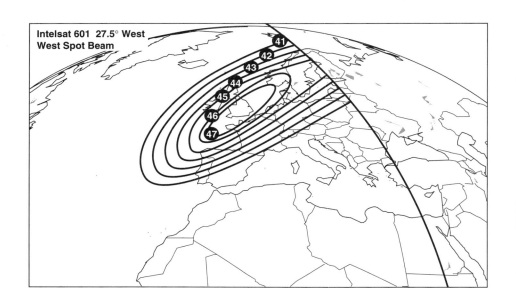

INTELSAT

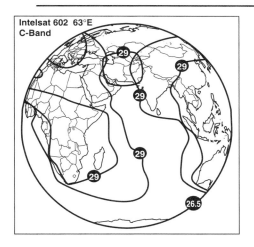
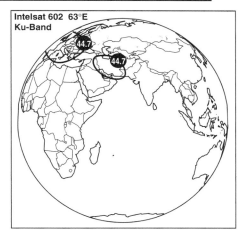
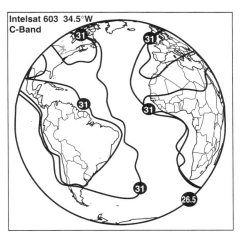
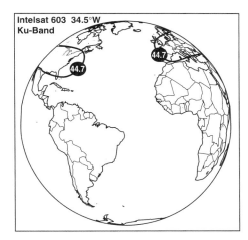
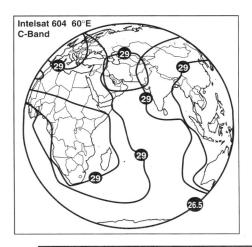
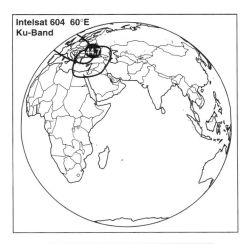

INTELSAT

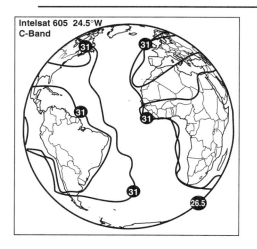
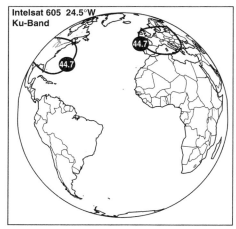
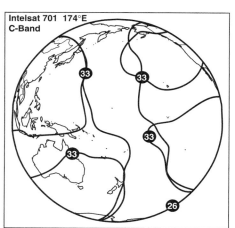
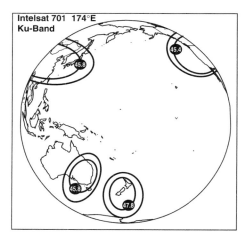
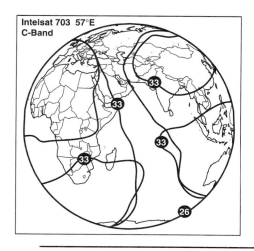
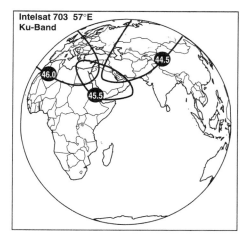

INTELSAT

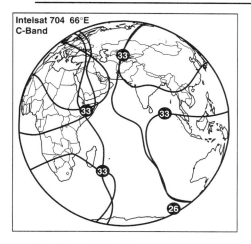

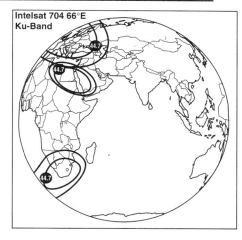

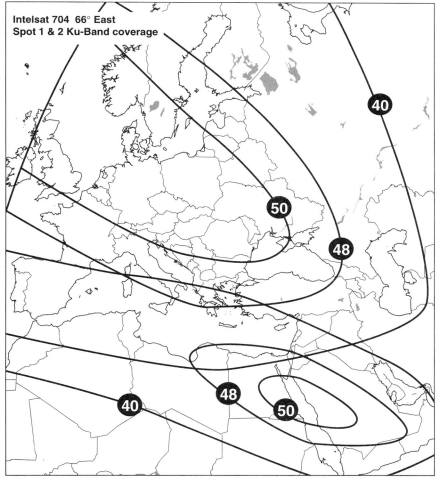

INTELSAT

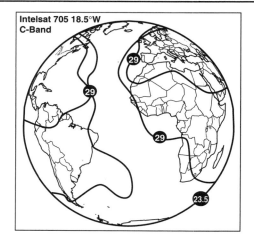

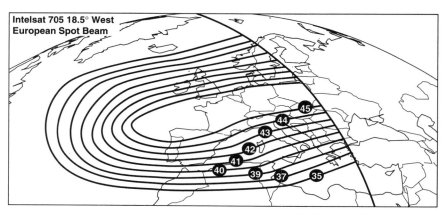

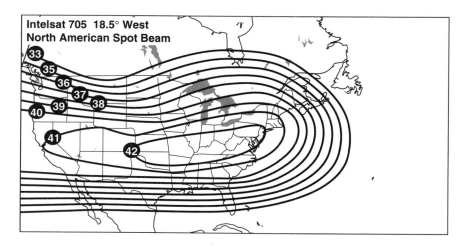

INTELSAT

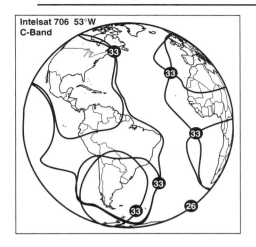
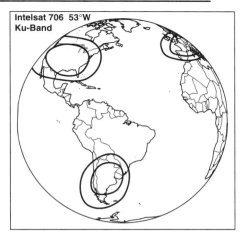
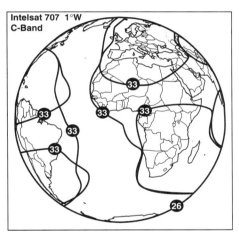
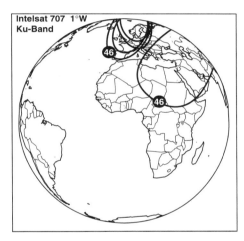
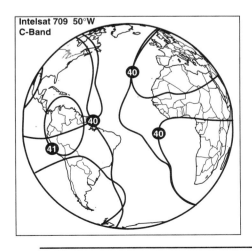
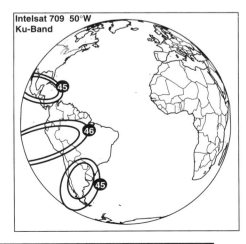

JCSAT/MEASAT▼

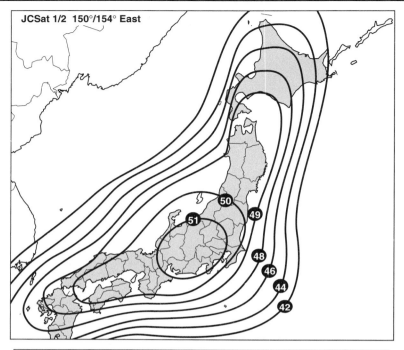

JCSat 1/2 150°/154° East

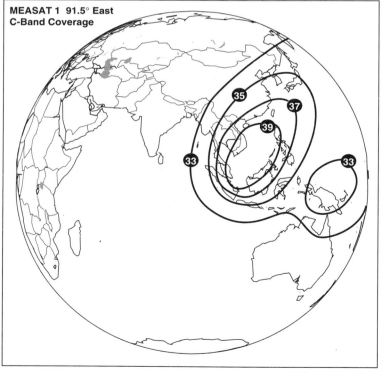

MEASAT 1 91.5° East
C-Band Coverage

OPTUS

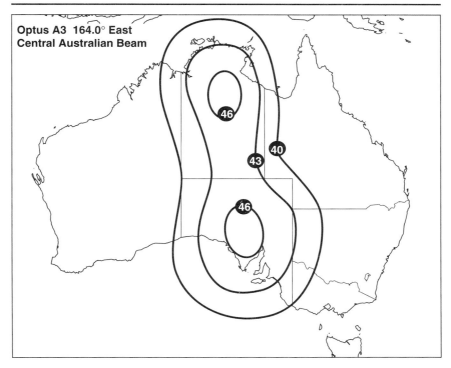

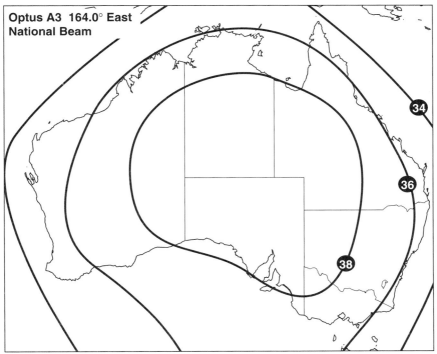

OPTUS

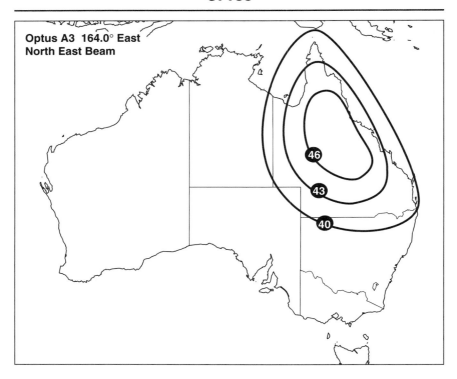

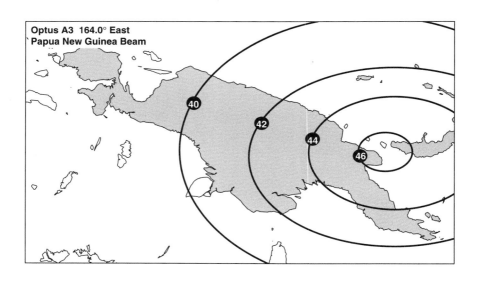

OPTUS

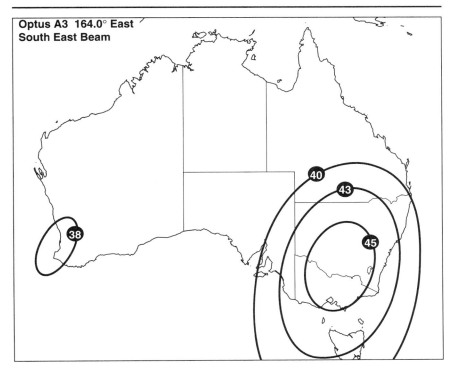

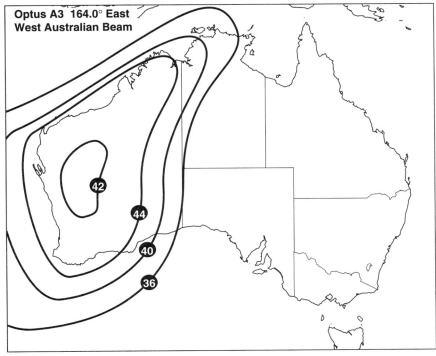

OPTUS

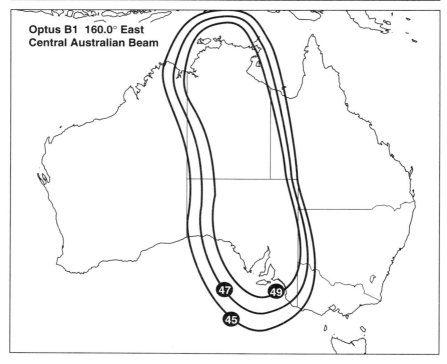

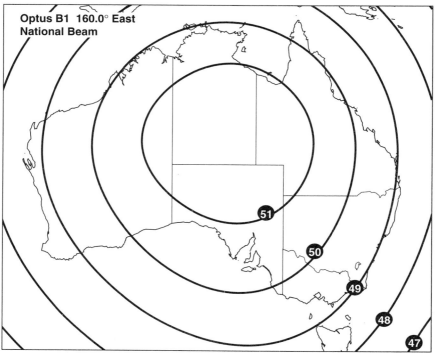

OPTUS

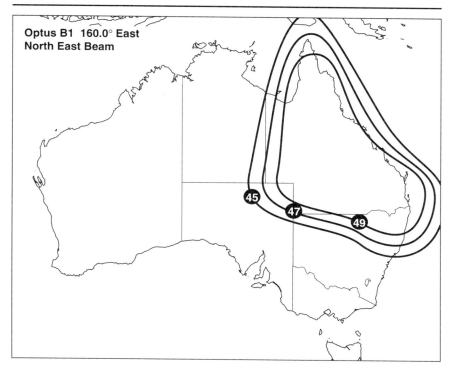

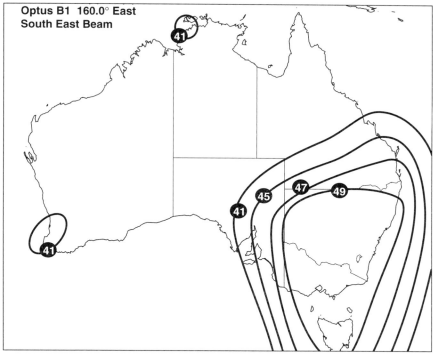

OPTUS

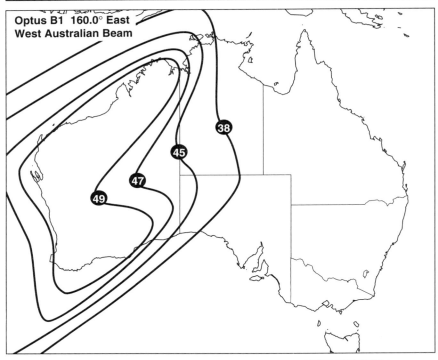

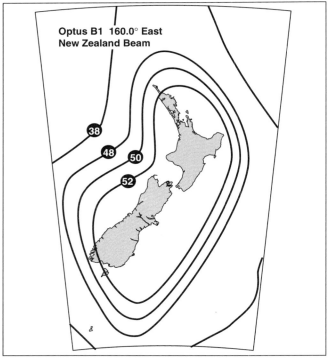

ORION

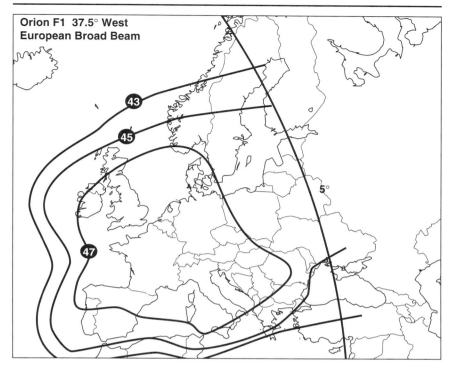

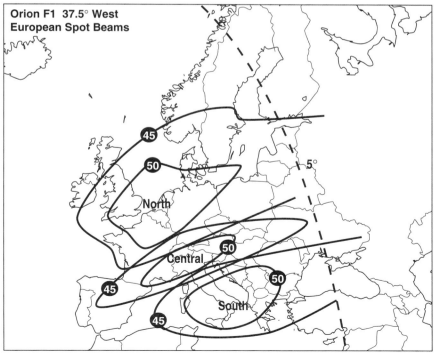

ORION

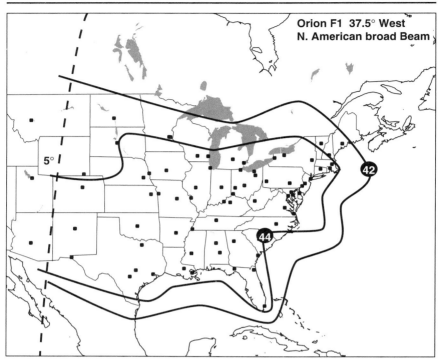

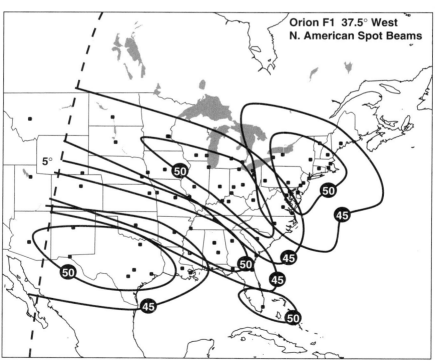

ORION

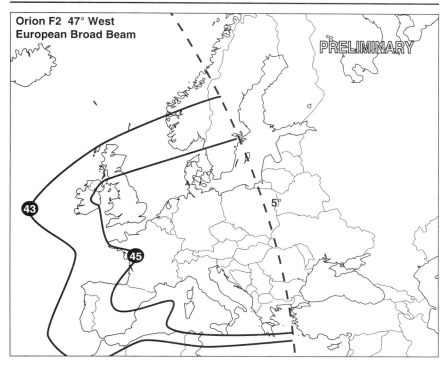

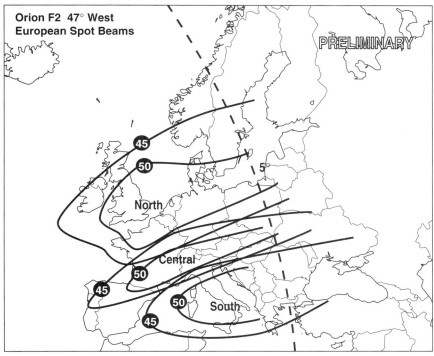

ORION

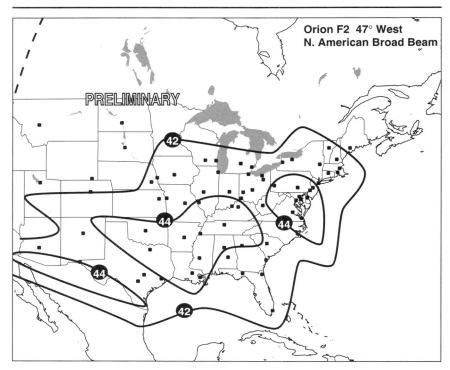

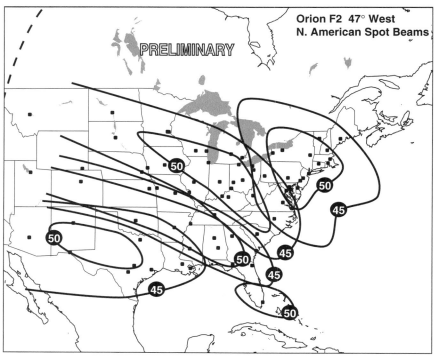

PALAPA

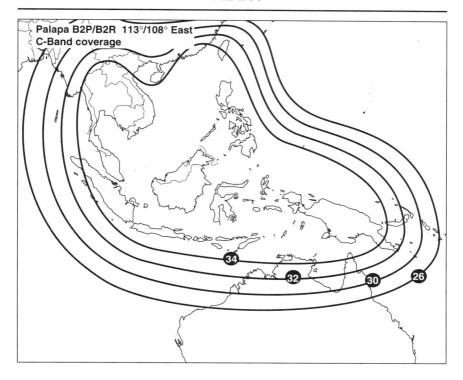

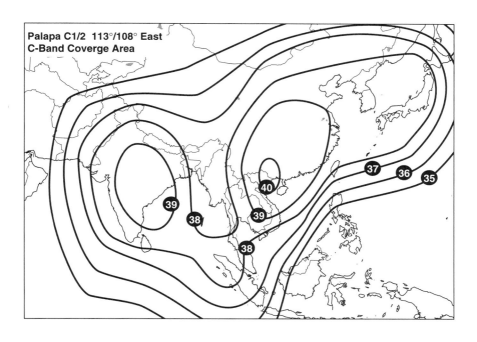

PALAPA

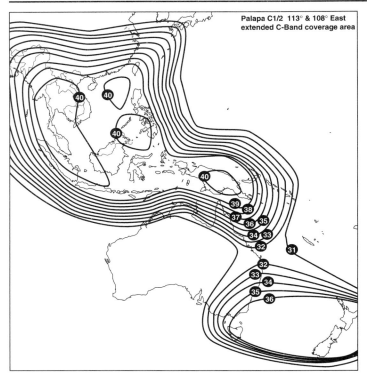

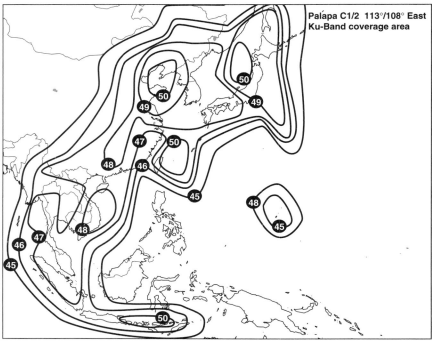

PANAMSAT

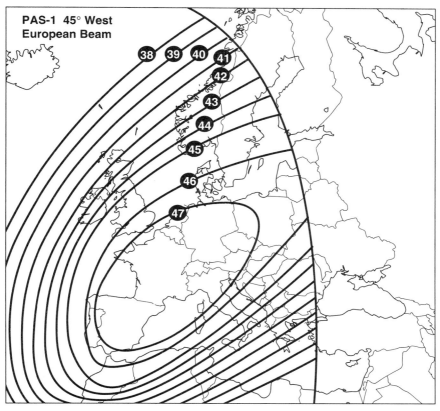

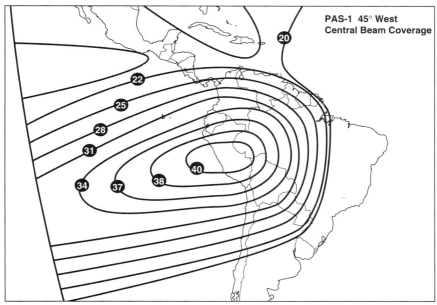

PANAMSAT

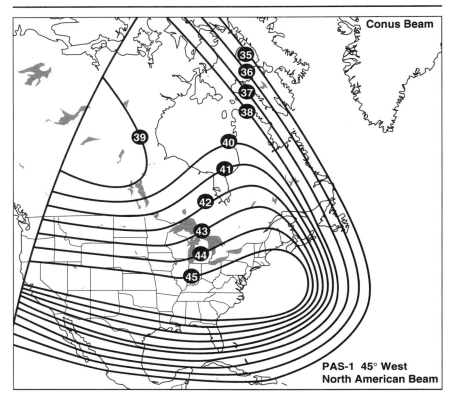

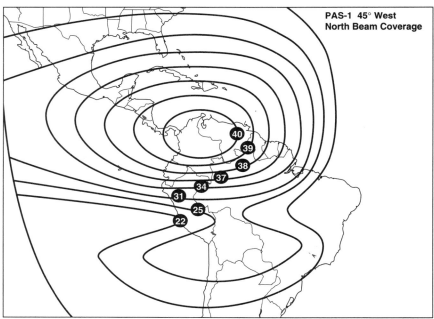

PANAMSAT

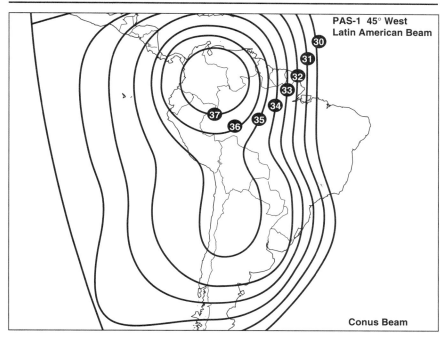

Conus Beam

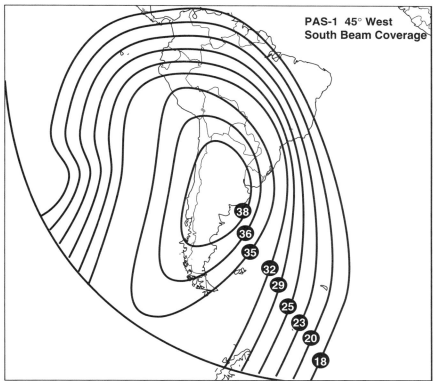

PANAMSAT

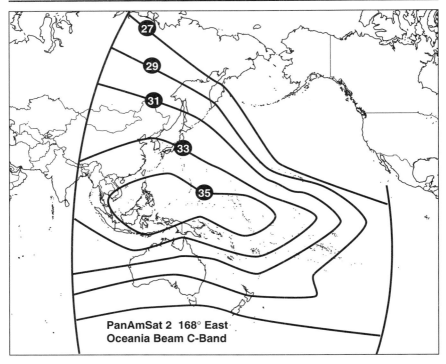

PanAmSat 2 168° East
Oceania Beam C-Band

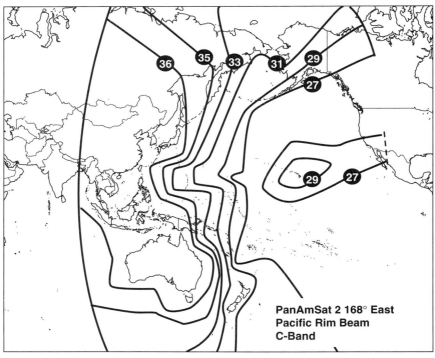

PanAmSat 2 168° East
Pacific Rim Beam
C-Band

PANAMSAT

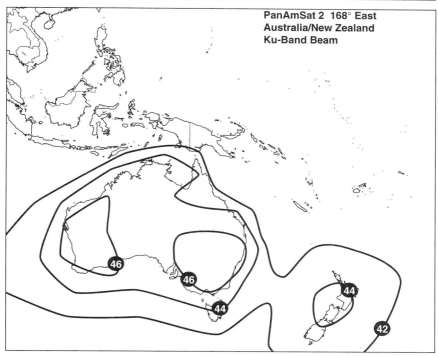

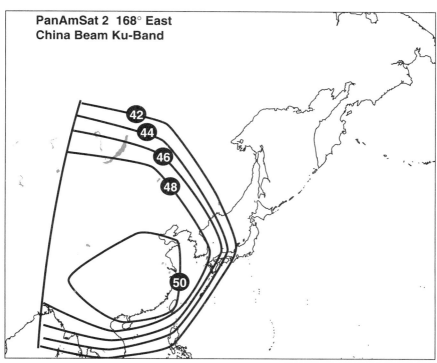

PANAMSAT

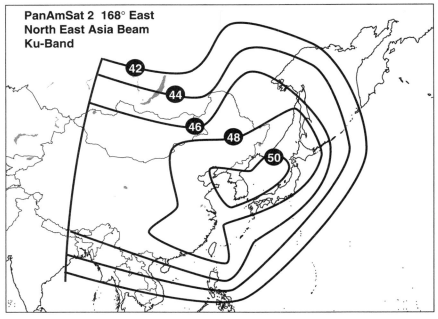

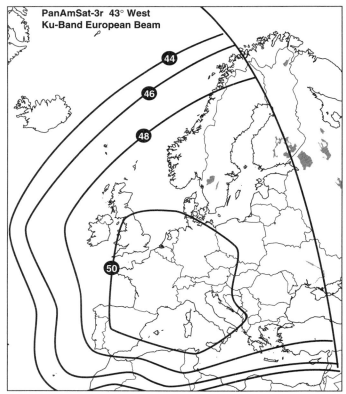

PANAMSAT

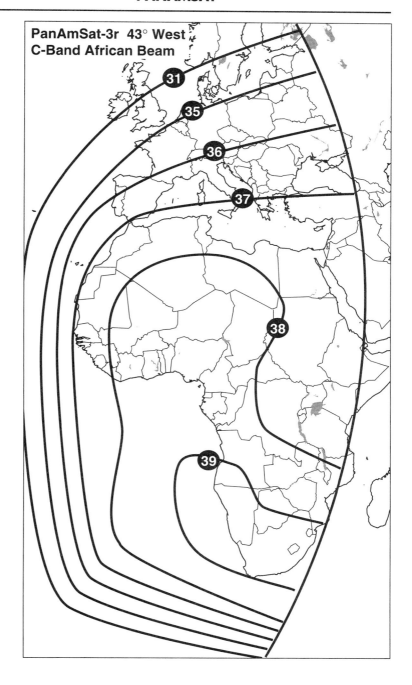

PANAMSAT

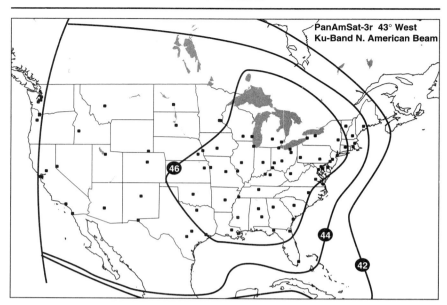

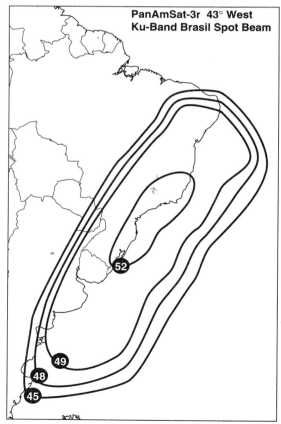

PANAMSAT

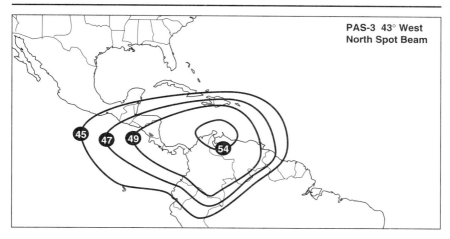

PANAMSAT

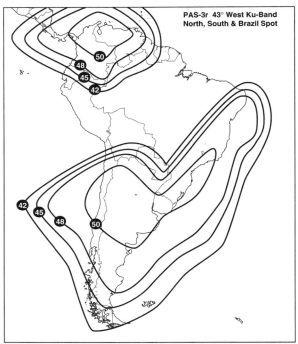

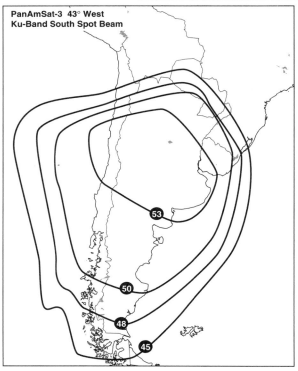

PANAMSAT

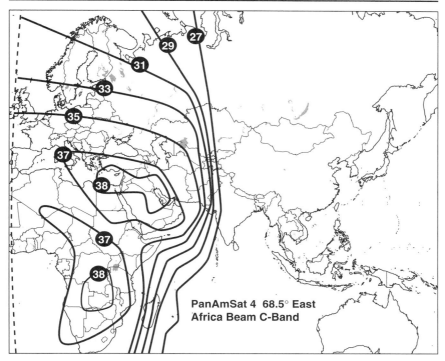

PanAmSat 4 68.5° East
Africa Beam C-Band

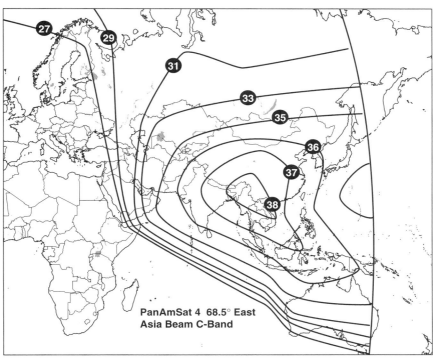

PanAmSat 4 68.5° East
Asia Beam C-Band

PANAMSAT

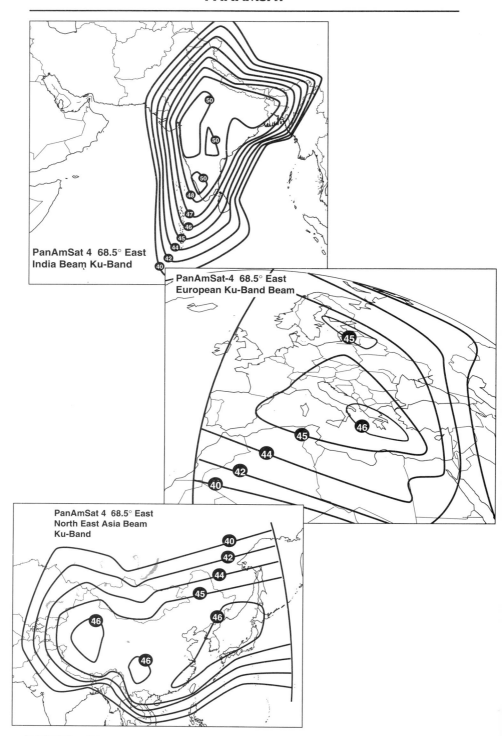

RIMSAT

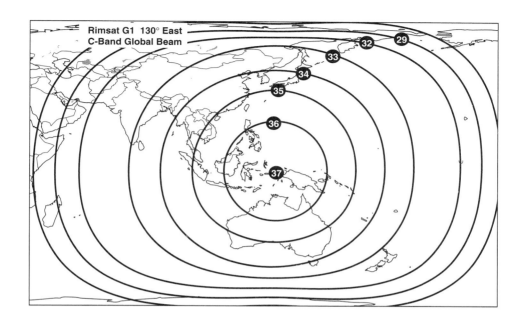

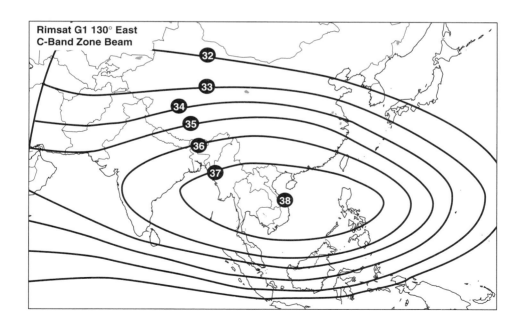

SATCOM

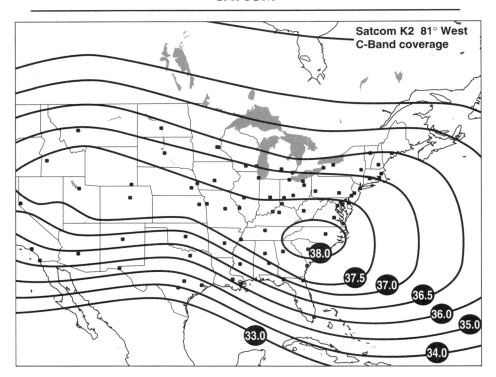

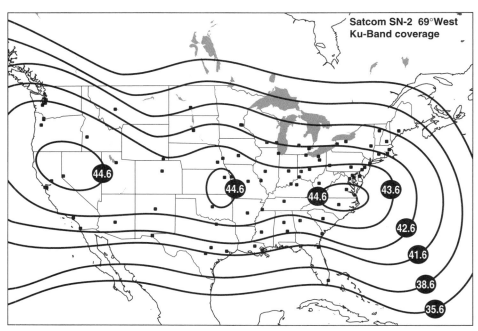

SBS

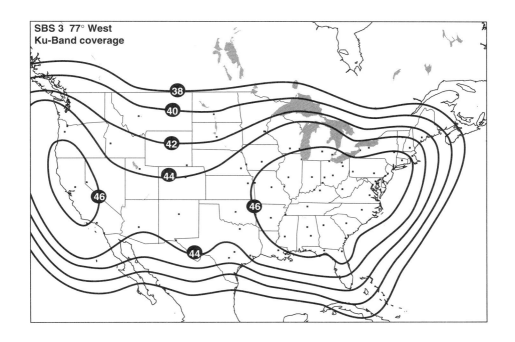

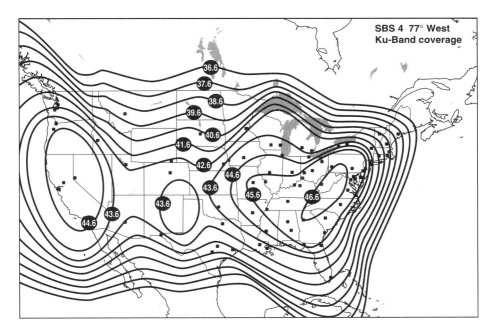

SBS

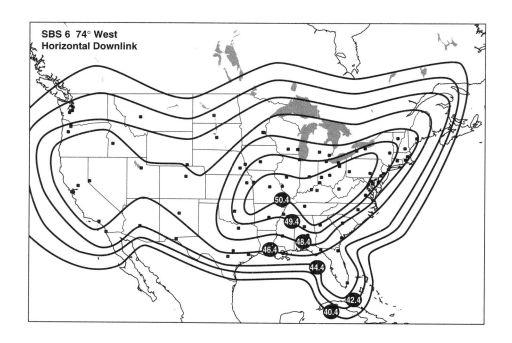

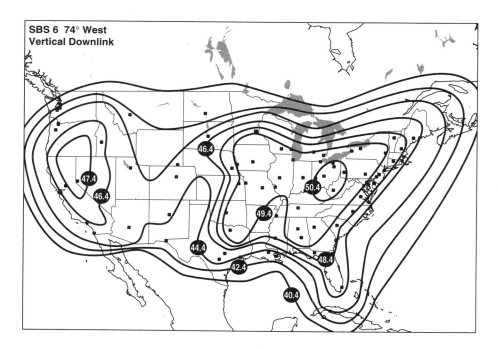

SPACENET

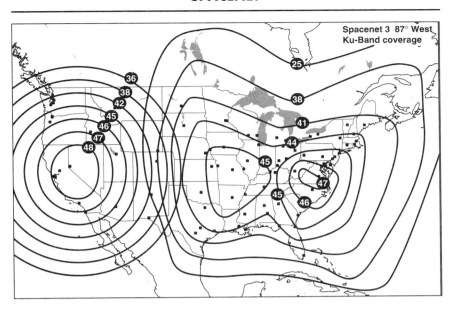

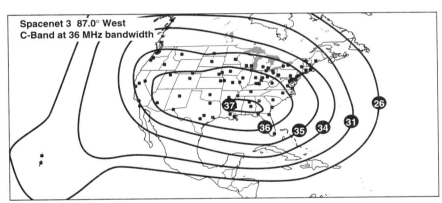

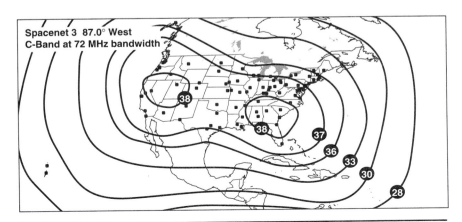

SUPERBIRD/TDF▼

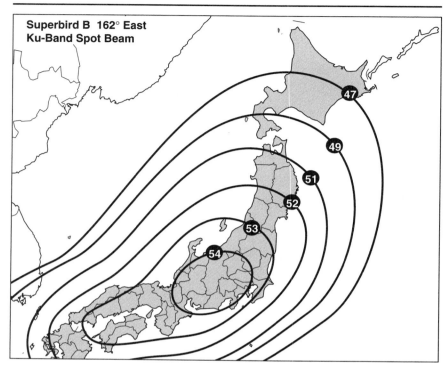

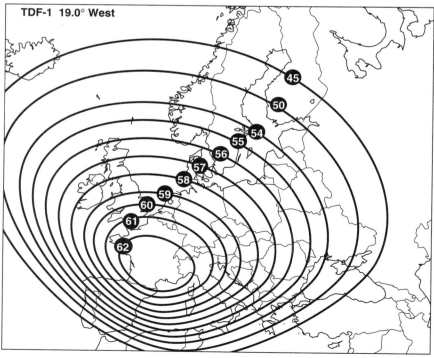

TDRSS/TELE-X▼

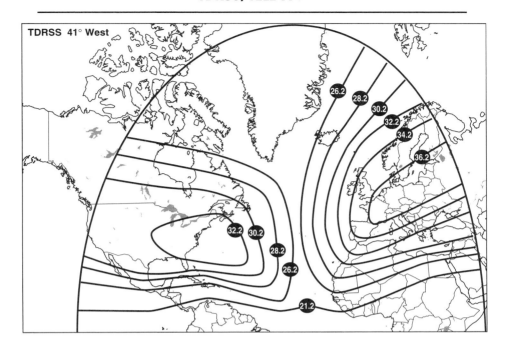

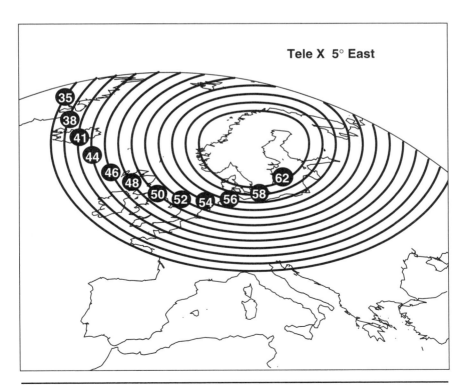

TÉLÉCOM

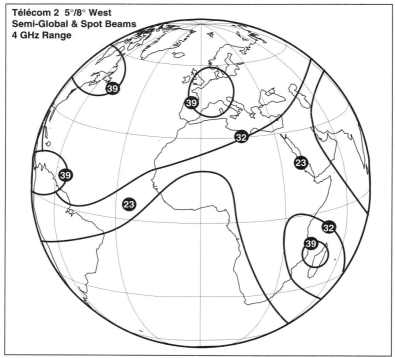

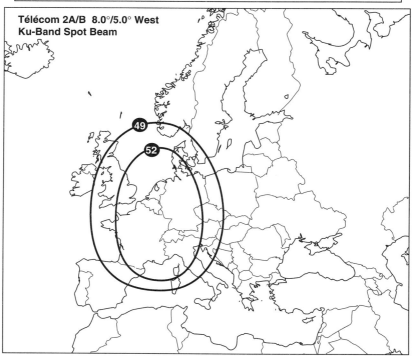

TELSTAR

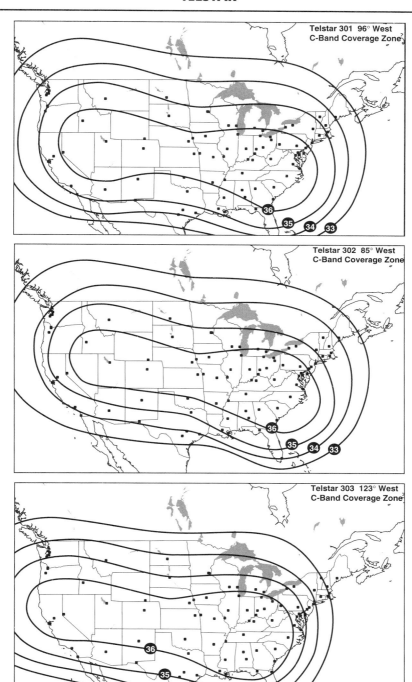

TELSTAR

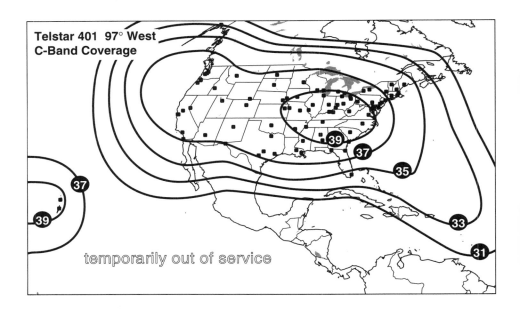

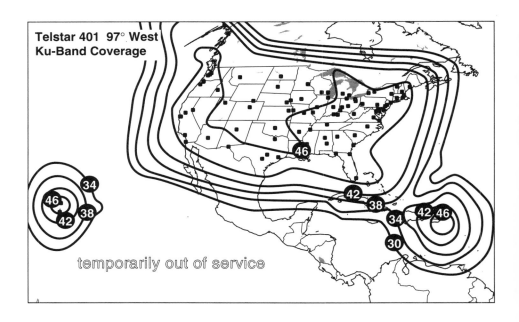

THAICOM

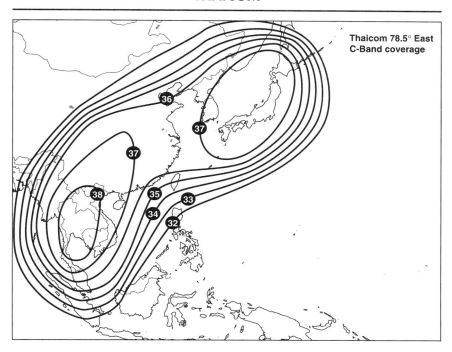

Thaicom 78.5° East
C-Band coverage

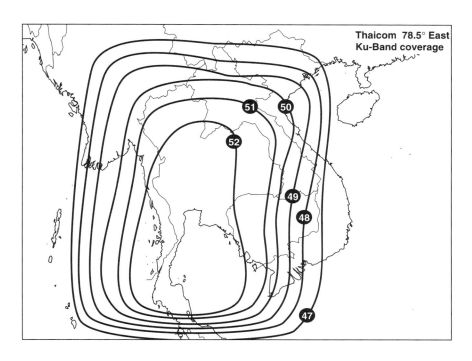

Thaicom 78.5° East
Ku-Band coverage

THOR/TV-SAT▼

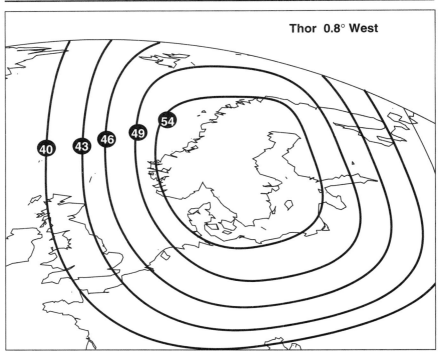

Thor 0.8° West

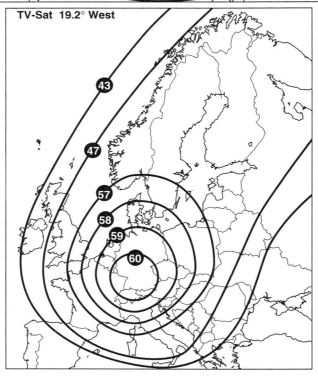

TV-Sat 19.2° West

TÜRKSAT

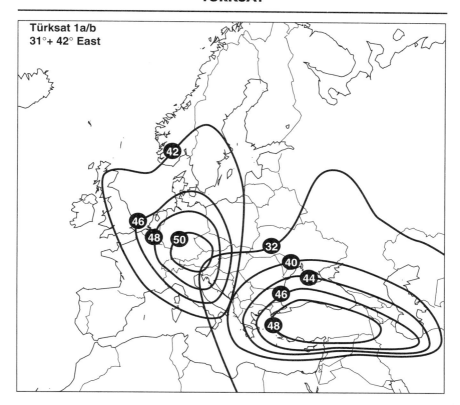

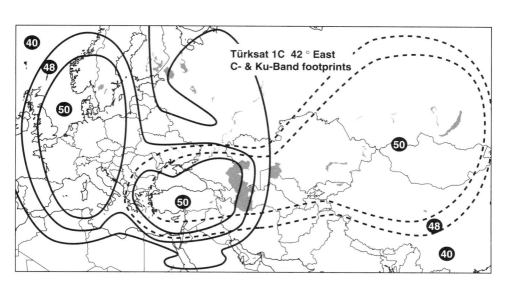

Satellite TV in Albania. Sights like this are common in many places in Tirana

CHAPTER 9
ITU REGION 1

TRANSPONDER LOADING

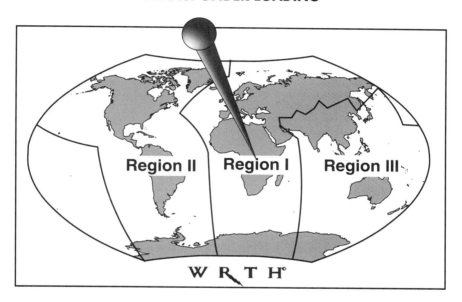

ITU REGION 1 STATIONS IN FREQUENCY ORDER

9.1. STATIONS IN FREQUENCY ORDER

PANAMSAT-4 68.5° EAST

Band	Freq.	Pol.	Tr.	Beam	Sce.	Program	Country	Language	System	Encr.	TT	A.M.	A.S.
C	3.790	Hor	3C	East Hemisph.	TV	Asia Business Channel	India	Hindi	PAL		No	6.80	
C	3.835	Ver	6L	East Hemisph.	TV	Home TV	India	Hindi	PAL		No	6.80/7.20	
C	3.845	Hor	5A	East Hemisph.	TV	ESPN	USA	English	PAL		No	6.50	A1
C	3.865	Hor	5U	East Hemisph.	TV	ESPN	USA	English	B-MAC		No		
C	3.865	Ver	5U	East Hemisph.	TV	BBC World	UK	English	PAL		No	6.60	
C	3.910	Hor	C7	East Hemisph.	TV	Sony Entertainment TV	India	Hindi	PAL		No	7.20	
C	4.040	Ver	C12	East Hemisph.	TV	Doordarshan TV	India	Hindi/English	PAL		No	6.30/6.80	
C	4.090	Hor	C13L	East Hemisph.	TV	CNN international	USA	English	PAL		No	6.80	
C	4.090	Hor	C13L	East Hemisph.	R	CNN R	USA	English			No	6.30	
C	4.113	Ver	C14U	Africa	TV	ConAir	USA	English	PAL				7.56/7.74
C	4.115	Hor	C13U	East Hemisph.	TV	TNT/Cartoon Network	USA	English	PAL		No	6.80	
C	4.157	Hor	C15L	East Hemisph.	O	occasional video					No	6.80	
C	4.157	Ver	C16L	East Hemisph.	O	occasional video					No	6.80	
C	4.182	Hor	C16U	East Hemisph.	TV	ATN	Africa	English	PAL				7.56/7.74
C	4.182	Ver	C16U	East Hemisph.	O	occasional video					No	6.80	
C	4.185	Ver	C15U	East Hemisph.	TV	MTV	Africa	English	PAL		No	6.50	
Ku3	12.538	Ver	13	AFS Spot	TV	M.Net South Africa	AFS	English	PAL		No	6.80	
Ku3	12.606	Ver	14	EUR Spot	TV	NHK Tokyo	Japan	Japanese	NTSC		No	7.02	
Ku3	12.664	Ver		AFS Spot	TV	SABC 1 (Afrikaans)	AFS	Afrikaans	PAL		No	7.56	6.66/6.84
Ku3	12.664	Ver		AFS Spot	TV	SABC 3 (Sindebele)	AFS	Sindebele	PAL		No	7.74	
Ku3	12.664	Ver		AFS Spot	TV	SABC 3 (Siswati)	AFS	Siswati	PAL		No	7.92	
Ku3	12.664	Ver		AFS Spot	TV	SABC 3 (Tshivenda)	AFS	Tshivenda	PAL		No	8.10	
Ku3	12.664	Ver		AFS Spot	TV	SABC 3 (Xitsonga)	AFS	Xitsonga	PAL		No	7.20	
Ku3	12.664	Ver		AFS Spot	R	SAFM	AFS	Afrikaans			No	7.38	
Ku3	12.664	Ver		AFS Spot	R	Lotus	AFS	Afrikaans			No	8.28	
Ku3	12.665	Ver	15L	AFS Spot	R	Youth	AFS				No	6.50	
Ku3	12.697	Ver	4KL	AFS Spot	TV	SABC TV1	AFS	English	PAL		No	7.02	6.66/6.84
Ku3	12.697	Ver	4KL	AFS Spot	TV	SABC 2	AFS	English	PAL		No	7.38	
Ku3	12.697	Ver	4KL	AFS Spot	TV	SABC 2	AFS	Sepedi	PAL		No	7.56	
Ku3	12.697	Ver	4KL	AFS Spot	TV	SABC 2	AFS	Sesotho	PAL		No	7.74	
Ku3	12.697	Ver	4KL	AFS Spot	TV	SABC 2	AFS	Setswana	PAL		No	7.92	
Ku3	12.697	Ver	4KL	AFS Spot	TV	SABC 2	AFS	Sixhosa	PAL		No	8.10	
Ku3	12.697	Ver	4KL	AFS Spot	TV	SABC 2	AFS	Sizulu	PAL		No	7.20	
Ku3	12.697	Hor	4KL	AFS Spot	R	Radio Sonder Grense	AFS	Afrikaans			No	8.28	
Ku3	12.698	Ver	15U	AFS Spot	R	Radio 2000	AFS	Afrikaans			No	6.50	
Ku3	12.724	Ver	16L	AFS Spot	TV	SABC CCV	AFS		PAL		No	7.02	6.66/6.84
Ku3	12.734	Ver	4KU	AFS Spot	TV	SABC NNTV	AFS		PAL		No	6.50	
Ku3	12.740	Ver	16U	AFS Spot	TV	ASTRA SPORT	AFS	Afrikaans	PAL		No		
Ku3				AFS Spot	TV	CDAT	AFS		PAL				

ITU REGION 1 STATIONS IN FREQUENCY ORDER

Band	Freq.	Pol.	Tr.	Beam	Sce.	Program	Country	Language	System	Encr.	TT	A.M.	A.S.
INTELSAT 704 66.0° EAST													
C	3.770	RHCP	11	West Hemisph.	O	occasional video	various		PAL/SEC/NTSC			6.60	
C	3.970	RHCP	14	West Hemisph.	O	occasional video	various		PAL/SEC/NTSC			6.60	
C	4.012	LHCP	15	North East Zone	O	occasional video	various		PAL			6.60	
C	4.100	LHCP	15	North East Zone	TV	NEPC TV	India	Hindi	PAL		No	6.60	
C	4.177	RHCP	38	Global	TV	C-Span	USA	English	PAL		No	6.60	
C	4.177	RHCP	38	Global	TV	WorldNet USIA	USA	English	PAL		No	6.60	
C	4.177	RHCP	38	Global	TV	Deutsche Welle TV	Germany	German/English/Spanish	PAL		No	6.60	
C	4.177	RHCP	38	Global	R	VOA Europe Network	USA	various				7.02	
C	4.177	RHCP	38	Global	R	VOA Ext. Sce.	USA	various				7.20	
C	4.177	RHCP	38	Global	R	VOA Ext. Sce.	USA	various				7.33	
C	4.177	RHCP	38	Global	R	VOA Ext. Sce.	USA	various				7.42	
C	4.177	RHCP	38	Global	R	VOA Ext. Sce.	USA	various				7.51	
C	4.177	RHCP	38	Global	R	Deutsche Welle R	Germany	various				7.60	
C	11.140	RHCP	63a	Global	O	occasional video	various		PAL/SECAM			6.60	
INTELSAT 602 63.0° EAST													
C	3.807	LHCP	4	South West Zone	TV	SABC-TV	South Africa	various	PAL	Irdeto	No	SiS	7.38/7.56
C	3.807	LHCP	4	South West Zone	R	Radio-5	South Africa	various				6.60	7.74/7.92
C	3.807	LHCP	4	South West Zone	R	Radio Metro	South Africa	various					
C	3.897	LHCP	12	South West Zone	TV	M-Net Int.	South Africa	various	PAL	Irdeto	No	SiS	
C	3.897	LHCP	4	South West Zone	R	Radio MF-702	South Africa	various				6.60	
C	4.055	LHCP	18	South East Zone	TV	Bop TV-1	Bophuthatswana	Setswana/English	PAL		No	6.65	
C	4.166	RHCP	38A	Global	O	occasional video	various		PAL/SEC/NTSC			6.60	
C	4.188	RHCP	38B	Global	O	occasional video	various		PAL/SEC/NTSC			6.60	
C	10.975	Hor	61A	West Spot	O	Telespazio	Italy		PAL			6.60	
Ku1	11.002	Ver	71	East Spot	TV	IRIB TV-2	Iran	Farsi	SECAM	Irdeto	No	6.80	
Ku1	11.002	Ver	71	East Spot	R	IRIB Radio-2 (int.)	Iran	Farsi				5.60/6.20	
Ku1	11.002	Ver	71	East Spot	R	IRIB Radio-1	Iran	various				5.95	
Ku1	11.011	Hor	61B	West Spot	O	former Rete 4			PAL			6.60	
Ku1	11.055	Hor	62	West Spot	O	former Cinquestelle			PAL			6.60	
Ku1	11.095	Hor	62	West Spot	R	Radiofonica	Italy	Italian	D2-MAC	EuroCrypt			
Ku1	11.100	Ver	73	East Spot	TV	IRIB TV-3	Iran	Farsi	SECAM		No	6.80	
Ku1	11.100	Ver	73	East Spot	R	unknown R						6.20	
Ku1	11.137	Hor	63A	West Spot	O	former Italia-1	Italy	Italian	PAL			6.60	
Ku1	11.155	Ver	73	East Spot	TV	IRIB TV-1	Iran	Farsi	SECAM		No	6.80	
Ku1	11.155	Ver	71	East Spot	R	IRIB Radio-1	Iran	various				6.20	
Ku1	11.173	Hor	63B	West Spot	O	former Canale-5			PAL			6.60	
INTELSAT 604 60.0° EAST													
C	4.166	RHCP	38A	Global	O	occasional video	various		PAL/SEC/NTSC			6.65	
C	4.188	RHCP	38B	Global	O	occasional video	various		PAL/SEC/NTSC			6.65	
INTELSAT 703 57.0° EAST													
C	3.920	RHCP	13	East Hemisph.	TV	ETV	Ethiopia	African dialect	PAL		No	6.60	
C	4.020	LHCP		East Hemisph.	TV	Sudan TV	Sudan	Arabic	PAL		No	6.60	

ITU REGION 1 STATIONS IN FREQUENCY ORDER

Band	Freq.	Pol.	Tr.	Beam	Sce.	Program	Country	Language	System	Encr.	TT	A.M.	A.S.
C	4.055	RHCP	13	East Hemisph.	TV	C-Span/Worldnet	USA	English	PAL		No	6.60	
C	4.059	LHCP		East Hemisph.	TV	NEPC-TV	India	Hindi	PAL		No	6.60	
C	4.133	LHCP		East Hemisph.	TV	TV 1-India	India	Hindi	PAL/SEC/NTSC		No	6.60	
C	4.166	RHCP	38A	Global	O	occasional video	various		PAL		No	6.65	
C	4.177	LHCP		East Hemisph.	TV	Muslim-TV MTA+	UK	Arabic	PAL/SEC/NTSC			6.60	
C	4.188	RHCP	38B	Global	O	occasional video	various		PAL			6.65	
Ku1	11.543	Hor		East Hemisph.	TV	Kasachstan 1 (private)	Kasachstan	Kasach	SECAM		No	7.00	
Ku1	11.543	Hor		East Hemisph.	R	Radio Almaty International	Kasachstan	Kasach	SECAM		No	7.50	
						GORIZONT 11/17 (STATIONAR 5) 53.0° EAST							
C	3.675	RHCP	6	Global	TV	ORT-1/Orbita 4	CIS	Russian	SECAM		No	7.0	
C	3.675	RHCP	6	Spot	TV	Afghanistan TV	Afghanistan	Pushtu	SECAM		No	7.0	
C	3.675	RHCP	6	Global	R	Radio Mayak	CIS	Russian				7.5	
Ku1	11.525	RHCP	K1	Spot	TV	CTC (STS)	CIS	Russian	SECAM		No	6.60	
						EUTELSAT I F1 48.0° EAST							
Ku1	10.990	Ver			O	TVCSM	Germany					6.65	
						TÜRKSAT 1B 42° EAST							
Ku1	10.965	Ver	4	West Spot	TV	ATV	Turkey	Turkish	PAL		No	6.65	
Ku1	10.977	Ver		East Spot	O		Turkey					6.60	
Ku1	11.030	Hor	13	East Spot	TV	Kanal D	Turkey	Turkish	PAL		No	6.60	
Ku1	11.030	Hor	13	East Spot	R	Radyo Kulup	Turkey	Turkish					7.02/7.20
Ku1	11.030	Hor	13	East Spot	R	Best FM	Turkey	Turkish					7.38/7.56
Ku1	11.030	Hor	13	East Spot	R	Radyo Forex	Turkey	Turkish					7.38/7.56
Ku1	11.073	Ver	14	East Spot	TV	Kanal-6	Turkey	Turkish					7.74/7.92
Ku1	11.100	Ver	12	West Spot	R	HBB	Turkey	Turkish	PAL		No	6.65	
Ku1	11.100	Ver	12	West Spot	TV	HBB FM	Turkey	Turkish	PAL		No	6.65	
Ku1	11.121	Ver	7B	East Spot	TV	Samanloyu TV	Turkey	Turkish					7.38/7.56
Ku1	11.144	Ver	8A	West Spot	TV	Kanal 7	Turkey	Turkish	PAL		No	6.65	
Ku1	11.144	Ver	8A	West Spot	R	Mamara FM	Turkey	Turkish	PAL		No	6.65	
Ku1	11.144	Ver	8A	West Spot	R	Tatlises R	Turkey	Turkish					7.02/7.20
Ku1	11.144	Ver	8A	West Spot	R	Best FM	Turkey	Turkish					7.74/7.92
Ku1	11.144	Ver	8A	East Spot	R	Morales FM	Turkey	Turkish					8.10/8.28
Ku1	11.173	Ver	16	West Spot	TV	TRT-1	Turkey	Turkish	PAL			7.38	
Ku1	11.181	Ver	8B	East Spot	O	former Kanal D	Turkey		PAL		No	6.65	
Ku1	11.462	Ver	9A	East Spot	TV	TRT-3	Turkey	Turkish	PAL		No	6.65	
Ku1	11.490	Ver	9B	East Spot	TV	TRT-4	Turkey	Turkish	PAL		No	6.65	
Ku1	11.555	Hor	10A	East Spot	TV	TRT Int.	Turkey	Turkish	PAL		No	6.65	
Ku1	11.569	Ver	2A	West Spot	TV	Euro D	Turkey	Turkish	PAL		No	6.60	
Ku1	11.569	Ver	2A	East Spot	R	Radyo D	Turkey	Turkish				7.02	
Ku1	11.640	Hor	11A	East Spot	O	occasional video	Turkey					6.80	
Ku1	11.678	Ver	3	Spot	O	future use	Turkey						
Ku1	11.678	Hor	11B	East Spot	O	occasional video	Turkey		PAL				

ITU REGION 1 STATIONS IN FREQUENCY ORDER

Band	Freq.	Pol.	Tr.	Beam	Sce.	Program	Country	Language	System	Encr.	TT	A.M.	A.S.
GORIZONT 12/22 40.0° EAST													
C	3.675	RHCP	6	Global	TV	TV Rossija 2	CIS	Russian	SECAM		No	7.5	
C	3.675	RHCP	6	Global	R	Radio Nostalgij	CIS	Russian					8.40/8.75
C	3.675	RHCP	6	Global	R	Radiostancija Majak	CIS	Russian				7.0	
C	3.675	RHCP	6	Global	R	Radio Golos Rossija	CIS	Russian				8.0	
C	3.930	RHCP	11	Global Beam	TV	RTP Internacional	Portugal	Portuguese	PAL		No	6.60/7.20	
C	3.930	RHCP	11	Global Beam	R	Radio UM	Portugal	Portuguese				7.02	
Ku1	11.525	LHCP	K1	Spot	O	occasional video	various		SECAM			7.0	
GALS 1/2 36.0° EAST													
Ku2	11.766	RHCP		West Spot	TV	Kazakhstan TV	Kazakhstan	Kazakh	SECAM		No	7.00	
Ku2	11.840	LHCP		West Spot	TV	TV-6 Moscow	C.I.S.	Russian	PAL			7.02	
Ku2	12.171	RHCP		West Spot	O							6.50	
Ku2	12.207	RHCP		West Spot	TV	HTB	Russia	Russian	SECAM			7.00	
ARABSAT 1C 30.5° EAST													
S	2.635	Ver	1A	South Zone	TV	Middle East Broadcasting Centre	UK	Arabic	PAL		No	6.60	
C	3.772	LHCP	4	Zone	O	occasional video	Saudi Arabia	Arabic	SCPC				
C	3.866	RHCP	9	Zone	O	occasional video						6.65	
C	3.955	RHCP	14	Zone	TV	EDTV	UAE	Arabic	PAL		Yes	6.60	
C	3.955	RHCP	14	Zone	R	Radio Dubai	UAE	Arabic				7.20	
C	3.955	RHCP	14	Zone	R	Radio Dubai	UAE	English				7.74	
C	4.062	LHCP	20	Zone	TV	Oman TV	Oman	Arabic	PAL		No	6.60	
C	4.078	LHCP	20	Zone	R	Oman R	Oman	Arabic	SCPC				4.078
C	4.107	LHCP	22	Zone	TV	Middle East Broadcast (MBC)	UK	Arabic	PAL		No	6.60	
C	4.107	LHCP	22	Zone	R	MBC R	UK	Arabic					7.38/7.56
C	4.125	RHCP	23	Zone	O	ASBU occasional video	various		PAL/SEC/NTSC		No	6.65	
C	4.140	LHCP	24	Zone	TV	JRTV Jordan	Jordan	Arabic	PAL		No	6.60	
C	4.181	RHCP	26	Zone	TV	Kuwait Space Channel	Kuwait	Arabic	PAL		No	6.60	
C	4.181	RHCP	26	Zone	R	Radio Kuwait (FM 89.5 & 98.9)	Kuwait	Arabic	PAL		No	5.80	
C	4.181	RHCP	26	Zone	R	FM Super Station (FM 99.7)	Kuwait	Arabic	PAL		No	7.40	
DFS-2 KOPERNIKUS 28.5° EAST													
Ku1	11.475	Hor	A1	Spot	TV	Cable Plus Filmový Kanál	Czech Republic	Czech	PAL	VideoCrypt II	No	6.65	7.02/7.20
Ku1	11.525	Ver	A2	Spot	TV	Premiéra TV	Czech Rep.	Czech	PAL	Syster	No	6.50	
Ku1	11.548	Hor	B1	Spot	O	occasional video (a.o. RTL West)	Germany	German	PAL		No	6.65	
Ku1	11.600	Ver	B2	Spot	O	occasional feeds	Germany	German			No	6.65	
Ku1	11.625	Hor	C1	Spot	O	occasional use	Germany	German				6.65	
Ku3	12.590	Ver	K3	Spot	O	Wir in Niedersachsen	Germany	German	PAL		No	6.65	
Ku3	12.625	Hor	K4	Spot	O	DBP Telekom	Germany		PAL			SiS	
Ku3	12.658	Ver	K5	Spot	O	RTL Nord feeds	Germany	German	PAL		No	6.65	7.02/7.20
Ku3	12.692	Hor	K6	Spot	O	RTL West feeds	Germany		PAL				
Ku3	12.725	Ver	7	Spot					PAL				
ARABSAT 2A 26.0° EAST													
C	3.720	RHCP	1	Zone	TV	Sharjah TV	Mauritania	Arabic	PAL		No	6.60	

163

ITU REGION 1 STATIONS IN FREQUENCY ORDER

Band	Freq.	Pol.	Tr.	Beam	Sce.	Program	Country	Language	System	Encr.	TT	A.M.	A.S.
C	3.740	LHCP	2	Zone	TV	LBC Lebanon	Lebanon	Arabic/French	PAL		No	6.60	
C	3.761	RHCP	3	Zone	TV	Egypt TV	Egypt	Arabic	PAL		No	6.60	
C	3.781	LHCP	4	Zone	TV	ART	Saudi Arabia	Arabic	PAL		No	6.60	
C	3.802	RHCP	5	Zone	TV	Nile TV	Egypt	Arabic	PAL		No	6.60	
C	3.822	LHCP	6	Zone	TV	Bahrain TV	Bahrain	Arabic	PAL		No	6.60	
C	3.843	RHCP	7	Zone	TV	CNN	USA	English	PAL		No	6.60	
C	3.863	LHCP	8	Zone	TV	Future Vision	Saudi Arabia	Arabic	PAL		No	6.60	
C	3.884	RHCP	9	Zone	TV	Orbit	Morocco	Arabic	SECAM		No	6.60	
C	3.905	LHCP	10	Zone	TV	Sudan TV	Sudan	Arabic	PAL		No	6.60	
C	3.905	LHCP	10	Zone	R	Radio Sudan	Sudan	Arabic				7.20	
C	3.925	RHCP	11	Zone	TV	Saudi Arabia TV 1	Saudi Arabia	Arabic	SECAM		No	6.60	
C	3.945	LHCP	12	Zone	TV	Canal France Internationale	France	French	PAL		No	6.60	
C	3.978	RHCP	13	Zone	TV	Saudi Arabia TV 2	Saudi Arabia	Arabic	SECAM		No	6.60	
C	3.998	LHCP	14	Zone	TV	Syrian TV	Syria	Arabic	PAL		No	6.60	
C	3.998	LHCP	14	Zone	R	Radio Damascus	Syria	Arabic				7.56	
C	4.043	RHCP	15	Zone	TV	Sara Vision		Arabic	PAL		No	6.60	
C	4.057	LHCP	16	Zone	TV	Dubai TV	UAE	Arabic	PAL		No	6.60	
C	4.057	LHCP	16	Zone	R	Radio Dubai	UAE	Arabic				7.20	
C	4.057	LHCP	16	Zone	R	Radio Dubai	UAE	English				7.74	
C	4.075	RHCP	17	Zone	TV	Abu Dhabi TV	UAE	Arabic	PAL		No	6.60	
C	4.098	LHCP	18	Zone	TV	MBC	UK	Arabic	PAL		No	6.60	7.38/7.56
C	4.098	LHCP	18	Zone	R	MBC FM	UK	Arabic					
C	4.139	LHCP	20	Zone	TV	Oman TV	Oman	Arabic	PAL		No	6.60	
C	4.166	RHCP	21	Zone	TV	Kuwait TV	Kuwait	Arabic	PAL		No	6.60	
C	4.166	LHCP	21	Zone	R	Radio Kuwait	Kuwait	Arabic				5.80	
C	4.166	RHCP	21	Zone	R	Superstation Kuwait	Kuwait	Arabic				7.56	
C	4.180	LHCP	22	Zone	TV	Yemen TV	Yemen	Arabic	PAL		No	6.60	
C	4.180	LHCP	22	Zone	R	Radio Yemen	Yemen	Arabic				7.20	
Ku3	12.521	Hor	1		TV	Al Jazeera Satellite Channel	Algeria	Arabic	PAL		No	6.60	
Ku3	12.536	Ver	2		TV	Saudi Arabia	Saudi Arabia	Arabic	PAL		No	6.60	
Ku3	12.562	Hor	3		TV	ART (1st Net)	Saudi Arabia	Arabic	PAL		No	5.80	
Ku3	12.577	Ver	4		TV	JRT Jordan	Jordan	Arabic	PAL		No	6.60	
Ku3	12.604	Hor	5		TV	Jordan TV	Jordan	Arabic	PAL		No	6.60	
Ku3	12.619	Ver	6		TV	Abu Dhabi TV	UAE	Arabic	PAL		No	6.60	
Ku3	12.646	Hor	7		TV	Kuwait TV	Kuwait	Arabic	PAL		No	6.60	
Ku3	12.661	Ver	8		TV	Saudia	Saudi Arabia	Arabic	PAL		No	6.60	
Ku3	12.685	Hor	9		TV	ART	Saudi Arabia	Arabic	PAL		No	6.60	
Ku3	12.700	Ver	10		R	Libya TV	Libya	Arabic	PAL		No	6.60	
Ku3	12.720	Hor	11		R	ART	Saudi Arabia	Arabic	PAL		No	6.60	
Ku3	12.735	Ver	12		R	MBC	UK	Arabic	PAL		No	6.60	

EUTELSAT I F4 25.5° EAST

Band	Freq.	Pol.	Beam	Sce.	Program	Country	Language	System	Encr.	TT	A.M.	A.S.
Ku1	11.136	Hor	West Spot	O	horse racing	UK	English	PAL		No	6.65	
Ku1	11.175	Hor	West Spot	O	horse racing	UK	English	PAL		No	6.65	
Ku1	11.175	Hor	West Spot	O	occasional use	UK	English	PAL		No	6.65	
Ku1	11.670	Hor	West Spot	O	occasional use	UK	English	PAL		No	6.65	

ITU REGION 1 STATIONS IN FREQUENCY ORDER

Band	Freq.	Pol.	Tr.	Beam	Sce.	Program	Country	Language	System	Encr.	TT	A.M.	A.S.
DFS-3 KOPERNIKUS 23.5° EAST													
Ku1	11.475	Hor	A1	Spot	TV	Sat-1	Germany	German	PAL		Yes	6.65	7.02/7.20
Ku1	11.525	Hor	A2	Spot	O	test pattern			PAL			6.65	
Ku1	11.550	Ver	B1	Spot	O	former ARTE						6.65	
Ku1	11.600	Ver	B2	Spot	O	occasional video			PAL			6.65	
Ku1	11.625	Hor	C1	Spot	O	occasional video			PAL			6.65	
Ku1	11.675	Hor	C2	Spot	O	occasional video			PAL			6.65	
Ku3	12.559	Hor	K2	Spot	O	occasional video			PAL			6.65	
Ku3	12.591	Ver	K3	Spot	O	occasional video			PAL			6.65	
Ku3	12.625	Hor	K4	Spot	R	Bayern-4 Klassik	Germany	German	DSR				DSR ch01 7.02/7.20
Ku3	12.625	Hor	K4	Spot	R	S2 Kultur	Germany	German	DSR				DSR ch02 7.02/7.20
Ku3	12.625	Hor	K4	Spot	R	RB2 Kulturell	Germany	German	DSR				DSR ch03 7.02/7.20
Ku3	12.625	Hor	K4	Spot	R	HR2 Kultur	Germany	German	DSR				DSR ch04 7.02/7.20
Ku3	12.625	Hor	K4	Spot	R	NDR-3	Germany	German	DSR				DSR ch05 7.02/7.20
Ku3	12.625	Hor	K4	Spot	R	Star*Sat R	Germany	German	DSR				DSR ch06 7.02/7.20
Ku3	12.625	Hor	K4	Spot	R	Deutschlandfunk	Germany	German	DSR				DSR ch07 7.02/7.20
Ku3	12.625	Hor	K4	Spot	R	WDR-3 Köln	Germany	German	DSR				DSR ch08 7.02/7.20
Ku3	12.625	Hor	K4	Spot	R	Deutschlandradio Berlin	Germany	German	DSR				DSR ch09 7.02/7.20
Ku3	12.625	Hor	K4	Spot	R	SR1 Saarland	Germany	German	DSR				DSR ch10 7.02/7.20
Ku3	12.625	Hor	K4	Spot	R	RPR-2	Germany	German	DSR				DSR ch11 7.02/7.20
Ku3	12.625	Hor	K4	Spot	R	Klassik R	Germany	German	DSR				DSR ch12 7.02/7.20
Ku3	12.625	Hor	K4	Spot	R	Der Oldiesender (RTL)	Germany	German	DSR				DSR ch13 7.02/7.20
Ku3	12.625	Hor	K4	Spot	R	Radioropa Info.	Germany	German	DSR				DSR ch14 7.02/7.20
Ku3	12.625	Hor	K4	Spot	R	MDR Sputnik	Germany	German	DSR				DSR ch15 7.02/7.20
Ku3	12.625	Hor	K4	Spot	R	Energy 93.3	Germany	German	DSR				DSR ch16 7.02/7.20
Ku3	12.692	Hor	K6	Spot	O	occasional video			PAL			6.65	
Ku3	12.725	Ver	K7	Spot	O	test pattern			PAL			6.65	
ASTRA 1D 19.2° EAST													
Ku1	10.714	Hor	49	Multi	TV	ARD/ZDF Kinderprogramm	Germany	German	PAL		Yes		7.02/7.20
Ku1	10.714	Hor	49	Multi	TV	ARTE	Germ./France	French/Ger.	PAL		Yes	7.38	7.02/7.20
Ku1	10.729	Ver	50	Multi	TV	CNBC	UK	English	PAL		Yes		7.02/7.20
Ku1	10.729	Ver	50	Multi	R	CMT R	UK	English				7.38	7.38/7.56
Ku1	10.743	Hor	51	Multi	TV	Country Music Europe	USA	English	PAL	VideoCrypt	No	6.50	7.02/7.20
Ku1	10.758	Ver	52	Multi	TV	QVC Deutschland	Germany	German	PAL		No	6.50	7.02/7.20
Ku1	10.773	Hor	53	Multi	TV	Chinese News Entertainment	China	Chinese	PAL		no	6.50	7.02/7.20
Ku1	10.773	Hor	53	Multi	TV	JSTV	Japan	Jap./English	PAL	VideoCrypt	Yes	6.50/7.38	7.02/7.20
Ku1	10.788	Ver	54	Multi	TV	Chinese Channel	UK	Chinese	PAL	Nagravision	Yes		7.02/7.20
Ku1	10.788	Ver	54	Multi	TV	Zee TV	UK	Asian	PAL	VideoCrypt I/II	No	6.50	7.02/7.20
Ku1	10.788	Ver	54	Multi	R	G-FM	UK	Hindi	ADR				
Ku1	10.802	Hor	55	Multi	TV	Teleclub	Switzerland	German	PAL	Syster	No		7.02/7.20
Ku1	10.802	Hor	55	Multi	R	Radio Evival	Switzerland	German				7.74	
Ku1	10.802	Hor	55	Multi	R	Swiss Radio International	Switzerland	English	ADR			7.92A	
Ku1	10.817	Ver	56	Multi	R	Swiss Radio International	Switzerland	various	ADR			7.92B	
Ku1	10.832	Hor	57	Multi	O	DVB (see digital stations listing)							
Ku1	10.832	Hor	57	Multi	O	ASTRA Promo Channel	UK	English	PAL		No		7.02/7.20

ITU REGION 1 STATIONS IN FREQUENCY ORDER

Band	Freq.	Pol.	Tr.	Beam	Sce.	Program	Country	Language	System	Encr.	TT	A.M.	A.S.
Ku1	10.847	Ver	58	Multi	TV	Granada Good Life	UK	English	PAL	VideoCrypt	No		7.02/7.20
Ku1	10.847	Ver	58	Multi	TV	The Computer Channel	UK	English	PAL	VideoCrypt	No		7.02/7.20
Ku1	10.847	Ver	58	Multi	R	RNW (Netwerk Europa)	Dutch	Dutch				7.38	
Ku1	10.847	Ver	58	Multi	R	RNW (Hola Holanda)	Dutch	Spanish				7.56	
Ku1	10.861	Hor	59	Multi	TV	Granada (Talk TV)	UK	English	PAL	VideoCrypt	Yes		7.02/7.20
Ku1	10.861	Hor	59	Multi	TV	Sky Scottish	UK	English	PAL	VideoCrypt	Yes		7.02/7.20
Ku1	10.876	Ver	60	Multi	TV	Sky Movies Gold	UK	English	PAL	VideoCrypt	Yes		7.02/7.20
Ku1	10.876	Ver	60	Multi	TV	The Racing Channel	UK	English	PAL	VideoCrypt	Yes		7.02/7.20
Ku1	10.876	Ver	60	Multi	TV	The Weather Channel	UK	English	PAL	VideoCrypt	Yes		7.02/7.20
Ku1	10.876	Ver	60	Multi	TV	Granada Good Life Channels	UK	English	PAL	VideoCrypt	Yes		7.02/7.20
Ku1	10.876	Ver	60	Multi	TV	WBTV (The Warner Channel)	UK	English	PAL	VideoCrypt	Yes		7.02/7.20
Ku1	10.876	Ver	60	Multi	TV	The Weather Channel	UK	English	PAL	VideoCrypt	Yes		7.02/7.20
Ku1	10.891	Hor	61	Multi	O	ProSieben Scheiz	Switzerland	German	PAL		Yes		7.02/7.20
Ku1	10.906	Ver	62	Multi	TV	H.O.T (Home Order TV)	UK	English	PAL		No		7.02/7.20
Ku1	10.920	Hor	63	Multi	TV	FilmNet (Polish)	Poland	English/Polish	PAL	VideoCrypt	Yes		7.02/7.20
Ku1	10.920	Ver	63	Multi	TV	The Adult Channel	UK	English	PAL	VideoCrypt	No		7.02/7.20
Ku1	10.920	Hor	63	Multi	R	Radio Vlaanderen/BRT NachtR	Belgium	Dutch				7.38	

ASTRA 1C 19.2° EAST

Band	Freq.	Pol.	Tr.	Beam	Sce.	Program	Country	Language	System	Encr.	TT	A.M.	A.S.
Ku1	10.935	Ver	64	Multi	TV	TM 3	Germany	German	PAL		Yes		7.38 (Ger.)/7.56 (Eng.)
Ku1	10.935	Ver	64	Multi	TV	What's In store	Netherlands	various	PAL		Yes		7.02/7.20 (Dutch)
Ku1	10.964	Hor	33	Multi	TV	ZDF	Germany	German	PAL		Yes		7.02/7.20
Ku1	10.964	Hor	33	Multi	R	Radio Sweden Int.	Sweden					7.38	
Ku1	10.964	Hor	33	Multi	R	Radio Sweden Int.	Sweden					7.56	
Ku1	10.964	Hor	33	Multi	R	Hit Radio FFH	Germany					8.10	
Ku1	10.979	Ver	34	Multi	TV	UK Living	UK	English	PAL	VideoCrypt	Yes		7.02/7.20
Ku1	10.979	Ver	34	Multi	TV	TVX The Fantasy Channel	UK	English	PAL	VideoCrypt	Yes		7.02/7.20
Ku1	10.979	Ver	34	Multi	R	BBC Radio One FM	UK	English					7.38/7.56
Ku1	10.979	Ver	34	Multi	R	BBC Radio 3	UK	English					7.74/7.92
Ku1	10.993	Ver	35	Multi	TV	The Family Channel	UK	English	PAL	VideoCrypt	Yes		7.02/7.20
Ku1	10.993	Ver	35	Multi	TV	TCC (The Children's Channel)	UK	English	PAL	VideoCrypt	No	6.50	7.02/7.20
Ku1	11.008	Ver	36	Multi	TV	MiniMax	Spain	Spanish	PAL	Nagravision	No		7.02/7.20
Ku1	11.023	Hor	37	Multi	TV	Cartoon Network	USA	various	PAL		Yes		7.38 (French)/7.56 (Swed)
Ku1	11.023	Hor	37	Multi	TV	TNT	USA	various	PAL		Yes		7.38 (French)/7.56 (Swed) 7.02/7.20
Ku1	11.038	Ver	38	Multi	TV	QVC	USA	English	PAL	clear/VCrypt	No		7.02/7.20
Ku1	11.038	Ver	38	Multi	R	TWR / ERF	USA / Germany	English / German				7.38	
Ku1	11.038	Ver	38	Multi	R	ERF 2	USA / Germany	English / German				7.56	
Ku1	11.052	Hor	39	Multi	TV	WDR 2	Germany	German	PAL		Yes		7.02/7.20
Ku1	11.052	Hor	39	Multi	R	Eins Live	Germany	German					7.38/7.56
Ku1	11.052	Hor	39	Multi	R	WDR 2	Germany	German	ADR	MUSICAM		6.12	
Ku1	11.052	Hor	39	Multi	R	WDR 2	Germany	German	ADR	MUSICAM		6.30	
Ku1	11.052	Hor	39	Multi	R	WDR 4	Germany	German	ADR	MUSICAM		6.66	
Ku1	11.052	Hor	39	Multi	R	WDR Radio 5	Germany	German	ADR	MUSICAM		6.84	

ITU REGION 1 STATIONS IN FREQUENCY ORDER

Band	Freq.	Pol.	Tr.	Beam	Sce.	Program	Country	Language	System	Encr.	TT	A.M.	A.S.
Ku1	11.067	Ver	40	Multi	TV	Cine Classics	Spain	Spanish	PAL	Nagravision	No		7.02/7.20
Ku1	11.082	Hor	41	Multi	TV	The Learning Channel (TLC)	UK	English	PAL	VideoCrypt	Yes		7.02/7.20
Ku1	11.082	Hor	41	Multi	TV	Home Shopping Network	UK	English	PAL	VideoCrypt	Yes	7.56 (Polish)	7.02/7.20
Ku1	11.082	Hor	41	Multi	TV	Discovery Channel	UK	English	PAL	VideoCrypt	Yes		7.02/7.20
Ku1	11.097	Ver	42	Multi	TV	Euro Business News	UK	various	PAL		No		7.02/7.20
Ku1	11.097	Ver	42	Multi	TV	Bravo	UK	English	PAL		No		7.02/7.20
Ku1	11.097	Ver	42	Multi	TV	Home Shopping Network	UK	English	PAL		No		7.02/7.20
Ku1	11.112	Hor	43	Multi	TV	Mitteldeutscher Rundfunk (MDR 3)	Germany	German	PAL		Yes		7.38/7.56
Ku1	11.112	Hor	43	Multi	R	MDR-Sputnik	Germany	German	ADR			6.30	
Ku1	11.112	Hor	43	Multi	R	MDR-Live	Germany	German	ADR			6.48	
Ku1	11.112	Hor	43	Multi	R	MDR-Info	Germany	German	ADR			6.66	
Ku1	11.112	Hor	43	Multi	R	MDR-Kultur	Germany	German	ADR			6.84	
Ku1	11.126	Ver	44	Multi	R	MDR-Sputnik	Mexico	Spanish	ADR				
Ku1	11.141	Hor	45	Multi	TV	Galavision	Germany	German	PAL		No		7.02/7.20
Ku1	11.141	Hor	45	Multi	TV	Bayerisches Fernsehen	Germany	German	ADR		Yes		7.02/7.20
Ku1	11.141	Hor	45	Multi	R	Bayern 1	Germany	German	ADR			6.12	
Ku1	11.141	Hor	45	Multi	R	Bayern 2	Germany	German	ADR			6.30	
Ku1	11.141	Hor	45	Multi	R	Bayern 3	Germany	German	ADR			6.48	
Ku1	11.141	Hor	45	Multi	R	Bayern 4 Klassik	Germany	German	ADR			6.66	
Ku1	11.141	Hor	45	Multi	R	Bayern 5 Aktuel	Austria	German	ADR			6.84	
Ku1	11.141	Hor	45	Multi	R	ORF Wien	Germany	German	ADR			8.10	
Ku1	11.156	Ver	46	Multi	R	Institute for Radio Techniques	USA	English	PAL	VideoCrypt	No	8.48	
Ku1	11.156	Ver	46	Multi	TV	Nickelodeon	USA	English	PAL	VideoCrypt	No		7.02/7.20
Ku1	11.156	Ver	46	Multi	TV	Paramount TV	France	various				7.38	7.02/7.20
Ku1	11.156	Ver	46	Multi	R	Radio France Internationale	France	various				7.56	
Ku1	11.156	Ver	46	Multi	R	Radio France Internationale 2	UK	English	PAL	VideoCrypt	Yes	6.50	7.02/7.20
Ku1	11.170	Hor	47	Multi	TV	Sky Sports Gold	UK	English	PAL	VideoCrypt	Yes	6.50	7.02/7.20
Ku1	11.170	Hor	47	Multi	TV	Sky Sports 2	UK	English	PAL	VideoCrypt	Yes	6.50	7.02/7.20
Ku1	11.170	Hor	47	Multi	TV	Sky Travel	UK	English	PAL	VideoCrypt	Yes	6.50	7.02/7.20
Ku1	11.170	Hor	47	Multi	TV	Sky Soap	UK	English	PAL	VideoCrypt	Yes	6.50	7.02/7.20
Ku1	11.170	Hor	47	Multi	TV	Christian Channel Europe	UK	English	PAL			7.38	
Ku1	11.170	Hor	47	Multi	TV	The Sci-Fi Channel	UK	English	PAL			7.56	
Ku1	11.170	Hor	47	Multi	TV	The History Channel	UK	English	PAL				
Ku1	11.170	Hor	47	Multi	R	GWR Classic Gold	UK	English	PAL				
Ku1	11.185	Ver	48	Multi	TV	ASDA FM	Germany	German	PAL		Yes		7.02/7.20
Ku1	11.185	Ver	48	Multi	TV	SWF/SDR (S-3)	UK	English	PAL	VideoCrypt	No		7.02/7.20
Ku1	11.185	Ver	48	Multi	TV	The Playboy Channel	Germany	German	ADR				6.12
Ku1	11.185	Ver	48	Multi	R	SDR-3	Germany	German	ADR				6.30
Ku1	11.185	Ver	48	Multi	R	SDR-1	Germany	German	ADR				6.48
Ku1	11.185	Ver	48	Multi	R	S2 Kultur	Germany	German	ADR				6.66
Ku1	11.185	Ver	48	Multi	R	SWF-1	Germany	German	ADR				6.84
Ku1	11.185	Ver	48	Multi	R	SWF-3							

ASTRA 1A 19.2° EAST

Band	Freq.	Pol.	Tr.	Beam	Sce.	Program	Country	Language	System	Encr.	TT	A.M.	A.S.
Ku1	11.214	Hor	1	Multi	TV	RTL-2	Germany	German	PAL		No		7.02/7.20
Ku1	11.214	Hor	1	Multi	R	Antenne Bayern	Germany	German					7.38/7.56

ITU REGION 1 STATIONS IN FREQUENCY ORDER

Band	Freq.	Pol.	Tr.	Beam	Sce.	Program	Country	Language	System	Encr.	TT	A.M.	A.S.
Ku1	11.229	Ver	2	Multi	TV	RTL Television	Germany	German	PAL		Yes		7.02/7.20
Ku1	11.229	Ver	2	Multi	R	Deutsche Welle	Germany	German					7.38/7.56
Ku1	11.229	Ver	2	Multi	R	Deutsche Welle	Germany	German				7.74	
Ku1	11.229	Ver	2	Multi	R	Deutsche Welle	Germany	various				7.92	
Ku1	11.244	Hor	3	Multi	TV	Granada Plus/Man & Motor	UK	English	PAL	VideoCrypt	Yes		7.02/7.20
Ku1	11.259	Ver	4	Multi	TV	Eurosport [English]	France	English	PAL		Yes	6.50	
Ku1	11.259	Ver	4	Multi	TV	Quantum Channel [English]	UK	English	PAL		Yes	6.50	
Ku1	11.259	Ver	4	Multi	TV	Eurosport [English]	France	English	PAL		Yes	7.02	
Ku1	11.259	Ver	4	Multi	TV	Eurosport [German]	France	German	PAL		Yes	7.20	
Ku1	11.259	Ver	4	Multi	TV	Quantum Channel [German]	UK	German	PAL		Yes	7.20	
Ku1	11.259	Ver	4	Multi	TV	Eurosport [Dutch]	France	Dutch	PAL		Yes	7.38	
Ku1	11.259	Ver	4	Multi	TV	Quantum Channel (Dutch)	UK	Dutch	PAL		Yes	7.38	
Ku1	11.259	Ver	4	Multi	TV	Quantum Channel (English)	UK	English	PAL		Yes	7.56	
Ku1	11.273	Hor	5	Multi	TV	VOX	Germany	German	PAL		Yes		7.02/7.20
Ku1	11.273	Hor	5	Multi	TV	Sell-a-Vision	Germany	German	PAL		Yes		7.02/7.20
Ku1	11.288	Ver	6	Multi	TV	Sat-1	Germany	German	PAL		Yes		7.02/7.20
Ku1	11.303	Hor	7	Multi	TV	Sky-2	UK	English	PAL	VideoCrypt	Yes		7.02/7.20
Ku1	11.303	Hor	7	Multi	TV	Fox Kids	UK	English	PAL	VideoCrypt	Yes		7.02/7.20
Ku1	11.318	Ver	8	Multi	TV	Sky One	UK	English	PAL	VideoCrypt	Yes		7.02/7.20
Ku1	11.318	Ver	8	Multi	R	Sky R	UK	English					7.38/7.56
Ku1	11.332	Hor	9	Multi	TV	Kabel-1	Germany	German	PAL		No		7.02/7.20
Ku1	11.332	Hor	9	Multi	R	Swiss Radio International	Switzerland	various				7.38	
Ku1	11.332	Hor	9	Multi	R	Swiss Radio International (English)	Switzerland	English				7.56	
Ku1	11.332	Hor	9	Multi	R	Radio Evva!	Switzerland	German				7.74	
Ku1	11.347	Ver	10	Multi	TV	3-Sat	Ger/Aus/Swit	German	PAL		Yes		7.02/7.20
Ku1	11.347	Ver	10	Multi	R	Deutschlandradio Köln	Germany	German	PAL				7.38/7.56
Ku1	11.347	Ver	10	Multi	R	D-Radio (Deutschlandradio Berlin)	Germany	German	PAL				7.74/7.92
Ku1	11.362	Hor	11	Multi	TV	Channel 5	UK	English	PAL	VideoCrypt	Yes		7.02/7.20
Ku1	11.377	Ver	12	Multi	TV	Sky News	UK	English	PAL		Yes		7.38/7.56
Ku1	11.377	Ver	12	Multi	R	Virgin 1215	UK	English	PAL				7.02/7.20
Ku1	11.391	Hor	13	Multi	TV	Super RTL	Germany	German	PAL		Yes		7.38/7.56
Ku1	11.391	Hor	13	Multi	R	RTL OldieSender	Germany	German					7.38/7.56
Ku1	11.406	Ver	14	Multi	TV	Pro-7	Germany	German	PAL		Yes		7.02/7.20
Ku1	11.406	Ver	14	Multi	R	Radio Horeb	Germany	German				7.38	
Ku1	11.421	Hor	15	Multi	TV	MTV	UK	English	PAL	VideoCrypt	Yes		7.02/7.20
Ku1	11.436	Ver	16	Multi	TV	Sky Movies	UK	English	PAL	VideoCrypt	Yes		7.02/7.20

ASTRA 1B 19.2° EAST

Band	Freq.	Pol.	Tr.	Beam	Sce.	Program	Country	Language	System	Encr.	TT	A.M.	A.S.
Ku1	11.464	Hor	17	Multi	TV	Premiere	Germany	German	PAL	Syster	Yes		7.02/7.20
Ku1	11.464	Hor	17	Multi	R	N-Joy R	Germany	German					7.38/7.56
Ku1	11.479	Ver	18	Multi	TV	The Movie Channel	UK	English	PAL	VideoCrypt	Yes		7.02/7.20
Ku1	11.479	Ver	18	Multi	TV	Sunrise R	UK	Hindi					
Ku1	11.493	Hor	19	Multi	TV	ARD-1	Germany	German	PAL		Yes		7.02/7.20
Ku1	11.493	Hor	19	Multi	R	ARD starpoint 1/2	Germany	German	ADR				6.12
Ku1	11.493	Hor	19	Multi	R	DLF	Germany	German	ADR				6.30

168

ITU REGION 1 STATIONS IN FREQUENCY ORDER

Band	Freq.	Pol.	Tr.	Beam	Sce.	Program	Country	Language	System	Encr.	TT	A.M.	A.S.
Ku1	11.493	Hor	19	Multi	R	DLR Berlin	Germany	German	ADR				6.48
Ku1	11.493	Hor	19	Multi	R	Deutsche Welle (German)	Germany	German	ADR				6.66
Ku1	11.493	Hor	19	Multi	R	Deutsche Welle (Eur. prgr1/2)	Germany	German	ADR				6.84
Ku1	11.493	Hor	19	Multi	R	SWF-3	Germany	German	ADR				7.38/7.56
Ku1	11.493	Hor	19	Multi	R	SR1 Eurowelle Saar	Germany	German	ADR				8.28
Ku1	11.508	Ver	20	Multi	TV	Sky Sports	UK	English	PAL	VideoCrypt	Yes		7.02/7.20
Ku1	11.508	Ver	20	Multi	R	DMX Instrumentals	UK	English	ADR	MUSICAM			7.92
Ku1	11.508	Ver	20	Multi	R	DMX Blues	UK	English	ADR	MUSICAM			8.10
Ku1	11.508	Ver	20	Multi	R	DMX Raggae	UK	English	ADR	MUSICAM			8.28
Ku1	11.508	Ver	20	Multi	R	DMX French language	UK	English	ADR	MUSICAM			8.46
Ku1	11.508	Ver	20	Multi	TV	United Christian Broadcasters	UK	English	PAL				
Ku1	11.523	Hor	21	Multi	TV	DSF Deutsches Sportfernsehen	Germany	German	PAL		No	7.56	7.02/7.20
Ku1	11.538	Ver	22	Multi	R	Radio Campanik	Germany	German	ADR				7.38/7.56
Ku1	11.538	Ver	22	Multi	TV	VH-1	UK	English	PAL		Yes		7.02/7.20
Ku1	11.538	Ver	22	Multi	R	World Radio Network (WRN 1)	UK	various				7.38	
Ku1	11.538	Ver	22	Multi	R	RTE Radio One	Ireland	English/Irish				7.56	
Ku1	11.538	Ver	22	Multi	R	America One	USA	English				7.74	
Ku1	11.538	Ver	22	Multi	R	Irish Satellite Radio Netw.	Ireland	Irish				7.92	
Ku1	11.553	Hor	23	Multi	TV	UK Gold	UK	English	PAL	VideoCrypt	Yes		7.02/7.20
Ku1	11.553	Hor	23	Multi	R	BBC World Service	UK	various				7.38	
Ku1	11.553	Hor	23	Multi	R	BBC Radio 4	UK	various				7.56	
Ku1	11.553	Hor	23	Multi	R	BBC Radio 2	UK	various				7.74	
Ku1	11.553	Hor	23	Multi	R	BBC 5 Live	UK	various				7.92	
Ku1	11.567	Ver	24	Multi	TV	BSkyB	UK	English	PAL	VideoCrypt	No		7.02/7.20
Ku1	11.582	Hor	25	Multi	TV	N-3 Nord-3	Germany	German	PAL		Yes		7.38/7.92
Ku1	11.582	Hor	25	Multi	R	NDR 4 SPE 2	Germany	German	PAL		Yes		7.74/7.92
Ku1	11.582	Hor	25	Multi	R	NDR 2	Germany	German	ADR	DMX	Yes		6.12
Ku1	11.582	Hor	25	Multi	R	NDR 3	Germany	German	ADR	DMX	Yes		6.30
Ku1	11.582	Hor	25	Multi	R	NDR 4	Germany	German	ADR	DMX	Yes		6.48
Ku1	11.582	Hor	25	Multi	R	N-Joy R	Germany	German	ADR	DMX	Yes		6.66
Ku1	11.582	Hor	25	Multi	R	NDR-2	Germany	German	ADR		Yes	7.74	6.84
Ku1	11.582	Hor	25	Multi	R	NDR-4	Germany	German	ADR		Yes	7.92	
Ku1	11.597	Ver	26	Multi	TV	The Disney Channel	UK	English	PAL	VideoCrypt	Yes		7.38/7.56
Ku1	11.597	Ver	26	Multi	TV	Sky Movies Gold	UK	English	PAL	VideoCrypt	Yes		7.02/7.20
Ku1	11.611	Hor	27	Multi	TV	Nickelodeon	UK	English	PAL		Yes		7.02/7.20
Ku1	11.611	Hor	27	Multi	TV	Sci-Fi Channel	UK	English	PAL		Yes		7.02/7.20
Ku1	11.611	Hor	27	Multi	TV	VH-1	Germany	German	PAL		Yes		7.38/7.56
Ku1	11.626	Ver	28	Multi	TV	CNN International	USA & UK	English	PAL		Yes		7.02/7.20
Ku1	11.626	Ver	28	Multi	R	CNN International (Spanish)	USA & UK	Spanish					
Ku1	11.626	Ver	28	Multi	TV	CNN R	USA & UK	English	PAL				
Ku1	11.641	Hor	29	Multi	TV	n-tv	Germany	German	PAL		No		7.02/7.20
Ku1	11.656	Ver	30	Multi	TV	Cinemania	Spain	Spanish	PAL	Syster	No		7.02/7.20
Ku1	11.656	Ver	30	Multi	R	Cadena Dial	Spain	Spanish			No		7.38/7.56
Ku1	11.656	Ver	30	Multi	R	Cadena Cero	Spain	Spanish			No		
Ku1	11.670	Hor	31	Multi	TV	Sky Sports 3	UK	English	PAL	VideoCrypt	Yes	7.74	7.02/7.20
Ku1	11.685	Ver	32	Multi	TV	Documania	Spain	Spanish	PAL	Syster	No		7.02/7.20

ITU REGION 1 STATIONS IN FREQUENCY ORDER

EUTELSAT II F3 16.0° EAST

Band	Freq.	Pol.	Tr.	Beam	Sce.	Program	Country	Language	System	Encr.	TT	A.M.	A.S.
Ku1	10.972	Ver	25	Wide	TV	RTM 1 (TV Marocaine)	Morocco	French	SECAM		No	6.60	
Ku1	10.972	Ver	25	Wide	R	RTM Berber	Morocco	Arabic				7.02	
Ku1	10.972	Ver	25	Wide	R	RTM Int.	Morocco	French				7.56	
Ku1	10.987	Ver	20	Wide	TV	Rendez-Vous	France	French	D2-MAC	EuroCrypt	Yes	6.65	ch 1
Ku1	10.987	Hor	20	Wide	TV	HRT Zagreb (Hvratska Televisija)	Croatia	Croatian	PAL			7.02	
Ku1	10.987	Hor	20	Wide	R	Radio Zagreb	Croatia	Croatian				6.60	
Ku1	11.003	Hor	26A	Wide	O	occasional video	Morocco		PAL			6.60	
Ku1	11.058	Ver	26A	Wide	O	occasional video	Turkey		PAL			6.60	
Ku1	11.063	Hor	21	Wide	O	occasional video			PAL			6.60	
Ku1	11.080	Hor	21	Wide	TV	Al Jazeera Satellite Channel	Qatar	Arabic	PAL		No	6.60	
Ku1	11.095	Ver	26B	Wide	TV	Algerian TV	Algeria	Arabic	PAL		No	6.60	
Ku1	11.095	Ver	26B	Wide	R	ENTV Radio 1 Arabic	Algeria	Arabic				7.02	
Ku1	11.095	Ver	26B	Wide	R	ENTV Radio 2 Kabyle	Algeria	Arabic				7.20	
Ku1	11.095	Ver	26B	Wide	R	ENTV La Châine 3	Algeria	Arabic				7.38	
Ku1	11.095	Ver	26B	Wide	R	Radio France Int.	France	Arabic				7.38	
Ku1	11.095	Ver	26B	Wide	R	Algerian Cultural R	Algeria	Arabic				7.56	
Ku1	11.140	Ver	27A	Wide	TV	Nile TV International	Egypt	Arabic			No	6.60	
Ku1	11.163	Hor	22B	Super	TV	Thaiwave	Thailand	Thai	PAL		No	6.60	
Ku1	11.163	Hor	22B	Super	TV	TV Eurotica		English	PAL	VideoCrypt	No	6.60	
Ku1	11.163	Hor	22B	Super	O	BT	UK		PAL		No	6.60	
Ku1	11.178	Ver	27B	Wide	TV	The Egyptian Space Channel	Egypt	Arabic	PAL		No	6.60	
Ku1	11.178	Ver	27	Wide	R	ERTU (general prgr)	Egypt	Arabic				7.02	
Ku1	11.178	Ver	27	Wide	R	Voice of the Arabs	Egypt	Arabic				7.20	
Ku1	11.178	Ver	27	Wide	R	Middle East R	Egypt	Arabic				7.38	
Ku1	11.178	Ver	27	Wide	R	ERTU (cultural)	Egypt	Arabic				7.56	
Ku1	11.556	Hor	32	Wide	TV	ART Europe	UK	Arabic	PAL		No	6.65	
Ku1	11.575	Ver	37	Wide	TV	Albanian TV (Shqiptar TV)	Albania	Albanian	PAL		No	6.65	
Ku1	11.575	Ver	37	Wide	TV	TelePace/Vatican TV	Italy/Vatican State		Latin	PAL		No	6.65
Ku1	11.575	Ver	37	Wide	TV	NTV	Russia	Russian	PAL		No	6.65	
Ku1	11.575	Ver	37	Wide	TV	TV România int.	Romania	Romanian	PAL		No	6.65	
Ku1	11.575	Ver	37	Wide	R	Radio Shqiptar	Albania	Albanian				7.20	
Ku1	11.595	Hor	33	Wide	TV	Duna 7	Hungary	Hungarian	PAL		No	6.50	7.02/7.20
Ku1	11.595	Hor	33	Wide	R	Bartok R	Hungary	various				7.02	
Ku1	11.595	Hor	33	Wide	R	Kussuth R	Hungary	various				7.38	
Ku1	11.595	Hor	33	Wide	R	Petöfi R	Hungary	various				7.38	
Ku1	11.616	Ver	38	Wide	O	occasional video			PAL			6.60	
Ku1	11.658	Ver	39	Wide	TV	RTT TV-7	Tunisia	Arabic	PAL		No	6.65	
Ku1	11.658	Ver	39	Wide	R	Tunis Radio Int.	Tunisia	Arabic				7.02	
Ku1	11.658	Ver	39	Wide	R	Radio Tunisia	Tunisia	Arabic				7.20	
Ku3	12.522	Hor	40	Wide	R	British Telecom	UK		PAL			6.60	
Ku3	12.583	Ver	46	Wide	O	vacant						6.60	
Ku3	12.604	Hor	42	Wide	O	vacant						6.60	
Ku3	12.625	Ver	47	Wide	O	vacant						6.60	
Ku3	12.646	Hor	43	Wide	O	vacant						6.60	
Ku3	12.667	Ver	48	Wide	O	vacant						6.60	

ITU REGION 1 STATIONS IN FREQUENCY ORDER

Band	Freq.	Pol.	Tr.	Beam	Sce.	Program	Country	Language	System	Encr.	TT	A.M.	A.S.
Ku3	12.708	Hor	44	Wide	O	vacant						6.60	
Ku3	12.708	Ver	49	Wide	O	vacant						6.60	

EUTELSAT II F1 & (HOTBIRD 1/2) 13.0° EAST

Band	Freq.	Pol.	Tr.	Beam	Sce.	Program	Country	Language	System	Encr.	TT	A.M.	A.S.
Ku1	10.972	Hor	20A	Super	TV	Viva 2	Germany	German	PAL		No	6.60	
Ku1	10.987	Ver	25	Wide	TV	NBC Super Channel [English]	UK	English	PAL		Yes	6.65	
Ku1	10.987	Ver	25	Wide	TV	NBC Super Channel [Dutch]	UK	Dutch	PAL		Yes	7.02	
Ku1	10.987	Ver	25	Wide	TV	NBC Super Channel [German]	UK	German	PAL		Yes	7.20	
Ku1	10.987	Ver	25	Wide	R	Virgin Megastore Music	UK	English	dig.				
Ku1	10.987	Ver	25	Wide	R	Granada FM	UK	English	dig.				
Ku1	10.987	Ver	25	Wide	R	World Radio Network 1	UK	English				7.38	6.30A
Ku1	10.987	Ver	25	Wide	R	Radio Sweden	Sweden	English				7.56	6.30B
Ku1	10.990	Hor	20	Super	O	Dutch PTT	Holland					6.60	
Ku1	11.005	Hor	20B	Super	TV	VIVA	Germany	German	PAL		No	6.65	7.02/7.20
Ku1	11.005	Hor	20B	Super	TV	ARTE	France	German	PAL		No	6.65	7.02/7.20
Ku1	11.055	Hor	21A	Super	TV	ARD/ZDF Kinderprogramm	Germany	German	PAL		No	6.65	7.02/7.20
Ku1	11.055	Hor	21A	Super	R	Radio Melodie	Germany	German					7.74/7.92
Ku1	11.080	Ver	26B	Multi	TV	Arte	France/Germany	German/French	PAL		No	6.65	
Ku1	11.080	Ver	26B	Multi	TV	La Cinquième	France	French	PAL		No	6.65	
Ku1	11.095	Ver	21B	Super	TV	RTL-2	Germany	German	PAL		Yes	6.60	7.02/7.20
Ku1	11.095	Ver	21B	Super	R	RFE/VOA	Germany/USA	Polish/English				8.10	
Ku1	11.148	Hor	22	Super	TV	Onyx TV	Germany	German/English/Spanish		PAL	No	6.65	
Ku1	11.162	Ver	27B	Wide	TV	Deutsche Welle TV	Germany	German				No	
Ku1	11.162	Ver	27B	Wide	R	Deutsche Welle [German]	Germany	German				7.74	
Ku1	11.162	Ver	27B	Wide	R	VOA Europe [English]	USA	English				7.92	
Ku1	11.162	Ver	27B	Wide	R	Deutsche Welle [European]	Germany	various				7.92	
Ku1	11.162	Ver	27B	Wide	R	VOA Europe [foreign]	USA	various				8.10	
Ku1	11.162	Ver	27B	Wide	R	Deutsche Welle [European]	Finland	various				8.46	
Ku1	11.162	Ver	27B	Wide	R	Radio Finland Int.	Germany	various					
Ku1	11.162	Hor	22B	Super	R	Deutsche Welle [Africa]	Turkey	Turkish				6.65	7.02/7.20
Ku1	11.181	Hor	22B	Super	TV	TRT Int.	Turkey	Turkish	PAL				
Ku1	11.181	Hor	1	Spot	R	Turkish R							
Ku1	11.221	Hor	1	Spot	TV	TVE-Internacional	Spain	Spanish	PAL		No	6.60	
Ku1	11.221	Hor	3	Spot	R	RNE Radio 1	Spain	Spanish				7.38	
Ku1	11.265	Hor	3	Spot	TV	Radio Exterior de España	Spain	Spanish				7.56	
Ku1	11.265	Hor	3	Spot	TV	European Business News (EBN)	UK	English	PAL		No	6.65	
Ku1	11.283	Hor	7	Spot	R	World Radio Network (WRN 1)	UK	English				7.20	
Ku1	11.304	Ver	5	Spot	TV	Radio Canada Int.	Canada	English/French				7.56	
Ku3	11.325	Ver	6	Wide	TV	TM3	UK	English	PAL		No	6.65	7.02/7.20
Ku1	11.325	Hor	6	Wide	TV	MCM Euromusique	France	French	PAL		No	6.60	7.02/7.20
Ku1	11.325	Hor	6	Wide	TV	TV 5 Internationale	France	French	PAL		No	6.60	
Ku1	11.325	Ver	6	Wide	R	France Info	France	French				7.20	
Ku1	11.325	Ver	6	Wide	R	France Inter	France	French				7.38	
Ku1	11.325	Ver	6	Wide	R	France Culture Europe	France	French				7.56	
Ku1	11.325	Ver	6	Spot	R	Radio Suisse Int.	Switzerland	French				7.74	
Ku1	11.366	Ver	8	Spot	TV	RAIUNO	Italy	Italian	PAL	Discrete	Yes	6.60	

ITU REGION 1 STATIONS IN FREQUENCY ORDER

Band	Freq.	Pol.	Tr.	Beam	Sce.	Program	Country	Language	System	Encr.	TT	A.M.	A.S.
Ku1	11.366	Ver	8	Spot	R	Radio Uno	Italy	Italian				7.38	
Ku1	11.366	Ver	8	Spot	R	Radio Due	Italy	Italian				7.56	
Ku1	11.387	Hor	9	Spot	TV	Eurosport [German]	France	German	PAL		Yes	6.65	
Ku1	11.387	Hor	9	Spot	TV	Quantum Channel	USA	English	PAL		Yes	6.65	
Ku1	11.387	Hor	9	Spot	TV	Eurosport [English]	France	English	PAL		Yes	7.02	
Ku1	11.387	Hor	9	Spot	TV	Eurosport [German]	France	German	PAL		Yes	7.02	
Ku1	11.387	Hor	9	Spot	TV	Eurosport [Dutch]	France	Dutch	PAL		Yes	7.38	
Ku1	11.387	Hor	9	Spot	TV	Eurosport [French]	France	French	PAL		Yes	7.56	
Ku1	11.408	Ver	10	Spot	TV	Canal Horizons	Senegal	French	SECAM	Syster	No	6.60	7.02/7.20
Ku1	11.408	Ver	10	Spot	R	Radio Italia	Italy	Italian					7.38/7.56
Ku1	11.408	Ver	10	Spot	R	Radio Maria	Italy	Italian				7.74	
Ku1	11.428	Hor	11	Spot	TV	Polsat	Poland	Polish	PAL		No	6.65	
Ku1	11.449	Ver	12	Spot	TV	RAIDUE	Italy	Italian	PAL	Discrete	Yes	6.60	
Ku1	11.449	Ver	12	Spot	R	Radio Due	Italy	Italian				7.38	
Ku1	11.449	Ver	12	Spot	R	Radio RAI Int.	Italy	Italian				7.56	
Ku1	11.472	Hor	13	Spot	TV	TV Polonia	Poland	Polish	PAL		No	6.65	
Ku1	11.472	Hor	13	Spot	R	PRT 1	Poland	Polish					7.38/7.56
Ku1	11.472	Hor	13	Spot	R	PRT 2	Poland	Polish					7.74/7.92
Ku1	11.472	Hor	13	Spot	R	PRT 3	Poland	Polish				8.10	
Ku1	11.472	Hor	13	Spot	R	PRT 5	Poland	Polish				8.28	
Ku1	11.492	Ver	14	Spot	R	RTL 7	Poland	Polish			No	6.65	
Ku1	11.513	Hor	15	Super	TV	Kanal + (Poland)	Poland	Polish	PAL		Yes	6.60	7.56
Ku1	11.534	Ver	16	Super	TV	RAITRE	Italy	Italian	PAL		No		7.02/7.20
Ku1	11.554	Hor	32	Super	TV	Middle East Broadcasting Centre	UK	Arabic	PAL		No	6.60	7.02/7.20
Ku1	11.554	Hor	32	Super	R	MBC R	UK	Arabic					7.38/7.56
Ku1	11.554	Hor	32	Super	R	World Radio Network (WRN 2)	UK	various				7.74	
Ku1	11.575	Ver	37	Super	TV	euroNews [German]	France	German	PAL		No	6.65	
Ku1	11.575	Ver	37	Super	TV	euroNews [English]	France	English	PAL		No	7.02	7.02/7.20
Ku1	11.575	Ver	37	Super	TV	euroNews [Spanish]	France	Spanish	PAL		No	7.20	7.38/7.56
Ku1	11.575	Ver	37	Super	TV	euroNews [French]	France	French	PAL		No	7.38	7.74/7.92
Ku1	11.575	Ver	37	Super	TV	euroNews [Italian]	France	Italian	PAL		No	7.56	
Ku1	11.596	Hor	33	Spot	R	Der Oldiesender	Germany	German				6.65	
Ku1	11.596	Hor	33	Spot	R	Deutschlandradio Köln	Germany	German					
Ku1	11.596	Hor	33	Spot	TV	VOX	Germany	German	PAL		Yes		
Ku1	11.616	Ver	38	Wide	TV	BBC World	UK	English				6.65	
Ku1	11.616	Ver	38	Wide	R	BBC World Service [English]	UK	English				7.38	
Ku1	11.616	Ver	38	Wide	R	BBC World Service [for Europe]	UK	English				7.56	
Ku1	11.616	Ver	38	Wide	R	BBC World Service	UK	English				7.74	
Ku1	11.616	Ver	38	Wide	R	BBC World Service	UK	English				7.92	
Ku1	11.616	Ver	38	Wide	R	BBC World Service	UK	English				8.10	
Ku1	11.727	Ver	50	Wide	TV	RTP International	Portugal	Portuguese	PAL		No	6.65	
Ku1	11.727	Ver	50	Wide	R	Radio Renascença 1	Portugal	Portuguese					7.38/7.56
Ku1	11.727	Ver	50	Wide	R	RFM/Radio Renascença 2	Portugal	Portuguese					7.74/7.92
Ku1	11.727	Ver	50	Wide	R	RDP Commercial Lisboa	Portugal	Portuguese					8.10/8.28
Ku1	11.727	Ver	50	Wide	R	Radio Portugal Int.	Portugal	Portuguese				7.02	
Ku1	11.727	Ver	50	Wide	R	RDP Antena 1	Portugal	Portuguese				7.20	

ITU REGION 1 STATIONS IN FREQUENCY ORDER

Band	Freq.	Pol.	Tr.	Beam	Sce.	Program	Country	Language	System	Encr.	TT	A.M.	A.S.
Ku1	11.745	Hor	51	Wide	TV	EDTV	UAE	Arabic	PAL		No	6.65	
Ku1	11.785	Hor	53	Wide	TV	Polonia 1	Poland	Polish	PAL		No	6.65	
Ku2	12.015	Hor	40	Super	TV	ART Europe	UK	Arabic	PAL		No	6.65	
Ku2	12.555	Hor	45	Wide	O	occasional video	various		PAL			6.60	
Ku3	12.563	Ver	41	Wide	O	EBU	various		PAL			6.60	
Ku3	12.563	Ver	41	Wide	O	WTN Worldwide Television News	UK		PAL	VideoCrypt	No	6.60/7.38	
Ku3	12.584	Ver	46	Wide	O	EBU	various	various	PAL			6.60	
Ku3	12.625	Ver	47	Wide	O	occasional video	various		PAL			6.60	
Ku3	12.667	Ver	48	Wide	O	occasional video	various		PAL			6.60	
Ku3	12.708	Ver	49	Wide	O	occasional video	various		PAL			6.60	
EUTELSAT II F2 10.0° EAST													
Ku1	10.972	Ver	25	Wide	TV	Vesa TV - Cable TV	Slovakia	Slovak	PAL		No	6.60	
Ku1	10.987	Hor	20A	Wide	TV	NTV	Russia	Russian	PAL		No	6.60	
Ku2	11.007	Ver	25B	Wide	O	occsional video	Italy		PAL			6.65	
Ku1	11.061	Hor	26A	Wide	O	occasional video	Italy		PAL			6.65	
Ku3	11.080	Ver	21	Wide	TV	Europe by Satellite			PAL		No	6.65	
Ku3	11.080	Hor	21	Wide	TV	BHT (Bosnian TV)	Bosnia Herzegovina		Serbian/CroatianPAL			No	
Ku1	11.095	Ver	26B	Wide	TV	TGRT	Turkey	Turkish	PAL		No	6.65	
Ku1	11.095	Hor	26B	Wide	R	TGRT FM (Huzur Radio)	Turkey	Turkish				7.20/7.38	
Ku3	11.095	Ver	26B	Wide	R	AKRA FM	Turkey	Turkish				8.20/8.38	
Ku1	11.136	Hor	27	Wide	O	EBU			PAL			6.65	
Ku1	11.158	Ver	27	Wide	O	various feeds	various		PAL			6.65	
Ku1	11.174	Hor	27	Wide	TV	AFN-Television, Frankfurt	USA	English	B-MAC [525 lines]		USA	No	
Ku1	11.594	Hor	33	Wide	TV	interSTAR	Greece	Greek	PAL		No	6.65	
Ku1	11.617	Ver	38	Wide	TV	ET-1/2/3	Turkey/Germany	Turkish	PAL		No	6.65	
Ku1	11.617	Hor	38	Wide	R	Metro FM	Turkey	Turkish				7.02/7.20	
Ku1	11.617	Ver	38	Wide	R	KRAL FM	Turkey	Turkish				7.38/7.56	
Ku1	11.617	Hor	38	Wide	R	Süper FM	Turkey	Turkish				8.10/8.28	
Ku1	11.658	Ver	39	Wide	O	occasional video			PAL			6.60	
EUTELSAT II F4 7.0° EAST													
Ku1	10.971	Hor	20a	Wide	O	EBU Channel A	various		PAL	SiS		SiS	
Ku1	10.987	Ver	25a	Wide	O	EBU Channel C	various		PAL	SiS			
Ku1	11.008	Hor	20b	Wide	O	EBU Channel B	various		PAL	SiS			
Ku1	11.009	Ver	25b	Wide	O	EBU Channel K	various		PAL	SiS			
Ku1	11.0266	Hor	25	Wide	R	Euroradio Channel R	various		SCPC				
Ku1	11.0399	Ver	26	Wide	R	Euroradio Channel T	various		SCPC				
Ku1	11.058	Ver	21	Wide	O	EBU Channel E	various		PAL	SiS			
Ku1	11.080	Ver	21	Wide	O	EBU Channel D	various		PAL	SiS			
Ku1	11.095	Hor	22A	Wide	O	EBU Channel F	various		PAL	SiS			
Ku1	11.146	Hor	22A	Wide	TV	PIK CYBC	Cyprus	Greek	PAL		No	6.60	
Ku1	11.146	Hor	22A	Wide	R	PIK Radio 3	Cyprus	Greek				7.20	
Ku1	11.178	Hor	22B	Wide	TV	RTS Beograd (RTV Srbija)	Yugoslavia	Serbian/English	PAL		Yes	6.65	
Ku1	11.178	Ver	22B	Wide	R	Radio Beograd	Yugoslavia	Serbian/English				7.02	
Ku1	11.575	Ver	37	Wide	O	occasional use						6.60	

ITU REGION 1 STATIONS IN FREQUENCY ORDER

Band	Freq.	Pol.	Tr.	Beam	Sce.	Program	Country	Language	System	Encr.	TT	A.M.	A.S.
Ku3	12.646	Hor	43	Wide	O	occasional video	various		PAL			6.60	
SIRIUS 5.2° EAST													
Ku2	11.785	RHCP	4	Spot	TV	VH1 Nordic	Sweden	Swedish	D2-MAC	EuroCrypt	Yes		ch 1
Ku2	11.785	RHCP	4	Spot	TV	Nova Shop	Sweden	Swedish	D2-MAC	EuroCrypt	Yes		ch 1
Ku2	11.861	RHCP	8	Spot	TV	ZTV (Sweden)	Sweden	Swedish/English	D2-MAC		Yes		ch 1
Ku2	11.861	RHCP	8	Spot	TV	Nickelodeon Sweden	Sweden	Swedish/English	D2-MAC		Yes		ch 1
Ku2	11.938	RHCP	12	Spot	TV	TV-4 (Sweden)	Sweden	Swedish/English	D2-MAC		Yes		ch 1
Ku2	12.015	RHCP	16	Spot	TV	TV 6 Sweden	Sweden	Swedish/English	D2-MAC		Yes		ch 1
TELE-X 5.0° EAST													
Ku2	12.475	LHCP	40	Spot	TV	Kanal 5	Sweden	Swedish	PAL		Yes	6.50	7.02/7.20
Ku2	12.475	LHCP	40	Spot	R	The Voice (of Scandinavia)	Sweden	Swedish					7.74/7.92
Ku2	12.475	LHCP	40	Spot	R	Radio Sweden/Radio Arlando	Sweden	Swedish				7.38	
Ku3	12.673	LHCP	C3	Spot	O	vacant							
TÉLÉCOM 2C 3.0° EAST													
Ku3	12.522	Ver	R1	Spot	R	French Digital Satellite R	France		ALCATEL				
Ku3	12.606	Ver	R3	Spot	O	Setanta Sport	Ireland	English	PAL		No	6.60	
Ku3	12.606	Ver	R3	Spot	O	TV 3 de Catalunya	Spain	Spanish	PAL			6.60	
Ku3	12.648	Ver	R4	Spot	O	BBC Orbit Arabic Service	UK	Arabic	PAL		No	5.80	
Ku3	12.690	Ver	R5	Spot	O	Canal Courses	France		D2-MAC	VideoCrypt	No		ch 1
Ku3	12.732	Ver	R5	Spot	O	occasional video	France		SECAM	EuroCrypt		5.80	
TELENOR 0.8° WEST													
Ku2	11.900	LHCP	10	Spot	TV	Cinema	Sweden	English	D-MAC	EuroCrypt	Yes	ch1	
Ku2	11.977	LHCP	14	Spot	TV	TV3+	Denmark	Danish	D2-MAC	EuroCrypt	No		ch 1
Ku2	12.054	LHCP	18	Spot	TV	SciFi Channel Nordic	Denmark	Danish	D2-MAC	EuroCrypt	No		
THOR 0.8° WEST													
Ku2	11.785	RHCP	7	Spot	TV	CNN Nordic	USA	English	D-MAC	EuroCrypt	Yes		ch1
Ku2	11.862	RHCP	8	Spot	TV	Eurosport	France	multilingual	D-MAC	EuroCrypt	Yes		ch 1
Ku2	11.938	RHCP	9	Spot	TV	Discovery Channel	UK	Englsiih	D-MAC	EuroCrypt	Yes	ch1	
Ku2	11.938	RHCP	9	Spot	TV	TCC The Children's Channel	UK	Englsiih	D-MAC	EuroCrypt	Yes		ch 1
Ku2	12.015	RHCP	16	Spot	TV	FilmNet 2	Sweden	English	D2-MAC	EuroCrypt	Yes		ch 1
Ku2	12.091	RHCP	20	Spot	TV	FilmNet 1 Nordic	Norway	English	D2-MAC	EuroCrypt	Yes		ch 1
INTELSAT 707 1.0° WEST													
C	3.840	RHCP	2	East Hemisph.	TV	TV5 Internationale/Afrique	France	French	SECAM		No		
C	3.860	RHCP	53	Zone	R	Radio Diffusion Nationale Tchadienne (RNT)	Chad	Chad	French / ethnic	SCPC			
C	3.900	RHCP	53	Zone	TV	AFRTS-SEB	USA	English	B-MAC	US-encryption	No	digital	
C	4.030	LHCP	24	East Hemisph.	TV	Libyan TV	Libya	Arabic	PAL		No	6.20	
C	4.140	LHCP	24	East Hemisph.	TV	Nile TV International	Egypt	Arabic	PAL		No	6.20	
C	4.140	LHCP	24	East Hemisph.	TV	Egyptian Satellite Channel	Egypt	Arabic	PAL		No	6.20	
C	4.175	RHCP	38	Global	TV	AFRTS	USA	English	B-MAC	US-encryption	No	digital	
C	4.185	LHCP	39	Global	TV	Yemeni Television	Rep of Yemen	Arabic	PAL		No	6.6	

ITU REGION 1 STATIONS IN FREQUENCY ORDER

Band	Freq.	Pol.	Tr.	Beam	Sce.	Program	Country	Language	System	Encr.	TT	A.M.	A.S.
C	4.185	LHCP	39	Global	R	Rep. of Yemen R	Rep. of Yemen	Arabic				7.20	
Ku1	11.016	Hor	61	West Spot	TV	TV Norge	Norway	Norwegian	PAL		Yes	6.60	
Ku1	11.017	Ver	71	East Spot	TV	IBA Channel-3	Israel	Hebrew	PAL		Yes	6.60/7.02	
Ku1	11.017	Ver	71	East Spot	R	Radio REQA	Israel	Hebrew				7.20	
Ku1	11.017	Ver	71	East Spot	R	Kol Israel Network B	Israel	Hebrew				7.38	
Ku1	11.017	Ver	71	East Spot	R	Kol Israel Network C	Israel	Hebrew				7.56	
Ku1	11.053	Ver	72	West Spot	TV	TV-1000	Sweden	Swedish	D-MAC	EuroCrypt	Yes		6.40/7.25
Ku1	11.086	Hor	62	West Spot	TV	TV-3	Norway	Norwegian	D2-MAC	EuroCrypt	Yes		6.85/8.20
Ku1	11.176	Ver	63	West Spot	TV	NRK 1	Norway	Norwegian	D-MAC	EuroCrypt S	Yes		
Ku1	11.178	Ver	73	East Spot	TV	Channel-2 TV Israel	Israel	English	PAL		Yes	6.60/7.02	
Ku1	11.178	Ver	73	East Spot	R	Voice of Music	Israel	English				7.20	
Ku1	11.178	Ver	73	East Spot	R	Israel R	Israel	English					7.38/7.56
Ku1	11.472	Ver	75	West Spot	TV	TV 3 Danmark	Denmark	Danish	D-MAC	EuroCrypt M	Yes		ch1
Ku1	11.482	Ver	75	West Spot	TV	NRK-2	Norway	Norwegian	D2-MAC	EuroCrypt	Yes		ch1
Ku1	11.553	Ver	65	West Spot	TV	TV-2	Norway	Norwegian	D2-MAC	EuroCrypt	Yes		ch1
Ku1	11.597	Ver	75L	West Spot	TV	TV 3 Sweden	Sweden	Swedish	D2-MAC	EuroCrypt	Yes		ch1
Ku1	11.665	Hor	69	West Spot	TV	DR-2	Denmark	Danish	D2-MAC				
Ku1	11.679	Hor	69	West Spot	TV	MTV Nordic	Denmark	Danish					

AMOS F1 4.0° WEST

Band	Freq.	Pol.	Tr.	Beam	Sce.	Program	Country	Language	System	Encr.	TT	A.M.	A.S.
Ku1	11.368	Hor	10L	West Spot	TV	ATV	Poland	Polish	PAL		No	6.80	
Ku1	11.345	Hor	9	West Spot	O	AMOS 1 test pattern	Israel		PAL				7.02/7.20

TÉLÉCOM 2B 5.0° WEST

Band	Freq.	Pol.	Tr.	Beam	Sce.	Program	Country	Language	System	Encr.	TT	A.M.	A.S.
C	3.710	LHCP	C1A	Semi Global	O	RFO/France-2 feeds	France	French	SECAM			5.80	
C	3.738	RHCP	C1B	Semi Global	O	RFO feeds	France	French	PAL/SEC/NTSC			6.60	
C	3.768	RHCP	C2A	Semi Global	O	Canal France feeds	France	French	PAL/SEC/NTSC			6.60	
C	3.793	RHCP	C2B	Semi Global	O	RFO/M6 feeds	France	French	SECAM			6.60	
Ku3	12.522	Ver	R1	Spot	TV	M-6 Metropole 6	France	French				5.80	
Ku3	12.522	Ver	R1	Spot	R	Frequence Mousquetaire	France	French					
Ku3	12.522	Ver	R1	Spot	TV	Europe-1	France	French					
Ku3	12.522	Ver	R1	Spot	R	UNICO Supermarché	France	French					
Ku3	12.543	Hor	K7	Spot	R	Fourviere FM	France	French				7.75	
Ku3	12.564	Ver	R2	Spot	TV	RTL-9	France/Germany	French/German	SECAM	SmartCrypt	No	8.65	7.75/8.65
Ku3	12.564	Ver	R2	Spot	TV	France-2	France	French	SECAM		Yes	6.60	
Ku3	12.564	Ver	R2	Spot	R	A. France & L'Essentiel	France	French				5.80	
Ku3	12.564	Ver	R2	Spot	R	Rire et Chansons	France	French					
Ku3	12.585	Ver	R8	Spot	R	Beur FM	France	French				6.40	
Ku3	12.606	Hor	R3	Spot	TV	La Chaîne Info (LCI)	France	French	SECAM	Syster	Yes	6.85	
Ku3	12.606	Hor	R3	Spot	TV	ARTE	France/Germany	French/German	SECAM		Yes	6.60	
Ku3	12.606	Hor	R3	Spot	R	La Cinquième	France	French				5.80	
Ku3	12.627	Hor	K9	Spot	TV	Classique FM	France	French					
Ku3	12.648	Ver	R4	Spot	TV	RTL-TVi	France	French	SECAM	SmartCrypt	No		7.75/8.65
Ku3	12.648	Ver	R4	Spot	TV	Tele Monte-Carlo (TMC)	Monte-Carlo	French	SECAM	Syster	Yes	6.60	
Ku3	12.648	Ver	R4	Spot	R	Europe 2	Monte-Carlo	French				5.80	7.25/8.20
Ku3	12.648	Ver	R4	Spot	R	Grand Magazines	Monte-Carlo	French				6.40	

175

ITU REGION 1 STATIONS IN FREQUENCY ORDER

Band	Freq.	Pol.	Tr.	Beam	Sce.	Program	Country	Language	System	Encr.	TT	A.M.	A.S.
Ku3	12.690	Ver	R5	Spot	TV	TF-1	France	French	SECAM		No	5.80	
Ku3	12.732	Ver	R6	Spot	O	VSAT transmissions	France		SECAM			5.80	

TÉLÉCOM 2A 8.0° WEST

C	3.715	LHCP	C1	Semi Global	O	RFO/France-2	France	French	SECAM		No	6.60	
C	3.738	LHCP	C1B	Semi Global	O	occasional video	France	French	PAL/SEC/NTSC			6.60	ch1
C	3.768	RHCP	C2B	Semi Global	TV	RFO Canal Permanent 2	France	French	PAL/SEC/NTSC			6.60	7.02/7.20
C	3.768	LHCP	C2B	Semi Global	O	occasional video	France	French	PAL/SEC/NTSC			6.60	
C	3.849	LHCP	C3	Semi Global	O	not in use	France	French					
C	3.907	LHCP	C4	Semi Global	O	not in use	France	French					
C	3.965	LHCP	C5A	Semi Global	O	not in use	France	French					
C	3.965	RHCP	C5B	Semi Global	O	not in use	France	French					
C	4.049	LHCP	C6A	Semi Global	O	not in use	France	French					
C	4.049	RHCP	C6B	Semi Global	O	not in use	France	French					
C	4.154	LHCP	C7A	Semi Global	O	not in use	France	French					
C	4.154	RHCP	C7B	Semi Global	O	not in use	France	French					
Ku3	12.522	Ver	R1	Spot	TV	Canal Plus	France	French	D2-MAC [16:9]	EuroCrypt	Yes	6.60	
Ku3	12.543	Hor	R7	Spot	TV	MCM Euromusique	France	French	SECAM	Syster	Yes	6.60	
Ku3	12.564	Ver	R2	Spot	TV	Paris Première	France	French	SECAM	Syster	Yes	6.60	
Ku3	12.585	Hor	R8	Spot	TV	Planète	France	French	SECAM	Syster	No	6.60	
Ku3	12.606	Ver	R3	Spot	TV	France Supervision	France	French	D2-MAC [16:9]	EuroCrypt	Yes	6.60	ch1
Ku3	12.606	Hor	R3	Spot	TV	France-2	France	French	D2-MAC [16:9]	EuroCrypt	Yes	6.60	ch1
Ku3	12.627	Ver	R9	Spot	TV	Ciné Cinéfil	France	French	SECAM	Syster	No	6.60	
Ku3	12.648	Hor	R4	Spot	TV	Canal Plus	France	French	SECAM	Syster	No	5.80	7.02/7.20
Ku3	12.669	Ver	R10	Spot	TV	Ciné Cinémas	France	French	D2-MAC [16:9]	EuroCrypt	No	6.60	
Ku3	12.689	Hor	R5	Spot	TV	Ciné Cinémas	France	French	SECAM	Syster	Yes	6.60	
Ku3	12.711	Ver	R11	Spot	TV	TV Sport Eurosport	France	French	SECAM	Syster	No	6.60	
Ku3	12.732	Hor	R6	Spot	TV	Canal (Jeunesse)	France	French	SECAM	Syster	No	6.60	
Ku3	12.732	Ver	R6	Spot	TV	Canal Jimmy	France	French	SECAM	Syster	No	6.60	

GORIZONT 26 11.0° WEST

C	3.675	RHCP	6	Spot	TV	Ostankino ORT Int.	CIS	Russian	SECAM		No	7.0	
C	3.675	RHCP	6	Spot	R	Radio Moscow Int / Radio Mayak	CIS	Russian			No	7.5	
Ku1	11.525	RHCP	K1	Spot	O	occasional video	CIS		PAL/SEC/NTSC		No	6.50/7.5	
Ku1	11.525	RHCP	K1	Spot	O	EBU Moscow	CIS		PAL/SEC/NTSC		No	7.5	

EXPRESS 2 14.0° WEST

C	3.675	RHCP	6	Spot	TV	Moscow-1	CIS	Russian	SECAM		No	7.0	
C	3.675	RHCP	6	Spot	TV	ORT Kanal 1	CIS	Russian	SECAM		No	7.0	
C	3.725	RHCP	9	Spot	R	Radio Rossija 4 Kanal	CIS	Russian	PAL		No	7.5	
C	3.825	RHCP	9	Global	TV	Muslim TV Ahmadiyya Int.	India	Arabic	SECAM		No	7.0	
C	4.025	RHCP	15	Semi Global	TV	APNA TV	India	Hindi	SECAM		No	6.85	
C	4.025	RHCP	15	Semi Global	TV	RTP Internacional	Portugal	Portuguese			No	7.80	
C	4.025	RHCP	15	Semi Global	R	RDP Internacional	Portugal	Portuguese			No	7.02	
C	4.025	RHCP	15	Semi Global	R	RDP Antena 1	Portugal	Portuguese			No	7.20	
C	4.075	RHCP	17	Spot	TV	NTV (Nezavisimoye TV)	Russia	Russian	SECAM		No	7.00	

ITU REGION 1 STATIONS IN FREQUENCY ORDER

Band	Freq.	Pol.	Tr.	Beam	Sce.	Program	Country	Language	System	Encr.	TT	A.M.	A.S.
INTELSAT 705 18.5° WEST													
C	4.075	RHCP	17	Spot	TV	NTV (Nezavisimoye TV)	Russia	Russian	SECAM		No	7.00	
Ku1	11.525	RHCP	K1	Spot	0	Reuters Moscow	CIS		PAL/SEC/NTSC			7.5	
C	4.166	RHCP	38A	Global	0	occasional video	various		PAL/SEC/NTSC			6.65	
C	4.188	RHCP	38B	Global	0	occasional video	various		PAL/SEC/NTSC			6.65	
Ku1	10.969	Hor	61A	West Spot	0	test pattern	Sweden	Swedish	D-MAC			ch1	
Ku1	10.975	Ver	71A	East Spot	0	RTI Milano	Italy	Italian	PAL			6.65	
Ku1	11.010	Ver	71B	East Spot	0	Telespazio	Italy	Italian	PAL			6.65	
Ku1	11.075	Ver	73A	East Spot	TV	MED TV	UK	Turkish/Kurdish/English	PAL			6.65	
Ku1	11.135	Ver	73A	East Spot	0	Telespazio	Italy	Italian	PAL			6.65	
Ku1	11.174	Ver	73B	East Spot	0	Telespazio	Italy	Italian	PAL			6.65	
Ku1	11.482	Ver	73B	East Spot	0	NBC New York	USA	English	PAL/NTSC			6.60	
INTELSAT 512 21.5° WEST													
C	4.055	RHCP	36	East Hemisph.	TV	Sky News	UK	English	PAL		No	6.60	
C	4.166	RHCP	38	Global	0	occasional video	UK					6.60/6.65	
C	4.188	RHCP	38	Global	0	occasional video	UK					6.60/6.65	
INTELSAT K 21.5° WEST													
Ku1	11.465	Ver	5	Euro	0	Reuters / RTL feeds	USA	English	PAL/NTSC/B-MAC				
Ku1	11.485	Hor	5A	Euro	0	n-tv feeds	Germany	German	PAL				
Ku1	11.498	Hor	5B	Euro	0	Reuters London	UK	English	PAL				6.60/ch1
Ku1	11.532	Ver	76L	Euro	0	Reuters Wash.	USA	English	PAL/NTSC/B-MAC			6.60	
Ku1	11.560	Ver	6A	Euro	0	occasional video	USA	English	PAL			6.60/738	
Ku1	11.560	Ver	6B	Euro	0	ABC News / Reuters Wash.	USA	English	PAL			6.60	
Ku1	11.560	Ver	6B	Euro	0	CBS News / Reuters Wash.	USA	English	NTSC				6.60/ch1
Ku1	11.590	Ver	7A	Euro	0	ESPN	USA	English	NTSC			6.60	
Ku1	11.648	Ver	8A	Euro	0	Reuters feeds	USA	English	PAL/NTSC/B-MAC			6.60	6.60/ch1
Ku1	11.652	Ver	8B	Euro	0	occasional video	USA	English	PAL/NTSC/B-MAC				6.60/ch1
Ku1	11.652	Ver	8B	Euro	0	Reuters feeds	USA	English	PAL/NTSC/B-MAC				6.60/ch1
Ku1	11.678	Hor	8B	Euro	0	occasional video	USA	English	PAL/NTSC/B-MAC				6.60/ch1
Ku1	11.682	Ver	8B	Euro	0	occasional video	USA	English	PAL/NTSC/B-MAC				6.60/ch1
Ku3	12.540	Ver	8B	Euro	0	occasional video	USA	English	PAL/NTSC/B-MAC				6.60/ch1
INTELSAT 605 24.5° WEST													
C	4.166	RHCP	38A	Global	TV	Channel Africa	South Africa	Afrikaans	PAL		No	6.60	
C	4.188	RHCP	38B	Global	0	occasional video	various					6.65	
INTELSAT 601 27.5° WEST													
C	3.650	RHCP	20	East Hemisph.	TV	MCM Afrique	South Africa	English	PAL		No	6.60	
C	3.743	RHCP	81	East Hemisph.	TV	WorldNet	USA	English	PAL		No	6.60	
C	3.743	RHCP	81	East Hemisph.	TV	C-Span	USA	English	PAL		No	6.60	
C	3.743	RHCP	81	East Hemisph.	TV	CNN International	USA	English	PAL	Irdeto	No	6.60	
C	3.743	RHCP	81	East Hemisph.	TV	Deutsche Welle TV	Germany	German/English/Spanish	PAL		No		6.60

ITU REGION 1 STATIONS IN FREQUENCY ORDER

Band	Freq.	Pol.	Tr.	Beam	Sce.	Program	Country	Language	System	Encr.	TT	A.M.	A.S.
C	3.743	RHCP	81	East Hemisph.	R	VOA [Europe]	USA	English				7.00	
C	3.743	RHCP	81	East Hemisph.	R	VOA [foreign 1]	USA	various				7.20	
C	3.743	RHCP	81	East Hemisph.	R	VOA [foreign 2]	USA	various				7.30	
C	3.743	RHCP	81	East Hemisph.	R	VOA [foreign 3]	USA	French				7.40	
C	3.743	RHCP	81	East Hemisph.	R	VOA [foreign 4]	USA	various				7.50	
C	3.743	RHCP	81	East Hemisph.	R	VOA [English]	USA	English				7.60	
C	3.795	LHCP	102L	South East Zone	TV	TV Angola	Angola	Portuguese	PAL		No	6.65	
C	3.803	RHCP	23A	East Hemisph.	TV	Canal Horizons	Senegal	French	SECAM	Syster	No	6.60	
C	3.884	LHCP	103A	East Hemisph.	TV	BOP TV	Senegal	French	SECAM	Syster	No	6.60	
C	3.884	LHCP	103A	East Hemisph.	TV	Radio Bop	Senegal	French					7.38/7.56
C	3.884	LHCP	103A	East Hemisph.	R	Mmbabatho R	Senegal	French					7.74/7.92
C	3.915	RHCP	23B	East Hemisph.	TV	Canale France Internationale	France	French	SECAM		No	5.80	
C	3.915	RHCP	23B	East Hemisph.	R	RFI Mondial en France	France	French				6.40	
C	3.915	RHCP	23B	East Hemisph.	R	RFI En Langues Européennes	France	various				6.85	
C	3.915	RHCP	23B	East Hemisph.	R	Radio SRI	Switzerland	various				7.75	
C	3.915	RHCP	23B	East Hemisph.	R	Radio Africa Nr.-1	Gabon	French				8.20	
C	3.915	RHCP	53	East Hemisph.	R	NHK Radio Japan Nippon	Japan	Japanese				8.65	
C	3.931	RHCP	24	West Hemisph.	TV	Bop TV Mmabatho	Bophuthatswana	various	PAL		No	6.60	
C	4.008	RHCP	24	West Hemisph.	TV	RTA-TV Algerian Television	Algeria	French	PAL		No	7.50	
C	4.008	RHCP	17	North East Zone	R	Algerian R	Algeria	French	SCPC				
C	4.048	RHCP	25	North East Zone	TV	CNN International	USA	English	PAL		No	6.65	
C	4.065	RHCP	25	North East Zone	TV	NTA Ch. 10	Nigeria	English/Hausa	PAL-NIL		No	digital	
Ku	10.995	Ver	71	West Spot	TV	BBC Prime	UK	English	D2-MAC	EuroCrypt	Yes	ch1	
Ku	11.016	Hor	61B	West Spot	O	IDB Atlantic Express/EBU	various		NTSC		No	6.60	
Ku	11.055	Hor	62A	West Spot	O	France Télécom	France	French	PAL		No	6.65	
Ku	11.135	Hor	63A	West Spot	O	test pattern			PAL			6.65	
Ku	11.175	Ver	73U	West Spot	TV	CMT Europe	USA	English	PAL		No	6.65	7.02/7.20
Ku	11.175	Ver	63B	West Spot	O	not in use							
Ku	11.475	Ver	74a	West Spot	O	EBU feeds			PAL		No	6.65	

HISPASAT 31.0° WEST

Band	Freq.	Pol.	Tr.	Beam	Sce.	Program	Country	Language	System	Encr.	TT	A.M.	A.S.
Ku1	11.538	Ver	7B	East Spot	O	occasional video	Spain	Spanish				6.60	
Ku1	11.597	Ver	8	East Spot	O	not in use	Spain	Spanish					
Ku1	11.668	Ver	9	East Spot	O	not in use	Spain	Spanish					
Ku2	12.015	Ver	15	East Spot	O	not in use	Spain	Spanish					
Ku2	12.149	LHCP	23	East Spot	TV	TeleDeporte	Spain	Spanish	PAL		No	6.60	7.38/7.56
Ku2	12.149	LHCP	23	East Spot	R	RNE Radio Uno	Spain	Spanish				6.60	
Ku2	12.149	LHCP	23	East Spot	R	RNE Radio 2	Spain	Spanish				7.74	
Ku2	12.149	LHCP	23	East Spot	R	RNE Radio Cinco	Spain	Spanish				7.92	
Ku2	12.227	LHCP	27	East Spot	TV	Canal Classico	Spain	Spanish	PAL		No	6.60	
Ku2	12.302	LHCP	31	East Spot	TV	Cinemania 2	Spain	Spanish	PAL		No	6.60	
Ku2	12.302	LHCP	31	East Spot	R	Cadena Principales	Spain	Spanish				6.60	
Ku2	12.379	LHCP	35	East Spot	TV	Telesat 5	Spain	Spanish	PAL		No	6.60	
Ku3	12.517	Hor	16	East Spot	O	not in use	Spain	Spanish					
Ku3	12.540	Hor	10	East Spot	O	not in use	Spain	Spanish					

ITU REGION 1 STATIONS IN FREQUENCY ORDER

Band	Freq.	Pol.	Tr.	Beam	Sce.	Program	Country	Language	System	Encr.	TT	A.M.	A.S.
Ku3	12.591	Ver	2	East Spot	TV	Canal Sur	Spain	Spanish	PAL		No	6.60	
Ku3	12.591	Ver	2	East Spot	TV	Canal On	Spain	Spanish	PAL		No	6.60	
Ku3	12.591	Hor	11	East Spot	O	not in use	Spain	Spanish					
Ku3	12.597	Hor	17	East Spot	O	not in use	Spain	Spanish					
Ku3	12.631	Ver	3	East Spot	TV	Tele-5	Spain	Spanish	PAL		No	6.60	
Ku3	12.631	Ver	3	East Spot	R	Radio Voz Galicia	Spain	Spanish				7.20	
Ku3	12.631	Ver	3	East Spot	R	Catalonia Principales Com	Spain	Spanish				7.74	
Ku3	12.671	Hor	13	East Spot	R	Antena Tres TV	Spain	Spanish					
Ku3	12.671	Hor	13	East Spot	TV	TV 3 de Catalunya	Spain	Spanish	DVB		No	6.60	
Ku3	12.711	Hor	14	East Spot	R	Cadena Cope Convencional	Spain	Spanish	PAL		No	7.02	
Ku3	12.711	Hor	14	East Spot	R	Radio Top 40	Spain	Spanish				7.56	

INTELSAT 603 34.5° WEST

Band	Freq.	Pol.	Tr.	Beam	Sce.	Program	Country	Language	System	Encr.	TT	A.M.	A.S.
C	4.166	RHCP	38A	Global	O	occasional video	USA		PAL/SEC/NTSC			6.65	
Ku1	11.003	Ver		Spot	TV	Muslim TV Ahmadiyya		Arabic	PAL		No	6.65	
Ku1	11.003	Ver		Spot	TV	Muslim TV Ahmadiyya		English	PAL		No	7.02	

ORION 1 37.5° WEST

Band	Freq.	Pol.	Tr.	Beam	Sce.	Program	Country	Language	System	Encr.	TT	A.M.	A.S.
Ku1	11.473	Hor	14	European Spot	O	vacant			PAL			6.60	
Ku1	11.483	Ver	5	European Spot	O	occasional video							
Ku1	11.494	Hor	15	European Spot	O	vacant			PAL			6.65	
Ku1	11.532	Hor	15		O	occasional video	USA	Dutch					
Ku1	11.540	Ver	6		O	AsiaNet	USA						
Ku1	11.596	Ver	15	European Spot	TV	Super Television Channel (USA)	USA		PAL			6.60	
Ku1	11.617	Hor	7		O	occasional video	USA	Romanian					6.60/7.20
Ku1	12.528	Hor	9		O	Orion Promo video	USA					6.60	
Ku3	12.585	Hor	3	European Spot	O	occasional video	USA					6.60	
Ku3	12.591	Ver			O	occasional video	USA					6.60	
Ku3	12.645	Hor			O	occasional video	USA					6.60	
Ku3	12.651	Ver			O	occasional video	USA					6.60	
Ku3	12.664	Ver			O	occasional video	USA					6.60	

TDRS ATLANTIC 41.0° WEST

Band	Freq.	Pol.	Tr.	Beam	Sce.	Program	Country	Language	System	Encr.	TT	A.M.	A.S.
C	3.800	Hor			O	occasional video	Mexico	Spanish	NTSC			6.2 / 6.80	
C	3.920	Hor	6		O	occasional video	Mexico	Spanish				6.2	
C	3.960	Hor			O	occasional video	Mexico	Spanish				6.2 / 6.80	
C	4.040	Hor			O	occasional video	Mexico	Spanish				6.2 / 6.80	
C	4.080	Hor			O	occasional video	Mexico	Spanish				6.2 / 6.80	
C	4.125	Hor			O	occasional video	Mexico	Spanish				6.2 / 6.80	

PANAMSAT F3R 43.0° WEST

Band	Freq.	Pol.	Tr.	Beam	Sce.	Program	Country	Language	System	Encr.	TT	A.M.	A.S.
C	4.100	Hor	14C	African Hemisph.	TV	ESPN	USA	English	B-MAC				ch 1
C	4.150	Hor	15A		O	occasional video							
C	4.180	Ver	15C	African Hemisph.	TV	CCTV4	China	Chinese	PAL	VideoCrypt	No		
Ku1	11.540	Ver			O	occasional video							
Ku3	12.560	Hor			O	occasional video					No	7.20	

ITU REGION 1 STATIONS IN FREQUENCY ORDER

Band	Freq.	Pol.	Tr.	Beam	Sce.	Program	Country	Language	System	Encr.	TT	A.M.	A.S.
Ku3	12.620	Ver			O	occasional video							
Ku3	12.680	Hor			O	occasional video							
Ku3	12.731	Hor			O	occasional video							
Ku3	12.750	Hor			O	occasional video							
PANAMSAT F1 45.0° WEST													
Ku1	11.461	Hor	19A	Spot	O	digital sce							
Ku1	11.515	Hor	19B	Spot	TV	Galavisión ECO	Mexico	Spanish	PAL		No		6.20/6.80
Ku1	11.515	Hor	19B	Spot	R	Kiss-FM, Mexico	Mexico	Spanish				7.40	
Ku1	11.595	Hor	20B	Spot	O	occasional video	varous		PAL/SEC/NTSC			6.60	
Ku1	11.639	Hor	21A	Spot	O	NBC feeds	USA	English	PAL/SEC/NTSC			6.60	
Ku1	11.676	Hor	21B	Spot	O	former NHK Tokyo	Japan	Japanese				6.60	
INTELSAT 709 50.0° WEST													
Ku1	11.010	Ver		Spot	O	NBC feeds for Super Channel	USA	English	PAL				
INTELSAT 706 53.0° WEST													
C	3.798	RHCP	12	East Hemisph.	TV	RTM-1 Morocco	Morocco	Arabic/French	SECAM		No	6.65	
C	3.990	RHCP	14	East Hemisph.	TV	M-2 Morocco	Morocco	Arabic/French	SECAM	Discrete-12	No	6.65	
C	3.990	RHCP	14	East Hemisph.	R	Radio Mediterranée Int. (Medi-1)	Morocco	Arabic/French					
Ku1	11.500	Ver	12	Spot	O	occasional video	Portugal	Portuguese	PAL			6.65	7.20/8.20

TV = Television R = Radio O = Others A.M. = Audio Mono A.S. = Audio Stereo

ITU REGION 1 DIGITAL STATIONS IN FREQUENCY ORDER

9.2. DIGITAL STATIONS IN FREQUENCY ORDER

Station	Frequency	Pol.	comments	Language	encr.	Symbol.	FEC	Tr.	Dish
KOPERNIKUS, DFS-2, 28,5° E									
Cable plus film	11.477,50	H	general prgr.	Czech		30.000	3/4	1A	90
Prima TV	11.477,50	H	general prgr.	Czech		30.000	3/4	1A	90
test	11.477,50	H	test pattern, TV-audio CT1	Czech		30.000	3/4	1A	90
RTL Lux 1	11.610,00	H	feed for Austria	German		8.350	1/2	90	
Sat 1/A-CH	11.621,00	H	general prgr.	German	Irdeto	8.447	1/2	90	
Sat 1/A-CH	11.643,00	H	general prgr.	German	Irdeto	8.447	1/2	90	
VT4-BXL Newsfeed	11.655,00	H	occ. video	Dutch		8.447	1/2	90	
VT4-BXL Newsfeed	11.666,00	H	occ. video	Dutch		8.447	3/4	90	
RTL Landermagazin HH	12.702,00	H	Regional TV wt. 18-18.30	German		5.632	3/4	90	
VT-4	12.506,00	V	general prgr.	Flemish		5.632	3/4	90	
Kanal Fem	12.514,00	V	general prgr.	Swedish		5.632	3/4	90	
Kanal Fem	12.522,00	V	general prgr.	Swedish		5.629	3/4	90	
Sat 1 Studio Hamburg	12.640,00	V	occ. video	German		5.629	3/4	90	
Sat 1 Bremen/Niedersachsen	12.647,00	V	occ. video	German		5.629	3/4	90	
Sat 1 NRW	12.655,00	V	occ. video	German		5.629	3/4	90	
Sat 1 Hessen/Rheinl.-Pfalz	12.661,00	V	occ. video	German		5.629	3/4	90	
Sat 1 Berlin	12.699,00	V	occ. video	German		5.629	3/4	90	
Sat 1 Berlin	12.674,00	V	occ. video	German		5.629	3/4	90	
Deutsche Telekom	12.573,00	V	occ. video	German		4.000	2/3	90	
KOPERNIKUS, DFS-3, 23,5° E									
Sat1	11.560,00	H	Tests	German		27.500	3/4	90	60
(TVCSM Usingen	11.560,00	H	test pattern	1000Hz. tone		27.500	3/4	90	60
Sat1	11.560,00	H	Tests	German		27.500	3/4	90	60
ATV2	11.560,00	H	atv	Turkish		27.500	3/4	90	60
TV Polonia	11.560,00	H	general prgr.	Polish		27.500	3/4	90	60
TVCSM Usingen	12.522,00	H	test pattern	1000Hz. tone		27.500	3/4	90	60
Dt. Telekom TVCSM Usingen	12.522,00	V	test pattern	1000Hz. tone		27.500	3/4	1K	90
Dt. Telekom TVCSM Usingen	12.522,00	V	test pattern	1000Hz. tone		27.500	3/4	1K	90
Kanal D	12.522,00	V	general prgr.	Turkish		27.500	3/4	1K	90
atv2	12.522,00	V	test pattern	Turkish		27.500	3/4	1K	90
Dt. Telekom TVCSM Usingen	12.522,00	V	test pattern	1000Hz. tone		27.500	3/4	1K	90
TV Polonia	12.522,00	V	?	Polish		27.500	3/4	1K	90
ADR	12.654,73	V	general prgr.	German		27.500	3/4	5K	90
ARD	12.654,73	V	general prgr.	German		27.500	3/4	5K	90
ZDF	12.654,73	V	general prgr.	German		27.500	3/4	5K	90
SAT 1	12.654,73	V	general prgr.	German		27.500	3/4	5K	90
RTL	12.654,73	V	general prgr.	German		27.500	3/4	5K	90
PRO 7	12.654,73	V	general prgr.	German		27.500	3/4	5K	90
Sell-a-Vision/Vox	12.654,73	V	general prgr.	German		27.500	3/4	5K	90

ITU REGION 1 DIGITAL STATIONS IN FREQUENCY ORDER

ASTRA, 19,2° E

Station	Frequency	Pol.	comments	Language	encr.	Symbol.	FEC	Tr.	Dish
DF1 Info-Kanal	10.817,50	V	Trailer	German	Irdeto	22.000	3/4	56	60
CineDom 1	10.817,50	V	Pay per View, Film 6 DM	German	Irdeto	22.000	3/4	56	60
CineDom 2	10.817,50	V	Pay per View, Film 6 DM	German	Irdeto	22.000	3/4	56	60
CineDom 3	10.817,50	V	Pay per View, Film 6 DM	German	Irdeto	22.000	3/4	56	60
CineDom 4	10.817,50	V	Pay per View, Film 6 DM	German	Irdeto	22.000	3/4	56	60
DSF Plus	11.719,50	H	Sport package (20-3)	German	Irdeto	27.500	3/4	65	60
Blue Movie	11.719,50	H	Erotic package Fr./Sa.night	German	Irdeto	27.500	3/4	65	60
Cine Action	11.797,50	H	Movie package	German	Irdeto	27.500	3/4	69	60
Star Kino	11.797,50	H	Movie package	German	Irdeto	27.500	3/4	69	60
Western Movies	11.797,50	H	Movie package	German	Irdeto	27.500	3/4	69	60
FilmPalast	11.797,50	H	Basic German package	German	Irdeto	27.500	3/4	69	60
Comedy & Co.	11.797,50	H	Basic German package	German	Irdeto	27.500	3/4	69	60
Krimi & Co.	11.797,50	H	Basic German package	German	Irdeto	27.500	3/4	69	60
Junior	11.797,50	H	basic package/06.30-19.30	German	Irdeto	27.500	3/4	69	60
K-toon	11.797,50	H	basic package/19.30-06.30	German	Irdeto	27.500	3/4	69	60
Herz & Co.	12.031,50	H	Basic German package	German	Irdeto	27.500	3/4	81	60
Discovery Channel	12.031,50	H	docs, science, educational	German	Irdeto	27.500	3/4	81	60
Clubhouse	12.031,50	H	Children's/basic package	German	Irdeto	27.500	3/4	81	60
Romantic Movies	12.031,50	H	Movie package/19.30-06.30	German	Irdeto	27.500	3/4	81	60
MTV	12.031,50	H	Music/basic package	Music	Irdeto	27.500	3/4	81	60
VH-1	12.031,50	H	Basic German package	German	Irdeto	27.500	3/4	81	60
15 DMX-audio prgrs.	12.031,50	H	basic package(music channel)		Irdeto	27.500	3/4	81	60
Sky News	12.070,50	H	Basic package English	English	Irdeto	27.500	3/4	83	60
CMT	12.070,50	H	Basic package English	English	Irdeto	27.500	3/4	83	60
DSF Golf	12.070,50	H	Sport package	German	Irdeto	27.500	3/4	83	60
NBC	12.070,50	H	Basic package English	English	Irdeto	27.500	3/4	83	60
CNBC	12.070,50	H	Basic package English	English	Irdeto	27.500	3/4	83	60
15 DMX-audio prgrs.	12.070,50	H	basic package (Music channels)	Music	Irdeto	27.500	3/4	83	60
Pro7 - IBC	12.051,00	V	Trailer	German	Irdeto	27.500	3/4	82	60
Classica	12.090,00	V	DF1 Extra	German	Irdeto	27.500	3/4	84	60
Heimatkanal	12.090,00	V	Basic German package	German	Irdeto	27.500	3/4	84	60
Cine Comedy	12.090,00	V	Movie package	German	Irdeto	27.500	3/4	84	60
BBC PRIME	12.090,00	V	special prgr.	English		27.500	3/4	84	60
Action1 @**	12.460,00	H	pay-per-View Tests	German	Irdeto	27.500	3/4	103	60
Action2 @**	12.460,00	H	pay-per-View Tests	German	Irdeto	27.500	3/4	103	60
Action3 @*¹	12.460,00	H	pay-per-View Tests	German	Irdeto	27.500	3/4	103	60
FN1NORDIC	11.953,50	H	movies	Scan.Ings./Finnish		27.500	3/4	77	60
PTV4	11.953,50	H	general prgr.	English		27.500	3/4	77	60
SSPNORDIC	11.953,50	H	sports channel	Scan.Ings./Finnish		27.500	3/4	77	60
HALNORDIC	11.953,50	H	series, movies	Scan. Ings/Eng		27.500	3/4	77	60
KAN5	11.953,50	H	general prgr.	Nor/Eng		27.500	3/4	77	60
FN2NORDIC	11.953,50	H	movies	Scan.Ings./Finnish		27.500	3/4	77	60
VT4	11.875,50	H	general prgr. (ab nachm.)	Flemish	Irdeto	27.500	3/4	73	60
BLOOMBERG	11.875,50	H	economy	English	Irdeto	27.500	3/4	73	60
DISCOVERY	11.875,50	H	docs, science, educational	English	Irdeto	27.500	3/4	73	60

ITU REGION 1 DIGITAL STATIONS IN FREQUENCY ORDER

Station	Frequency	Pol.	comments	Language	encr.	Symbol.	FEC	Tr.	Dish
CMT	11.875,50	H	Country Music	Dutch	Irdeto	27.500	3/4	73	60
ADLT	11.875,50	H	Erotic Channel (01.00-05.00)	Dutch	Irdeto	27.500	3/4	73	60
MSAT	11.875,50	H	Filme, Serien, Unterhaltung	ung		27.500	3/4	73	60
TV NORGE	11.875,50	H	general prgr.	norw		27.500	3/4	73	60
TCC	11.875,50	H	children s	English		27.500	3/4	73	60
20 DMX-audio prgrs.	11.875,50	H	DMX-audio channels	music		27.500	3/4	73	60
FN1 HOLLAND	12.012,00	V	movies	German/Dutch/Eng		27.500	3/4	80	60
FN2 BELGIUM	12.012,00	V	movies	English/Dutch		27.500	3/4	80	60
HALMBENELUX	12.012,00	V	series/movies	English/Dutch		27.500	3/4	80	60
FN1 BELGIUM	12.012,00	V	movies	English/Dutch		27.500	3/4	80	60
RTL4	12.012,00	V	general prgr.	Dutch		27.500	3/4	80	60
Kink FM	12.012,00	V	Radio	Dutch		27.500	3/4	80	60
OLDI	12.012,00	V	Radio/RTL d.b.Oldies	German		27.500	3/4	80	60
RTL5	12.012,00	V	general prgr.	Dutch		27.500	3/4	80	60
EURT	12.012,00	V	Erotic (00.00-05.00)	multilingual		27.500	3/4	80	60
NBC	12.265,50	H	economy	English	Irdeto	27.500	3/4	93	60
CNBC	12.265,50	H	stockmarkets	English	Irdeto	27.500	3/4	93	60
TRAV	12.265,50	H	travel	English	Irdeto	27.500	3/4	93	60
EBN	12.265,50	H	Business channel	English	Irdeto	27.500	3/4	93	60
PERFORMANCE	12.265,50	H	20.00-02.00/classic	English	Irdeto	27.500	3/4	93	60
BTP	12.265,50	H	24 h. Jazz	English/Dutch	Irdeto	27.500	3/4	93	60
WEATHER	12.265,50	H	weather channel	Music	Irdeto	27.500	3/4	93	60
20 DMX-Kanale	12.265,50	H	DMX-audio	English	Irdeto	27.500	3/4	93	60
PULSE movie	12.343,50	H	test (Engl.)	English	Irdeto	27.500	3/4	97	60
PULSE movie	12.343,50	H	test (Engl.)	English	Irdeto	27.500	3/4	97	60
PULSE movie	12.343,50	H	test (Engl.)	English	Irdeto	27.500	3/4	97	60
PULSE movie	12.343,50	H	test (Engl.)	English	Irdeto	27.500	3/4	97	60
PULSE movie	12.343,50	H	test (Engl.)	English	Irdeto	27.500	3/4	97	60
PULSE movie	12.343,50	H	test (Engl.)	English	Irdeto	27.500	3/4	97	60
PULSE movie	12.343,50	H	test (Engl.)	English	Irdeto	27.500	3/4	97	60
PULSE movie	12.343,50	H	test (Engl.)	English	Irdeto	27.500	3/4	97	60
PULSE movie	12.343,50	H	test (Engl.)	English	Irdeto	27.500	3/4	97	60
SBS6	12.343,50	H	general prgr.	Dutch	Irdeto	27.500	3/4	97	60
VERONICA	12.441,00	V	general prgr.	Dutch	Irdeto	27.500	3/4	102	60
Hitradio Veronica	12.441,00	V	?	1000Hz. tone		27.500	3/4	102	60
TEST 1	12.441,00	V	Radio	Dutch	Irdeto	27.500	3/4	102	60
TEST2	12.441,00	V	movies	French	Irdeto	27.500	3/4	102	60
FN2 HOLLAND	12.441,00	V	movies	French	Irdeto	27.500	3/4	102	60
CINECINEMAS 16/9	11.739,00	V	general prgr.	French	Irdeto	27.500	3/4	66	60
CANAL+ 16/9	11.739,00	V	?	French	Irdeto	27.500	3/4	66	60
RTPI	11.739,00	V	?	French	Irdeto	27.500	3/4	66	60
VOYAGE	11.739,00	V	open window		Irdeto	27.500	3/4	66	60
KIOSQUE 4	11.739,00	V			Irdeto	27.500	3/4	66	60

ITU REGION 1 DIGITAL STATIONS IN FREQUENCY ORDER

Station	Frequency	Pol.	comments	Language	encr.	Symbol.	FEC	Tr.	Dish
KIOSQUE 5	11.739,00	V	open window	French	Irdeto	27.500	3/4	66	60
KIOSQUE 6	11.739,00	V	open window	French	Irdeto	27.500	3/4	66	60
MTV	11.739,00	V	music channel	English	Irdeto	27.500	3/4	66	60
EUROSPORT	11.778,00	V	sports channel	French	Irdeto	27.500	3/4	68	60
MONTE CARLO TMC	11.778,00	V	general prgr.	French	Irdeto	27.500	3/4	68	60
PLANETE	11.778,00	V	science	French	Irdeto	27.500	3/4	68	60
PARIS PREMIERE	11.778,00	V	arts & theatre	French	Irdeto	27.500	3/4	68	60
#	11.778,00	V	French	French	Irdeto	27.500	3/4	68	60
MCM	11.778,00	V	music channel	French	Irdeto	27.500	3/4	68	60
LCI	11.778,00	V	news channel	French	Irdeto	27.500	3/4	68	60
Canal J	11.778,00	V	children's (tags)	French	Irdeto	27.500	3/4	68	60
MULTIMUSIC 1	11.778,00	V	10 Radios (eins Offen)	French	Irdeto	27.500	3/4	68	60
CINE-CINEMAS	11.817,00	V	movies	French	Irdeto	27.500	3/4	70	60
CINE-CINEMAS I	11.817,00	V	movies	French	Irdeto	27.500	3/4	70	60
CINE-CINEMAS II	11.817,00	V	movies	French	Irdeto	27.500	3/4	70	60
CINE-CINEFIL	11.817,00	V	oldies movies	French	Irdeto	27.500	3/4	70	60
MOSAIQUE	11.817,00	V	Trailer	French	Irdeto	27.500	3/4	70	60
CANAL JIMMY	11.817,00	V	youth (evenings)	French	Irdeto	27.500	3/4	70	60
MUZZIK	11.817,00	V	music channel	French	Irdeto	27.500	3/4	70	60
LA CHAINE METEO	11.817,00	V	weather channel	French	Irdeto	27.500	3/4	70	60
LES RADIOS	11.817,00	V	10 Rad.&int.Menu, Hector	French	Irdeto	27.500	3/4	70	60
MULTIMUSIC 2	11.817,00	V	5 Radios	French	Irdeto	27.500	3/4	70	60
CANAL+	11.856,00	V	general prgr. (open window)	French	Irdeto	27.500	3/4	72	60
CANAL+ BLEU	11.856,00	V	general prgr.	French	Irdeto	27.500	3/4	72	60
CANAL+ JAUNE	11.856,00	V	open window	French	Irdeto	27.500	3/4	72	60
KIOSQUE 1	11.856,00	V	open window	French	Irdeto	27.500	3/4	72	60
KIOSQUE 2	11.856,00	V	preview channel	French	Irdeto	27.500	3/4	72	60
KIOSQUE 3	11.856,00	V	open window	French	Irdeto	27.500	3/4	72	60
GUIDE TV	11.856,00	V	movies/cartoons	French		27.500	3/4	72	60
TNT	11.836,50	H	cartoons	English		27.500	3/4	71	60
Cartoon Network	11.836,50	H	London feeds/1-5 Chin. Ch.	English		27.500	3/4	71	60
TVBS-Europe	11.836,50	H	travel & tourist info.	Chinese		27.500	3/4	71	60
Travel	11.836,50	H	news channel	English		27.500	3/4	71	60
CNN Int.	11.836,50	H	Radio (news)	English		27.500	3/4	71	60
CNN Radio	11.836,50	H	Tests	English		27.500	3/4	71	60
RTL Television CH	11.914,50	H	general prgr.	German	Irdeto	27.500	3/4	75	60
RTL Television A	11.914,50	H	Radio	German	Irdeto	27.500	3/4	75	60
RTLradio Oldiesender	11.914,50	H	movies	German		27.500	3/4	75	60
CINEMANIA	11.934,00	V	movies	Spanish		27.500	3/4	76	60
CINEMANIA 2	11.934,00	V	st. logo	Spanish		27.500	3/4	76	60
TAQUILLA 3	11.934,00	V	st. logo	1000Hz. tone		27.500	3/4	76	60
TAQUILLA 4	11.934,00	V	st. logo	1000Hz. tone		27.500	3/4	76	60
TAQUILLA 5	11.934,00	V	science & nature	1000Hz. tone		27.500	3/4	76	60
DISCOVERY	11.934,00	V	movies	Spanish	Irdeto	27.500	3/4	76	60
CANAL+	11.973,00	V		Spanish	Irdeto	27.500	3/4	78	60

ITU REGION 1 DIGITAL STATIONS IN FREQUENCY ORDER

Station	Frequency	Pol.	comments	Language	encr.	Symbol.	FEC	Tr.	Dish
CANAL+ AZUL	11.973,00	V	movies	Spanish	Irdeto	27.500	3/4	78	60
CANAL+ ROJO	11.973,00	V	movies	Spanish	Irdeto	27.500	3/4	78	60
MOSAICO	11.973,00	V	Trailer	Spanish	Irdeto	27.500	3/4	78	60
SPORTMANIA	11.973,00	V	sports channel	Spanish	Irdeto	27.500	3/4	78	60
RADIO	11.973,00	V	5 Radios	Spanish	Irdeto	27.500	3/4	78	60
AUDIOMANIA	11.973,00	V	6 Radios	Spanish	Irdeto	27.500	3/4	78	60
MULTICLASICA	11.973,00	V	Radio	Spanish	Irdeto	27.500	3/4	78	60
Taquilla Test	11.973,00	V	Trailer	Spanish		27.500	3/4	78	60
GUIA TV	11.973,00	V	prgr. guide	Spanish		27.500	3/4	78	60
TAQUILLA 6	12.246,00	V	st. logo	1000Hz. tone		27.500	3/4	92	60
TAQUILLA 7	12.246,00	V	st. logo	1000Hz. tone		27.500	3/4	92	60
TAQUILLA 8	12.246,00	V	st. logo	1000Hz. tone		27.500	3/4	92	60
SCR TEST	12.246,00	V	Multimedia-Trailer	Spanish		27.500	3/4	92	60
DOCUMANIA	12.285,00	V	documataries	Spanish	Irdeto	27.500	3/4	94	60
MINIMAX/ALBUM	12.285,00	V	children's	Spanish	Irdeto	27.500	3/4	94	60
CINECLASSICS [@*]	12.285,00	V	movies (oldies)	Spanish		27.500	3/4	94	60
SEASONS	12.285,00	V	Tests (movies)	Spanish		27.500	3/4	94	60
TAQUILLA 0	12.285,00	V	Tests (movies)	Spanish		27.500	3/4	94	60
TAQUILLA 1	12.285,00	V	Tests (movies)	Spanish		27.500	3/4	94	60
TAQUILLA 2	12.285,00	V	Tests (movies)	Spanish		27.500	3/4	94	60
premiere 1	12.109,50	H	general prgr.	German		27.500	3/4	85	60
premiere 2	12.109,50	H	timeshifted (afternoon prgr)	German		27.500	3/4	85	60
PPV 1	12.109,50	H	pay-per-view (Film 6 DM)	German	Irdeto	27.500	3/4	85	60
PPV 2	12.109,50	H	pay-per-view (Film 6 DM)	German	Irdeto	27.500	3/4	85	60
PPV 3	12.109,50	H	pay-per-view (Film 6 DM)	German	Irdeto	27.500	3/4	85	60
PPV 4	12.109,50	H	pay-per-view (Film 6 DM)	German	Irdeto	27.500	3/4	85	60
PEP	12.109,50	H	prgr. guide	German		27.500	3/4	85	60
ASTRA INFO	12.168,00	H	(Promo Video) open window	various		27.500	3/4	88	60
SAT1	12.168,00	V	general prgr.	German	Irdeto	27.500	3/4	88	60
KABEL1	12.168,00	V	general prgr.	German	Irdeto	27.500	3/4	88	60
DSF	12.168,00	V	sports channel	German	Irdeto	27.500	3/4	88	60
ZDF	12.168,00	V	general prgr.	German		27.500	3/4	88	60
ARD	12.168,00	V	general prgr.	German		27.500	3/4	88	60
TRAVEL	12.168,00	V	travel/open window	English		27.500	3/4	88	60
SPECTACLE	12.129,00	V	?	French		27.500	3/4	86	60
KIOSQUE 7	12.129,00	V	open window	French	Irdeto	27.500	3/4	86	60
KIOSQUE 8	12.129,00	V	open window	French	Irdeto	27.500	3/4	86	60
KIOSQUE 9	12.129,00	V	open window	French	Irdeto	27.500	3/4	86	60
SEASONS	12.129,00	V	fishing channel	French	Irdeto	27.500	3/4	86	60
FRANCE COURSES	12.129,00	V	?	French	Irdeto	27.500	3/4	86	60
DT 90-1 SST	12.207,00	V	?	French		27.500	3/4	90	60
DT 90-2 C2P	12.207,00	V	?	French		27.500	3/4	90	60
MULTITHEM-1	12.207,00	V	open window	French	Irdeto	27.500	3/4	90	60
BLOOMBERG TV	12.207,00	V	economy channel	French	Irdeto	27.500	3/4	90	60
DT 90-5 PPV1	12.207,00	V	?	French		27.500	3/4	90	60
DT 90-6 PPV2	12.207,00	V	?	French		27.500	3/4	90	60

ITU REGION 1 DIGITAL STATIONS IN FREQUENCY ORDER

Station	Frequency	Pol.	comments	Language	encr.	Symbol.	FEC	Tr.	Dish
DT 90-7 PPV3	12.207,00	V	Trailer	French	Irdeto	27.500	3/4	90	60
DT 90-8 WEB	12.207,00	V	tests (tape loop) open window	English		27.500	3/4	90	60
DT 90-9 AS1	12.207,00	V	tests (tape loop) open window	English	Irdeto	27.500	3/4	90	60
DT 90-10 RADIO1	12.207,00	V	10 audio channels	English	Irdeto	27.500	3/4	90	60
DT 90-11 RADIO2	12.207,00	V	4 audio channels	French	Irdeto	27.500	3/4	90	60
DT 90-12 AS2	12.207,00	V	tests (tape loop)	English		27.500	3/4	90	60
DT90-13 CH1	12.207,00	V	tests (tape loop)	English		27.500	3/4	90	60
DT 90-14 CH2	12.207,00	V	tests (tape loop)	English		27.500	3/4	90	60
DT 90-15 CH3	12.207,00	V	tests (tape loop)	English		27.500	3/4	90	60
DT 90-16 CH4	12.207,00	V	tests (tape loop)	French		27.500	3/4	90	60
DT 90-17 CH5	12.207,00	V	?	French	Irdeto	27.500	3/4	90	60
EUTELSAT II F3 16° EAST									
rita28 video music [@**]	11.005,00	H	?	?	Irdeto	5.642	3/4	90	90
rete tmc2 [@**]	11.005,00	H	?	?	Irdeto	5.632	3/4	90	90
TV10 [@**]	11.014,00	H	Dutch entertainment	Dutch		5.629	3/4	90	90
Radio 10 Gold [@**]	11.014,00	H	Radio.	Dutch		5.629	3/4	90	90
TMF [@**]	11.024,00	H	music channel	Dutch		5.629	3/4	90	90
TMC [@**]	11.052,00	H	?	?	Irdeto	5.629	3/4	90	90
Test [@**]	11.052,00	H	?	?	Irdeto	5.926	3/4	90	90
Wizja TV [@*]	11.060,00	H	general prgr.	Polish		5.632	3/4	90	90
ENEX-Feed [@*]	12.519,00	V	occ. video	English		5.632	3/4	120	90
ENEX-Feed [@*]	12.528,00	V	occ. video	English		5.632	3/4	120	90
ENEX-Feed [@*]	12.546,00	V	occ. video	English		5.632	3/4	120	90
Number ONE TV [@*]	12.599,00	V	music channel	Turkish	?	?	3/4	120	90
EUTELSAT II F-1/HOTBIRD 1/2 13° E									
T9 [@**]	11.345,00	H	?	Italian	Irdeto	27.500	2/3	7	90
T10 [@**]	11.345,00	H	?	Italian	Irdeto	27.500	2/3	7	90
T11 [@**]	11.345,00	H	?	Italian	Irdeto	27.500	2/3	7	90
T12 [@**]	11.345,00	H	?	Italian	Irdeto	27.500	2/3	7	90
T13 [@**]	11.345,00	H	?	Italian	Irdeto	27.500	2/3	7	90
T14 [@**]	11.345,00	H	?	Italian	Irdeto	27.500	2/3	7	90
DX16-20 [@**]	11.345,00	H	5 DMX audio channels music	Italian	Irdeto	27.500	2/3	7	90
T15 [@**]	11.283,00	V	?	Italian	Irdeto	27.500	2/3	4	90
T16 [@**]	11.283,00	V	?	Italian	Irdeto	27.500	2/3	4	90
T17 [@**]	11.283,00	V	?	Italian	Irdeto	27.500	2/3	4	90
T18 [@**]	11.283,00	V	?	Italian	Irdeto	27.500	2/3	4	90
T19 [@**]	11.283,00	V	?	Italian	Irdeto	27.500	2/3	4	90
T20 [@**]	11.283,00	V	?	Italian	Irdeto	27.500	2/3	4	90
RDON [@**]	11.283,00	V	Radio	Italian	Irdeto	27.500	2/3	4	90
RITA [@**]	11.283,00	V	Radio	Italian	Irdeto	27.500	2/3	4	90
TMC	11.958,00	V	Music channel	Italian/Eng	Irdeto	27.491	3/4	62	70
TMC2	11.958,00	V	Music channel	Italian/Eng	Irdeto	27.491	3/4	62	70
BLTV	11.958,00	V	stockmarket/science	Italian/Eng	Irdeto	27.491	3/4	62	70

ITU REGION 1 DIGITAL STATIONS IN FREQUENCY ORDER

Station	Frequency	Pol.	comments	Language	encr.	Symbol.	FEC	Tr.	Dish
DSCV	11.958.00	V	nature/science	English	Irdeto	27,491	3/4	62	70
HALL	11.958.00	V		Italian/Eng	Irdeto	27,491	3/4	62	70
TOON	11.958.00	V	cartoons	Italian/Eng	Irdeto	27,491	3/4	62	70
CNN	11.958.00	V	news channel	English	Irdeto	27,491	3/4	62	70
BBC	11.958.00	V	foreign sce.	English	Irdeto	27,491	3/4	62	70
MTEO	11.996.00	V	weather channel	Italian/Eng	Irdeto	27,491	3/4	64	70
MTV	11.996.00	V	music channel	Italian/Eng	Irdeto	27,491	3/4	64	70
BOJ	11.996.00	V	?	Italian/Eng	Irdeto	27,491	3/4	64	70
PIU3	11.996.00	V	?	Italian/Eng	Irdeto	27,491	3/4	64	70
PRO2	11.996.00	V	test pattern	Italian/Eng	Irdeto	27,491	3/4	64	70
PRO3	11.996.00	V	test pattern	Italian/Eng	Irdeto	27,491	3/4	64	70
PRO4	11.996.00	V	?	Italian/Eng	Irdeto	27,491	3/4	64	70
T1	11.996.00	V	test pattern	Italian/Eng	Irdeto	27,491	3/4	64	70
T2	11.996.00	V	test pattern	Italian/Eng	Irdeto	27,491	3/4	64	70
DX07-10	11.996.00	V	4 DMX-audio channels	music	Irdeto	27,491	3/4	64	70
PIU1	12.034.00	V		Italian	Irdeto	27,491	3/4	66	70
PLUS	12.034.00	V	?	Italian	Irdeto	27,491	3/4	66	70
PIU2	12.034.00	V	?	Italian	Irdeto	27,491	3/4	66	70
SPT2	12.034.00	V	?	Italian	Irdeto	27,491	3/4	66	70
PR01	12.034.00	V	Trailer	Italian	Irdeto	27,491	3/4	66	70
DX01-06	12.034.00	V	6 DMX-audio channels	music	Irdeto	27,491	3/4	66	70
T3	12.072.00	V	st. logo	Italian	Irdeto	27,491	3/4	68	70
T4	12.072.00	V	st. logo	Italian	Irdeto	27,491	3/4	68	70
T5	12.072.00	V	st. logo	Italian	Irdeto	27,491	3/4	68	70
T6	12.072.00	V	st. logo	Italian	Irdeto	27,491	3/4	68	70
T7	12.072.00	V	st. logo	Italian	Irdeto	27,491	3/4	68	70
T8	12.072.00	V	5 DMX-audio channels	music	Irdeto	27,491	3/4	68	70
DX11-15	12.072.00	V	foreign sce. (audio mono A)	Italian		27,491	3/4	68	70
RAI International	11.464.00	V	(audio mono B)	multilingual		4,357	3/4	12V	90
Rai Radio Intern.	11.464.00	V	general prgr.	Italian		4,357	3/4	12V	90
RAIUNO	11.804.00	V	general prgr.	Italian		27,500	2/3	54	70
RAIDUE	11.804.20	V	general prgr.	Italian		27,500	2/3	54	70
RAITRE	11.804.20	V	Kath. program (occ.)	Italian		27,500	2/3	54	70
TELEPACE	11.804.20	V	Radio	Italian		27,500	2/3	54	70
RADIOUNO	11.804.20	V	Radio	Italian		27,500	2/3	54	70
RADIODUE	11.804.20	V	Radio	Italian		27,500	2/3	54	70
RADIOTRE	11.804.20	V		Italian		27,500	2/3	54	70
I1 [@**]	11.919.28	V	general prgr.	Italian	Irdeto	27,500	2/3	60	70
C 5 [@**]	11.919.28	V	general prgr.	Italian	Irdeto	27,500	2/3	60	70
R 4 [@**]	11.919.28	V	general prgr.	Italian	Irdeto	27,500	2/3	60	70
TEST [@**]	11.919.28	V	Tests	Italian	Irdeto	27,500	2/3	60	70
TSTMultivision 1 [@**]	12.543.00	V	Pay-per-view Tests	French		27,500	3/4	46	90
TSTMultivision 2 [@**]	12.543.00	V	Pay-per-view Tests	French		27,500	3/4	46	90
TSTMultivision 3 [@**]	12.543.00	V	Pay-per-view Tests	French		27,500	3/4	46	90
TSTMultivision 4 [@**]	12.543.00	V	Pay-per-view Tests	French		27,500	3/4	46	90
VH-1 Export [@**]	12.543.00	V	music channel	English		27,500	3/4	46	90

ITU REGION 1 DIGITAL STATIONS IN FREQUENCY ORDER

Station	Frequency	Pol.	comments	Language	encr.	Symbol.	FEC	Tr.	Dish
TPSTESTLACINQUIEME [@**]	12.543,00	V	Tests	French	Irdeto	27.500	3/4	46	90
TF1	12.583,00	V	general prgr.	French	Irdeto	27.500	3/4	46	90
France 2	12.583,00	V	general prgr.	French	Irdeto	27.500	3/4	46	90
France 3	12.583,00	V	general prgr.	French	Irdeto	27.500	3/4	46	90
Canal Promo	12.583,00	V	trailer	French		27.500	3/4	46	90
Arte / La Cinquieme	12.583,00	V	German/French cultural prgr.	French/German		27.500	3/4	46	90
M6	12.583,00	V	general prgr.	French	Irdeto	27.500	3/4	46	90
Serie Club	12.583,00	V	series!	French	Irdeto	27.500	3/4	46	90
RTL9	12.583,00	V	general prgr.	French	Irdeto	27.500	3/4	46	90
LCI	12.708,00	V	news channel	French	Irdeto	27.500	3/4	49	90
Eurosport	12.708,00	V	sports channel	French	Irdeto	27.500	3/4	49	90
CNN	12.708,00	V	news channel	English	Irdeto	27.500	3/4	49	90
BBC World	12.708,00	V	news channel	English	Irdeto	27.500	3/4	49	90
BBC Prime	12.708,00	V	Info & docs.	English	Irdeto	27.500	3/4	49	90
TV5	12.708,00	V	general prgr.	French	Irdeto	27.500	3/4	49	90
TV7	12.708,00	V	general prgr.	Arabic	Irdeto	27.500	3/4	49	90
CANAL ASSEMBLEE	12.708,00	V	parliaments channel	French	Irdeto	27.500	3/4	49	90
Radio Classique	12.708,00	V	Radio Classique	French	Irdeto	27.500	3/4	49	90
Cinestar 1	11.938,00	H	Movies pay-per-view	French	Irdeto	27.500	3/4	61	70
Cinestar 2	11.938,00	H	Movies pay-per-view	French	Irdeto	27.500	3/4	61	70
Cinetoile	11.938,00	H	?	French	Irdeto	27.500	3/4	61	70
Teletoon	11.938,00	H	cartoons	French	Irdeto	27.500	3/4	61	70
Odyssee	11.938,00	H	Naturkanal (ab 11 Uhr)	French	Irdeto	27.500	3/4	61	70
Teva	11.938,00	H	women	French	Irdeto	27.500	3/4	61	70
Supervision	11.938,00	H	general prgr. 16:9	French	Irdeto	27.500	3/4	61	70
FUN TV	11.938,00	H	youth & music channel	French	Irdeto	27.500	3/4	61	70
FESTIVAL	11.938,00	H	movies	French	Irdeto	27.500	3/4	61	90
Multivision 1	12.091,00	H	movies	French	Irdeto	27.500	3/4	69	70
Multivision 2	12.091,00	H	movies	French	Irdeto	27.500	3/4	69	70
Multivision 3	12.091,00	H	movies	French	Irdeto	27.500	3/4	69	70
Multivision 4	12.091,00	H	movies	French	Irdeto	27.500	3/4	69	70
Multivision 5	12.091,00	H	movies	French	Irdeto	27.500	3/4	69	70
Multivision 6	12.091,00	H	movies	French	Irdeto	27.500	3/4	69	70
FRANCE COURSES	12.091,00	H	sports & cars channel	French	Irdeto	27.500	3/4	69	70
CANAL AUTO	12.091,00	H		French	Irdeto	27.500	3/4	69	70
METEO EXPRESS	12.091,00	H	interactive weather ch.	French	Irdeto	27.500	3/4	69	70
AB 1 [@**]	11.678,00	H	Trailer	French		27.500	3/4	34 H	90
CHEVAL [@**]	11.678,00	H	?	French		27.500	3/4	34 H	90
AB CARTOONS	11.678,00	H	cartoons	French		27.500	3/4	34 H	90
CHASSE & PECHE	11.678,00	H	?	French		27.500	3/4	34 H	90
VIVE LA VIE	11.678,00	H	?	French		27.500	3/4	34 H	90
CINE PALACE	11.678,00	H	movies	French		27.500	3/4	34 H	90
ROMANCE	11.678,00	H	romantic	French		27.500	3/4	34 H	90
AUTOMOBILE / XXL [@**]	11.678,00	H	(erotics at night)	French	Irdeto	27.500	3/4	34 H	90
MUSIQUE CLASSIQUE	11.678,00	H	music channel	French		27.500	3/4	34 H	90
AB SAT	12.521,41	H	Trailer	French	Irdeto	27.500	3/4	40H	90

ITU REGION 1 DIGITAL STATIONS IN FREQUENCY ORDER

Station	Frequency	Pol.	comments	Language	encr.	Symbol.	FEC	Tr.	Dish
AB SPORTS	12.521,41	H	sports channel	French	Irdeto	27.500	3/4	40H	90
ANIMAUX	12.521,41	H	?	French	Irdeto	27.500	3/4	40H	90
ENCYCLOPEDIA	12.521,41	H	science	French	Irdeto	27.500	3/4	40H	90
ESCALES	12.521,41	H	?	French	Irdeto	27.500	3/4	40H	90
RIRE	12.521,41	H	various	French	Irdeto	27.500	3/4	40H	90
POLAR	12.521,41	H	?	French	Irdeto	27.500	3/4	40H	90
ACTION	12.521,41	H	series	French	Irdeto	27.500	3/4	40H	90
MELODY	12.521,41	H	music channel	French		27.500	3/4	40H	90
VH-1	11.241,50	V	music channel	German		27.500	3/4	2	90
MTV 3 [@**]	11.241,50	V	music channel	English/Dutch		27.500	3/4	2	90
Bloomberg [@**]	11.241,50	V	economy channel	English	Irdeto	27.500	3/4	2	90
Sci-Fi [@**]	11.241,50	V	Science F	English		27.500	3/4	2	90
MTV 2	11.241,50	V	music channel	German		27.500	3/4	2	90
WRN [@**]	11.241,50	V	various int. radio scs.	English		27.500	3/4	2	90
MTV 1	11.241,50	V	music channel	English		27.500	3/4	2	90
VH-1 Export	11.241,50	V	music channel	English		27.500	3/4	2	90
FNet	11.823,00	H	movies	Greek		25.384	3/4	55	90
SSport	11.823,00	H	sports channel	English/Greek		25.384	3/4	55	90
Test 1	11.823,00	H	Tests			25.384	3/4	55	90
Test 2	11.823,00	H	Tests			25.384	3/4	55	90
Test 3	11.823,00	H	Tests			25.384	3/4	55	90
Test 4	11.823,00	H	Tests			25.384	3/4	55	90
Test 5	11.823,00	H	Tests			25.384	3/4	55	90
ART-EUROPE [@**]	12.015,00	H	general prgr.	Arabic		27.500	3/4	55	90
ART-EUROPE [@**]	12.015,00	H	general prgr.	Arabic		27.500	3/4	65	90
ART-MOVIE [@**]	12.015,00	H	art 4	Arabic		27.500	3/4	65	90
ART-MUSIC [@**]	12.015,00	H	art 5			27.500	3/4	65	90
TEST- SPORT [@**]	12.015,00	H	test pattern	1000Hz. tone		27.500	3/4	65	90
LBC SAT [@**]	12.015,00	H	general prgr.	Arabic		27.500	3/4	65	90
MCM Euromusique	11.289,00	H	music channel	French		27.500	3/4	65	90
RTL Television [@**]	11.610,00	H	general prgr.	German		5.140	2/3	90	90
APTV [@**]	12.549,00	H	Feed/sporad.	English		5.697	3/4	33H	90
WTN LONDON [@**]	12.558,00	H	Feed/sporad.	English		5.632	3/4	41A	90
WTN New York [@**]	12.567,00	H	Feed/sporad.	English		5.632	3/4	41A	90

EUTELSAT II F-2, 10° E

Station	Frequency	Pol.	comments	Language	encr.	Symbol.	FEC	Tr.	Dish
Antena-1 [@**]	11.024,25	H	general prgr.	Romanian	Irdeto	5.631	3/4	90	
Programm pentu cablisti [@**]	11.024,25	H	general prgr.	Romanian	Irdeto	5.631	3/4	90	
Advent-1 [@**]	11.631,00	V	general prgr. (KRAL)	Turkish	Irdeto	5.631	3/4	38V	90

TELE X, 5° E

Station	Frequency	Pol.	comments	Language	encr.	Symbol.	FEC	Tr.	Dish
TNT [@**]	12.322,00	LHCP	feeds for Astra			30.000	7/8	60	
Cartoon Network [@**]	12.322,00	LHCP	feeds for Astra			30.000	7/8	60	
TVBS-Europe [@**]	12.322,00	LHCP	feeds for Astra			30.000	7/8	60	
Travel [@**]	12.322,00	LHCP	feeds for Astra			30.000	7/8	60	
CNN Int. [@**]	12.322,00	LHCP	feeds for Astra			30.000	7/8	60	

ITU REGION 1 DIGITAL STATIONS IN FREQUENCY ORDER

Station	Frequency	Pol.	comments	Language	encr.	Symbol.	FEC	Tr.	Dish
CNN Radio9+++	12.322,00	LHCP	feeds for Astra	various		30.000	7/8	60	
?	12.668,00	LHCP	?			23.562	3/4	60	
INTELSAT 707, 1° W									
?	10.974,00	H	?	various	Irdeto	25.776	3/4	90	90
Denmarks Radio 1 [@**]	11.592,00	H	DR1	Danish		17.494	3/4	69	90
Denmarks Radio 2 [@**]	11.592,00	H	DR2	Danish		17.494	3/4	69	90
Sportskanalen [@**]	11.592,00	H	parallel DR2	Danish		17.494	3/4	69	90
Radio 2 [@**]	11.592,00	H	until 18 hr, TV-audio DR2	Danish		17.494	3/4	69	90
TELECOM 2B, 5° W									
FESTIVAL	12.669,00	H	feed	French	Irdeto	27.500	3/4	90	
INTELSAT K, 21,5° W									
RFTV	11.550,00	V	occ. feeds	English		5.632	3/4	90	
Brightstar V6u	11.558,00	V	occ. feeds	English		5.632	3/4	90	
RTV News	11.566,00	V	occ. feeds	English		5.632	3/4	90	
Reuters Moscov	11.551,00	V	occ. feeds	English		5.632	3/4	90	
RTV WNS	11.559,00	H		English		5.632	3/4	90	
3ABN	12.615,00	V	church	English		20.000	1/2	90	
3ABN back up	12.615,00	V	church	English		20.000	1/2	90	
INTELSAT 601, 27,5° W									
IPPV	10.973,00	H	?	Dutch	Irdeto	27.500	3/4	160	
PLS	10.973,00	H	?	Dutch	Irdeto	27.500	3/4	150	
PLS	10.973,00	H	?	Dutch	Irdeto	27.500	3/4	150	
PLS	10.973,00	H		Dutch	Irdeto	27.500	3/4	150	
ALT	10.973,00	H		Dutch		27.500	3/4	150	
PROMO	10.973,00	H	Trailer (half pict.)	English		27.500	3/4	150	
NL1 [@**]	10.973,00	H	?	Dutch	Irdeto	27.500	3/4	150	
LANG	10.973,00	H	?	Dutch	Irdeto	27.500	3/4	150	
HLL	10.973,00	H		Dutch	Irdeto	27.500	3/4	150	
WEA	10.973,00	H	weather channel	English	Irdeto	27.500	3/4	150	
EBN	10.973,00	H	economy channel	English	Irdeto	27.500	3/4	150	
NBC	10.973,00	H	economy/stockmarkets	English		27.500	3/4	150	
Channel 1	11.506,00	H	id	?		6.197	7/8	64	150
Channel 2	11.506,00	H	id	?		6.197	7/8	64	150
MC Europe	11.580,00	V	audio package, 31 stations			6.111	3/4	150	
DVB Mode Test*]	11.660,00	V	Discovery Channel	English		27.500	3/4	150	
HISPASAT 1B, 30° W									
Galavision	12.591,00	H	general prgr.	Spanish		27.500	3/4	11	90
TVE2	12.591,00	H	general prgr.	Spanish		27.500	3/4	11	90
TVE1	12.591,00	H	general prgr.	Spanish		27.500	3/4	11	90
Canal Clasico	12.591,00	H	cultural channel	Spanish		27.500	3/4	11	90
Teledeporte	12.591,00	H	sports channel	Spanish		27.500	3/4	11	90

ITU REGION 1 DIGITAL STATIONS IN FREQUENCY ORDER

Station	Frequency	Pol.	comments	Language	encr.	Symbol.	FEC	Tr.	Dish
TVE Int	12.591,00	H	foreign sce.	Spanish		27.500	3/4	11	90
TVG	12.591,00	H	Regional-TV, Galicija	Spanish		27.500	3/4	11	90
Telemadrid	12.591,00	H	Regional TV, Madrid	Spanish		27.500	3/4	11	90
Mosaico	12.591,00	H	Trailer	Spanish		27.500	3/4	11	90
Galeusca	12.631,00	H	entertainment	Spanish		27.500	3/4	12	90
TVE int	12.631,00	H	foreign sce. (no audio)	Spanish		27.500	3/4	12	90
TVG	12.631,00	H	TV Galicija (no audio)	Spanish		27.500	3/4	12	90
Telemadrid	12.631,00	H	Regional TV, Madrid	Spanish		27.500	3/4	12	90
Teledeporte	12.631,00	H	sports channel	Spanish		27.500	3/4	12	90
Telefonica	12.631,00	H	test pattern	Spanish		27.500	3/4	12	90
Canal 9	12.631,00	H	Regional TV, Valencia	Spanish		27.500	3/4	12	90
TV3 Sat	12.631,00	H	TV3 de Catalunya, 20-24 hr	Spanish		27.500	3/4	3/4	90
Telenoticias	12.456,00	LHCP	news channel	English/Sp		24.000	3/4	12	90
Fiesta/EBN	12.456,00	LHCP	entertainment/economy	Spanish		24.000	3/4	12	90
Cine Color	12.456,00	LHCP	movies	Spanish		24.000	5/6	39	90
Cine Siempre	12.456,00	LHCP	movies	Spanish		24.000	5/6	39	90
Discovery [@*]	12.456,00	LHCP	science/nature	English/Sp		24.000	5/6	39	90
VH-1	11.538,00	H	music channel	English		27.500	5/6	39	90
CANAL+	11.538,00	H	Tests	Spanish	Irdeto	27.500	5/6	16	90
CANAL+ AZUL	11.538,00	H	Tests	Spanish	Irdeto	27.500	3/4	16	90
CANAL+ ROJO	11.538,00	H	Tests	Spanish	Irdeto	27.500	3/4	16	90
MOSAICO	11.538,00	H	Trailer	Spanish		27.500	3/4	16	90
SPORTMANIA	11.538,00	H	sports channel	Spanish		27.500	3/4	16	90
TAQUILLA TEST	11.538,00	H	Tests	Spanish		27.500	3/4	16	90
GUIA TEST	11.538,00	H	(1) 40 PRINCIPALIS	Spanish		27.500	3/4	16	90
RADIO [@*]	11.538,00	H	(2) CADENA DIAL span	Spanish		27.500	3/4	16	90
			(3) M-80	Spanish			3/4	16	90
			(4) RADIOLE	Spanish			3/4	16	90
			(5) SINFO RADIO	Spanish			3/4	16	90
AUDIOMANIA	11.538,00	H	(1) POP-ROCK	Spanish		27.500	3/4	16	90
MULTICLASSICA	11.538,00	H	Radio ?	Spanish		27.500	3/4	16	90

ORION 1, 37,5° W

Station	Frequency	Pol.	comments	Language	encr.	Symbol.	FEC	Tr.	Dish
?	11.543,00	H	feeds			6.000	3/4	90	
Channel On [@*]	11.605,00	H	feeds ?			18.898	7/8	90	
Performance the Arts.... [@*]	11.605,00	H	feeds			18.898	7/8	90	
Apna TV [@*]	11.605,00	H	feeds			18.898	7/8	90	
Weather Network [@*]	11.605,00	H	feeds	English		18.898	7/8	90	
NTL Orion Test Channel [@*]	11.605,00	H	feeds	English		18.898	7/8	90	
Channel On [@*]	11.625,00	V	?	Urdu		18.901	7/8	90	
Performance the Arts.... [@*]	11.625,00	V	?	English		18.901	7/8	90	
Apna TV [@*]	11.625,00	V	general prgr.	English	Irdeto	18.901	7/8	90	
Weather Network [@*]	11.625,00	V	weather channel			18.901	7/8	90	
NTL Orion Test Channel [@*]	11.625,00	V	news			18.901	7/8	90	
Service 1	11.680,00		?	?		6.000	3/4	90	

191

ITU REGION 1 DIGITAL STATIONS IN FREQUENCY ORDER

Station	Frequency	Pol.	comments	Language	encr.	Symbol.	FEC	Tr.	Dish
A3	12.746,00	V	feed for Astra	Hungarian		3.615	3/4	90	

Frequency = TV-Frequency in Mhz
Pol. = Polarisationsebene (H=Horizontal, V=Vertical, LHCP= Left Hand Circular Polarized)
@*† = reception only through PID codes
@**= reception only through manual freq. entry
Language = station language
encr. = Irdeto encrypted
Tr. = Transpondernummer
Dish= minimal dish size

CHAPTER 10
ITU REGION 2

TRANSPONDER LOADING

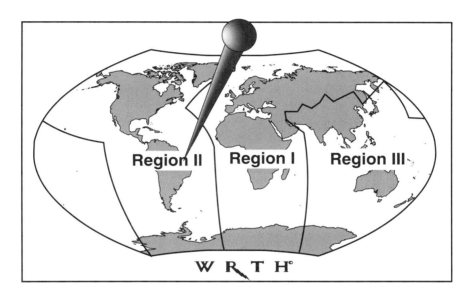

ITU REGION 2 STATIONS IN FREQUENCY ORDER

10.1. STATIONS IN FREQUENCY ORDER

Band	Freq.	Pol.	Tr.	Beam	Sce.	Program	Country	Language	System	Encr.	TT	A.M.	A.S.
INTELSAT 707 1.0° WEST													
C-	4.175	RHCP	24	Global	TV	AFRTS (U.S.)	USA	English	powervu		No		
C-	4.175	RHCP	24	Global	R	AFRTS Radio	USA	English				7.41	digital
GORIZONT 26 11.0° WEST													
C-	3.675	RHCP	6	Global	TV	ORT TV-1	Russia	Russian	SECAM		No	7.00	
C-	3.825	RHCP	6	Global	O	EBU Moscow	Russia	Russian	SECAM		No	7.00	
EXPRESS 2 14.0° WEST													
C-	3.675	RHCP	1	Global	TV	ORT TV-1	Russia	Russian	SECAM		No	7.00	
C-	3.675	RHCP	6	Global	R	Radio Rossija	Russia	Russian				7.50	
C-	3.825	RHCP	9	Global	TV	Muslim TV	Russia	Russian	PAL		No	6.60	
C-	3.825	RHCP	9	Global	TV	APNA TV	India	Hindi	PAL		No	6.60	
C-	3.825	RHCP	9	Global	TV	Canal France Int.	France	French	PAL		No	6.60	
C-	4.025	RHCP	15	Global	TV	RTP Internacional	Portugal	Portuguese	SECAM		No	6.60	
C-	4.025	RHCP	15	Global	R	RTP foreign sce.	Portugal	Portuguese					7.02/7.74
C-	4.025	RHCP	15	Global	R	Radio Portugal	Portugal	Portuguese					7.38/7.56
C-	4.125	RHCP	17	Global	TV	Channel 1 Moscow	Russia	Russian	SECAM		No	7.0	
C-	4.125	RHCP	17	Global	R	Voice of Russia World Service	C.I.S.	Russian				7.5	
INTELSAT K 21.5° WEST													
Ku1-	11.485	Hor	5	NA Spot	O	Eurocast feeds	USA	English	D2-MAC				
Ku1-	11.545	Hor	6	NA Spot	O	Reuters feeds	Germany	German	NTSC			6.60	
Ku1-	11.605	Hor	7	N & Lat. Am. Spot	TV	Deutsche Welle TV (German)	Germany	various	PAL		Yes	7.02	
Ku1-	11.605	Hor	7	N & Lat. Am. Spot	TV	Deutsche Welle TV (foreign)	Germany	various	PAL		Yes	7.20	
Ku1-	11.605	Hor	7	N & Lat. Am. Spot	R	Deutsche Welle	Germany	various				7.38	
Ku1-	11.605	Hor	7	N & Lat. Am. Spot	R	Deutsche Welle	Germany	various				7.56	
Ku1-	11.605	Hor	7	N & Lat. Am. Spot	R	Swiss Radio International	Switzerland	various				7.74	
Ku1-	11.665	Hor	8	NA Spot	O	AP feeds	USA	English				8.10	
Ku2-	11.735	Hor	1	NA Spot	TV	Europlus/Teleplus	Italy	Italian/German	D2-MAC	EuroCrypt	Yes	6.60	ch 1
Ku1-	11.795	Hor	2	NA Spot	O	Reuters news feeds	USA	English	PAL			6.60	
Ku1-	11.855	Hor	3	NA Spot	O	Reuters news feeds	USA	English	PAL			6.60	
Ku1-	11.915	Hor	4	NA Spot	TV	UAE TV Dubai	UAE	Arabic	NTSC		No	6.60	
Ku1-	11.915	Hor	4	NA Spot	R	Radio Otto	Italy	Italian				6.20	
Ku1-	11.915	Hor	4	NA Spot	R	Radio Dubai	UAE	Arabic				7.42	
Ku1-	11.915	Hor	4	NA Spot	R	Italian Radio	Italy	Italian				8.00	
INTELSAT 605 24.5° WEST													
C-	4.166	RHCP	23	Global	O	occasional video	USA		PAL/SEC/NTSC			6.60	
C-	4.188	RHCP	24	Global	O	occasional video	USA		PAL/SEC/NTSC			6.60	

ITU REGION 2 STATIONS IN FREQUENCY ORDER

Band	Freq.	Pol.	Tr.	Beam	Sce.	Program	Country	Language	System	Encr.	TT	A.M.	A.S.
						INTELSAT 601 27.5° WEST							
C-	3.761	LHCP	3	NW Zone	TV	TF1	France	French	PAL		No	5.80	
C-	3.900	RHCP	10	West Hemisph.	TV	Cartoon Network	USA	English	B-MAC			6.60	
C-	3.920	RHCP	11	West Hemisph.	O	NBC feeds	UK	English	NTSC			6.60	
C-	3.980	LHCP	14	West Hemisph.	O	ABC London	USA	English	NTSC			6.60	
C-	3.995	RHCP	15	West Hemisph.	TV	USIA WorldNet	USA	various	NTSC		No	6.80 / 5.94	
C-	3.995	RHCP	15	West Hemisph.	R	VOA	USA	various				5.90/6.10	
C-	3.995	RHCP	15	West Hemisph.	R	VOA	USA	various				7.32	
C-	4.015	LHCP	16	NW Zone	O	CBS London / Brigtstar	UK	English	PAL			6.60	
C-	4.015	LHCP	16	NW Zone	O	BBC WSTV	UK	English	PAL			6.60	
C-	4.055	LHCP	18	Global	O	CNN London	USA	English	PAL			6.60	
						HISPASAT 30.0° WEST							
Ku2-	12.015	RHCP	15	Spot	TV	Hispavision	Spain	Spanish	NTSC		No	5.8	
Ku2-	12.015	RHCP	15	Spot	R	Radio Nacional de España	Spain	Spanish				7.38	
Ku2-	12.015	RHCP	15	Spot	R	Radio Exterior	Spain	Spanish				7.56	
Ku2-	12.078	RHCP	19	Spot	TV	TVE internacional	Spain	Spanish	NTSC		No	5.8	
						INTELSAT 603 4.5° WEST							
C-	4.166	RHCP	23	Global	O	occasional video	USA		PAL/SEC/NTSC			6.60	
C-	4.188	RHCP	24	Global	O	occasional video	USA		PAL/SEC/NTSC			6.60	
						ORION F1 37.5° WEST							
Ku1-	11.617	Ver	7	Spot	TV	Super Television Channel	(USA)		PAL/SEC/NTSC				6.60/7.20
Ku2-	11.800	Ver		Spot	O	feeds (Teleport)	Germany	Romanian					
						TDRSS 41.0° WEST							
C-	3.720	Hor	1	Spot	TV	SSVC	UK	English	PAL	Cryptovision	No	5.80/7.20	
C-	3.720	Hor	1	Spot	R	BFBS 2 Radio	UK	English				6.12	
C-	3.720	Hor	1	Spot	R	BFBS 3 Radio	UK	English				6.30	
C-	3.720	Hor	1	Spot	R	BFBS 1 Radio	UK	English				7.02	
C-	3.720	Hor	1	Spot	R	BBC WS (Ukrainian)	UK	Ukrainian				7.20	
C-	3.720	Hor	1	Spot	R	BBC WS (Europe)	UK	various				7.56	
C-	3.720	Hor	1	Spot	R	BBC WS (foreign)	UK	various				7.74	
C-	3.720	Hor	1	Spot	R	BBC WS (foreign)	UK	various				7.92	
C-	3.720	Hor	1	Spot	R	BBC WS (foreign)	UK	various				8.10	
C-	3.800	Hor	3	Spot	TV	occasional video	Mexico	Spanish	NTSC			6.2 / 6.80	
C-	3.880	Hor	9	Spot	TV	C-Span	USA	English	PAL		No	6.8	
C-	3.920	Hor	11	Spot	TV	America-1	USA	English	NTSC		No	6.2	
C-	3.960	Hor	6	Spot	O	occasional video	Mexico	Spanish				6.2 / 6.80	
C-	3.960	Hor	7	Spot	O	Chalfont Teleport	USA	English				6.2/6.8	
C-	4.040	Hor	9	Spot	O	occasional video	Mexico	Spanish				6.2/6.8	
C-	4.040	Hor	7	Spot	TV	NASA Select TV	USA	English	NTSC			6.2 / 6.80	
C-	4.080	Hor	10	Spot	O	occasional video	Mexico	Spanish				6.2 / 6.80	
C-	4.120	Hor	11	Spot	O	occasional video	Mexico	Spanish	NTSC			6.2 / 6.80	

ITU REGION 2 STATIONS IN FREQUENCY ORDER

Band	Freq.	Pol.	Tr.	Beam	Sce.	Program	Country	Language	System	Encr.	TT	A.M.	A.S.
PANAMSAT F1 45.0° WEST													
C-	3.720	Hor	1	LA Spot	TV	CNN Int.	USA	English	B-MAC		Yes	ch1	
C-	3.730	Hor	2	North Beam	O	Omnivision			NTSC		No	6.8	
C-	3.750	Ver	2a	SA Spot	TV	TV Naçionale de Chile (ch 7)	Chile	Spanish	NTSC	Leitch			
C-	3.750	Hor	7b	North Beam	O	future video							
C-	3.750	Hor	3	North Beam	O	Omnivision			NTSC			6.2	
C-	3.769	Ver	2b	SA Spot	TV	TV Naçionale de Chile (ch 10)	Chile	Spanish	NTSC		No		
C-	3.771	Hor	8a	CA Spot	TV	Peru TV Ch. 5	Peru	Spanish	NTSC	Leitch	No	6.8	
C-	3.789	Hor	8b	CA Spot	TV	Peru TV Ch. 4	Peru	Spanish	NTSC		No	6.8	
C-	3.798	Ver	5	LA Spot	TV	TNT Internaçional	Turkey	Spanish	B-MAC		No	ch1	
C-	3.840	Ver	4	SA Spot	TV	HBO/Cinemax	USA	Spanish	digital		No		
C-	3.851	Hor	10a	CA Spot	TV	Peru TV Ch. 13	Peru	Spanish	NTSC		No	6.8	
C-	3.871	Ver	10b	CA Spot	TV	Peru TV Ch. 2	Peru	Spanish	NTSC		No	6.8	
C-	3.883	Ver	5	LA Spot	TV	ECO-Televisa	Mexico	Spanish	digital		No		
C-	3.900	Hor	11	NA Spot	TV	HBO/Cinemax	USA	Spanish	digital		No		
C-	3.910	Hor	6a	SA Spot	TV	Telefe	Argentina	Spanish	PAL		No	6.8	
C-	3.929	Ver	6b	SA Spot	TV	Space	Argentina	Spanish	PAL		No	6.8	
C-	3.940	Hor	12	CA Spot	TV	Peru TV	Argentina	Spanish	PAL		No	6.8	
C-	3.998	Hor	13b	CA Spot	O	RAI/WTN/VisNews	UK/Italy	various	PAL/SEC/NTSC		No	6.8	
C-	4.119	Ver	15a	CA Spot	TV	ESPN International	USA	Spanish	B-MAC				
C-	4.140	Hor	18a	CA Spot	TV	Canal Sur	Peru	Spanish	digital		No		
Ku2-	11.724	Hor	22a	NA Conus	O	Fuji TV / WTN London	Japan	Japanese	NTSC		No	6.2/6.8	
Ku2-	11.760	Hor	22b	NA Conus	TV	RTV Beograd	Croatia	Croat	NTSC		Yes	6.2	
Ku2-	11.760	Hor	22b	NA Conus	R	Radio Beograd 1	Croatia	Croat				6.8	
Ku2-	11.808	Hor	23a	NA Conus	TV	BBC World Service	UK	Eng/Span/Por	SCPC				
INTELSAT 709 50.0° WEST													
C-	3.820	RHCP	3		O	feeds	USA						
C-	3.920	RHCP	1		O	feeds	USA						
C-	3.962	RHCP	7	Zone	TV	RCTV (Venezuela 7)	Venezuela	Spanish	NTSC		No	6.80	
C-	4.143	RHCP	11	Zone	TV	Television Boliviana	Bolivia	Spanish	NTSC		No	6.80	
INTELSAT 706 53.0° WEST													
C-	3.985	RHCP	7	Zone	TV	BBC World	UK	English	NTSC				
C-	4.015	RHCP	8	Zone	O	CBC London feeds	UK	English	NTSC		No		
C-	4.160	RHCP	12	Zone	TV	Argentina TV a Color	UK	English	PAL				
SATCOM SN 2 69.0° WEST													
C-	3.720	Ver	1	Conus	TV	SportsChannel New York	USA	English	NTSC	VideoCipher II+	No		
C-	3.720	Ver	1	Conus	R	Superstation WLIR 1300 AM NY	USA	English				7.60	
C-	3.740	Ver	13L	Conus	TV	GEMS TV	USA	English	NTSC		No		
C-	3.760	Hor	2	Conus	TV	WorldNet	USA	English	NTSC	VideoCipher II+	No	6.8	6.8
C-	3.760	Hor	2	Conus	TV	Deutsche Welle TV	Germany	various	NTSC		No	6.8	6.8
C-	3.760	Hor	2	Conus	TV	United States Info. Agency	USA	English	NTSC		No	6.8	5.41
C-	3.760	Hor	2	Conus	R	VOA	USA	various				5.92	
C-	3.760	Hor	2	Conus	R	VOA	USA	various				6.12	

ITU REGION 2 STATIONS IN FREQUENCY ORDER

Band	Freq.	Pol.	Tr.	Beam	Sce.	Program	Country	Language	System	Encr.	TT	A.M.	A.S.
C-	3.760	Hor	2	Conus	R	VOA	USA	various				7.30	
C-	3.760	Hor	2	Conus	R	VOA	USA	various				7.56	
C-	3.760	Hor	2	Conus	R	VOA	USA	English				7.63	
C-	3.780	Ver	13U	Conus	TV	System United for Retransm.	Latin America	Spanish	NTSC		No	6.2/6.8	
C-	3.780	Ver	13U	Conus	R	La Voz de la Resistancia	US	Spanish				5.80	
C-	3.780	Ver	13U	Conus	R	Radio France Int. (RFI)	France	various				7.38	
C-	3.780	Ver	13U	Conus	R	WCMQ-FM Hialeah, FL	USA	Spanish					7.74/7.92
C-	3.800	Hor	3	Conus	R	NASA Engenering Channel	USA	English	NTSC			6.2/6.8	
C-	3.840	Hor	4	Conus	O	occasional video	USA		NTSC		No	6.8	
C-	3.880	Ver	5	Conus	TV	NASA Select Channel	USA	English				6.8	
C-	3.920	Hor	6	Conus	TV	SportsChannel Philadelphia	US	English	NTSC	VideoCipher II+		7.60	
C-	3.920	Hor	6	Conus	R	Radio Tropical	Haiti	Haitian Creole				No	6.8
C-	3.920	Ver	6	Conus	R	WNWK Newark, NY	US	various				8.30	
C-	4.000	Ver	8	Conus	O	occasional video	USA	English	NTSC			6.2/6.8	
C-	4.040	Ver	9	Conus	O	occasional video	USA	English	NTSC			6.2/6.8	
C-	4.100	Ver	17U	Conus	TV	AFRTS	USA	English	B-MAC		No		
C-	4.100	Ver	17U	Conus	R	AFRTS Radio Sce.	USA	English				7.35	
C-	4.120	Ver	11	Conus	TV	SportsChannel New England	USA	English	NTSC	VideoCipher II+		No	
C-	4.120	Hor	11	Conus	R	Christian Music Sce.	USA	English	NTSC	VideoCipher II+		No	
C-	4.140	Ver	18L	Conus	TV	Newsport	US	English	NTSC	VideoCipher II+		No	
C-	4.180	Ver	18U	Conus	TV	SportsChannel New York Plus	USA		NTSC	Leitch	Yes	6.2/6.8	
Ku2-	11.900	Ver	20L	Conus	O	CNN/TV ASAHI feeds	USA		NTSC			6.8	5.58
Ku2-	12.060	Ver	23	Conus	TV	The Kentucky Netw.	USA	English	NTSC			6.8	6.27.6
Ku2-	12.140	Hor	24U	Conus	O	occasional video	USA						5.58

GALAXY 6 74.0° WEST

Band	Freq.	Pol.	Tr.	Beam	Sce.	Program	Country	Language	System	Encr.	TT	A.M.	A.S.
C-	3.720	Hor	1	Conus	O	horse racing	USA	English	B-MAC				
C-	3.760	Hor	3	Conus	R	digital radio	USA		SCPC				
C-	3.780	Ver	4	Conus	O	Sports/dogs/horse racing	USA		NTSC/B-MAC				
C-	3.800	Hor	5	Conus	TV	NHK Japan	Japan	Japanese	NTSC	VideoCipher I		6.2/6.8	
C-	3.820	Ver	6	Conus	TV	TV-Japan	Japan	Japanese	NTSC			6.2/6.8	
C-	3.820	Hor	6	Conus	TV	TV-Japan	Japan	English	NTSC			6.2	
C-	3.840	Ver	7	Conus	O	Sports/dogs/horse racing	USA		B-MAC			6.8	
C-	3.860	Hor	8	Conus	O	Sports/dogs/horse racing	USA		B-MAC				
C-	3.880	Ver	9	Conus	TV	MuchMusic	Canada	English	NTSC	VideoCipher II+		No	
C-	3.900	Hor	10	Conus	TV	Arab Netw. America (ANA)	USA	Arabic	NTSC			6.2/6.8	
C-	3.900	Ver	10	Conus	R	ANA Radio Netw.	USA	Arabic				5.80	
C-	3.900	Hor	10	Conus	R	WNTL-AM Indian Head	USA	English				5.80	
C-	3.920	Ver	11	Conus	O	Sports/dogs/horse racing	USA		B-MAC				
C-	3.940	Ver	12	Conus	TV	TV Asia	USA		NTSC	VideoCipher II+		No	
C-	3.960	Hor	13	Conus	TV	Independent Film Channel	USA	English	NTSC	VideoCipher II+		No	
C-	3.980	Ver	14	Conus	TV	Cornerstone TV	USA	English	NTSC			6.2/6.8	
C-	3.980	Hor	14	Conus	R	KHNC-AM	USA	English				5.80	
C-	3.980	Ver	14	Conus	R	Talk Radio feeds	USA	English				7.58	
C-	4.000	Hor	15	Conus	TV	Midwest Sports Channel	USA	English	NTSC	VCII+		No	
C-	4.000	Hor	15	Conus	R	WCCO-AM, Minneapolis	USA	English				6.2	

ITU REGION 2 STATIONS IN FREQUENCY ORDER

Band	Freq.	Pol.	Tr.	Beam	Sce.	Program	Country	Language	System	Encr.	TT	A.M.	A.S.
C-	4.040	Hor	17	Conus	O	Tokyo Broadcasting Sce.	USA		NTSC			6.2/6.8	
C-	4.060	Ver	18	Conus	TV	Merchandise and Entertainm. TV (MET)	USA	English	NTSC		No	6.2/6.8	
C-	4.080	Hor	19	Conus	TV	The University Netw.	USA	English	NTSC			6.2/6.8	
C-	4.100	Ver	20	Conus	O	CNN Headline News	USA		NTSC	VCII+			
C-	4.120	Hor	21	Conus	O	Sports/dogs/horse racing	USA		B-MAC/NTSC				
C-	4.140	Ver	22	Conus	O	Sports/dogs/horse racing	USA		B-MAC/NTSC				
C-	4.160	Hor	23	Conus	TV	Worship TV	USA	English	NTSC		No	6.2/6.8	
C-	4.180	Ver	24	Conus	O	Sports/dogs/horse racing	USA		NTSC			6.2/6.8	
SBS 6 74.0° WEST													
Ku2-	11.717	Hor	1	Conus	O	BBC 9pm News	UK	English	PAL				
Ku2-	11.774	Hor	3	Conus	O	Northfield Horse Racing	USA	English	NTSC			6.8	
Ku2-	11.963	Hor	11	Conus	O	Conus Comm.	USA	English	NTSC			6.2/6.8	
Ku2-	11.985	Ver	12	Conus	O	Conus Comm.	USA	English	NTSC			6.2/6.8	
Ku2-	12.019	Hor	13	Conus	O	Conus Comm.	USA	English	NTSC			6.2/6.8	
Ku2-	12.043	Ver	14	Conus	O	occasional video	USA	English	NTSC			6.2/6.8	
Ku2-	12.092	Ver	16	Conus	TV	MCET Educational Netw.	USA	English	NTSC		No	6.2/6.8	
Ku2-	12.110	Hor	17	Conus	O	occasional video	USA	English	NTSC		No	6.2/6.8	
Ku2-	12.174	Hor	18	Conus	O	occasional video	USA	English	NTSC			6.2/6.8	
SBS 4 77.0° WEST													
Ku2-	11.780	Hor	2	Conus	O	NBC Network feeds	USA	English	NTSC			6.2/6.8	
Ku2-	11.823	Hor	3	Conus	O	NBC Network feeds	USA	English	NTSC			6.2/6.8	
Ku2-	11.872	Hor	4	Conus	O	NBC Network feeds	USA	English	NTSC			6.2/6.8	
Ku2-	11.921	Hor	5	Conus	O	NBC Network feeds	USA	English	NTSC			6.2/6.8	
Ku2-	11.970	Hor	6	Conus	O	NBC Network feeds	USA	English	NTSC			6.2/6.8	
Ku2-	12.019	Hor	7	Conus	O	NBC Network feeds	USA	English	NTSC			6.2/6.8	
Ku2-	12.068	Hor	8	Conus	O	NBC Network feeds	USA	English	NTSC			6.2/6.8	
Ku2-	12.117	Hor	9	Conus	O	NBC Network feeds	USA	English	NTSC			6.2/6.8	
Ku2-	12.166	Hor	10	Conus	O	NBC Network feeds	USA	English	NTSC			6.2/6.8	
TELSTAR 302 85.0° WEST													
C-	3.720	Ver	1	Conus	O	Sports/dogs/horse racing	USA	English	B-MAC/NTSC			6.2/6.8	
C-	3.760	Ver	2	Conus	TV	Wholesale Shopping Netw.	USA	English	NTSC		No	6.2/6.8	
C-	3.800	Ver	3	Conus	TV	The Baseball Netw. (TBN)	USA	English	NTSC	Leitch		6.2/6.5	
C-	3.800	Ver	3	Conus	O	Syndication Show feeds	USA	English	NTSC			6.2/6.5	
C-	3.960	Ver	13	Conus	TV	FLIX movie services	USA	English	NTSC	VCII+	No		
C-	3.960	Hor	7	Conus	TV	Turner Vision Promo Sce.	USA	English	NTSC			6.2/6.8	
C-	4.000	Ver	15	Conus	TV	Exxxtreme/Climaxxx Promo Ch.	USA	English	NTSC	VCII+			
C-	4.020	Hor	16	Conus	O	Red Mile horse racing	USA	English	NTSC		No	6.2/6.8	
C-	4.040	Ver	17	Conus	TV	TV Erotica	USA	English	NTSC	VCII+	No		
C-	4.060	Hor	18	Conus	TV	Around the World after Dark	USA	English	NTSC	VCII+	No		
C-	4.060	Ver	18	Conus	O	TV Erotica feeds	USA	English	NTSC				
C-	4.080	Hor	19	Conus	O	Sports/dogs/horse racing	USA	English	B-MAC/NTSC			6.2/6.8	
C-	4.080	Ver	20	Conus	O	Sports/dogs/horse racing	USA	English	B-MAC/NTSC				
C-	4.100	Hor	20	Conus	TV	La Cadena de Miagro	USA	Spanish	NTSC	Leitch	No	6.8	

ITU REGION 2 STATIONS IN FREQUENCY ORDER

Band	Freq.	Pol.	Tr.	Beam	Sce.	Program	Country	Language	System	Encr.	TT	A.M.	A.S.
C-	4.120	Ver	21	Conus	TV	Skyvision Home Shopping Ch.	USA	English	NTSC		No	6.2/6.8	
C-	4.120	Ver	21	Conus	R	Tech Talk Radio	USA	English	NTSC			5.80	
C-	4.140	Hor	22	Conus	O	occasional video	USA					6.2/6.8	

GE AMERICOM K2 85.0° WEST

Band	Freq.	Pol.	Tr.	Beam	Sce.	Program	Country	Language	System	Encr.	TT	A.M.	A.S.
Ku2-	11.729	Ver	1	Conus	TV	Primestar DTH	USA	English	Digicipher	USA	No		
Ku2-	11.758	Ver	2	Conus	TV	Primestar DTH	USA	English	Digicipher	USA	No		
Ku2-	11.788	Hor	3	Conus	TV	Primestar DTH	USA	English	Digicipher	USA	No		
Ku2-	11.817	Ver	4	Conus	TV	Primestar DTH	USA	English	Digicipher	USA	No		
Ku2-	11.847	Hor	5	Conus	TV	Primestar DTH	USA	English	Digicipher	USA	No		
Ku2-	11.876	Ver	6	Conus	TV	Primestar DTH	USA	English	Digicipher	USA	No		
Ku2-	11.906	Hor	7	Conus	TV	Primestar DTH	USA	English	Digicipher	USA	No		
Ku2-	11.935	Ver	8	Conus	TV	Primestar DTH	USA	English	Digicipher	USA	No		
Ku2-	11.965	Hor	9	Conus	TV	Primestar DTH	USA	English	Digicipher	USA	No		
Ku2-	11.994	Ver	10	Conus	TV	Primestar DTH	USA	English	Digicipher	USA	No		
Ku2-	12.024	Hor	11	Conus	TV	Primestar DTH	USA	English	Digicipher	USA	No		
Ku2-	12.053	Ver	12	Conus	TV	Primestar DTH	USA	English	Digicipher	USA	No		
Ku2-	12.083	Hor	13	Conus	TV	Primestar DTH	USA	English	Digicipher	USA	No		
Ku2-	12.143	Ver	15	Conus	TV	Primestar DTH	USA	English	Digicipher	USA	No		
Ku2-	12.171	Hor	16	Conus	TV	Primestar DTH	USA	English	Digicipher	USA	No		

SATCOM SN3R 87.0° WEST

Band	Freq.	Pol.	Tr.	Beam	Sce.	Program	Country	Language	System	Encr.	TT	A.M.	A.S.
C-	3.720	Hor	1	Conus	R	various radio	USA	English	SCPC		No	6.8/6.12	5.76/5.94
C-	3.740	Ver	13L	Conus	TV	Nebraska Educational TV	USA	English	NTSC			6.48	
C-	3.740	Hor	13L	Conus	R	Nebraska Talking Book Netw.	USA	English					5.76/5.94
C-	3.740	Ver	13L	Conus	R	KUCV-FM	USA	English					
C-	3.760	Hor	2	Conus	TV	WSBK, Boston	USA	English	NTSC	VCII+	No	6.2	
C-	3.760	Ver	13U	Conus	R	WROL-AM	USA	Spanish					
C-	3.780	Hor	13U	Conus	TV	KUCV-FM	USA	English	NTSC	VCII+	No	6.8/6.12	5.76/5.94
C-	3.780	Ver	13U	Conus	R	Univision	USA	Spanish					
C-	3.800	Hor	3	Conus	TV	Yesterday USA	USA	English				6.8	
C-	3.800	Ver	3	Conus	R	USA Radio Netw.	USA	English	NTSC	VCII+	No	6.8	5.76/6.12
C-	3.800	Hor	3	Conus	TV	WPIX, New York	USA	English					
C-	3.880	Hor	5	Conus	R	CNN Radio Netw.	USA	English					
C-	3.880	Hor	5	Conus	R	Unistar	USA	English					
C-	3.880	Hor	5	Conus	R	American Urban Radio Netw.	USA	English				5.62	
C-	3.880	Hor	5	Conus	O	CNN Contract Channel	USA	English	NTSC		No		5.76/5.94
C-	3.920	Hor	6	Conus	R	various radio	USA	English	SCPC				6.30/6.48
C-	3.960	Hor	7	Conus	TV	CNN	USA	English	SCPC			6.8/6.2	
C-	3.980	Hor	16L	Conus	TV	KTLA, Los Angeles	USA	English	NTSC	Leitch			
C-	4.000	Hor	8	Conus	R	Ambassador Insp. Radio	USA	English	SCPC	VCII+	No	5.96	
C-	4.000	Hor	8	Conus	R	Radio Sedeye Iran	USA	Farsi				6.20	
C-	4.000	Hor	8	Conus	R	Ambassador Insp. Radio	USA	English				6.48	
C-	4.020	Ver	16U	Conus	TV	CNN International	USA	English	NTSC	Leitch			
C-	4.040	Hor	9	Conus	R	various radio	USA	English	SCPC				

ITU REGION 2 STATIONS IN FREQUENCY ORDER

Band	Freq.	Pol.	Tr.	Beam	Sce.	Program	Country	Language	System	Encr.	TT	A.M.	A.S.
C-	4.060	Ver	17L	Conus	TV	Sellevision	USA	English	NTSC		No	5.04	
C-	4.060	Ver	17L	Conus	R	EZ Listening	USA	English				5.22	
C-	4.060	Ver	17L	Conus	R	commercials	USA	English				5.40	
C-	4.060	Ver	17L	Conus	R	EZ Listening	USA	English				5.55	
C-	4.060	Ver	17L	Conus	R	commercials	USA	English				5.76	
C-	4.060	Ver	17L	Conus	R	Safeway Markets	USA	English				5.94	
C-	4.060	Ver	17L	Conus	R	commercials	USA	English				6.12	
C-	4.060	Ver	17L	Conus	R	C&W	USA	English				6.30	
C-	4.060	Ver	17L	Conus	R	commercials	USA	English				6.48	
C-	4.060	Ver	17L	Conus	R	Soft music	USA	English				7.22	
C-	4.060	Ver	17L	Conus	R	Soft music	USA	English				7.40	
C-	4.060	Ver	17L	Conus	R	EZ Listening	USA	English				7.58	
C-	4.060	Ver	17L	Conus	R	Spanish sce.	USA	Spanish				7.76	
C-	4.060	Ver	17L	Conus	R	commercials	USA	English					
C-	4.080	Hor	10	Conus	TV	Sport South	USA	English	NTSC	VCII+	No	6.8	
C-	4.120	Hor	11	Conus	TV	Pro-Am sports, Detroit	USA	English	NTSC	VCII+	No	6.8/6.2	
C-	4.160	Hor	12	Conus	TV	Home Team sports, Baltimore	USA	English	NTSC	VCII+	No	6.8/6.2	
C-	4.180	Hor	18U	Conus	TV	American One	USA	English	NTSC		No		
Ku2-	12.060	Hor	23	West Spot	TV	Oragon Ed Net	USA	English	NTSC		No		
Ku2-	12.140	Hor	24	East Spot	O	NYNet	USA	English	NTSC		No		

TELSTAR 402 89.0° WEST

Band	Freq.	Pol.	Tr.	Beam	Sce.	Program	Country	Language	System	Encr.	TT	A.M.	A.S.
C-	3.760	Ver	3	Spot	TV	The Babe Network	USA	English	NTSC		No	6.8	
C-	3.780	Hor	4	Spot	TV	Shop at Home	USA	English	NTSC		No	6.8	
C-	3.800	Ver	5	Spot	O	Fox Network feeds	USA	English	NTSC		No	6.2/6.8	
C-	3.820	Hor	6	Spot	TV	The X Channel	USA	English	NTSC	VCII+	No		
C-	3.860	Ver	8	Spot	TV	TV 69	USA	English	NTSC	VCII+	No		
C-	3.860	Ver	8	Spot	TV	Fantasy Cafe TV	USA	English	NTSC	VCII+	No		
C-	3.860	Ver	8	Spot	O	Sunday football feeds	USA	English	NTSC	Leitch	No		
C-	3.900	Ver	10	Spot	TV	XXXPlore	USA	English	NTSC	VCII+	No		
C-	3.920	Hor	11	Spot	TV	The Outdoor Channel	USA	English	NTSC		No		6.2/6.8
C-	3.920	Ver	11	Spot	R	Yesterday USA	USA	English	NTSC		No		5.8
C-	3.940	Ver	12	Spot	TV	American Collectibles Netw.	USA	English	NTSC				6.2/6.8
C-	3.960	Hor	13	Spot	O	Fox Network feeds	USA	English	NTSC				6.2/6.8
C-	3.980	Ver	14	Spot	O	ABC feeds	USA	English	NTSC				6.2/6.8
C-	4.020	Hor	16	Spot	TV	Eurotica	USA	English	NTSC	VCII+	No		
C-	4.040	Ver	17	Spot	O	Fox Network feeds	USA	English	NTSC		No		
C-	4.060	Hor	18	Spot	TV	HRT Croatia	Croatia	Croatian	NTSC		No	5.8	
C-	4.060	Hor	18	Spot	TV	Rete Otto	Italy	Italian	NTSC		No	7.78	
C-	4.060	Hor	18	Spot	R	Antenna Greece	Greece	Greek	NTSC		No		
C-	4.080	Ver	19	Spot	R	Russian TV Network	USA	Russian	NTSC		No		6.2/6.8
C-	4.080	Ver	19	Spot	TV	Nat. Jewish TV Network	USA	Hebrew	NTSC		No		6.2/6.8
C-	4.100	Hor	20	Spot	TV	GOP TV	USA	English	NTSC		No		6.2/6.8
C-	4.100	Hor	20	Spot	O	Superior Lifestock Auction	USA	English	NTSC		No		6.2/6.8
C-	4.120	Ver	21	Spot	O	feeds	USA	English	NTSC		No		6.2/6.8
C-	4.140	Ver	22	Spot	O	feeds	USA	English	NTSC		No		6.2/6.8

ITU REGION 2 STATIONS IN FREQUENCY ORDER

Band	Freq.	Pol.	Tr.	Beam	Sce.	Program	Country	Language	System	Encr.	TT	A.M.	A.S.
C-	4.160	Hor	23	Spot	TV	La Cadena de Milagro	USA	Spanish	NTSC		No		6.2/6.8
C-	4.180	Ver	24	Spot	TV	Panda America (Home Shopping)	USA	English	NTSC		No		6.2/6.8
Ku2-	11.984	Hor	10	Spot	TV	Alphastar DBS System	USA	English	COMPR.		No		
Ku2-	12.033	Ver	11	Spot	TV	Alphastar DBS System	USA	English	COMPR.		No		
Ku2-	12.046	Hor	12	Spot	TV	Alphastar DBS System	USA	English	COMPR.		No		
Ku2-	12.095	Ver	13	Spot	TV	Alphastar DBS System	USA	English	COMPR.		No		
Ku2-	12.108	Hor	14	Spot	TV	Alphastar DBS System	USA	English	COMPR.		No		
Ku2-	12.170	Hor	16	Spot	TV	Alphastar DBS System	USA	English	COMPR.		No		

GALAXY 7 91.0° WEST

Band	Freq.	Pol.	Tr.	Beam	Sce.	Program	Country	Language	System	Encr.	TT	A.M.	A.S.
C-	3.740	Ver	2	Conus	O	CBS Contract Channel	USA	English	NTSC			6.2/6.8	
C-	3.760	Ver	2	Conus	TV	Action Pay-per-View	USA	English/Spanish	NTSC	VCII+	No	6.2/6.8	
C-	3.780	Ver	4	Conus	O	occasional video	USA	English	NTSC			6.2/6.8	
C-	3.800	Hor	5	Conus	O	occasional video	USA	English	NTSC			6.2/6.8	
C-	3.820	Hor	6	Conus	TV	Game Show Network	USA	English	NTSC	VCII+			
C-	3.840	Ver	7	Conus	TV	The Golf Channel	UK	English	NTSC	VCII+			
C-	3.840	Hor	7	Conus	TV	BBC Breakfast News	UK	English	NTSC			6.8	
C-	3.860	Ver	8	Conus	TV	HBO II (East)	USA	English	NTSC	VCII+			
C-	3.880	Hor	9	Conus	TV	RAI	Italy	Italian	NTSC			6.8	
C-	3.900	Ver	10	Conus	TV	United Arab Emirates TV (UAE)	UAE	Arabic	NTSC		No	6.2/6.8	
C-	3.900	Ver	10	Conus	R	Radio Maria	UAE	Arabic				5.80	
C-	3.900	Hor	10	Conus	R	Radio Dubai	UAE	Arabic				7.48	
C-	3.900	Ver	10	Conus	R	Religious Music	USA	English				8.03	
C-	3.920	Hor	11	Conus	TV	Estacion Montello	USA	Spanish	NTSC			6.2/6.8	
C-	3.940	Ver	12	Conus	TV	The International Channel	USA	English	NTSC	VCII+			
C-	3.980	Ver	14	Conus	TV	HBO II (West)	UAE	Arabic	NTSC	VCII+			
C-	4.000	Hor	15	Conus	TV	TCI Preview Channel UA	USA	English	NTSC	VCII+			
C-	4.040	Ver	17	Conus	TV	Via TV (interactive TV)	USA	English	NTSC				
C-	4.060	Hor	18	Conus	TV	Turner Vision Promo Channel	USA	English	NTSC				
C-	4.060	Ver	18	Conus	TV	SkyVision Promo Channel	USA	English	NTSC				
C-	4.060	Ver	18	Conus	O	CBS Contract Channel	USA	English	NTSC				
C-	4.060	Hor	18	Conus	O	BBC Breakfast News	USA	English	NTSC	VCII+			
C-	4.080	Hor	19	Conus	R	WCBS-AM, Newsradio 88, NY	USA	English				7.38	
C-	4.080	Ver	19	Conus	O	CBS East	USA	English	NTSC		No	6.2/6.8	
C-	4.100	Ver	20	Conus	TV	National Empowerment TV Netw.	USA	English	NTSC		No	6.2/6.8	
C-	4.120	Hor	21	Conus	TV	La Cadena di Milagro	USA	Spanish	NTSC			6.8	
C-	4.140	Ver	22	Conus	TV	NewsTalk Television	USA	English	NTSC		No	6.2/6.8	
C-	4.160	Hor	23	Conus	TV	fX Movies	USA	English	NTSC				
C-	4.180	Ver	24	Conus	TV	HBO III (East)	USA	English	NTSC				
Ku2-	11.720	Ver	1K	Conus	TV	G.O.P. TV	USA	English	NTSC		No	6.8/6.2	
Ku2-	11.810	Hor	6	Conus	TV	Classic Sports Network	USA	English	NTSC		No	6.8/6.2	
Ku2-	11.900	Hor	10K	Conus	TV	Hospitality TV	USA	English	B-MAC				
Ku2-	11.900	Ver	10K	Conus	TV	The People's Network	USA	English	NTSC		No	6.8/6.2	
Ku2-	11.930	Hor	15K	Conus	O	Asian American Satellite Netw. (TAN)	USA	Korean	NTSC				
Ku2-	11.990	Hor	15K	Conus	TV	Muslim Television	USA	English	NTSC		No	6.8/6.2	
Ku2-	11.990	Ver	15	Conus	O	The Asian Netw. (TAN)	USA	Korean	NTSC				

ITU REGION 2 STATIONS IN FREQUENCY ORDER

Band	Freq.	Pol.	Tr.	Beam	Sce.	Program	Country	Language	System	Encr.	TT	A.M.	A.S.
Ku2-	12.020	Ver	16	Conus	TV	Quorum multi-level marketing	USA	Korean	NTSC		No	6.8/6.2	
Ku2-	12.065	Hor	17K	Conus	TV	Automotive Sat. Training Netw.	USA	English	B-MAC	USA	No	6.2/6.8	
Ku2-	12.065	Hor	17K	Conus	TV	National Weather Netw.	USA	English	NTSC		No	6.8	
Ku2-	12.065	Hor	17K	Conus	TV	Airport News and Transportation Netw.	USA	English	NTSC		No	6.2/6.8	
Ku2-	12.080	Ver	19K	Conus	TV	The People's Network	USA	English	NTSC		No		
Ku2-	12.110	Ver	21	Conus	TV	TCI TV	USA	English	B-MAC		No		
Ku2-	12.140	Ver	22	Conus	TV	Real Estate TV Network	USA	English	NTSC		No	6.2/6.8	

GALAXY 3 93.5° WEST

Band	Freq.	Pol.	Tr.	Beam	Sce.	Program	Country	Language	System	Encr.	TT	A.M.	A.S.
C-	3.720	Hor	1	Conus	TV	TVN Theatre 1	USA	English	NTSC	VCII+	No		
C-	3.740	Ver	2	Conus	TV	TVN Theatre 2	USA	English	NTSC	VCII+	No		
C-	3.760	Hor	3	Conus	TV	TVN Theatre 3	USA	English	NTSC	VCII+	No		
C-	3.780	Ver	4	Conus	TV	TVN Theatre 4	USA	English	NTSC	VCII+	No		
C-	3.800	Hor	5	Conus	TV	TVN Theatre 5	USA	English	NTSC	VCII+	No		
C-	3.820	Ver	6	Conus	TV	TVN Theatre 6	USA	English	NTSC	VCII+	No		
C-	3.840	Hor	7	Conus	TV	TVN Theatre 7	USA	English	NTSC	VCII+	No		
C-	3.860	Ver	8	Conus	TV	TVN Theatre 8	USA	English	NTSC	VCII+	No		
C-	3.880	Hor	9	Conus	TV	TVN Theatre 9	USA	English	NTSC		No		
C-	3.900	Ver	10	Conus	TV	TVN Theatre 10	USA	English	NTSC		No	6.8	
C-	3.920	Hor	11	Conus	TV	America's Collectibles Netw.	USA	English	NTSC		No	6.8	
C-	3.940	Ver	12	Conus	TV	RAI	Italy	Italian	NTSC		No		
C-	4.000	Hor	15	Conus	TV	Gospel Music TV	USA	English	NTSC	VCII+	No		
C-	4.020	Ver	16	Conus	TV	HBO East 2	USA	English	NTSC	VCII+	No		
C-	4.040	Hor	17	Conus	TV	Cinemax East 2	USA	English	NTSC	VCII+	No		6.2/6.8
C-	4.060	Ver	18	Conus	TV	Infomerica TV	USA	English	NTSC		No		
C-	4.080	Hor	19	Conus	TV	HBO West 3	USA	English	NTSC	VCII+	No		
C-	4.100	Ver	20	Conus	TV	Infomerica TV	USA	English	NTSC	VCII+	No	6.2/6.8	
C-	4.120	Hor	21	Conus	TV	Three Angels Broadcasting	USA	English	NTSC		No	6.8	
C-	4.160	Ver	23	Conus	0	sports/dogs/horse racing	USA	English	NTSC		No	6.2/6.8	
C-	4.180	Hor	24	Conus	TV	Galaxy Lat. America DBS	USA	various	various	various	No		
Ku2-	11.720	Hor	1	Spot	TV	Galaxy Lat. America DBS	USA	various	digital		No		
Ku2-	11.750	Ver	2	Spot	TV	Galaxy Lat. America DBS	USA	various	digital		No		
Ku2-	11.750	Hor	3	Spot	TV	Galaxy Lat. America DBS	USA	various	digital		No		
Ku2-	11.780	Hor	4	Spot	TV	Galaxy Lat. America DBS	USA	various	digital		No		
Ku2-	11.810	Ver	5	Spot	TV	Galaxy Lat. America DBS	USA	various	digital		No		
Ku2-	11.810	Hor	6	Spot	TV	Galaxy Lat. America DBS	USA	various	digital		No		
Ku2-	11.840	Hor	7	Spot	TV	Galaxy Lat. America DBS	USA	various	digital		No		
Ku2-	11.870	Ver	8	Spot	TV	Galaxy Lat. America DBS	USA	various	digital		No		
Ku2-	11.870	Hor	9	Spot	TV	Galaxy Lat. America DBS	USA	various	digital		No		
Ku2-	11.900	Hor	10	Spot	TV	Galaxy Lat. America DBS	USA	various	digital		No		
Ku2-	11.930	Ver	11	Spot	TV	Galaxy Lat. America DBS	USA	various	digital		No		
Ku2-	11.930	Hor	12	Spot	TV	Galaxy Lat. America DBS	USA	various	digital		No		
Ku2-	11.960	Hor	13	Spot	TV	Galaxy Lat. America DBS	USA	various	digital		No		
Ku2-	11.990	Ver	14	Spot	TV	Galaxy Lat. America DBS	USA	various	digital		No		
Ku2-	11.990	Hor	15	Spot	TV	Galaxy Lat. America DBS	USA	various	digital		No		
Ku2-	12.020	Hor	16	Spot	TV	Galaxy Lat. America DBS	USA	various	digital		No		

ITU REGION 2 STATIONS IN FREQUENCY ORDER

Band	Freq.	Pol.	Tr.	Beam	Sce.	Program	Country	Language	System	Encr.	TT	A.M.	A.S.
Ku2-	12.050	Ver	17	Spot	TV	Galaxy Lat. America DBS	USA	various	digital		No		
Ku2-	12.050	Hor	18	Spot	TV	Galaxy Lat. America DBS	USA	various	digital		No		
Ku2-	12.080	Ver	19	Spot	TV	Galaxy Lat. America DBS	USA	various	digital		No		
Ku2-	12.110	Hor	20	Spot	TV	Galaxy Lat. America DBS	USA	various	digital		No		
Ku2-	12.110	Ver	21	Spot	TV	Galaxy Lat. America DBS	USA	various	digital		No		
Ku2-	12.140	Hor	22	Spot	TV	Galaxy Lat. America DBS	USA	various	digital		No		
Ku2-	12.170	Ver	23	Spot	TV	Galaxy Lat. America DBS	USA	various	digital		No		
Ku2-	12.170	Hor	24	Spot	TV	Galaxy Lat. America DBS	USA	various	digital		No		

TELSTAR 401 97.0° WEST (temporarily out of service)

Band	Freq.	Pol.	Tr.	Beam	Sce.	Program	Country	Language	System	Encr.	TT	A.M.	A.S.
C-	3.720	Ver	1	Conus	TV	Canadian Exxtacy	Canada	English / French	NTSC	VCII+		6.2/6.8	
C-	3.720	Ver	2	Conus	TV	VTC Satellite Network	Canada	English / French	NTSC		No	6.2/6.8	
C-	3.760	Ver	3	Conus	O	Paramount / Keystone	USA	English	NTSC			6.2/6.8	
C-	3.780	Hor	4	Conus	O	Fox Netw. feeds	USA	English	NTSC	Leitch		6.2/6.8	
C-	3.800	Ver	5	Conus	O	Syndication show feeds	USA	English	NTSC			6.2/6.8	
C-	3.800	Ver	5	Conus	O	North Carolina Open Net	USA	English	NTSC			6.2/6.8	
C-	3.820	Ver	6	Conus	O	Buena Vista	USA	English	NTSC			6.2/6.8	
C-	3.840	Hor	7	Conus	TV	Fox Nwtwork feeds	USA	English	NTSC			6.2/6.8	
C-	3.860	Hor	8	Conus	TV	Public Broadcasting Sce. (PBS)	USA	English	NTSC	clear/Leitch	No		
C-	3.880	Ver	9	Conus	O	FOX Network feeds	USA	English	NTSC			6.2/6.8	
C-	3.900	Ver	10	Conus	O	FOX Network feeds	USA	English	NTSC			6.2/6.8	
C-	3.920	Hor	11	Conus	O	ABC feeds	USA	English	NTSC			6.2/6.8	
C-	3.940	Hor	12	Conus	O	ABC feeds	USA	English	NTSC			6.2/6.8	
C-	3.960	Ver	13	Conus	O	FOX (East)	USA	English	NTSC			6.8	
C-	3.980	Ver	14	Conus	O	FOX (West)	USA	English	NTSC	VCII+			
C-	4.000	Hor	15	Conus	TV	Exxtasy 2	Canada	English	NTSC	various		6.2/6.8	
C-	4.020	Hor	16	Conus	O	sports/dogs/horse racing	USA	English	various	various		6.2/6.8	
C-	4.040	Ver	17	Conus	O	sports/dogs/horse racing	USA	English	various	various			
C-	4.060	Ver	18	Conus	TV	United Paramount Network (UPN)	USA	English	NTSC				
C-	4.080	Hor	19	Conus	TV	The Baseball Network	USA	English	NTSC	Leitch	No	6.2/6.8	
C-	4.080	Hor	19	Conus	O	occasional video	USA	English	NTSC				
C-	4.100	Ver	20	Conus	O	ABC	USA	English	NTSC				
C-	4.120	Ver	21	Conus	O	ABC	USA	English	NTSC				
C-	4.140	Hor	22	Conus	O	ABC	USA	English	NTSC				
C-	4.160	Hor	23	Conus	O	NASA TV highlights	USA	English	NTSC				
C-	4.180	Ver	24	Conus	O	S. Carolina Educational TV	USA	English	NTSC				
C-	11.790	Ver	3A	Conus	TV	SRC Educational Netw.	USA	English	NTSC		No	6.2/6.8	
Ku2-	11.855	Hor	6	Conus	TV	PBS (schedule C)	USA	English	NTSC		No	6.2/6.8	
Ku2-	11.902	Ver	7	Conus	TV	PBS (schedule D)	USA	English	NTSC		No	6.2/6.8	
Ku2-	11.914	Hor	8	Conus	TV	Georgia Public TV (GPTV)	USA	English	NTSC		No	6.2/6.8	
Ku2-	12.092	Hor	14A	Conus	R	Georgia Radio Reading Sce. (GRRS)	USA	English	NTSC			5.76	
Ku2-	12.092	Hor	14A	Conus	R	Peach State Public Radio	USA	English	NTSC				5.40/5.58
Ku2-	12.123	Hor	14B	Conus	TV	Georgia Public TV (GPTV)	USA	English	NTSC		No	6.2/6.8	
Ku2-	12.157	Ver	15	Conus	O	ABC feeds	USA	English	NTSC		No	6.2/6.8	

ITU REGION 2 STATIONS IN FREQUENCY ORDER

Band	Freq.	Pol.	Tr.	Beam	Sce.	Program	Country	Language	System	Encr.	TT	A.M.	A.S.
Ku2-	12.175	Ver	16	Conus	O	ABC feeds	USA	English				6.2/6.8	
GALAXY 4 99.0° WEST													
C-	3.720	Hor	1	Conus	R	various radio	USA	English	SCPC				
C-	3.740	Ver	2	Conus	R	various radio	USA	English	SCPC				
C-	3.760	Hor	3	Conus	R	various radio	USA	English	SCPC				
C-	3.780	Ver	4	Conus	R	various radio	USA	English	SCPC				
C-	3.800	Hor	5	Conus	O	occasional video	USA	English	NTSC				
C-	3.820	Ver	5	Conus	TV	Shepherd's Chapel	USA	English	NTSC		No	6.8	
C-	3.820	Hor	6	Conus	R	WCRP, Guayama, Puerto Rico	Puerto Rico	Spanish				6.53	
C-	3.820	Ver	6	Conus	R	KBVA-FM, Bella Vista	Puerto Rico	Spanish					5.58/5.76
C-	3.820	Ver	6	Conus	R	Easy Listening	USA	English					6.20/7.78
C-	3.840	Hor	7	Conus	O	occasional video	USA	English	NTSC				
C-	3.880	Hor	9	Conus	O	occasional video	USA	English	NTSC				
C-	3.900	Ver	10	Conus	TV	WABC, New York	USA	English	NTSC	VCII+	No		
C-	3.920	Hor	11	Conus	O	occasional video	USA	English	NTSC				
C-	3.940	Ver	12	Conus	O	occasional video	USA	English	various	various			
C-	3.960	Hor	13	Conus	O	Syndication Show feeds	USA	Russian	NTSC			6.20	
C-	3.980	Ver	14	Conus	TV	WRAL, Raleigh, NC	USA	Spanish	NTSC	VCII+	No	7.42	6.2/6.8
C-	4.000	Hor	15	Conus	TV	World Harvest TV	USA	English	NTSC		No	7.46	6.2/6.8
C-	4.000	Hor	15	Conus	R	World Harvest Radio, Indiana	USA	English			No	7.55	
C-	4.000	Hor	15	Conus	R	World Harvest Radio, Indiana	USA	English				7.64	
C-	4.000	Hor	15	Conus	R	World Harvest Radio, Indiana	USA	English			No	7.70	
C-	4.000	Hor	15	Conus	R	World Harvest Radio (Shortwave)	USA	English					5.57/5.8
C-	4.000	Hor	15	Conus	R	WHME	USA	English	NTSC				
C-	4.020	Ver	16	Conus	O	occasional video	USA	English	NTSC				
C-	4.040	Hor	17	Conus	O	occasional video	USA	English	NTSC				
C-	4.060	Ver	18	Conus	O	occasional video	USA	English	NTSC				
C-	4.080	Hor	19	Conus	O	occasional video	USA	English	NTSC				
C-	4.100	Ver	20	Conus	R	WCBS-FM New York	USA	English	NTSC				
C-	4.100	Ver	20	Conus	O	occasional video	USA	English	NTSC				
C-	4.120	Hor	21	Conus	TV	WXIA, Atlanta	USA	English	NTSC	VCII+	No	7.40	
C-	4.140	Ver	22	Conus	O	occasional video	USA	English	NTSC				
C-	4.180	Ver	24	Conus	TV	Jong Ten	Taiwan	Taiwanese	NTSC				
Ku2-	11.840	Hor	7	Conus	TV	WMNB Russian Language Station	USA	Russian	NTSC (inverted video)				
Ku2-	11.930	Hor	12	Conus	O	CBS NewsNet	USA	English	SECAM			6.2/6.8	
Ku2-	12.032	Ver	17	Conus	TV	Hong Kong TVB Jade Channel	Hong Kong	English	NTSC		No	6.8/6.2	5.41
Ku2-	12.050	Hor	18	Conus	TV	The Filipino Channel	Philippines	English	NTSC	Oak Orion	No	6.8/6.2	
Ku2-	12.170	Hor	24	Conus							No	6.8/6.2	
SPACENET 4 101.0° WEST													
C-	3.720	Ver	1	Conus	TV	Encore 2	USA	English	NTSC	VCII+	No		
C-	3.740	Hor	7	Conus	TV	WHDH-TV, Boston, MA	USA	English	NTSC	VCII+	No		
C-	3.780	Hor	8	Conus	TV	WUSA-TV, Washington, DC	USA	English	NTSC	VCII+	No		
C-	3.780	Ver	4	Conus	TV	Westerns Encore 3	USA	English	NTSC	VCII+	No		

ITU REGION 2 STATIONS IN FREQUENCY ORDER

Band	Freq.	Pol.	Tr.	Beam	Sce.	Program	Country	Language	System	Encr.	TT	A.M.	A.S.
C-	3.820	Hor	9	Conus	TV	KNBC-TV, LA	USA	English	NTSC	VCII+	No		
C-	3.860	Hor	10	Conus	TV	KOMO-TV, Seattle	USA	English	NTSC	VCII+	No		
C-	3.860	Hor	8	Conus	TV	KOMO-TV Seattle	USA	English	NTSC	VCII+	No		
C-	3.900	Hor	11	Conus	TV	WFLD-TV, Chicago	USA	English	NTSC	VCII+	No		
C-	3.900	Ver	10	Conus	TV	FOXNet - Prime Time	USA	English	NTSC	VCII+	No		
C-	3.920	Ver	11	Conus	TV	Encore 8	USA	English	NTSC	VCII+	No		
C-	3.920	Hor	11	Conus	TV	Encore 8	USA	Spanish	NTSC		No		
C-	3.940	Hor	12	Conus	TV	HTV Hispanic music videos	USA	English	NTSC		No		6.2/6.8
C-	3.980	Hor	14	Conus	TV	Nat. Prgr. Sce. Preview channel	USA	English	NTSC		No		6.2/6.8
C-	4.060	Hor	18	Conus	TV	Encore 8	USA	English	NTSC	VCII+	No		
C-	4.180	Hor	24	Conus	TV	KPIX TV, San Francisco	USA	English	NTSC	VCII+	No		

DBS 1 101.0 WEST

Band	Freq.	Pol.	Tr.	Beam	Sce.	Program	Country	Language	System	Encr.	TT	A.M.	A.S.
Ku2-	12.2388	CP	2	Conus	TV	DIRECTV	USA	English	MPEG-2				
Ku2-	12.2677	CP	4	Conus	TV	DIRECTV	USA	English	MPEG-2				
Ku2-	12.2969	CP	6	Conus	TV	DIRECTV	USA	English	MPEG-2				
Ku2-	12.3260	CP	8	Conus	TV	DIRECTV	USA	English	MPEG-2				
Ku2-	12.3552	CP	10	Conus	TV	DIRECTV	USA	English	MPEG-2				
Ku2-	12.3843	CP	12	Conus	TV	DIRECTV	USA	English	MPEG-2				
Ku2-	12.4135	CP	14	Conus	TV	DIRECTV	USA	English	MPEG-2				
Ku2-	12.4427	CP	16	Conus	TV	DIRECTV	USA	English	MPEG-2				
Ku2-	12.4718	CP	18	Conus	TV	DIRECTV	USA	English	MPEG-2				
Ku2-	12.5010	CP	20	Conus	TV	DIRECTV	USA	English	MPEG-2				
Ku2-	12.5301	CP	22	Conus	TV	DIRECTV	USA	English	MPEG-2				
Ku2-	12.5593	CP	24	Conus	TV	USSB	USA	English	MPEG-2				
Ku2-	12.5885	CP	26	Conus	TV	USSB	USA	English	MPEG-2				
Ku2-	12.6176	CP	28	Conus	TV	USSB	USA	English	MPEG-2				
Ku2-	12.6478	CP	30	Conus	TV	USSB	USA	English	MPEG-2				
Ku2-	12.6759	CP	32	Conus	TV	USSB	USA	English	MPEG-2				

AMERICOM GE 1 103.0° WEST

Band	Freq.	Pol.	Tr.	Beam	Sce.	Program	Country	Language	System	Encr.	TT	A.M.	A.S.
C-	3.780	Ver	3	Spot	TV	Sportschannel Ohio	USA	English	NTSC	VCII+	No		
C-	3.820	Ver	6	Spot	TV	WNBC New York	USA	English	NTSC	VCII+	No	5.80	
C-	3.820	Ver	6	Spot	R	WCNJ FM, Hazlet, NY	USA	English					
C-	3.840	Hor	6	Spot	TV	Cornerstone TV	USA	English	NTSC		No	5.8	6.2/6.8
C-	3.840	Hor	6	Spot	R	American Freedom Netw.	USA	English	N			7.56	
C-	3.860	Ver	6	Spot	R	WW Freedom Radio Network.	USA	English	NTSC	VCII+	No		
C-	3.860	Hor	8	Spot	TV	Sportschannel Chicago	USA	English	NTSC	VCII+	No		
C-	3.880	Ver	9	Spot	TV	Sportsouth (regional)	USA	English	NTSC	VCII+	No		
C-	3.900	Hor	10	Spot	TV	WJLA, Washington	USA	English	NTSC	VCII+	No		
C-	3.980	Ver	14	Spot	TV	Sportschannel New England	USA	English	NTSC	VCII+	No		
C-	4.020	Hor	16	Spot	TV	Sportschannel Pacific	USA	English	NTSC	VCII+	No		
C-	4.040	Hor	17	Spot	TV	Sportschannel alternatives	USA	English	NTSC	VCII+	No		
C-	4.040	Hor	19	Spot	TV	Nat. Enpowerment TV Netw.	USA	English	NTSC		No		
C-	4.180	Ver	24	Spot	TV	WRAL, Raleigh, NC	USA	English	NTSC	VCII+	No		6.2/6.8
Ku2-	11.760	Hor	3	Spot	TV	NBC Contract Channel (Eastern)	USA	English	NTSC		No		6.2/6.8

ITU REGION 2 STATIONS IN FREQUENCY ORDER

Band	Freq.	Pol.	Tr.	Beam	Sce.	Program	Country	Language	System	Encr.	TT	A.M.	A.S.
Ku2-	11.840	Hor	7	Spot	TV	NBC Contract Channel (Pacific)	USA	English	NTSC		No		6.2/6.8
Ku2-	11.880	Hor	9	Spot	TV	NBC Contract Channel (Mountain)	USA	English	NTSC		No		6.2/6.8
Ku2-	12.000	Hor	15	Spot	TV	NBC Contract Channel	USA	English	NTSC		No		6.2/6.8
Ku2-	12.100	Ver	20	Spot	TV	CycleSat commercials	USA	English	NTSC		No		6.2/6.8
Ku2-	12.120	Hor	21	Spot	TV	NBC Contract Channel	USA	English	NTSC		No		6.2/6.8
Ku2-	12.140	Hor	22	Spot	TV	CCC (Chinese Comm. Channel)	USA	Chinese	NTSC		No		6.2/6.8
Ku2-	12.160	Hor	23	Spot	TV	NBC Contract Channel	USA	English	NTSC		No		6.2/6.8

GSTAR 4 105.0° WEST

Band	Freq.	Pol.	Tr.	Beam	Sce.	Program	Country	Language	System	Encr.	TT	A.M.	A.S.
Ku2-	11.852	Hor	3	Conus	TV	CNN Newsource	USA	English	NTSC	Leitch	No	6.8	6.8
Ku2-	11.974	Hor	5	Conus	O	occasional video	USA	English	NTSC			6.2/6.8	
Ku2-	12.035	Hor	14	Conus	O	occasional video	USA	English	NTSC	Leitch		6.2/6.8	
Ku2-	12.096	Hor	7	Conus	O	CNN Newsource	USA	English	NTSC			6.8	

ANIK E2 107.3° WEST

Band	Freq.	Pol.	Tr.	Beam	Sce.	Program	Country	Language	System	Encr.	TT	A.M.	A.S.
C-	3.720	Hor	1	Spot	TV	CBMT, Montreal/CBC	Canada	French	NTSC		No		
C-	3.720	Hor	1	Spot	R	CBC Radio affiliates	Canada	English				5.78	
C-	3.720	Hor	1	Spot	R	CBM AM	Canada	French				6.12	
C-	3.720	Hor	1	Spot	R	CBC Radio	Canada	French				7.38	5.38/5.58
C-	3.720	Hor	1	Spot	R	CBC Radio	Canada	English					
C-	3.820	Ver	7	Spot	TV	CBC Newsworld	Canada	English	NTSC		No	6.8	
C-	3.820	Ver	7	Spot	R	CBC Radio (Eastern)	Canada	English				5.40	
C-	3.820	Ver	7	Spot	R	CBC Radio (Atlantic)	Canada	English				5.58	
C-	3.820	Ver	7	Spot	R	Voice Print	Canada	English				7.44	
C-	3.820	Ver	7	Spot	R	CBC FM	Canada	English					5.78/5.96
C-	3.820	Ver	7	Spot	R	CBC Stereo Radio (Atlantic)	Canada	English					6.12/6.30
C-	3.820	Ver	7	Spot	O	CBCM occasional video	Canada	English	NTSC		No		
C-	3.840	Hor	9	Spot	TV	CBC North	Canada	English	NTSC		No		
C-	3.880	Hor	7	Spot	TV	CBCM	Canada	French	NTSC				6.2/6.8
C-	3.960	Hor	17	Spot	O	various feeds	Canada	French	NTSC				6.2/6.8
C-	4.040	Ver	18	Spot	TV	Video Catalog Channel	Canada	English	NTSC	VCII+			
C-	4.060	Ver	22	Spot	TV	Venus Adult	Canada	French	NTSC		No		
C-	4.140	Hor	23	Spot	O	CBC feeds	Canada	French	NTSC		No	6.8	6.2/6.8
C-	4.160	Ver	24	Spot	TV	CTV Television Netw.	Canada	French/English	NTSC		No		5.76/5.94
C-	4.180	Ver	24	Spot	TV	Ontario Legislature	Canada	French	NTSC		No	5.41/6.00	
Ku2-	11.939	Hor	24	Spot	TV	La Chaine French	Canada	French	NTSC		No	5.41	5.76/5.94
Ku2-	11.974	Hor	25	Spot	TV	TV Ontario	Canada	English	NTSC		No	5.41	
Ku2-	12.000	Hor	26	Spot	R	CJRT FM, Toronto	Canada	English	NTSC				5.76/5.94
Ku2-	12.000	Hor	26	Spot	R	Northern Native Radio	Canada	English	NTSC		No		6.43/6.53
Ku2-	12.096	Hor	29	Spot	TV	Atlantic Satellite Netw. Halifax	Canada	French	NTSC		No		5.76/5.94
C-	38.60	Ver	8	Spot	TV	Global TV	Canada		NTSC	Leitch			6.43/6.53
													5.8/6.2

SOLIDARIDAD F1 109.2 WEST

Band	Freq.	Pol.	Tr.	Beam	Sce.	Program	Country	Language	System	Encr.	TT	A.M.	A.S.
C-	3.760	Ver	2N	Spot	R	various radio	Mexico	Spanish	SCPC				
C-	3.840	Ver	4N	Spot	TV	XEQ-TV 9	Mexico	Spanish	NTSC		No	6.2	
C-	3.840	Ver	4N	Spot	R	XEW-FM, WFM, Mexico	Mexico	Spanish				7.38	

ITU REGION 2 STATIONS IN FREQUENCY ORDER

Band	Freq.	Pol.	Tr.	Beam	Sce.	Program	Country	Language	System	Encr.	TT	A.M.	A.S.
C-	3.900	Hor	3W/L	Spot	0	Gobierno de la Republica	Mexico	English	NTSC			6.2	
MORELOS F2 116.7° WEST													
C-	3.860	Ver	4N	Spot	TV	XHGC-TV (Canal 5)	Mexico	Spanish	NTSC		No	6.2	
C-	3.860	Ver	4N	Spot	TV	Q-CVC	Mexico	Spanish	NTSC		No	6.2	
C-	3.860	Ver	4N	Spot	R	XEWA-FM, Mexico City	Mexico	Spanish				7.38	
C-	3.900	Ver	5N	Spot	O	SEP (Education Ministry)	Mexico	Spanish	NTSC			6.2	
C-	3.920	Hor	3W/U	Spot	TV	XEIPN-TV Canal 11, Mexico City	Mexico	Spanish	NTSC			6.2	
C-	3.980	Ver	7N	Spot	TV	XEW-TV, Mexico City	Mexico	Spanish	NTSC	clear/Leitch	No	6.2	
C-	3.980	Ver	7N	Spot	R	XEX-FM 101.7, Mexico City	Mexico	Spanish			Yes	7.38	
C-	4.020	Ver	8N	Spot	TV	XHIMT, Canal 22	Mexico	Spanish	NTSC			6.2	
C-	4.080	Hor	5W/U	Spot	O	TeleNoticias	Spain	Spanish	NTSC			6.2	
C-	4.120	Ver	6W	Spot	TV	XHIMT-TV, TV7	Mexico	Spanish	NTSC		No	6.2	
C-	4.120	Ver	6W	Spot	TV	Telecasa	Mexico	Spanish	NTSC		No	6.2	
C-	4.180	Ver	12N	Spot	TV	Canal 13 (XHDF-TV)	Mexico	Spanish	NTSC		Yes	6.2	
ECHOSTAR 1/2 119.0° WEST													
Ku2-					TV	Echostar DISH DBS			MPEG-2				
GALAXY 9 123.0° WEST													
C-	3.720	Ver	1	West Hemisph.	TV	BBC Breakfast News	UK	English	NTSC		No		
C-	3.760	Ver	3	West Hemisph.	TV	NHK Tokyo	Japan	Japanese	NTSC		No		
C-	3.980	Hor	14	West Hemisph.	TV	Sundance Channel	USA	English	NTSC	VCII+	No		
C-	4.000	Ver	15	West Hemisph.	TV	Showtime West	USA	English	NTSC	VCII+	No		
C-	4.040	Hor	17	West Hemisph.	TV	Nickelodeon (West)	USA	English	NTSC	VCII+	No		
C-	4.060	Ver	18	West Hemisph.	TV	Movie Channel (West)	USA	English	NTSC	VCII+	No		
C-	4.080	Hor	19	West Hemisph.	TV	MTV (West)	USA	English	NTSC	VCII+	No		6.2/6.8
C-	4.120	Ver	21	West Hemisph.	TV	ESPN	USA	English	NTSC	VCII+	No		6.2/6.8
C-	4.160	Ver	23	West Hemisph.	TV	Computer Television Netw.	USA	English	NTSC		No		
SBS 5 123.0° WEST													
Ku2-	11.898	Ver	12	Conus	TV	WMNB	USA	Russian	NTSC (inverted)		No	5.4	
Ku2-	12.166	Hor	10	Conus	O	occasional video	USA	English	B-MAC				
GALAXY 5 125.0° WEST													
C-	3.720	Hor	1	Conus	TV	Disney Channel (East)	USA	English	NTSC	VCII+	No	6.2/6.8	5.58/5.76
C-	3.740	Ver	2	Conus	TV	Playboy at Night	USA	English	NTSC	VCII+	No	6.8	
C-	3.740	Ver	2	Conus	TV	Satellite City TV	USA	English					
C-	3.760	Hor	3	Conus	R	KLON FM, Longbeach, CA	USA	English					
C-	3.760	Hor	3	Conus	TV	Trinity Broadcasting Network	USA	English	NTSC		No	6.2	5.58/5.78
C-	3.780	Ver	4	Conus	R	Trinity Broadcasting (Spanish)	USA	Spanish					
C-	3.800	Hor	5	Conus	TV	Sci-Fi Channel	USA	English	NTSC	VCII+	No		
C-	3.800	Hor	5	Conus	TV	CNN Headline News	USA	English	NTSC	VCII+	No		6.2/7.58
C-	3.820	Ver	6	Conus	R	CNN Radio	USA	English					
C-	3.820	Ver	6	Conus	TV	WTBS, Atlanta	USA	English	NTSC	VCII+	No	6.2	
C-	3.820	Ver	6	Conus	R	WRN (World Radio Network 2)	USA	English					

ITU REGION 2 STATIONS IN FREQUENCY ORDER

Band	Freq.	Pol.	Tr.	Beam	Sce.	Program	Country	Language	System	Encr.	TT	A.M.	A.S.
C-	3.820	Ver	6	Conus	R	Brother Stair	USA	English				6.44	
C-	3.820	Ver	6	Conus	R	WRN (World Radio Network 1)	USA	English				6.8	
C-	3.840	Hor	7	Conus	TV	WGN, Chicago Superstation	USA	English	NTSC	VCII+	No	6.8	6.3/6.48
C-	3.840	Hor	7	Conus	R	Yesterday USA	USA	English					6.30/6.48
C-	3.840	Hor	7	Conus	R	WFMT, Chicago	USA	English					
C-	3.840	Hor	7	Conus	R	WFMT, Chicago	USA	English					
C-	3.860	Ver	8	Conus	TV	Home Box Office (West)	USA	English	NTSC	VCII+	No		
C-	3.880	Hor	9	Conus	TV	ESPN	USA	English	NTSC	VCII+	No		
C-	3.900	Ver	10	Conus	TV	MOR Music Television	USA	English	NTSC		No	6.8	5.58/5.78
C-	3.920	Hor	11	Conus	TV	The Family Channel	USA	English	NTSC	VCII+			
C-	3.920	Hor	11	Conus	R	CBN Radio Netw.	USA	English					
C-	3.940	Ver	12	Conus	TV	Discovery Channel	USA	English	NTSC	VCII+	Yes	6.8	6.3/6.48
C-	3.940	Ver	12	Conus	R	Barker Channel	USA	English			No		
C-	3.960	Hor	13	Conus	TV	Consumer News & Business Channel (CNBC)	USA	English	NTSC	VCII+	No	6.8	
C-	3.960	Hor	13	Conus	R	Barker Channel	USA	English	NTSC	VCII+	No		
C-	3.980	Ver	14	Conus	TV	ESPN 2	USA	English	NTSC	VCII+	No		
C-	4.000	Hor	15	Conus	TV	Home Box Office (East)	USA	English	NTSC	VCII+	No		
C-	4.020	Ver	16	Conus	TV	Cinemax (West)	USA	English	NTSC	VCII+	No		
C-	4.040	Hor	17	Conus	TV	Turner Network Television (TNT)	USA	English	NTSC	VCII+			7.38/7.56
C-	4.060	Ver	18	Conus	TV	The Nashville Network	USA	English	NTSC				
C-	4.060	Ver	18	Conus	R	WSM-AM, Nashville	USA	English					
C-	4.080	Hor	19	Conus	TV	USA Network East	USA	English	NTSC	VCII+	No	6.8	5.58/5.78
C-	4.100	Ver	20	Conus	TV	Black Entertainment TV (BET)	USA	English	NTSC		No	6.8	
C-	4.120	Hor	21	Conus	TV	Knowledge TV	USA	English	NTSC		No		5.04/7.74
C-	4.120	Hor	21	Conus	R	SuperAudio (American Country Favorites)	USA	English					5.22/5.4
C-	4.120	Hor	21	Conus	R	SuperAudio (Prime Demo)	USA	English					5.58/5.78
C-	4.120	Hor	21	Conus	R	SuperAudio (Soft Sounds)	USA	English					5.94/6.12
C-	4.120	Hor	21	Conus	R	SuperAudio (Light Rock)	USA	English					6.3/6.48
C-	4.120	Hor	21	Conus	R	SuperAudio (Classical Collections)	USA	English					7.38/7.56
C-	4.120	Hor	21	Conus	R	SuperAudio (New Age of Jazz)	USA	English					8.11/8.31
C-	4.120	Hor	21	Conus	R	SuperAudio (Classic Hits)	USA	English					
C-	4.140	Ver	22	Conus	TV	CNN Headline News	USA	English	NTSC	VCII+	No	6.3	
C-	4.140	Ver	22	Conus	R	CNN Radio	USA	English					
C-	4.160	Hor	23	Conus	TV	Arts & Entertainment TV	USA	English	NTSC		No	6.8	
C-	4.160	Hor	23	Conus	R	Barker Channel	USA	English					

GSTAR 2 125.0° WEST

Band	Freq.	Pol.	Tr.	Beam	Sce.	Program	Country	Language	System	Encr.	TT	A.M.	A.S.
Ku2-	11.866	Ver	11	Spot	O	occasional video	USA	English	NTSC			6.2/6.8	
Ku2-	11.927	Ver	12	Spot	O	occasional video	USA	English	NTSC			6.2/6.8	
Ku2-	11.988	Ver	13	Spot	TV	CNN Airport Channel	USA	English	NTSC			6.80	
Ku2-	12.035	Hor	6	Spot	TV	CNN International	USA	English	NTSC	Leitch			
Ku2-	12.035	Hor	6	Spot	R	CNN Radio	USA	English					
Ku2-	12.049	Ver	14	Spot	O	occasional video	USA	English	NTSC			6.3	
Ku2-	12.096	Hor	7	Spot	O	occasional video	USA	English	NTSC		No	6.2/6.8	
Ku2-	12.110	Ver	15	Spot	O	occasional video	USA	English	NTSC		No	6.2/6.8	
Ku2-	12.157	Hor	8	Spot	O	occasional video	USA	English	NTSC			6.2/6.8	
Ku2-	12.171	Ver	16	Spot	O	CourtTV (backhauls)	USA	English	NTSC			6.8	

ITU REGION 2 STATIONS IN FREQUENCY ORDER

Band	Freq.	Pol.	Tr.	Beam	Sce.	Program	Country	Language	System	Encr.	TT	A.M.	A.S.
						SATCOM C3 131.0° WEST							
C-	3.720	Ver	1	Conus	TV	The Family Channel (West)	USA	English	NTSC	VCII+	No	6.20	
C-	3.720	Ver	1	Conus	R	CBN Radio	USA	English			No	6.2/6.8	
C-	3.740	Hor	2	Conus	TV	The Learning Channel	USA	English	NTSC		No	6.8	
C-	3.760	Ver	3	Conus	TV	Viewer's Choice	USA	English	NTSC	VCII+	No		
C-	3.760	Ver	3	Conus	R	Barker Channel	USA	English			No		
C-	3.780	Hor	4	Conus	TV	Lifetime West	USA	English	NTSC	VCII+	No	6.8	
C-	3.780	Hor	4	Conus	R	Barker Channel	USA	English			No	6.2/6.8	
C-	3.800	Ver	5	Conus	TV	Odyssey Network	USA	English	NTSC		No	6.8	
C-	3.840	Ver	7	Conus	TV	C-SPAN 1	USA	English	NTSC		No	5.22	
C-	3.840	Ver	7	Conus	R	C-SPAN audio sce 1 (a.o. VOA)	USA	English				5.41	
C-	3.840	Ver	7	Conus	R	C-SPAN audio sce 2 (a.o. BBC)	USA	English				5.58	
C-	3.840	Ver	7	Conus	R	C-SPAN 1 ASAP	USA	English					
C-	3.860	Hor	8	Conus	TV	QVC Fashion Channel	USA	English	NTSC		No	6.8/6.2	
C-	3.880	Ver	9	Conus	R	Digital Cable Radio	USA	English					
C-	3.900	Hor	10	Conus	TV	Home Shopping Club 2	USA	English	NTSC	VCII+	No	6.8/6.2	
C-	3.920	Ver	11	Conus	TV	Prime Network	USA	English	NTSC		No		
C-	3.940	Hor	12	Conus	R	digital cable radio	USA	English	DSR		No		
C-	3.960	Ver	13	Conus	TV	Weather Channel	USA	English	NTSC	VCII+	No		
C-	3.980	Hor	14	Conus	TV	New England Sports Channel	USA	English	NTSC	VCII+	No		
C-	4.000	Ver	15	Conus	TV	Showtime East	USA	English	NTSC	VCII+	No		
C-	4.000	Ver	15	Conus	R	Barker Channel	USA	English			No	6.8	
C-	4.020	Hor	16	Conus	TV	Music Television (West)	USA	English	NTSC		No	6.8	
C-	4.020	Hor	16	Conus	R	Barker Channel	USA	English			No		
C-	4.040	Ver	17	Conus	TV	The Movie Channel	USA	English	NTSC	VCII+	No	6.8	
C-	4.040	Ver	17	Conus	R	Barker Channel	USA	English			No		
C-	4.060	Hor	18	Conus	TV	Nickelodeon (West)	USA	English	NTSC		No	6.8	
C-	4.060	Hor	18	Conus	R	Barker Channel	USA	English			No		
C-	4.080	Ver	19	Conus	TV	digitally compressed video	USA	English	DIGICIPHER		No		
C-	4.100	Hor	20	Conus	TV	Infomercial Channel	USA	English	NTSC		No	6.2/6.8	
C-	4.120	Ver	21	Conus	TV	Comedy Central (East)	USA	English	NTSC	VCII+	No		
C-	4.140	Hor	22	Conus	TV	American TV Netw.	USA	English	NTSC		No	6.2/6.8	
C-	4.160	Ver	23	Conus	TV	El Entertainment	USA	English	NTSC		No	6.2/6.8	
C-	4.160	Ver	23	Conus	R	CBN News	USA	English			No	7.23	
C-	4.180	Hor	24	Conus	R	Digital Music Express	USA	English	NTSC		No		
						GALAXY 1 133.0° WEST							
C-	3.720	Hor	1	Conus	TV	Comedy Central	USA	English	NTSC	VCII+	No	6.8	
C-	3.760	Hor	3	Conus	TV	Encore	USA	English	NTSC	VCII+	No	6.8	
C-	3.820	Hor	6	Conus	TV	Z-Music	USA	English	NTSC		No	6.8	
C-	3.840	Ver	7	Conus	TV	Disney Channel (West)	USA	English	NTSC	VCII+	No	6.8	
C-	3.860	Hor	8	Conus	TV	The Cartoon Network	USA	English	NTSC	VCII+	No		
C-	3.880	Ver	9	Conus	TV	Shop-at-Home Network	USA	English	NTSC		No		
C-	3.900	Hor	10	Conus	TV	America's Talking	USA	English	NTSC		No		
C-	3.920	Ver	11	Conus	TV	Eternal Word TV	USA	English	NTSC		No		
C-	3.940	Hor	12	Conus	TV	Valuevision	USA	English	NTSC		No		6.2/6.8

ITU REGION 2 STATIONS IN FREQUENCY ORDER

Band	Freq.	Pol.	Tr.	Beam	Sce.	Program	Country	Language	System	Encr.	TT	A.M.	A.S.
C-	3.980	Ver	14	Conus	TV	Shop-at-Home	USA	English	NTSC		No	6.8	
C-	3.980	Ver	14	Conus	TV	ESPN Blackout Channel	USA	English	NTSC		No	6.8	
C-	4.000	Ver	15	Conus	TV	CNN Int.	USA	English	NTSC		No		
C-	4.020	Hor	16	Conus	TV	Turner Classic Movies	USA	English	NTSC	VCII+	No	6.2/6.8	
C-	4.040	Hor	17	Conus	TV	The New Inspirational Netw.	USA	English	NTSC	VCII+	No	6.2/6.8	
C-	4.080	Ver	19	Conus	TV	Cinemax (East)	USA	English	NTSC	VCII+	No	6.80	5.58/5.78
C-	4.100	Ver	20	Conus	TV	Home and Garden TV Netw.	USA	Spanish	NTSC		No		
C-	4.120	Hor	21	Conus	TV	USA Network West	USA	English	NTSC	VCII+	No		6.2/6.8
C-	4.140	Ver	22	Conus	TV	Nostalgia Channel	USA	English	NTSC		No		
C-	4.160	Hor	23	Conus	TV	Cinemax II (East)	USA	English	NTSC	VCII+	No		
C-	4.180	Ver	24	Conus	TV	Global Shopping Network	USA	Spanish	NTSC		No	6.3/6.48	
SATCOM C4 135.0° WEST													
C-	3.720	Ver	1	Conus	TV	American Movie Classics	USA	English	NTSC	VCII+	No		
C-	3.760	Ver	3	Conus	TV	Nickelodeon (East)	USA	English	NTSC	VCII+	No	6.8	
C-	3.760	Hor	3	Conus	R	Barker Channel	USA	English	NTSC		No		
C-	3.780	Hor	4	Conus	TV	Lifetime (East)	USA	English	NTSC	VCII+	No	6.8	
C-	3.780	Hor	4	Conus	R	Barker Channel	USA	English	NTSC		No	6.8	
C-	3.800	Ver	5	Conus	TV	Deutsche Welle TV	Germany	various	NTSC		No	7.02	
C-	3.800	Hor	5	Conus	R	Deutsche Welle	Germany	various	NTSC		No		
C-	3.820	Hor	6	Conus	TV	Madison Square Garden	USA	English	NTSC	VCII+	No	6.2	
C-	3.820	Hor	6	Conus	R	WQCD-FM (CD 101.9), NY	USA	English	NTSC		No		
C-	3.840	Ver	7	Conus	TV	Bravo	USA	English	NTSC	VCII+	No	6.2/6.8	
C-	3.860	Hor	8	Conus	TV	PrevuGuide	USA	English	NTSC		No	5.90	
C-	3.860	Hor	8	Conus	R	United Video Background Music	USA	English	NTSC		No		7.38/7.56
C-	3.860	Hor	8	Conus	R	KILA, Las Vegas	USA	English	NTSC		No		
C-	3.880	Ver	9	Conus	TV	QVC, Home Shopping Netw.	USA	English	NTSC		No	6.2/6.8	
C-	3.900	Hor	10	Conus	TV	Home Shopping Network 1	USA	English	NTSC		No	6.2/6.8	
C-	3.900	Hor	10	Conus	R	In Touch	USA	English			No	7.87	
C-	3.900	Hor	10	Conus	R	Business Radio Network	USA	English			No	8.06	
C-	3.900	Hor	10	Conus	R	WUSF-FM	USA	English			No	8.26	
C-	3.940	Hor	12	Conus	TV	Nustar	USA	English	NTSC		No	6.8	
C-	3.960	Hor	13	Conus	TV	The Travel Channel	USA	English	NTSC		No	6.8	
C-	3.980	Ver	14	Conus	TV	Cable Health Club	USA	English	NTSC		No	6.8	
C-	4.000	Ver	15	Conus	TV	WWOR-TV	USA	English	NTSC	VCII+	No		
C-	4.020	Hor	16	Conus	R	WQXR-FM	USA	English	NTSC		No		6.30/6.48
C-	4.040	Ver	17	Conus	TV	Request TV 1	USA	English	NTSC	VCII+	No		
C-	4.040	Ver	17	Conus	TV	Music TV (East)	USA	English	NTSC	VCII+	No		
C-	4.080	Ver	19	Conus	TV	C-Span 2 ASAP	USA	English	NTSC		No	5.58	
C-	4.100	Ver	20	Conus	TV	Showtime (West)	USA	English	NTSC	VCII+	No		
C-	4.100	Hor	20	Conus	R	Barker Channel	USA	English			No		
C-	4.120	Ver	21	Conus	TV	Discovery (East)	USA	English	NTSC	VCII+	No	6.8	
C-	4.120	Ver	21	Conus	R	Barker Channel	USA	English			No		
C-	4.140	Hor	22	Conus	TV	The Movie Channel (West)	USA	English	NTSC	VCII+	No	6.8	
C-	4.140	Hor	22	Conus	R	Barker Channel	USA	English			No		
C-	4.160	Ver	23	Conus	TV	VH-1	USA	English	NTSC	VCII+	No	6.8	

ITU REGION 2 STATIONS IN FREQUENCY ORDER

Band	Freq.	Pol.	Tr.	Beam	Sce.	Program	Country	Language	System	Encr.	TT	A.M.	A.S.
C-	4.180	Hor	24	Conus	TV	Country Music TV	USA	English	NTSC	VCII+	No		
						SATCOM C1 137.0° WEST							
C-	3.720	Hor	1	Conus	TV	SportsChannel Chicago Plus	USA	English	NTSC	VCII+	No	6.2/6.8	
C-	3.720	Hor	1	Conus	TV	SportsChannel Cincinnati	USA	English	NTSC	VCII+	No		
C-	3.720	Hor	1	Conus	TV	SportsChannel Hawaii	USA	English	NTSC		No		
C-	3.740	Ver	2	Conus	TV	KMGH, Denver	USA	English	NTSC	VCII+	No	7.50	
C-	3.740	Ver	2	Conus	R	For The People Netw.	USA	English	NTSC		No		
C-	3.760	Hor	3	Conus	TV	KRMA, Denver	USA	English	NTSC	VCII+	No	5.58	
C-	3.760	Hor	3	Conus	R	Colorado Talking Book	USA	English	NTSC		No	5.58	
C-	3.760	Hor	3	Conus	R	BBC World Radio	USA	English	NTSC		No		
C-	3.780	Ver	4	Conus	TV	SportsChannel Pacific	USA	English	NTSC	VCII+	No	5.8	
C-	3.780	Ver	4	Conus	R	KJAZ	USA	English	NTSC		No		
C-	3.800	Hor	5	Conus	TV	KDVR, Denver (FOX affiliate)	USA	English	NTSC	VCII+	No	5.58	
C-	3.800	Hor	5	Conus	R	Talk Radio Network	USA	English	NTSC		No		
C-	3.820	Ver	6	Conus	TV	KMGH, Denver	USA	English	NTSC	VCII+	No	5.8	
C-	3.820	Ver	6	Conus	R	KSL-AM, Salt Lake City	USA	English	NTSC		No		
C-	3.840	Hor	7	Conus	TV	Prime Ticket, California	USA	English	NTSC	VCII+	No	6.8	
C-	3.860	Ver	8	Conus	TV	NBC	USA	English	NTSC	VCII+	No	6.2/6.8	
C-	3.880	Hor	9	Conus	TV	SportsChannel Florida	USA	English	NTSC	VCII+	No		
C-	3.880	Hor	9	Conus	TV	Sunshine Backout Channel	USA	English	NTSC	VCII+	No		
C-	3.880	Hor	9	Conus	TV	SportsChannel Chicago	USA	English	NTSC	VCII+	No		
C-	3.900	Ver	10	Conus	TV	Home Sports Entertainment	USA	English	NTSC	VCII+	No		
C-	3.920	Hor	11	Conus	TV	Network One (N1)	USA	English	NTSC		No	7.38	
C-	3.960	Hor	13	Conus	TV	SportsChannel Chicago	USA	English	NTSC	VCII+	No	7.56	
C-	3.980	Ver	14	Conus	TV	KCNC, Denver	USA	English	NTSC	VCII+	No		
C-	4.000	Hor	15	Conus	TV	SportsChannel Ohio, Flori., Cinci.	USA	English	NTSC	VCII+	No		
C-	4.000	Hor	15	Conus	R	RAI Italy Radio Netw.	Italy	Italian	NTSC		No		
C-	4.000	Ver	15	Conus	TV	Sun Radio Network	USA	English	NTSC		No		
C-	4.020	Ver	16	Conus	TV	Newsport	USA	English	NTSC	VCII+	No		
C-	4.040	Hor	17	Conus	TV	HSE2	USA	English	NTSC	VCII+	No	6.8	6.2/6.8
C-	4.040	Hor	17	Conus	TV	Prime Sports Intermountain (West)	USA	English	NTSC	VCII+	No	6.2/6.8	
C-	4.040	Hor	17	Conus	TV	Cal-Span	USA	Russian	NTSC		No		
C-	4.060	Ver	18	Conus	TV	Prime Sports Showcase	USA	English	NTSC	VCII+	No		
C-	4.080	Hor	19	Conus	O	FOX Network	USA	English	NTSC	VCII+	No		
C-	4.100	Ver	20	Conus	TV	The International Channel	USA	English	NTSC	VCII+	No		
C-	4.140	Ver	22	Conus	TV	Prime Sports Northwest	USA	English	NTSC	VCII+	No		
C-	4.160	Hor	23	Conus	TV	KWGN, Denver	USA	English	NTSC	VCII+	No		
C-	4.180	Ver	24	Conus	TV	Sunshine Network	USA	English	NTSC	VCII+	No		
						SATCOM C5 139.0° WEST							
C-	3.760	Ver	3	Conus	R	various digital radio	USA		SCPC		No	6.8	
C-	3.880	Ver	9	Conus	R	various digital radio	USA		SCPC		No	6.48	
C-	4.100	Ver	21	Conus	R	various digital radio	USA		SCPC				
C-	4.180	Hor	24	Conus	TV	Alaska Satellite TV Project	USA	English	NTSC		No		
C-	4.180	Hor	24	Conus	R	In Touch	USA	English					

ITU REGION 2 STATIONS IN FREQUENCY ORDER

Band	Freq.	Pol.	Tr.	Beam	Sce.	Program	Country	Language	System	Encr.	TT	A.M.	A.S.
C-	4.180	Hor	24	Conus	R	KSKA-FM, Anchorage	USA	English					7.38/7.56

TV = Television   **R** = Radio   **O** = Others   **VC** = VideCipher

ITU REGION 2 DIGITAL STATIONS IN FREQUENCY ORDER

10.2. DIGITAL STATIONS IN FREQUENCY ORDER

GE Americom K2 85° West
Ku-Band

Tr.	Freq	Block IF
T02(V)	11758.5	1008.5
T03(H)	11788.0	1038
T04(V)	11817.5	1067.5
T05(H)	11847.0	1097
T06(V)	11876.5	1126.5
T07(H)	11906.0	1156
T08(V)	11935.5	1185.5
T09(H)	11965.0	1215
T10(V)	11994.5	1244.5
T11(H)	12024.0	1274
T12(V)	12053.5	1303.5
T13(H)	12083.0	1333
T14(V)	12112.5	1362.5
T15(H)	12142.0	1392
T16(V)	12171.5	1421.5

Primestar Services (transponders 2-16)

1 HBO East
2 HBO East 2
3 HBO East 3
7 Cinemax East
8 Cinemax 2
13 TV Japan (English)
14 TV Japan (Japanese)
15 Future Service
17 Future Service
19 Future Service
27 STARZ!
31 Encore3 - Westerns
32 Encore4 - Mystery
33 Encore
34 The Disney Channel East
35 The Disney Channel West
40 The Golf Channel
47 C-SPAN
48 CNBC
49 The Weather Channel
50 CNN International (occ)
51 Cable News Network
52 CNN Headline News
52 (data) Ingenious News service
55 Prevue Channel
56 Future Service
58 Turner Network TV
59 Turner Classic Movies
60 TV Land
61 Comedy Central
63 Turner Broadcasting System
65 The Discovery Channel
66 The Learning Channel
68 Arts & Entertainment
70 USA Network
71 Sci-Fi Channel
72 The Family Channel
73 The Cartoon Network
74 Nickelodeon/Nick at Nite
75 E! Entertainment TV
76 Lifetime
77 The Nashville Network
78 Country Music TV
80 MTV
83 Odyssey
84 QVC (occ)
111 WHDH-TV (NBC) Boston
112 WSB-TV (ABC) Atlanta
117 WUSA-TV (CBS) Washington
120 KTVU-TV (FOX) Oakland/S.F.
124 WHYY-TV (PBS) Philadelphia
131 ESPN
132 ESPN2
133 ESPN 2
137 Classic Sports Network
138 Mega+1 - sports
141 New England Sports Network
142 Madison Square Garden Network
Home and Garden TV
143 Empire Sports Network
144 Fox Sports Pittsburgh
145 Home Team Sports
146 SportSouth
147 Sunshine Network
148 Pro Am Sports System (PASS)
149 Future Service
152 Fox Sports Midwest
153 Fox Sports Rocky Mountain
154 Fox Sports Southwest
155 Fox Sports Intermountain West
156 Fox Sports Northwest
157 Fox Sports Arizona
158 Fox Sports West
159 Midwest Sports Channel
181 HBO en Espanol
182 HBO2 en Espanol
183 HBO3 en Espanol
187 Cinemax Selecciones
188 Cinemax Selecciones
190 Univision
201 PPV - Viewer's Choice
202 PPV - Request 1
203 PPV - Request 5
204 PPV - Hot Choice
205 PPV - Continuous Hits 1
207 PPV - Continuous Hits 3
208 PPV - Request 2
210 PPV - Request 4
221 Playboy (occ)
301 Superradio - Classic Hits
302 Superradio - America's Country Fav.
303 Superradio - Light 'N Lively Rock
304 Superradio - Soft Sounds
305 Superradio - Classic Collections
306 Superradio - New Age Jazz
311 DMX Audio - Lite Jazz

ITU REGION 2 DIGITAL STATIONS IN FREQUENCY ORDER

312 DMX Audio - Classic Rock
313 DMX Audio - 70's Oldies
314 DMX Audio - Adult Contemporary
315 DMX Audio - Hottest Hits
316 DMX Audio - Modern Country
317 DMX Audio - Traditional Blues
318 DMX Audio - Salsa
527 Testing - Test Channel

Telstar 402 89 °West

Tr.	Freq	Block IF
T01(V)	11730	980
T02(H)	11743	993
T03(V)	11790	1040
T04(H)	11803	1053
T05(V)	11850	1100
T06(H)	11863	1113
T07(V)	11910	1160
T08(H)	11923	1173
T09(V)	11971	1221
T10(H)	11984	1234
T11(V)	12033	1283
T12(H)	12046	1296
T13(V)	12095	1345
T14(H)	12108	1358
T15(V)	12157	1407
T16(H)	12170	1420

T402R-16(KU) COMPR Alphastar DBS System

Alphastar services:

1 A&E Network
2 ABC (WJLA Washington DC)
3 Alpha Preview Channel
4 Asian Television Network (ATV)
5 C-Span 1 (US House)
6 C-Span 2 (US Senate)
7 Cartoon Network
8 CBS (WRAL Raleigh, NC)
9 Cinemax East
10 Cinemax 2
11 Cinemax West
12 Classic Sports Network
13 CNBC
14 CNN
15 CNN International/CNN fn
16 Comedy Central
17 Country Music Television
18 Court TV
19 Discovery Channel
20 Disney Channel (E)
21 Disney Channel (W)
22 E! Entertainment Television
23 Egyptian Satellite Channel
24 Encore
25 Encore Plus
26 ESPN
27 ESPN2
28 Family Channel
29 FOX Network (Foxnet)
30 Fox Sports Southwest
31 Fox Sports Northwest
32 Fox Sports Rocky Mountain/Midwest/Intermountain West
33 Fox Sports West
34 Fox Sports Pittsburgh
35 Golf Channel
36 HBO East
37 HBO East 2
38 HBO West 2
39 HBO East 3
40 HBO West
41 Headline News
42 History Channel
43 Learning Channel
44 Lifetime
45 Madison Square Garden
46 MSNBC
47 MTV
48 Nashville Network
49 NBC (WNBC New York)
50 NewsTalk Television
51 Nickelodeon / Nick at Nite
52 Nile Drama
53 Nile TV
54 PBS Network (National)
55 Playboy TV
56 10 PPV Channels
57 Sci-Fi Channel
58 Showtime
59 Showtime 2
60 Showtime West
61 Starz!
62 Sundance Film Channel
63 Sunshine Network
64 TBS Atlanta
65 The Movie Channel
66 The Movie Channel West
67 Turner Classic Movies
68 Turner Network Television (TNT)
69 TV Land
70 USA Network
71 Venus (adult)
72 VH-1
73 Weather Channel
74 WGN Chicago
+ 30 DMX Audio Channels

Galaxy 3R 95°West

Tr.	Freq	Block IF
T01(H)	11720	970
T02(V)	11750	1000
T03(H)	11750	1000
T04(H)	11780	1030
T05(V)	11810	1060
T06(H)	11810	1060
T07(V)	11840	1090

ITU REGION 2 DIGITAL STATIONS IN FREQUENCY ORDER

T08(V)	11870	1120
T09(H)	11870	1120
T10(H)	11900	1150
T11(V)	11930	1180
T12(H)	11930	1180
T13(H)	11960	1210
T14(V)	11990	1240
T15(H)	11990	1240
T16(H)	12020	1270
T17(V)	12050	1300
T18(H)	12050	1300
T19(H)	12080	1330
T20(V)	12110	1360
T21(H)	12110	1360
T22(H)	12140	1390
T23(V)	12170	1420
T24(H)	12170	1420

Galaxy Latin America (all 24 Ku transponders):

100 DTV Coming Attractions, programming
102 TNT Latin America
104 TeleUno All the Spelling Series
106 USA
108 SONY Entertainment TV (SET)
110 WBTV (The Warner Channel)
112 MAS Mexican Channel
118 GEMS The Women channel
130 TVE Television Espanola
132 Antena 3 Espana
134 RAI Italia
138 DW From Germany
140 RTPI from Portugal
156 TVN Chile
166 TV Azteca Canal 7 Mexico
172 TV Azteca Canal 13 Mexico
180 Cartoon Network
182 ZAZ cartoons and children programming
184 LOCOMOTION cartoons
194 MTV Latino
210 ESPN International
212 ESPN 2
232 CBS Telenoticias
234 BBC World Service
236 CNN International
252 Bloomberg
262 Travel Channel
272 Discovery Channel
300 DTV
302 MultiPremier Movies
304 Films & Arts
306 MultiCinema Movies
310 Cine Latino Movies
330 HBO OLE WEST
332 HBO OLE EAST
333 Music (radio station)
334 HBO OLE 2
336 CINEMAX WEST
338 CINEMAX EAST
400 DTV
402-436 CINE DIRECT (DirectTicket)
500 DTV
502-536 DIRECT EVENT (NFL & MLB Packages)
592 AdultVision
600-601 DTV
602-660 Music (30 radio stations without commercials)

DBS-1/2/3 101.2° West
DirecTV uses transponders 1-15, 16-23, 25, 27, 29 and 31

USSB uses transponders 24,26,28,30, and 32

(100-899 DIRECTV) 277 Encore 7 - WAM!
100 Direct Ticket Previews
101 Direct Ticket Special Events
102 Direct Ticket PPV Channel
103 Direct Ticket PPV Channel
104 Direct Ticket PPV Channel
105 Direct Ticket PPV Channel
108 Direct Ticket PPV Channel
109 Direct Ticket PPV Channel
110 Direct Ticket PPV Channel
111 Direct Ticket PPV Channel
112 Direct Ticket PPV Channel
113 Direct Ticket PPV Channel
114 Direct Ticket PPV Channel
115 Direct Ticket PPV Channel
116 Direct Ticket PPV Channel
117 Direct Ticket PPV Channel
118 Direct Ticket PPV Channel
119 Direct Ticket PPV Channel
120 Direct Ticket PPV Channel
121 Direct Ticket PPV Channel
122 Direct Ticket PPV Channel
123 Direct Ticket PPV Channel
124 Direct Ticket PPV Channel
125 Direct Ticket PPV Channel
127 Direct Ticket PPV Channel
128 Direct Ticket PPV Channel
129 Direct Ticket PPV Channel
130 Direct Ticket PPV Channel
132 Direct Ticket PPV Channel
133 Direct Ticket PPV Channel
134 Direct Ticket PPV Channel
135 Direct Ticket PPV Channel
138 Direct Ticket PPV Channel
139 Direct Ticket PPV Channel
140 Direct Ticket PPV Channel
141 Direct Ticket PPV Channel
142 Direct Ticket PPV Channel
143 Direct Ticket PPV Channel
145 Direct Ticket PPV Channel
146 Direct Ticket PPV Channel

ITU REGION 2 DIGITAL STATIONS IN FREQUENCY ORDER

149 Direct Ticket PPV Channel
150 Direct Ticket PPV Channel
151 Direct Ticket PPV Channel
154 Direct Ticket PPV Channel
155 Direct Ticket PPV Channel
156 Direct Ticket PPV Channel
158 Direct Ticket PPV Channel
160 Direct Ticket PPV Channel
162 Direct Ticket PPV Channel
164 Direct Ticket PPV Channel
170 Direct Ticket PPV Channel
175 Direct Ticket PPV Channel
180 Direct Ticket PPV Channel
181 Direct Ticket PPV Channel
182 Direct Ticket PPV Channel
185 Direct Ticket PPV Channel
188 Direct Ticket PPV Channel
190 Direct Ticket PPV Channel
201 DirecTV Events Calendar
202 Cable News Network
203 Court Television
204 CNN Headline News
206 ESPN
207 ESPN (Alternate)
208 ESPN2
212 Turner Network Television
213 Home Shopping Network
214 Home and Garden TV
215 E! Entertainment Television
216 MuchMusic U.S. version
217 Black Entertainment TV (BET)
219 American Movie Classics
220 Turner Classic Movies
221 Arts and Entertainment (A&E)
222 The History Channel
223 The Disney Channel (East)
224 The Disney Channel (West)
225 The Discovery Channel
226 The Learning Channel
227 Cartoon Network
229 USA Network (East)
230 TRIO
232 Family Channel (East)
233 Superstation TBS
235 The Nashville Network
236 Country Music Television
240 Sci-Fi Channel
242 C-SPAN 1
243 C-SPAN 2
245 Bloomberg Direct (Bloomberg
246 CNBC
247 MSNBC
248 The Weather Channel
249 Fox News Channel
250 CBC Newsworld International
252 CNN International / CNNfn
254 America's Health Network
258 Bravo
266 Independent Film Channel
269 STARZ! - West

270 STARZ! - East
271 Encore - 60's, 70's, 80's movies
272 Encore 2 - Love Stories
273 Encore 3 - Westerns
274 Encore 4 - Mystery
275 Encore 5 - Action
276 Encore 6 - True Stories and Drama
989 MTV (East)
278 Plex Encore 1
282 WRAL-TV, Raleigh, NC (CBS)
283 KPIX-TV, San Francisco (CBS)
284 WNBC-TV, New York (NBC)
285 KNBC-TV, Los Angeles (NBC)
286 PBS National Feed
287 WJLA-TV, Washington (ABC)
288 KOMO-TV, Seattle (ABC)
289 FOXNet (FOX)
297 Informational Channel
298 TV ASIA (leaves DSS Dec 31st)
299 In-store Dealer Info Network
300 DirecTV Sports Offer
301 Special Events Calendar
302 Special Events Calendar
303 NewSport
304 Golf Channel
305 Classic Sports Network
306 Speedvision
307 Outdoor Life Network
309 SPORTSCHANNEL New England
310 Madison Square Garden
311 New England Sports Network (NESN)
312 SPORTSCHANNEL New York
313 Empire Sports Network
314 SPORTSCHANNEL Philadelphia
315 Fox Sports Pittsburgh
316 Home Team Sports (HTS)
317 SportSouth
318 Sunshine Network
319 SPORTSCHANNEL Florida
320 Pro Am Sports (PASS)
321 SPORTSCHANNEL Ohio
322 SPORTSCHANNEL Cincinnati
323 SPORTSCHANNEL Chicago
324 Midwest Sports Channel (MSC)
325 Fox Sports Southwest
326 Fox Sports Intermountain West/Midwest/Rocky Mountain
329 Fox Sports Arizona
330 Fox Sports Northwest
331 Fox Sports West
332 SPORTSCHANNEL Pacific
401 Spice
402 Playboy At Night
501 AUDIO - Hit list
502 AUDIO - Dance
503 AUDIO - Hip Hop
504 AUDIO - Urban Beat
505 AUDIO - Reggae
506 AUDIO - Blues

ITU REGION 2 DIGITAL STATIONS IN FREQUENCY ORDER

507 AUDIO - Jazz
508 AUDIO - Jazz Plus
509 AUDIO - Contemp. Jazz
510 AUDIO - New Age
511 AUDIO - Eclectic Rock
512 AUDIO - Modern Rock
513 AUDIO - Classic Rock
514 AUDIO - Rock Plus
515 AUDIO - Metal
516 AUDIO - Solid Gold Oldies
517 AUDIO - Soft Rock
518 AUDIO - Love Songs
519 AUDIO - Progressive Country
520 AUDIO - Contemp. Country
521 AUDIO - Country Gold
522 AUDIO - Big Band
523 AUDIO - Easy Listening
524 AUDIO - Classic Favorites
525 AUDIO - Classics in Concert
526 AUDIO - Contemp. Christian
527 AUDIO - Gospel
528 AUDIO - For Kids Only
529 AUDIO - Sounds of the Season
530 AUDIO - Spectrum I
531 AUDIO - Spectrum II
550 AUDIO - Lite Classical
551 AUDIO - EZ - Vocals
552 AUDIO - Soft Album Mix
553 AUDIO - The Trend
554 AUDIO - Tropical
555 AUDIO - Mexicana
599 AUDIO - NRTC (realtor channel)
757 Microsoft TV

(900-999 USSB)
899 USSB Promos
900 USSB Pay-per-view
960 TV Land
963 All News Channel Business)
967 Lifetime (East)
968 Nickelodeon (East)
970 FLIX
973 Cinemax (East)
974 Cinemax 2
975 Cinemax (West)
977 The Movie Channel (East)
978 The Movie Channel (West)
980 HBO (East)
981 HBO 2 (East)
982 HBO 3
983 HBO (West)
984 HBO 2 (West)
985 Showtime (East)
986 Showtime 2
987 Showtime (West)
990 Comedy Central
995 Sundance Channel
999 USSB Promos

Echostar-1/2 119.0° West
Echostar DISH uses transponders 1, 3, 5, 7, 9, 11, 13, 15, 17,1 9 and 21

100 DISH Network Channel
102 USA Network
104 Comedy Central
106 TV Land
108 Lifetime
110 TV Food Network
112 Home and Garden Network
114 E! Entertainment Television
116 Game Show Network
118 Arts and Entertainment
120 History Channel
122 Sci-Fi Channel
124 Black Entertainment Television
132 Turner Classic Movies
138 Turner Network Television
140 ESPN
141 ESPN Alternate
142 ESPN2
143 ESPN2 Alternate
144 ESPNews
160 M-TV
162 VH-1
166 Country Music Television
168 The Nashville Network
170 Nickelodeon
172 The Disney Channel
176 The Cartoon Network
178 The Learning Channel
180 The Family Channel
182 The Discovery Channel
184 Animal Planet
200 Cable News Network
202 Headline News
204 Court TV
206 CNN International / CNNfn
208 CNBC
210 CSpan
212 CSPan 2
214 The Weather Channel
216 National Empowerment Television
220 The Travel Channel
226 QVC Shopping Network
230 TBS
232 KTLA, Los Angeles
234 WPIX, New York
236 WSBK, Boston
240 WGN, Chicago
241 WNBC-TV, New York (NBC)
242 KNBC-TV, Los Angeles (NBC)
243 WRAL-TV, Raleigh (CBS)
244 KPIX-TV, San Francisco (CBS)
245 WJLA-TV, Washington (ABC)
246 KOMO-TV, Seattle (ABC)
247 FOXNet
249 PBS
260 Trinity Broadcasting Net

ITU REGION 2 DIGITAL STATIONS IN FREQUENCY ORDER

261 Eternal Word TV Network
267 American Family Radio (Sky Angel)
268 Calvary Chapel Radio (Sky Angel)
269 Bob Jones University Radio (Sky Angel)
270 The Worship Channel (Sky Angel)
271 Praise TV (Sky Angel)
272 FamilyNet (Sky Angel)
273 Cornerstone Television (Sky Angel)
274 100 Plus Ministries (Sky Angel)
275 Dominion Variety and International/ Home School Channel (Sky Angel)
300 HBO East
301 HBO 2 East
302 HBO 3 East
303 HBO West
304 HBO 2 West
310 Showtime East
311 Showtime West
312 Showtime East 2
318 Sundance Channel
319 FLIX
320 Cinemax East
321 Cinemax East 2
322 Cinemax West
330 The Movie Channel East
331 The Movie Channel West
401 The Golf Channel
412 Madison Square Garden (MSG)
414 Fox Sports Rocky Mountain
416 Fox Sports Southwest
417 Fox Sports West
418 Fox Sports Midwest
420 SportSouth
422 Sunshine Network
424 Home Team Sports
426 Fox Sports Northwest
428 Fox Sports Pittsburgh
430 Pro-Am Sports (PASS)
432 Empire Sports Network
434 New England Sports Network
436 Midwest Sports Channel
500 PPV 1 (events)
501 PPV 2
502 PPV 3
503 PPV 4
504 PPV 5
505 PPV 6
506 PPV 7

507 PPV 8
508 PPV 9
509 PPV 10
600 RAI (Italy)
602 ART (Arabic)
604 Antenna TV (Greek)
620 MTV Latino
622 Univision
624 Galavision
626 Fox Sports Americas
628 Telemundo
700 Preview Channel (Dealers)
900 Business TV
901 Business TV

DISH CD(TM)
950 Young Country
951 Country Gold
952 Country Currents
953 Jukebox Gold
954 70's Song Book
955 Adult Favorites
956 Adult Contemporary
957 Adult Alternative
958 HitLine
959 Classic Rock
960 The Edge
961 Power Rock
962 Non-stop Hip Hop
963 Urban Beat
964 Latin Styles
965 Fiesta Mexicana
966 Eurostyle
967 Jazz Traditions
968 Contemporary Jazz Flavors
969 Expressions
970 Contemporary Instrumentals
971 Concert Classics
972 Light Classical
973 Easy Instrumentals
974 Big Band Era
975 Contemporary Christian
976 KidZone
977 LDS Radio Network
978 Blues
979 Reggae
980 New Age

CHAPTER 11
ITU REGION 3

TRANSPONDER LOADING

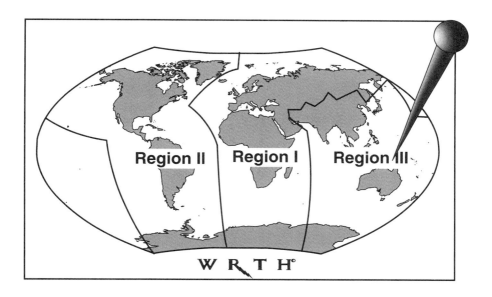

ITU REGION 3 STATIONS IN FREQUENCY ORDER

Band	Freq.	Pol.	Tr.	Beam	Sce.	Program	Country	Language	System	Encr.	TT	A.M.	A.S.
TDRS-F5 185.7° EAST													
C-	3.845	Hor			TV	Fuji TV Network	Japan	Japanese	dig. compressed		No		
C-	3.987	Hor			TV	BBC WSTV	UK	English	dig. compressed		Yes		
INTELSAT 511 180.0° EAST													
C-	3.720	RHCP	11	W. Hemisph.	O	occasional video	Australia	English	Vidiplex		No	6.6	
C-	3.765	RHCP	11	W. Hemisph.	TV	Network 10 Australia	Australia	English	Vidiplex		No	6.65	
C-	3.876	RHCP	13	W. Hemisph.	TV	Channel 7 Australia	Australia	English	PAL	encrypted	No	6.65	
C-	3.894	LHCP	10	W. Hemisph.	TV	ABC Australia	Australia	English	Vidiplex		No	6.65	
C-	3.925	RHCP	13	W. Hemisph.	TV	NHK Tokyo	Japan	Japanese	NTSC		No	6.65	
C-	3.930	RHCP	13	W. Hemisph.	TV	Channel 9 Australia	Australia	English			No	6.65	
C-	3.975	LHCP	14	W. Hemisph.	TV	WorldNet/C-Span/Deutsche Welle TV	USA	English			No	7.00	
C-	3.975	RHCP	14	W. Hemisph.	R	VOA (English/Spanish)	USA	English / Spanish				7.20	
C-	3.975	RHCP	14	W. Hemisph.	R	VOA (French/Spanish)	USA	French / Spanish				7.35	
C-	3.975	RHCP	14	W. Hemisph.	R	VOA (Vietnamese/Tibetan etc.)	USA	Vietnamese / Tibetan				7.45	
C-	3.975	RHCP	14	W. Hemisph.	R	VOA (Lao/Korean)	USA	Lao / Korean				7.53	
C-	3.975	RHCP	14	W. Hemisph.	R	VOA (Chinese)	USA	Chinese				7.60	
C-	3.975	RHCP	14	W. Hemisph.	R	VOA (English)	USA	English				6.6	
C-	4.015	RHCP	44	W. Hemisph.	O	NHK/TV Tokyo/ABC	USA	English	NTSC		No	6.65	
C-	4.045	RHCP	35	Global	TV	RFO Tahiti	France	French	SECAM		No	6.65	
C-	4.135	LHCP	37	Global	TV	Nine Netw. Australia	Australia	English	PAL		No	6.65	
C-	4.135	RHCP	36	Global	O	occasional video	varies		PAL				
C-	4.167	LHCP	38	Global	O	occasional video	New Zealand	English	PAL				
C-	4.188	RHCP	38	Global	TV	TV New Zealand	New Zealand	English	digital				
Ku1-	11.135	Hor	79	E. Hemisph.	O	NHK feeds	Japan	Japanese	PAL			6.8/7.8	
Ku1-	11.508	Hor	79	E. Hemisph.	TV	TV Asahi	Japan	Japanese	NTSC			6.8/7.8	
INTELSAT 702 177.0° EAST													
C-	4.166	RHCP	38	Global	O	occasional video	Japan	Japanese	NTSC			6.65	
C-	4.177	LHCP	39	Global	TV	AFRTS	USA	English	B-MAC		Yes		
C-	4.177	RHCP	39	Global	R	AFRTS Radio	USA	English				7.4	
C-	4.187	LHCP	40	Global	O	occasional video	Taiwan	Taiwanese Chinese	NTSC			6.65	
INTELSAT 701 174.0° EAST													
C-	4.166	RHCP	38A	Global	O	occasional video						6.65	
C-	4.188	RHCP	38B	Global	O	occasional video						6.65	
PANAMSAT-2 169.0° EAST													
C-	3.723	Hor		China Beam	TV	Chinese TV Network	China	Chinese	dig. compression		No		
C-	3.723	Hor		China Beam	TV	Asia Business News	China	Chinese	dig. compression		No		
C-	3.780	Hor		China Beam	TV	Discovery Channel	USA	English	B-MAC	EuroCrypt	No		
C-	3.804	Ver		China Beam	TV	MTV Mandarin	China	Mandarin	B-MAC	EuroCrypt	No		
C-	3.862	Ver		China Beam	TV	ESPN International	USA	English	PAL		No	6.8	
C-	3.901	Ver		China Beam	TV	Country Music TV	USA	English	dig. compression		No		
C-	3.901	Ver		China Beam	TV	BBC World Service TV	USA	English	dig. compression		No		
C-	3.925	Hor		China Beam	TV	Worldnet	USA	English	PAL		No	6.8	

ITU REGION 3 STATIONS IN FREQUENCY ORDER

Band	Freq.	Pol.	Tr.	Beam	Sce.	Program	Country	Language	System	Encr.	TT	A.M.	A.S.
C-	3.967	Hor		China Beam	TV	CNN Int	USA	English	PAL		No	6.3	
C-	3.967	Hor		China Beam	R	CNN Radio	USA	English					
C-	3.990	Ver		China Beam	TV	Prime International	USA	English	dig. compression		No	6.6/7.8	6.2/6.8
C-	4.030	Hor		China Beam	TV	NHK Int. TV	Japan	Japanese	NTSC		No	6.6/7.8	
C-	4.090	Hor		China Beam	TV	ABS/CBN	AUS/CHN	various	dig. compression		No	6.2/6.8	
C-	4.093	Ver		China Beam	TV	NBC Asia	USA	English	dig. compression		No	6.2/6.8	
C-	4.150	Ver		China Beam	TV	TNT/Cartoon Netw.	USA	English	dig. compression		No		
C-	4.150	Ver		China Beam	TV	Chinese TV Network	USA	English	dig. compression		No		
C-	4.150	Ver		China Beam	TV	Asia Business News	USA	English	dig. compression		No		
Ku2-	12.410	Ver		China Beam	TV	TVBS	Ciina	Chinese	PAL				
Ku3-	12.524	Ver		China Beam	TV	Chinese TV Network	Taiwan	English					6.2/6.8
Ku3-	12.704	Hor		China Beam	TV	Meishi Entertainment TV	Ciina	Chinese	dig. compression				
Ku3-	12.730	Ver		China Beam	TV	Channel KTV	Ciina						

SUPERBIRD B1 162.0° EAST

Band	Freq.	Pol.	Tr.	Beam	Sce.	Program	Country	Language	System	Encr.	TT	A.M.	A.S.
Ku2-	12.310	Hor	2B	Spot	O	Superbird B color bar	Japan	Japanese	NTSC		No	dig.	
Ku2-	12.330	Ver	2B	Spot	O	Channel Yokohama	Japan	Japanese	NTSC		No	dig.	
Ku2-	12.370	Ver	2B	Spot	TV	Family Theatre	Japan	Japanese	NTSC		No		
Ku2-	12.410	Ver	6A	Spot	TV	Cable Soft Netw.	Japan	Japanese	NTSC	Skyport	No		PCM
Ku2-	12.430	Hor	5B	Spot	TV	Cable Soft Network	Japan	Japese	NTSC	Skyport	No		PCM
Ku2-	12.450	Ver	7	Spot	TV	Super Channel	USA	English	NTSC	Skyport	No		PCM
Ku2-	12.450	Ver	7	Spot	TV	Weather Channel	USA	English	NTSC	Skyport	No		PCM
Ku2-	12.490	Ver	8	Spot	TV	Nihon Cable TV Netw.	Japan	Japanese	NTSC		No		PCM
Ku2-	12.530	Ver	8	Spot	TV	Life Design Channel	Japan	Japanese	NTSC		No		PCM
Ku2-	12.550	Ver	19	Spot	TV	Green Channel	Japan	Japanese	NTSC		No		PCM
Ku2-	12.570	Ver	11	Spot	TV	Space Vision Network	Japan	Japanese	NTSC	Skyport	No		PCM
Ku2-	12.610	Ver	13	Spot	TV	Asahi New Star	Japan	Japanese	NTSC	Skyport	No		dig.
Ku2-	12.650	Ver	15	Spot	TV	Japan Cable Television	Japan	Japanese	NTSC	Skyport	No		PCM
Ku2-	12.690	Ver	19	Spot	TV	MTV Japan (Music Channel)	UK	English	NTSC	Skyport	No		PCM
Ku2-	12.710	Hor	19	Spot	TV	Playboy Channel	Japan	Japanese	NTSC	Skyport	No		PCM
Ku2-	12.730	Ver	19	Spot	TV	Star Channel	Japan	Japanese	NTSC	Skyport	No		PCM

OPTUS B1 160.0° EAST

Band	Freq.	Pol.	Tr.	Beam	Sce.	Program	Country	Language	System	Encr.	TT	A.M.	A.S.
Ku2-	12.313	Hor	1	NA	O	CAA Communications	Australia	English	PAL		No		
Ku2-	12.375	Hor	10	NA	O	occasional video	Australia	English	PAL				
Ku2-	12.394	Ver	1	NA	O	occasional video	Australia	English	PAL/NTSC				
Ku2-	12.421	Ver	3L	NA	O	7 Network (SPS)	Australia	English	E-PAL	VAMP	No		7.38/7.56
Ku2-	12.438	Ver	3U	NA	TV	TV Oceania	Japan	Japanese	PAL	VideoCrypt			
Ku2-	12.456	Ver	11	NA	O	Network 10 (interchange)	Australia	English	E-PAL	VAMP			
Ku2-	12.482	Ver	4L	NA	O	Network 10	Australia	English	PAL	EuroCrypt			
Ku2-	12.533	Ver	4U	NA	O	SBS Specialized Broadcasting	Australia	English	B-MAC				7.38/7.56
Ku2-	12.548	Ver	5	SE	O	ABC TV (interchange)	Australia	English	B-MAC				7.38/7.56
Ku2-	12.597	Ver	13L	NA	TV	SkyTV	Australia	English	PAL	EuroCrypt	No		7.38/7.56
Ku2-	12.626	Hor	6	NE	TV	ABC TV HACBSS	Australia	English	B-MAC	EuroCrypt			
Ku2-	12.661	Ver	7	SE	TV	ABC HACBSS	Australia	English	B-MAC	EuroCrypt			

ITU REGION 3 STATIONS IN FREQUENCY ORDER

Band	Freq.	Pol.	Tr.	Beam	Sce.	Program	Country	Language	System	Encr.	TT	A.M.	A.S.
Ku2-	12.688	Hor	15	NE	TV	Queensland Television	Australia	English	B-MAC	EuroCrypt	Yes		
Ku2-	12.725	Ver	8	NA	O	Sky Special Event	Australia	English	B-MAC				
SUPERBIRD A 158.0° EAST												
Ku3-	12.290	Ver	2A	Spot	TV	Feisuo Satellite TV	Japan	Japanese	dig. compression			7.75/6.2	
Ku3-	12.290	Ver	2A	Spot	TV	Kuoshin Satellite TV	Japan	Japanese	dig. compression			7.75/6.2	
Ku3-	12.290	Ver	2A	Spot	TV	Friendly TV	Japan	Japanese	dig. compression			7.75/6.2	
Ku3-	12.300	Ver	2A	Spot	O	Tokyo Broadcasting System	Japan	Japanese	NTSC			7.75/6.2	
Ku3-	12.320	Hor	2B	Spot	O	Tokyo Broadcasting System	Japan	Japanese	NTSC			7.75/6.2	
Ku3-	12.350	Hor	3B	Spot	O	Nippon TV Network	Japan	Japanese	NTSC			7.75/6.2	
Ku3-	12.380	Hor	5A	Spot	O	Fuji TV Network	Japan	Japanese	NTSC			7.75/6.2	
Ku3-	12.520	Hor	13	Spot	O	TV Tepco	Japan	Japanese	NTSC			6.2	
Ku3-	12.610	Hor	13	Spot	O	Superbird A test patern	Japan	Japanese	NTSC			6.2	
OPTUS A3 156.0° EAST												
Ku2-	12.341	Ver	2	WA	TV	Golden West Network	Australia	English	B-MAC	EuroCrypt	Yes		
Ku2-	12.436	Hor	11	NE	TV	TV Oceania	Japan	Japanese	PAL	VideoCrypt	No		7.38/7.56
Ku2-	12.501	Hor	12	CA	TV	ABC HACBSS	Australia	English	B-MAC	VideoCrypt	Yes		
Ku2-	12.533	Ver	5	SE	O	occasional video	Australia	English	B-MAC				
Ku2-	12.629	Hor	14	CA	O	RCTS (Imparja)	Australia	English	B-MAC	VideoCrypt	Yes		
Ku2-	12.661	Ver	7	WA	TV	Enterprise Channel	Australia	English	B-MAC		Yes		
Ku2-	12.661	Hor	7	WA	TV	TV Oceania	Japan	Japanese	PAL		No		7.38/7.56
Ku2-	12.693	Hor	15	CA	TV	ABC HACBSS	Australia	English	B-MAC	VideoCrypt	Yes		
Ku2-	12.725	Ver	8	WA	TV	ABC HACBSS	Australia	English	B-MAC	EuroCrypt	Yes		
JCSAT 2 154.0° EAST												
Ku2-	12.268	Ver	1	Spot	TV	Japan Leisure Channel	Japan	Japanese	NTSC		No		PCM
Ku2-	12.283	Hor	2	Spot	TV	Japan Religious Channel	Japan	Japanese	NTSC		No		PCM
Ku2-	12.283	Hor	2	Spot	TV	Cable TV Access Channel	Japan	Japanese	NTSC		No		PCM
Ku2-	12.298	Ver	3	Spot	TV	Business News Network	Japan	Japanese	NTSC		No		PCM
Ku2-	12.313	Hor	4	Spot	TV	Nikkei Satellite News	Japan	Japanese	NTSC		No		
Ku2-	12.328	Ver	5	Spot	O	JCSat 2 test patern	Japan	Japanese	NTSC			6.2	
Ku2-	12.343	Hor	6	Spot	TV	Diamond Channel	Japan	Japanese	NTSC	M-System	No	dig.	dig.
Ku2-	12.358	Ver	6	Spot	TV	M Channel (JMTV)	Japan	Japanese	NTSC	M-System	No	dig.	dig.
Ku2-	12.373	Hor	8	Spot	TV	Green Channel	Japan	Japanese	NTSC		No		PCM
Ku2-	12.403	Hor	10	Spot	TV	M Channel (JMTV)	Japan	Japanese	NTSC	M-System	No		
Ku2-	12.403	Ver	10	Spot	TV	Rainbow Channel	Japan	Japanese	NTSC		No		PCM
Ku2-	12.448	Hor	13	Spot	TV	Midnight Blue	Japan	Japanese	NTSC		No		
Ku2-	12.493	Hor	16	Spot	TV	Satellite Culture Japan	Japan	Japanese	NTSC	M-System	No	dig.	dig.
Ku2-	12.523	Hor	18	Spot	TV	Space Shower TV	Japan	Japanese	NTSC	COATEC	No	6.2	PCM
Ku2-	12.538	Ver	21	Spot	TV	Keirin Channel	Japan	Japanese	NTSC		No		PCM
Ku2-	12.553	Hor	21	Spot	TV	Kids Station	Japan	Japanese	NTSC		No		PCM
Ku2-	12.583	Hor	22	Spot	TV	Japan Sports Channel	Japan	Japanese	NTSC	COATEC			PCM
Ku2-	12.598	Ver	23	Spot	R	PCM Zipang	Japan	Japanese	NTSC	COATEC			PCM
Ku2-	12.598	Ver	23	Spot	R	R Sky	Japan	Japanese	NTSC	COATEC			PCM
Ku2-	12.613	Hor	24	Spot	TV	Satellite ABC	Japan	Japanese	NTSC	COATEC	No		PCM

ITU REGION 3 STATIONS IN FREQUENCY ORDER

Band	Freq.	Pol.	Tr.	Beam	Sce.	Program	Country	Language	System	Encr.	TT	A.M.	A.S.
Ku2-	12.628	Ver	25	Spot	R	Music Bird / Satellite Music	Japan	Japanese	NTSC	COATEC	No		PCM
Ku2-	12.643	Hor	23	Spot	TV	Satellite Theatre	Japan	Japanese	NTSC	COATEC	No		PCM
Ku2-	12.673	Hor	24	Spot	TV	BBC WSTV	UK	various	NTSC	COATEC			PCM
Ku2-	12.688	Hor	25	Spot	O	test pattern			NTSC				PCM

JCSAT 1 150.0° EAST

Band	Freq.	Pol.	Tr.	Beam	Sce.	Program	Country	Language	System	Encr.	TT	A.M.	A.S.
Ku2-	12.283	Hor	2	Spot	O	TV Asiha (SNG)	Japan	Japanese	NTSC			6.2	
Ku2-	12.328	Ver	5	Spot	O	test pattern JCSat 1	Japan	Japanese	NTSC			6.2	
Ku2-	12.373	Hor	8	Spot	O	NHK (SNG)	Japan	Japanese	NTSC				PCM
Ku2-	12.403	Hor	10	Spot	O	NHK (SNG)	Japan	Japanese	NTSC				PCM
Ku2-	12.463	Ver	15	Spot	TV	P-Sat	Japan	Japanese	NTSC		No		PCM
Ku2-	12.658	Ver	15	Spot	O	JCSat 1 color bar	Japan	Japanese	NTSC				PCM

STATSIONAR 16 145.0° EAST

Band	Freq.	Pol.	Tr.	Beam	Sce.	Program	Country	Language	System	Encr.	TT	A.M.	A.S.
C-	3.875	RHCP	6	Spot	TV	Dubl-1	CIS	Russian	SECAM		No	7.5	
C-	3.875	RHCP	6	Spot	R	R Rossia	CIS	Russian				7.0	
Ku1-	11.525	RHCP	6	Spot	R	Sakha TV	CIS	Russian	SECAM			7.0	

RIMSAT-2 (GORIZONT 30) 142.5 EAST

Band	Freq.	Pol.	Tr.	Beam	Sce.	Program	Country	Language	System	Encr.	TT	A.M.	A.S.
C-	3.675	LHCP		Spot	TV	Music Asia	India	Hindi	PAL		No	6.6	
C-	3.725	LHCP		Spot	TV	Ray TV	Shri Lanka	Tamil	PAL		No	6.6	
C-	3.825	LHCP		Spot	TV	Vi.Jay TV	Shri Lanka	Tamil	PAL		No	6.6	
C-	3.875	LHCP		Spot	TV	EM-TV	Papua New Guinea	English	English	PAL	No		6.6

GORIZONT 18 140.0° EAST

Band	Freq.	Pol.	Tr.	Beam	Sce.	Program	Country	Language	System	Encr.	TT	A.M.	A.S.
C-	3.675	RHCP	6	Spot	TV	Orbita-1	CIS	Russian	SECAM		No	7.0	
C-	3.675	RHCP	6	Spot	R	R Mayak	CIS	Russian				7.5	
C-	3.725	RHCP	10	Global	TV	NTV	CIS	Russian	SECAM		No	7.5	

APSTAR-1 138.0° EAST

Band	Freq.	Pol.	Tr.	Beam	Sce.	Program	Country	Language	System	Encr.	TT	A.M.	A.S.
C-	3.636	Ver		China Beam	TV	Unique Business Channel	China	Chinese	dig. compression		No		
C-	3.636	Ver		China Beam	TV	CBHS Hour	China	Chinese	dig. compression		No		
C-	3.636	Ver		China Beam	TV	Rainbow Channel	China	Chinese	dig. compression		No		
C-	3.636	Ver		China Beam	TV	TV 4 Channel	China	Chinese	dig. compression		No		
C-	3.666	Ver		China Beam	TV	Chinese Satellite TV (CSTV)	China	Chinese	dig. compression		No		
C-	3.666	Ver		China Beam	TV	CSTV Music Channel	China	Chinese	dig. compression		No		
C-	3.666	Ver		China Beam	TV	CSTV News Channel	China	Chinese	dig. compression		No		
C-	3.666	Ver		China Beam	TV	Taiwan Satellite TV	Taiwan	English	PAL		No		
C-	3.800	Hor		China Beam	TV	CETV Shandong	China	Chinese	PAL		No	6.6	
C-	3.840	Ver		China Beam	TV	China Educational TV-1	China	Chinese	PAL		No	6.6	
C-	3.860	Hor		China Beam	TV	MTV Mandarin	China	Chinese	B-MAC	EuroCrypt	No		
C-	3.880	Ver		China Beam	TV	China Educational TV-2	China	Chinese	PAL		No	6.6	
C-	3.900	Ver		China Beam	TV	TVBS	China	Chinese	dig. compression		No		dig
C-	3.990	Ver		China Beam	TV	CNN International	USA	English	PAL		No	6.8	
C-	3.980	Ver		China Beam	R	CNN Radio	USA	English	PAL		No	6.3	
C-	4.020	Ver		China Beam	TV	TNT/Cartoon Network	USA	English	B-MAC		No	6.2	

ITU REGION 3 STATIONS IN FREQUENCY ORDER

Band	Freq.	Pol.	Tr.	Beam	Sce.	Program	Country	Language	System	Encr.	TT	A.M.	A.S.
C-	4.040	Hor		China Beam	TV	Walt Disney TV	USA	English	B-MAC	EuroCrypt	No		dig.
C-	4.060	Hor		China Beam	TV	HBO Asia	USA	English	B-MAC	EuroCrypt	No		dig.
C-	4.100	Ver		China Beam	TV	ESPN International	USA	English	B-MAC	EuroCrypt	No		dig.
C-	4.140	Ver		China Beam	TV	Discovery Channel	USA	English	B-MAC	EuroCrypt	No		dig.
C-	4.160	Hor		China Beam	TV	China Entertainment TV	China	Chinese	PAL		No	6.6	
C-	4.180	Ver		China Beam	TV	PHTV-Sanlih Channel	China	Chinese	dig. compression		No		
C-	4.180	Ver		China Beam	TV	PHTV-Toei Channel	China	Chinese	dig. compression		No		
C-	4.180	Ver		China Beam	TV	PHTV-Information Channel	China	Chinese	dig. compression		No		

APSTAR-1A 134.0° EAST

Band	Freq.	Pol.	Tr.	Beam	Sce.	Program	Country	Language	System	Encr.	TT	A.M.	A.S.
C-	3.820	Ver		China Beam	TV	China Central TV 2	China	Chinese	PAL		No	6.6	
C-	3.860	Ver		China Beam	TV	China Central TV 1	China	Chinese	PAL		No	6.6	
C-	4.020	Ver		China Beam	TV	Zhejiang TV Station	China	Chinese	PAL		No	6.6	
C-	4.040	Hor		China Beam	TV	Xizang TV Station	China	Chinese	PAL		No	6.6	
C-	4.080	Hor		China Beam	TV	Sichuan TV Station	China	Chinese	PAL		No	6.6	
C-	4.100	Ver		China Beam	TV	Shandong TV Staton 1	China	Chinese	PAL		No	6.6	
C-	4.120	Hor		China Beam	TV	Xinjiang TV Staton 1	China	Chinese	PAL		No	6.6	
C-	4.180	Ver		China Beam	TV	China Central TV 2	China	Chinese	PAL		No	6.6	

GORIZONT-G41 130.0° EAST

Band	Freq.	Pol.	Tr.	Beam	Sce.	Program	Country	Language	System	Encr.	TT	A.M.	A.S.
C-	3.775	LHCP		Zone	TV	Lao National TV	Laos		PAL		No	6.60	

JCSAT-3 128.0° EAST

Band	Freq.	Pol.	Tr.	Beam	Sce.	Program	Country	Language	System	Encr.	TT	A.M.	A.S.
C-	3.980	Hor		Spot	0	JCSat test pattern	Japan		PAL		No	6.8	PCM
Ku2-	12.280	Hor		Spot	0	JCSat test pattern	Japan		NTSC		No		PCM
Ku2-	12.386	Hor		Spot	0	JCSat test pattern	Japan		NTSC		No		
Ku2-	12.523	Hor		Spot	TV	Car Information TV	Japan	Japanese	dig. compression		No		dig.
Ku2-	12.523	Hor		Spot	TV	Satellite ABC	Japan	Japanese	dig. compression		No		dig.
Ku2-	12.538	Ver		Spot	TV	Pioneer Music Satellite	Japan	Japanese	dig. compression		No		dig.
Ku2-	12.538	Ver		Spot	TV	Travel Channel	Japan	Japanese	dig. compression		No		dig.
Ku2-	12.538	Ver		Spot	TV	Voice of the Earth	Japan	Japanese	dig. compression		No		dig.
Ku2-	12.538	Ver		Spot	TV	Mondo 21	Japan	Japanese	dig. compression		No		dig.
Ku2-	12.568	Ver		Spot	TV	World Entertainment	Japan	Japanese	dig. compression		No		dig.
Ku2-	12.568	Ver		Spot	TV	BGV Channel	Japan	Japanese	dig. compression		No		dig.
Ku2-	12.568	Ver		Spot	TV	Karaoke Channel	Japan	Japanese	dig. compression		No		dig.
Ku2-	12.568	Ver		Spot	TV	Satellite Culture Japan	Japan	Japanese	dig. compression		No		dig.
Ku2-	12.568	Ver		Spot	TV	Bloomberg Info. TV	Japan	Japanese	dig. compression		No		dig.
Ku2-	12.568	Ver		Spot	TV	Digital Tampa 501	Japan	Japanese	dig. compression		No		dig.
Ku2-	12.568	Ver		Spot	TV	Digital Tampa 502	Japan	Japanese	dig. compression		No		dig.
Ku2-	12.568	Ver		Spot	TV	Sound with Radio	Japan	Japanese	dig. compression		No		dig.
Ku2-	12.598	Ver		Spot	TV	Nihon TV Network	Japan	Japanese	dig. compression		No		dig.
Ku2-	12.598	Hor		Spot	TV	Space Vision Network	Japan	Japanese	dig. compression		No		dig.
Ku2-	12.643	Ver		Spot	TV	Satellite News	Japan	Japanese	dig. compression		No		dig.
Ku2-	12.643	Ver		Spot	TV	Chinese TV Network	Japan	Japanese	dig. compression		No		dig.
Ku2-	12.643	Ver		Spot	TV	Korea Vision	Korea	Chinese	dig. compression		No		dig.
Ku2-	12.658	Hor		Spot	TV	Kids Station	Japan	Japanese	dig. compression		No		dig.

ITU REGION 3 STATIONS IN FREQUENCY ORDER

Band	Freq.	Pol.	Tr.	Beam	Sce.	Program	Country	Language	System	Encr.	TT	A.M.	A.S.
Ku2-	12.658	Ver		Spot	TV	Prefec Mulch	Japan	Japanese	dig. compression		No		dig.
Ku2-	12.658	Ver		Spot	TV	Prefec Today	Japan	Japanese	dig. compression		No		dig.
Ku2-	12.688	Ver		Spot	TV	Kikkei Satellite News	Japan	Japanese	dig. compression		No		dig.
Ku2-	12.688	Ver		Spot	TV	Cinefil Imagica	Japan	Japanese	dig. compression		No		dig.
Ku2-	12.688	Ver		Spot	TV	Theatre Television	Japan	Japanese	dig. compression		No		dig.
Ku2-	12.703	Hor		Spot	TV	KN Television	Japan	Japanese	dig. compression		No		dig.
Ku2-	12.703	Hor		Spot	TV	IBC TV Network	Japan	Japanese	dig. compression		No		dig.
Ku2-	12.718	Ver		Spot	TV	Shopping Channel	Japan	Japanese	dig. compression		No		dig.
KOREASAT 1 116.0° EAST													
Ku2-	11.823	LHCP		Spot	TV	KBS Satellite TV 1	Korea	Korean	PAL		No		
KOREASAT 1 116.0° EAST													
Ku2-	11.823	LHCP		Spot	TV	KBS Satellite TV 2	Korea	Korean	PAL		No		
Ku2-	12.550	Ver		Spot	O	occasional video	Korea	Korean	PAL				
Ku2-	12.590	Ver		Spot	O	occasional video	Korea	Korean					
Ku2-	12.630	Ver		Spot	O	occasional video	Korea	Korean					
Ku2-	12.670	Ver		Spot	O	occasional video	Korea	Korean					
CHINASAT 5 115.5° EAST													
C-	3.760	Hor		Spot	TV	Zhejiang TV Station	China	Chinese	PAL		No	6.6	
C-	3.803	Hor		Spot	TV	China Central TV-2	China	Chinese	PAL		No	6.6	
C-	3.883	Hor		Spot	TV	China Central TV-1	China	Chinese	PAL		No	6.6	
PALAPA B2P 113.0° EAST													
C-	3.620	Hor	4	Spot	TV	CNBC Asia	USA	English	PAL	EuroCrypt	No	6.6	
C-	3.720	Hor	4	Spot	TV	Discovery Channel	USA	English	B-MAC	EuroCrypt	No	6.6	
C-	3.740	Ver	5	Spot	TV	Rajawari Citra Televisi Indonesia (RCTI)	Indonesia	Indonesian	B-MAC	EuroCrypt	No	6.6	
C-	3.820	Hor	6	Spot	TV	TV-1 (Malaysia)	Malaysia	English	PAL		No	6.6	
C-	3.840	Ver	7	Spot	TV	TVRI	Indonesia	Indonesian	PAL		No	6.8	
C-	3.880	Hor	9	Spot	TV	ATVI Australia	Australia	English	PAL		No	7.20	6.3/6.48
C-	3.880	Hor	9	Spot	R	R Australia	Australia	English	PAL		No	6.80	
C-	3.900	Ver	10	Spot	TV	TV-3 Malaysia	Malaysia	English	PAL		No	6.80	
C-	3.920	Ver	11	Spot	TV	GMA Philippines	Philippines	English	B-MAC	EuroCrypt	Yes		
C-	3.960	Ver	13	Spot	TV	ABS-CBN Philippines	Philippines	English	PAL		Yes	6.8	
C-	3.980	Ver	14	Spot	TV	CNN International	USA	English	B-MAC		No		
C-	4.000	Ver	15	Spot	TV	HBO Asia	USA	English	PAL		No	6.8	
C-	4.020	Hor	16	Spot	TV	AN-TEVE	Indonesia	Indonesian	dig. compression		No		
C-	4.050	Ver	17	Spot	TV	KBP Peoples Network	Philippines	English	PAL		No		
C-	4.080	Hor	19	Spot	TV	Televisi Pendidikan Indonesia (TPI)	Indonesia	Indonesian	B-MAC	EuroCrypt	No	6.8	
C-	4.100	Ver	20	Spot	TV	ESPN International	USA	English	PAL		Yes		
C-	4.120	Hor	21	Spot	TV	MTV Asia	UK	English	PAL		No	6.6	
C-	4.140	Ver	24	Spot	TV	Singapore Int. TV	Singapore	English	PAL		No	6.8	
C-	4.140	Ver	22	Spot	TV	R TV Brunei	Brunei	English	PAL		No		6.2/6.8
C-	4.160	Hor	22	Spot	TV	Canal France Int.	France	French	PAL		No	6.6	
C-	4.160	Ver	23	Spot	TV	Canal France International	France	French	PAL		No		

225

ITU REGION 3 STATIONS IN FREQUENCY ORDER

Band	Freq.	Pol.	Tr.	Beam	Sce.	Program	Country	Language	System	Encr.	TT	A.M.	A.S.
C-	4.180	Hor	23	Spot	TV	Star TV	Australia	English	PAL		No		
BS-3B 110.0° EAST													
Ku2-	11.804	RHCP	5	Spot	TV	Japan Satellite Broadcasting Co.	Japan	Japanese	NTSC	COATEC	No		PCM
Ku2-	11.804	RHCP	5	Spot	R	St. Giga	Japan	Japanese	NTSC				PCM
Ku2-	11.880	RHCP	9	Spot	O	HDTV test transmission	Japan	Japanese	MUSE HDTV				
PALAPA B2R 108.0° EAST													
C-	4.000	Hor	15	Spot	TV	TVRI Indonesia	Indonesia	Indonesian	PAL		No	6.8	
ASIASAT 1 (HONG KONG) 105.5° EAST													
C-	3.760	Hor	2	NB	TV	Guizhou TV Station	Mongolia	Mongolian	SECAM		No	6.6	
C-	3.800	Hor	3	NB	TV	Prime Sports (Mandarin)	Hong Kong	Mandarin	NTSC		No	5.94	
C-	3.800	Hor	3	NB	TV	Prime Sports	Hong Kong	English	PAL		No		5.58/5.76
C-	3.840	Hor	4	NB	TV	MTV Asia	Hong Kong	Mandarin	NTSC-M		No		5.58/5.76
C-	3.840	Hor	4	NB	R	BBC World Service	UK	various	PAL		No	5.94/6.12	
C-	3.860	Ver	11	SB	TV	Prime Sports (Mandarin)	Hong Kong	Mandarin	PAL		No	7.2	
C-	3.860	Ver	11	SB	TV	Prime Sports	Hong Kong	English	PAL		No		6.3/6.48
C-	3.880	Hor	5	NB	TV	Star Movies	China	Mandarin	PAL		No		5.58/5.76
C-	3.900	Ver	5	SB	TV	MTV Asia	Hong Kong	Mandarin	PAL		No		6.30/6.48
C-	3.900	Ver	5	SB	R	BBC World Service	UK	various	PAL		No	7.20	
C-	3.920	Hor	6	NB	TV	Chinese Channel (Mandarin)	Hong Kong	Mandarin	NTSC		No	6.20	
C-	3.920	Hor	6	NB	TV	Chinese Channel	UK	English	NTSC		No		5.58/576
C-	3.920	Hor	6	NB	R	BBC World Service (Chinese)	UK	Chinese	PAL		No		5.94/6.12
C-	3.940	Ver	5	SB	TV	BBC Asia (mandarin)	UK	Mandarin	PAL		No	7.20	
C-	3.940	Ver	5	SB	TV	BBC Asia	UK	English	PAL		No		6.30/6.48
C-	3.960	Hor	7	NB	TV	Star Plus	Hong Kong	English	NTSC		No		5.58/5.76
C-	3.960	Hor	7	NB	R	BBC World Service (Arabic)	UK	Arabic	PAL		No		5.94/6.12
C-	3.980	Ver	7	SB	TV	Zee TV (mandarin)	Hong Kong	Mandarin	PAL		No	6.8	
C-	3.980	Ver	7	SB	TV	Zee TV (English)	Hong Kong	English	PAL		No		6.30/6.48
C-	3.980	Ver	7	SB	R	BBC World Service	UK	various	PAL		No	7.20	
C-	4.020	Hor	8	SB	TV	Star TV Plus	Hong Kong	Mandarin	PAL		No	7.20	
C-	4.020	Hor	8	SB	R	BBC World Service	UK	various	NTSC		No	6.6	
C-	4.040	Ver	9	NB	TV	Yunnan TV Station-1	Hong Kong	Chinese	NTSC		No	6.6/7.5	
C-	4.040	Ver	9	NB	TV	Guizhou TV Station-1	Hong Kong	Chinese	PAL		No	7.5	
C-	4.040	Ver	9	NB	R	Yunnan PBS	Hong Kong	Chinese	PAL		No	7.5	
C-	4.040	Ver	9	NB	R	Guizhou PBS	Hong Kong	Chinese	PAL		No		
C-	4.060	Ver	9	SB	TV	Star TV Chinese	Hong Kong	Chinese	PAL		No		6.30/6.48
C-	4.100	Ver	10	SB	TV	Pakistan TV	Pakistan	Urdu	PAL		No	6.6	
C-	4.120	Hor	11	NB	TV	CCTV-4	China	Chinese	NTSC		No	6.2	
C-	4.140	Ver	11	SB	TV	Myanmar TV	Burma	Burmese	PAL		No	6.6	
C-	4.180	Ver	11	SB	TV	Star TV	Burma	Burmese	NTSC		No	6.6	
GORIZONT 23 103.0° EAST													
C-	3.675	RHCP	6	Global	TV	Dub'l IV	CIS	Russian	SECAM		No	7.0	
C-	3.876	RHCP	8	Global	TV	ART	Russia	Arabic	PAL		No	7.5	

ITU REGION 3 STATIONS IN FREQUENCY ORDER

Band	Freq.	Pol.	Tr.	Beam	Sce.	Program	Country	Language	System	Encr.	TT	A.M.	A.S.
ASIASAT 2 (HONG KONG) 100.5° EAST													
C-	3.640	Hor		SB	TV	Egyptian Satellite Channel	Egypt	Arabic	PAL		No	6.6	
C-	3.660	Ver		SB	TV	TV Shopping Network	China	Chinese	PAL		No	6.2/6.8	
C-	3.680	Hor		SB	TV	TV Mongol	Mongolia	Mongolian	SECAM	Syster	No	6.6	
C-	3.680	Hor		SB	TV	WorldNet	Mongolia	Mongolian	PAL		No	6.6	
C-	3.720	Hor		SB	TV	Henan Satellite TV			PAL		No	6.6	
C-	3.760	Hor		SB	TV	Star TV	China	Chinese	PAL		No		6.30/6.48
C-	3.840	Hor		SB	TV	Guangdong Satellite TV	China	Chinese	PAL		No	6.6	
C-	3.900	Ver		SB	TV	Star Plus Japan	Japan	Japanese	dig. compression		No		
C-	3.900	Ver		SB	TV	Viva Channel	Japan	Japanese	dig. compression		No		
C-	3.900	Ver		SB	TV	Sky News	Japan	Japanese	dig. compression		No		
C-	3.900	Ver		SB	TV	BBC World Service TV	Japan	Japanese	dig. compression		No		
C-	3.960	Hor		SB	TV	China Central TV 4	China	Chinese	PAL		No	6.65	
C-	3.980	Ver		SB	TV	RTP International	Portugal	Portuguese	PAL		No	6.65	
C-	3.980	Ver		SB	R	RTP Antena 1	Portugal	Portuguese			No	7.02	
C-	4.000	Ver		SB	TV	Deutsche Welle TV	Germany	various	dig. compression		No		
C-	4.000	Ver		SB	TV	TV 5	France	French	dig. compression		No		
C-	4.000	Ver		SB	TV	MCM	France	French	dig. compression		No		
C-	4.000	Ver		SB	TV	RTE Int.	Spain	Spanish	dig. compression		No		
C-	4.000	Ver		SB	R	World Radio Network	UK	various	dig. compression		No		
GORIZONT 19 96.5° EAST													
C-	3.675	RHCP	6	Spot	TV	Orbita II	CIS	Russian	SECAM		No	7.0	
C-	3.675	RHCP	6	Spot	R	R Mayak	CIS	Russian			No	7.5	
C-	3.825	RHCP	9	North Hemisph.	TV	CCTV-4	China	Chinese	PAL		No	6.6	
C-	3.875	RHCP	10	North Hemisph.	TV	Azerbaijan Radio TV	Azerbaijan	Azerbaijani	SECAM		No	7.4	
INSAT 2B 94.0° EAST													
S-	2.575	LHCP	S1	Spot	TV	Doordarshan TV	India	Indian	PAL		No	5.5	
S-	2.615	LHCP	S1	Spot	TV	Doordarshan TV	India	Indian	PAL		No	5.5	
C-	3.865	Ver	C3	Spot	TV	DD Channel 7	India	Indian	PAL		No	6.6	
C-	4.120	Ver	C10	Spot	TV	DD Channel 1	India	Indian	PAL		No	5.5	
C-	4.140	Ver	C11	Spot	TV	DD Channel 10	India	Indian	PAL		No	5.5	
C-	4.160	Ver	C12	Spot	TV	DD Channel 2	India	Indian	PAL		No	5.8	
MEASAT-1 91.5° EAST													
C-	3.710	Hor		Spot	TV	Dai Truyen Hinh Vietnam	Vietnam	Vietnamese	PAL		No	6.5	
C-	4.00	Hor		Spot	O	occasional video	Vietnam	Vietnamese	PAL		No	6.5	
GORIZONT 20 90.0° EAST													
C-	3.675	RHCP	6	Spot	TV	Dub'l II	CIS	Russian	SECAM		No	7.0	
C-	3.675	RHCP	6	Spot	R	R Rossia	CIS	Russian			No	7.5	
C-	3.875	RHCP	10	Spot	R	R Mayak	CIS	Russian			No	7.5	
C-	3.916	RHCP	11	Spot	O	Orbita (feeds)	CIS	Russian	SECAM		No	7.0	
C-	3.916	RHCP	11	Spot	O	Dub'l II (feeds)	CIS	Russian	SECAM		No	7.5	

ITU REGION 3 STATIONS IN FREQUENCY ORDER

Band	Freq.	Pol.	Tr.	Beam	Sce.	Program	Country	Language	System	Encr.	TT	A.M.	A.S.
INSAT 1D 82.9° EAST													
S-	2.575	LHCP	S1	Spot	TV	Doordarshan TV	India	Indian	PAL		No	5.5	
S-	2.615	LHCP	S1	Spot	TV	Community TV	India	Indian	PAL		No	5.5	
GORIZONT 16/25 80.0° EAST													
C-	3.675	RHCP	6	Spot	TV	Dub'l IV	CIS	Russian	SECAM		No	7.0	
C-	3.675	RHCP	6	Spot	R	Rstancija Mayak (RSM)	CIS	Russian	SECAM		No	7.5	
C-	3.875	RHCP	10	North Hemisph.	TV	TV 6 Mockba	C.I.S.	Russian	PAL		No	7.5	
C-	3.875	Hor		North Hemisph.	TV	VTV-4 (Vietnam)	Vietnam	Vietnamese			No	7.5	
C-	3.875	RHCP	10	North Hemisph.	R	R Hanoi	Vietnam	Vietnamese	PAL		No	7.0	
THAICOM-1 78.5° EAST													
C-	3.760	Ver		Spot	TV	Army TV (ch. 5)	Thailand	Thai	PAL		No		5.5/6.6
C-	3.870	Ver		Spot	TV	Channel 11 Thailand	Thailand	Thai	PAL		No		5.5/6.6
C-	3.890	Ver		Spot	O	Channel 11 Thailand (feeds)	Thailand	Thai	PAL		No		5.5/6.6
C-	3.950	Ver		Spot	TV	Dai Truyen Hinh Vietnam	Vietnam	Vietnamese	PAL		No	7.5	
C-	3.750	Hor		Spot	TV	Channel 7 Thailand	Thailand	Thai	PAL		No		5.5/6.6
C-	3.950	Hor		Spot	TV	Channel 3 Thailand	Thailand	Thai	PAL		No		5.5/6.6
C-	3.970	Hor		Spot	TV	Channel 9 Thailand	Thailand	Thai	PAL		No		5.5/6.6
INSAT 2A 74.0° EAST													
C-	4.115	Hor	21	Spot	TV	Doordarshan TV	India	Indian	PAL		No	5.5	
PANAMSAT-4 68.5° EAST													
C-	3.785	Hor		Asian	TV	Asia Business TV	Japan	Japanese	PAL		No	6.8	
C-	3.785	Ver		Asian	TV	Discovery Channel	India	English	PAL		No	6.8	
C-	3.800	Ver		Asian	TV	BBC World Service TV	UK	English	PAL		No	6.6	
C-	3.836	Ver		Asian	TV	China Central TV 4	China	Chinese	PAL		No	6.8	
C-	3.865	Hor		Asian	TV	ESPN	USA	English	B-MAC		No		
C-	3.905	Ver		Asian	TV	Sony Entertainment TV	Japan	Japanese	NTSC		No	6.65	
C-	3.910	Hor		Asian	TV	Sony Entertainment Network	USA	English	PAL		No	7.2	
C-	4.030	Ver		Asian	TV	Doordarshan TV	India	English/Hindi/Urdu	PAL		No	6.80	
C-	4.085	Hor		Asian	TV	CNN International	USA	English	PAL		No	6.8	
C-	4.115	Hor		Asian	TV	TNT/Cartoon Network	USA	English	PAL		No	6.65	
C-	4.179	Ver		Asian	TV	Asia TV Network	India	Hindi	PAL		No	6.3/6.6	
C-	4.185	Hor		Asian	TV	MTV Asia	India	English	PAL		No		7.55/7.75
Ku3-	12.600	Ver		Asian	TV	NHK TV-Japan	Japan	Japanese	NTSC		No	6.65	
INTELSAT 704 66.0° EAST													
C-	4.055	RHCP	15	E. Hemisph.	TV	Canal France Internationale	France	French	SECAM		No	5.8	
INTELSAT 602 63.0° EAST													
C-	4.168	LHCP	38	Global	O	occasional video			PAL/SEC/NTSC			6.6	
C-	4.188	LHCP	38	Global	O	occasional video			PAL/SEC/NTSC			6.6	

ITU REGION 3 STATIONS IN FREQUENCY ORDER

Band	Freq.	Pol.	Tr.	Beam	Sce.	Program	Country	Language	System	Encr.	TT	A.M.	A.S.
							INTELSAT 703 57.0° EAST						
C-	3.750	RHCP	51	East Zone	TV	Sun Music TV (Tamil Sce.)	Sri Lanka	Tamil	PAL		No	6.3/6.6	
C-	3.750	RHCP	51	East Zone	TV	Gemini TV	Sri Lanka		PAL		No	6.3/6.6	
C-	3.750	RHCP	51	East Zone	TV	Money TV			PAL		No	6.3/6.6	
C-	3.808	RHCP	51	East Zone	TV	Sun TV	Sri Lanka	Tamil	PAL		No	6.6	
C-	3.980	RHCP	51	East Zone	TV	Asianet	India	Malayalam	PAL		No	6.6	
C-	4.050	LHCP	51	East Zone	TV	World Net	India	English	PAL		No	6.6	
C-	4.065	LHCP	51	East Zone	TV	NEPC TV	India	English	PAL		No	6.6	
C-	4.135	LHCP	51	East Zone	TV	TVI	India	English	PAL		No	6.6	
C-	4.178	LHCP	51	East Zone	TV	Muslim TV Ahmadiyya Int.			PAL		No	6.6	

TV = Television   R = Radio   O = Others

229

CHAPTER 12

STATIONS IN ALPHABETICAL ORDER

12.1. ITU REGION 1 TV

12.2. ITU REGION 1 RADIO

12.3. ITU REGION 2 TV

12.4. ITU REGION 2 RADIO

12.5. ITU REGION 3 TV

12.6. ITU REGION 3 RADIO

12.1. ITU REGION 1 TV IN ALPHABETICAL ORDER

Station	Country	Satellite	Position	Band	Frequency	Pol.	Tr	Beam
3-Sat	Ger/Aus/Swit	ASTRA 1A	19.2° East	Ku1-Band	11.347	Ver	10	Multi
Abu Dhabi TV	UAE	Arabsat 2A	26.0° East	C-Band	4.075	RHCP	17	Zone
Abu Dhabi TV	UAE	Arabsat 2A	26.0° East	Ku3-Band	12.619	Ver	6	Zone
Afghanistan TV	Afghanistan	Gorizont 11/17 (Stationar 5)	53.0° East	C-Band	3.675	RHCP	6	Spot
AFN-Television, Frankfurt	USA	Eutelsat II f2	10.0° East	Ku1-Band	11.174	Ver	27	Wide
AFRTS	USA	Intelsat 707	1.0° West	C-Band	4.175	RHCP	38	Global
AFRTS-SEB	USA	Intelsat 707	1.0° West	C-Band	3.900	RHCP	53	Zone
Al Jazeera Satellite Channel	Qatar	Eutelsat II f3	16.0° East	Ku1-Band	11.080	Hor	21	Wide
Al Jazeera Satellite Channel	Algeria	Arabsat 2A	26.0° East	Ku3-Band	12.521	Hor	1	Zone
Albanian TV (Shqiptar TV)	Albania	Eutelsat II f3	16.0° East	Ku1-Band	11.575	Ver	37	Wide
Algerian TV	Algeria	Eutelsat II f3	16.0° East	Ku3-Band	11.095	Hor	26B	Wide
Antena Tres TV	Spain	Hispasat	31.0° West	Ku3-Band	12.671	Hor	13	East Spot
ARD-1	Germany	ASTRA 1B	19.2° East	Ku1-Band	11.493	Ver	19	Multi
ARD/ZDF Kinderprogramm	Germany	ASTRA 1D	19.2° East	Ku1-Band	10.714	Hor	49	Multi
ARD/ZDF Kinderprogramm	Germany	Eutelsat II f1 (Hotbird 1/2)	13.0° East	Ku1-Band	11.055	Ver	21A	Super
ART	Saudi Arabia	Arabsat 2A	26.0° East	C-Band	3.781	LHCP	4	Zone
ART	Saudi Arabia	Arabsat 2A	26.0° East	Ku3-Band	12.685	Hor	9	Zone
ART (1st Net)	Saudi Arabia	Arabsat 2A	26.0° East	Ku3-Band	12.562	Hor	3	Zone
ART Europe	UK	Eutelsat II f1 (Hotbird 1/2)	13.0° East	Ku2-Band	12.015	Hor	40	Super
ART Europe	UK	Eutelsat II f3	16.0° East	Ku1-Band	11.556	Hor	32	Wide
ARTE	Germany	Eutelsat II f1 (Hotbird 1/2)	13.0° East	Ku3-Band	11.055	Hor	21A	Super
ARTE	France/Germany	Télécom 2B	5.0° West	C-Band	12.606	Ver	R3	Spot
ARTE	France/Germany	ASTRA 1D	19.2° East	Ku1-Band	10.714	Hor	49	Multi
ARTE	France/Germany	Eutelsat II f1 (Hotbird 1/2)	13.0° East	Ku1-Band	11.080	Ver	26B	Multi
Asia Business Channel		PanAmSat-4	68.5° East	C-Band	3.790	Hor	3C	East Hemisph.
AsiaNet		Orion 1	37.5° West	Ku1-Band	11.596	Hor	15	European Spot
ASTRA SPORT	AFS	PanAmSat-4	68.5° East	Ku3-Band	12.734	Ver	4KU	AFS Spot
ATN	Africa	PanAmSat-4	68.5° East	C-Band	4.182	Hor	C16U	East Hemisph.
ATV	Turkey	Türksat 1B	42° East	Ku1-Band	10.965	Ver	4	West Spot
ATV	Poland	AMOS F1	4.0° West	Ku1-Band	11.368	Hor	10L	West Spot
Bahrain TV	Bahrain	Arabsat 2A	26.0° East	C-Band	3.822	LHCP	6	Zone
Bayerisches Fernsehen	Germany	ASTRA 1C	19.2° East	Ku1-Band	11.141	Hor	45	Multi
BBC Orbit Arabic Service		Télécom 2C	3.0° East	Ku3-Band	12.648	Ver	R4	Spot
BBC Prime	UK	Intelsat 601	27.5° West	Ku1-Band	10.995	Hor	71	West Spot
BBC World	UK	Eutelsat II f1 (Hotbird 1/2)	13.0° East	C-Band	11.616	Ver	38	Wide
BBC World	UK	PanAmSat-4	68.5° East	Ku1-Band	3.865	Ver	5U	East Hemisph.
BHT (Bosnian TV)	Bosnia Herzegovina	Eutelsat II f2	10.0° East	Ku1-Band	11.080	Hor	21	Wide
BOP TV	Senegal	Intelsat 601	27.5° West	C-Band	3.884	LHCP	103A	East Hemisph.
Bop TV Mmabatho	Bophuthatswana	Intelsat 601	27.5° West	C-Band	3.931	RHCP	53	East Hemisph.
Bop TV-1	Bophuthatswana	Intelsat 602	63.0° East	C-Band	4.055	LHCP	18	South East Zone

ITU REGION 1 TV IN ALPHABETICAL ORDER

Station	Country	Satellite	Position	Band	Frequency	Pol.	Tr	Beam
Bravo	UK	ASTRA 1C	19.2°East	Ku1-Band	11.097	Ver	42	Multi
BSkyB	UK	ASTRA 1B	19.2°East	Ku1-Band	11.567	Ver	24	Multi
C-Span	USA	Intelsat 704	66.0°East	C-Band	4.177	RHCP	38	Global
C-Span	USA	Intelsat 601	27.5°West	C-Band	3.743	RHCP	81	East Hemisph.
C-Span/Worldnet	USA	Intelsat 703	57.0°East	C-Band	4.055	RHCP	13	East Hemisph.
Cable Plus Filmovy Kanal	Czech Republic	DFS-2 Kopernikus	28.5°East	Ku2-Band	11.475	Hor	A1	Spot
Canal Classico	Spain	Hispasat	31.0°West	Ku2-Band	12.227	LHCP	27	East Spot
Canal France Internationale	France	Arabsat 2A	26.0°East	C-Band	3.945	LHCP	12	Zone
Canal Horizons	Senegal	Intelsat 601	27.5°West	C-Band	3.803	RHCP	23A	East Hemisph.
Canal Horizons	Senegal	Eutelsat II f1(Hotbird 1/2)	13.0°East	Ku1-Band	11.408	Ver	10	Spot
Canal Jeunesse)	France	Télécom 2A	8.0°West	Ku3-Band	12.732	Ver	R6	Spot
Canal Jimmy	France	Télécom 2A	8.0°West	Ku3-Band	12.732	Ver	R6	Spot
Canal On	Spain	Hispasat	31.0°West	Ku3-Band	12.591	Ver	2	East Spot
Canal Plus	France	Télécom 2A	8.0°West	Ku3-Band	12.522	Ver	R1	Spot
Canal Plus	France	Télécom 2A	8.0°West	Ku3-Band	12.648	Ver	R4	Spot
Canal Sur	Spain	Hispasat	31.0°West	Ku3-Band	12.591	Ver	2	East Spot
Canale France Internationale	France	Intelsat 601	27.5°West	C-Band	3.915	Hor	23B	East Hemisph.
Cartoon Network	USA	ASTRA 1C	19.2°East	Ku1-Band	11.023	Hor	37	Multi
CCTV4	China	PanAmSat f3r	43.0°East	C-Band	4.180	Hor	15C	African Hemisph.
CDAT	AFS	PanAmSat-4	68.5°East	Ku3-Band	12.740	Ver	16U	AFS Spot
Channel 5	UK	ASTRA 1A	19.2°East	Ku1-Band	11.362	Ver	11	Multi
Channel Africa	South Africa	Intelsat 605	24.5°West	C-Band	4.166	RHCP	38A	Global
Channel-2 TV Israel	Israel	Intelsat 707	1.0°West	Ku1-Band	11.178	Ver	73	East Spot
Chinese Channel	UK	ASTRA 1D	19.2°East	Ku1-Band	10.788	Ver	54	Multi
Chinese News Entertainment	China	ASTRA 1C	19.2°East	Ku1-Band	10.773	Ver	53	Multi
Christian Channel Europe	UK	ASTRA 1D	19.2°West	Ku1-Band	10.770	Ver	47	Multi
Ciné Cinéfil	France	Télécom 2A	8.0°West	Ku3-Band	12.627	Hor	R9	Spot
Ciné Cinémas	France	Télécom 2A	8.0°West	Ku3-Band	12.689	Hor	R10	Spot
Ciné Cinémas	France	Télécom 2A	8.0°West	Ku3-Band	12.689	Hor	R5	Spot
Ciné Classics	Spain	ASTRA 1C	19.2°East	Ku1-Band	11.067	Ver	40	Multi
Cinema	Sweden	TeleNor	0.8°West	Ku2-Band	11.900	LHCP	10	Spot
Cinemania	Spain	ASTRA 1B	19.2°East	Ku1-Band	11.656	Ver	30	Multi
Cinemania 2	Spain	Hispasat	31.0°West	Ku2-Band	12.302	LHCP	31	East Spot
CMT Europe	USA	Intelsat 601	27.5°West	Ku1-Band	11.175	Ver	73U	West Spot
CNBC	UK	ASTRA 1D	19.2°East	Ku1-Band	10.729	Ver	50	Multi
CNN International	USA	Arabsat 2A	26.0°East	C-Band	3.843	RHCP	7	Zone
CNN International	USA	ASTRA 1B	19.2°East	Ku1-Band	11.626	Ver	28	Multi
CNN International	USA	Intelsat 601	27.5°West	C-Band	3.743	RHCP	81	East Hemisph.
CNN International	USA	Intelsat 601	27.5°West	C-Band	4.048	RHCP	17	North East Zone
CNN International	USA	PanAmSat-4	68.5°East	C-Band	4.090	RHCP	C13L	East Hemisph.
CNN Nordic	USA	Thor	0.8°West	Ku2-Band	11.785	Hor	7	Spot
ConAir	USA	PanAmSat-4	68.5°East	C-Band	4.113	Ver	C14U	Africa
Country Music Europe	USA	ASTRA 1D	19.2°East	Ku1-Band	10.743	Hor	51	Multi
CTC (STS)	CIS	Gorizont 11/17 (Stationar 5)	53.0°East	Ku1-Band	11.525	LHCP	K1	Spot
Deutsche Welle TV	Germany	Intelsat 704	66.0°East	C-Band	4.177	RHCP	38	Global
Deutsche Welle TV	Germany	Eutelsat II f1(Hotbird 1/2)	13.0°East	Ku1-Band	11.162	Ver	27B	Wide

ITU REGION 1 TV IN ALPHABETICAL ORDER

Station	Country	Satellite	Position	Band	Frequency	Pol.	Tr	Beam
Deutsche Welle TV	Germany	Intelsat 601	27.5° West	C-Band	3.743	RHCP	81	East Hemisph.
Discovery Channel	UK	ASTRA 1C	19.2° East	Ku1-Band	11.082	Hor	41	Multi
Discovery Channel	UK	Thor	0.8° West	Ku2-Band	11.938	RHCP	9	Spot
Documania	Spain	ASTRA 1B	19.2° East	Ku1-Band	11.685	Ver	32	Multi
DR-2	Denmark	Intelsat 707	1.0° West	Ku1-Band	11.665	Ver	69	West Spot
DSF Deutsches Sportfernsehen	Germany	ASTRA 1B	19.2° East	Ku1-Band	11.523	Hor	21	Multi
Dubai TV	UAE	Arabsat 2A	26.0° East	C-Band	4.057	LHCP	16	Zone
Duna 7	Hungary	Eutelsat II f3	16.0° East	Ku1-Band	11.595	Hor	33	Wide
EDTV	UAE	Arabsat 1C	30.5° East	C-Band	3.955	RHCP	14	Zone
EDTV	UAE	Eutelsat II f1(Hotbird 1/2)	13.0° East	Ku1-Band	11.745	Hor	51	Wide
Egypt TV	Egypt	Arabsat 2A	26.0° East	C-Band	3.761	RHCP	3	Zone
Egyptian Satellite Channel	Egypt	Intelsat 707	1.0° West	Ku1-Band	4.140	LHCP	24	East Hemisph.
ESPN	USA	PanAmSat-4	68.5° East	C-Band	3.865	Hor	5U	East Hemisph.
ESPN	USA	PanAmSat-4	68.5° East	C-Band	3.845	Hor	5A	East Hemisph.
ESPN	USA	PanAmSat f3r	43.0° East	C-Band	4.100	Hor	14C	African Hemisph.
ET-1/2/3	Greece	Eutelsat II f2	10.0° East	Ku1-Band	11.594	RHCP	33	Wide
ETV	Ethiopia	Intelsat 703	57.0° East	C-Band	3.920	RHCP	13	East Hemisph.
Euro Business News	UK	ASTRA 1C	19.2° East	Ku1-Band	11.097	Ver	42	Multi
Euro D	Turkey	Türksat 1B	42° East	Ku1-Band	11.569	Ver	2A	West Spot
euroNews	France	Eutelsat II f1(Hotbird 1/2)	13.0° East	Ku1-Band	11.575	Ver	37	Super
Europe by Satellite	UK	Eutelsat II f2	10.0° East	Ku1-Band	11.080	Hor	21	Wide
European Business News (EBN)	UK	Eutelsat II f1(Hotbird 1/2)	13.0° East	Ku2-Band	11.265	Hor	3	Spot
Eurosport	France	Thor	0.8° West	Ku3-Band	11.862	RHCP	8	Spot
Eurosport	France	ASTRA 1A	19.2° East	Ku1-Band	11.259	Hor	4	Multi
Eurosport	France	Eutelsat II f1(Hotbird 1/2)	13.0° East	Ku1-Band	11.387	Ver	9	Spot
FilmNet 1 Nordic	Norway	Thor	0.8° West	Ku2-Band	12.091	RHCP	20	Spot
FilmNet 2	Sweden	Thor	0.8° West	Ku2-Band	12.015	RHCP	16	Zone
FilmNet (Polish)	Poland	ASTRA 1D	19.2° East	Ku1-Band	10.920	Hor	63	Multi
Fox Kids	UK	ASTRA 1A	19.2° East	Ku1-Band	11.303	Ver	7	Multi
France Supervision	France	Télécom 2A	8.0° West	Ku3-Band	12.606	Ver	R3	Spot
France-2	France	Télécom 2B	5.0° West	Ku3-Band	12.564	Ver	R2	Spot
France-2	France	Télécom 2A	8.0° West	Ku3-Band	12.606	Ver	R3	Spot
Future Vision	Saudi Arabia	Arabsat 2A	26.0° East	C-Band	3.863	LHCP	8	Zone
Galavisión	Mexico	ASTRA 1C	19.2° East	Ku1-Band	11.126	Ver	44	Multi
Galavisión ECO	Mexico	PanAmSat f1	45.0° West	Ku1-Band	11.515	Ver	19B	Spot
Granada Good Life	UK	ASTRA 1D	19.2° East	Ku1-Band	10.847	Ver	58	Multi
Granada Good Life Channels	UK	ASTRA 1D	19.2° East	Ku1-Band	10.876	Ver	60	Multi
Granada Plus/Man & Motor	UK	ASTRA 1A	19.2° East	Ku1-Band	11.244	Hor	3	Multi
Granada (Talk TV)	UK	ASTRA 1D	19.2° East	Ku1-Band	10.861	Ver	59	Multi
H.O.T (Home Order TV)	UK	ASTRA 1D	19.2° East	Ku1-Band	10.906	Ver	62	Multi
HBB	Turkey	Türksat 1B	42° East	Ku1-Band	11.100	Ver	12	West Spot
Home Shopping Network	UK	ASTRA 1C	19.2° East	Ku1-Band	11.082	Hor	41	Multi
Home Shopping Network	UK	ASTRA 1C	19.2° East	Ku1-Band	11.097	Ver	42	Multi
Home TV	India	PanAmSat-4	68.5° East	C-Band	3.835	Hor	6L	East Hemisph.
HRT Zagreb (Hrvatska Televisija)	Croatia	Eutelsat II f3	16.0° East	Ku1-Band	10.987	Hor	20	Wide
HTB	Russia	Gals 1/2	36.0° East	Ku2-Band	12.207	RHCP		West Spot

233

ITU REGION 1 TV IN ALPHABETICAL ORDER

Station	Country	Satellite	Position		Band	Frequency	Pol.	Tr	Beam
IBA Channel-3	Israel	Intelsat 707	1.0°	West	Ku1-Band	11.017	Ver	71	East Spot
interSTAR	Turkey/Germany	Eutelsat II f2	10.0°	East	Ku1-Band	11.617	Ver	38	Wide
IRIB TV-1	Iran	Intelsat 602	63.0°	East	Ku1-Band	11.155	Ver	73	East Spot
IRIB TV-2	Iran	Intelsat 602	63.0°	East	Ku1-Band	11.002	Ver	71	East Spot
IRIB TV-3	Iran	Intelsat 602	63.0°	East	Ku1-Band	11.100	Ver	73	East Spot
Jordan TV		Arabsat 2A	26.0°	East	Ku3-Band	12.604	Ver	5	Zone
JRT Jordan	Jordan	Arabsat 2A	26.0°	East	C-Band	12.577	Ver	4	Zone
JRTV Jordan	Jordan	Arabsat 1C	30.5°	East	C-Band	4.140	LHCP	24	Zone
JSTV	Japan	ASTRA 1D	19.2°	East	Ku1-Band	10.773	Hor	53	Multi
Kabel-1	Germany	ASTRA 1A	19.2°	East	Ku1-Band	11.332	Hor	9	Multi
Kanal 5	Sweden	Tele-X	5.0°	East	Ku2-Band	12.475	LHCP	40	Spot
Kanal 7	Turkey	Türksat 1B	42°	East	Ku1-Band	11.144	Ver	8A	West Spot
Kanal D	Turkey	Türksat 1B	42°	East	Ku1-Band	11.030	Hor	13	East Spot
Kanal + (Poland)	Poland	Eutelsat II f1(Hottbird 1/2)	13.0°	East	Ku1-Band	11.513	Hor	15	Super
Kanal-6	Turkey	Türksat 1B	42°	East	Ku1-Band	11.073	Hor	14	East Spot
Kasachstan 1 (private)	Kasachstan	Intelsat 703	57.0°	East	Ku1-Band	11.543	RHCP		East Hemisph.
Kazakhstan TV	Kazakhstan	Gals 1/2	36.0°	East	Ku2-Band	11.766	RHCP		West Spot
Kuwait Space Channel	Kuwait	Arabsat 1C	30.5°	East	C-Band	4.181	RHCP	26	Zone
Kuwait TV	Kuwait	Arabsat 2A	26.0°	East	C-Band	4.166	RHCP	21	Zone
Kuwait TV	Kuwait	Arabsat 2A	26.0°	East	Ku3-Band	12.646	Hor	7	Zone
La Chaîne Info (LCI)	France	Télécom 2B	5.0°	West	Ku3-Band	12.585	Ver	R8	Spot
La Cinquième	France	Télécom 2B	5.0°	West	Ku3-Band	12.606	Ver	R3	Spot
La Cinquième	France	Eutelsat II f1(Hotbird 1/2)	13.0°	East	Ku1-Band	11.080	Ver	26B	Multi
LBC Lebanon	Lebanon	Arabsat 2A	26.0°	East	C-Band	3.740	LHCP	2	Zone
Libyan TV	Libya	Intelsat 707	1.0°	West	C-Band	4.030	Ver	24	East Hemisph.
M.Net South Africa	AFS	PanAmSat-4	68.5°	East	Ku3-Band	12.538	Ver	13	AFS Spot
M-2 Morocco	Morocco	Intelsat 706	53.0°	West	C-Band	3.990	RHCP	14	East Hemisph.
M-6 Métropole 6	France	Télécom 2B	5.0°	West	Ku3-Band	12.522	Ver	R1	Spot
M-Net Int.	South Africa	Intelsat 602	63.0°	East	C-Band	3.897	LHCP	12	South West Zone
MBC	UK	Arabsat 2A	26.0°	East	C-Band	4.098	LHCP	18	Zone
MCM Afrique	South Africa	Intelsat 601	27.5°	West	Ku3-Band	3.650	RHCP	20	East Hemisph.
MCM Euromusique	France	Eutelsat II f1(Hotbird 1/2)	13.0°	East	Ku1-Band	11.304	Hor	5	Spot
MCM Euromusique	France	Télécom 2A	8.0°	West	Ku3-Band	12.543	Hor	R7	East Spot
MED TV	UK	Intelsat 705	18.5°	West	Ku1-Band	11.075	Ver	73A	Zone
Middle East Broadcast (MBC)	UK	Arabsat 1C	30.5°	East	C-Band	4.107	LHCP	22	South Zone
Middle East Broadcasting Centre	UK	Arabsat 1C	30.5°	East	S-Band	2.635	Ver	1A	Super
Middle East Broadcasting Centre	UK	Eutelsat II f1(Hotbird 1/2)	13.0°	East	Ku1-Band	11.554	Hor	32	Multi
MiniMax	Spain	ASTRA 1C	19.2°	East	Ku1-Band	11.008	Hor	36	Multi
Mitteldeutscher Rundfunk (MDR 3)	Germany	ASTRA 1C	19.2°	East	Ku1-Band	11.112	Hor	43	Multi
Moscow-1	CIS	Express 2	14.0°	West	C-Band	3.675	RHCP	6	Spot
MTV	UK	ASTRA 1A	19.2°	East	Ku1-Band	11.421	Ver	15	Multi
MTV	Africa	PanAmSat-4	68.5°	East	C-Band	4.185	Ver	C15U	East Hemisph.
MTV Nordic	Denmark	Intelsat 707	1.0°	West	Ku1-Band	11.679	Hor	69	West Spot
Muslim TV Ahmadiyya		Intelsat 603	34.5°	West	Ku1-Band	11.003	Ver		Spot
Muslim TV Ahmadiyya Int.		Express 2	14.0°	West	C-Band	3.725	RHCP	9	Global
Muslim-TV MTA+	UK	Intelsat 703	57.0°	East	C-Band	4.177	LHCP		East Hemisph.

234

ITU REGION 1 TV IN ALPHABETICAL ORDER

Station	Country	Satellite	Position		Band	Frequency	Pol.	Tr	Beam
N-3 Nord-3	Germany	ASTRA 1B	19.2°	East	Ku1-Band	11.582	Hor	25	Multi
n-tv	Germany	ASTRA 1B	19.2°	East	Ku1-Band	11.641	Hor	29	Multi
NBC Super Channel	UK	Eutelsat II f1 (Hotbird 1/2)	13.0°	East	Ku1-Band	10.987	Ver	25	Wide
NEPC TV	India	Intelsat 704	66.0°	East	C-Band	4.100	LHCP	15	North East Zone
NEPC-TV	Japan	Intelsat 703	57.0°	East	C-Band	4.059	LHCP		East Hemisph.
NHK Tokyo	Japan	PanAmSat-4	68.5°	East	Ku3-Band	12.606	Ver	14	EUR Spot
Nickelodeon	USA	ASTRA 1C	19.2°	East	Ku1-Band	11.156	Hor	46	Multi
Nickelodeon	Germany	ASTRA 1B	19.2°	East	Ku1-Band	11.611	Hor	27	Multi
Nile TV	Egypt	Arabsat 2A	26.0°	East	C-Band	3.802	RHCP	5	Zone
Nile TV International	Egypt	Intelsat 707	1.0°	West	C-Band	4.140	LHCP	24	East Hemisph.
Nile TV International	Egypt	Eutelsat II f3	16.0°	East	Ku1-Band	11.140	Ver	27A	Wide
Nockelodeon Sweden	Sweden	Sirius	5.2°	East	Ku2-Band	11.861	RHCP	8	Spot
Nova Shop	Sweden	Sirius	5.2°	East	Ku2-Band	11.785	RHCP	4	Spot
NRK 1	Norway	Intelsat 707	1.0°	West	Ku1-Band	11.176	Ver	63	West Spot
NRK-2	Norway	Intelsat 707	1.0°	West	Ku1-Band	11.482	Hor		West Spot
NTA Ch. 10	Nigeria	Intelsat 601	27.5°	West	C-Band	4.065	RHCP	25	North East Zone
NTV	Russia	Eutelsat II f2	10.0°	East	Ku1-Band	10.987	Hor	20A	Wide
NTV	Russia	Eutelsat II f3	16.0°	East	Ku1-Band	11.575	Ver	37	Wide
NTV (Nezavisimoye TV)	Russia	Express 2	14.0°	West	C-Band	4.075	RHCP	17	Spot
Oman TV	Oman	Arabsat 1C	30.5°	East	C-Band	4.062	LHCP	20	Zone
Oman TV	Oman	Arabsat 2A	26.0°	East	C-Band	4.139	LHCP	20	Zone
Onyx TV		Eutelsat II f1 (Hotbird 1/2)	13.0°	East	Ku1-Band	11.148	Hor	22	Super
Orbit	Morocco	Arabsat 2A	26.0°	East	C-Band	3.884	RHCP	9	Zone
ORT-1	CIS	Express 2	14.0°	West	C-Band	3.675	RHCP	6	Spot
ORT-1/Orbita 4	CIS	Gorizont 11/17 (Stationar 5)	53.0°	East	C-Band	3.675	LHCP	6	Global
Ostankino ORT Int.	CIS	Gorizont 26	11.0°	West	C-Band	3.675	RHCP	6	Spot
Paramount TV	USA	ASTRA 1C	19.2°	East	Ku1-Band	11.156	Ver	46	Multi
Paris Première	France	Télécom 2A	8.0°	West	Ku3-Band	12.564	Hor	R2	Spot
PIK CYBC	Cyprus	Eutelsat II f4	7.0°	East	Ku1-Band	11.146	Hor	22A	Wide
Planete	France	Télécom 2A	8.0°	West	Ku3-Band	12.585	Hor	R8	Spot
Polonia 1	Poland	Eutelsat II f1 (Hotbird 1/2)	13.0°	East	Ku1-Band	11.785	Hor	53	Wide
Polsat	Poland	Eutelsat II f1 (Hotbird 1/2)	13.0°	East	Ku1-Band	11.428	Hor	11	Super
Premiera TV	Czech Rep.	DFS-2 Kopernikus	28.5°	East	Ku1-Band	11.525	Hor	A2	Spot
Premiere	Germany	ASTRA 1B	19.2°	East	Ku1-Band	11.464	Hor	17	Multi
Pro-7	Germany	ASTRA 1A	19.2°	East	Ku1-Band	11.406	Ver	14	Multi
Quantum Channel	USA	Eutelsat II f1 (Hotbird 1/2)	13.0°	East	Ku1-Band	11.387	Hor	9	Spot
Quantum Channel	UK	ASTRA 1A	19.2°	East	Ku1-Band	11.259	Ver	4	Multi
QVC	USA	ASTRA 1C	19.2°	East	Ku1-Band	11.038	Ver	38	Multi
QVC Deutschland	Germany	ASTRA 1D	19.2°	East	Ku1-Band	10.758	Ver	52	Multi
RAIDUE	Italy	Eutelsat II f1 (Hotbird 1/2)	13.0°	East	Ku1-Band	11.449	Hor	12	Spot
RAITRE	Italy	Eutelsat II f1 (Hotbird 1/2)	13.0°	East	Ku1-Band	11.534	Ver	16	Super
RAIUNO	Italy	Eutelsat II f3	16.0°	East	Ku1-Band	11.366	Ver	8	Spot
Rendez-Vous	France	Télécom 2A	8.0°	West	Ku1-Band	10.987	Hor	20	Wide
RFO Canal Permanent 2	France	Intelsat 601	27.5°	West	C-Band	3.768	RHCP	C2B	Semi Global
RTA-TV Algerian Television	Algeria	Intelsat 601	27.5°	West	C-Band	4.008	RHCP	24	West Hemisph.
RTL 7	Poland	Eutelsat II f1 (Hotbird 1/2)	13.0°	East	Ku1-Band	11.492	Ver	14	Spot

ITU REGION 1 TV IN ALPHABETICAL ORDER

Station	Country	Satellite	Position		Band	Frequency	Pol.	Tr	Beam
RTL Television	Germany	ASTRA 1A	19.2°	East	Ku1-Band	11.229	Ver	2	Multi
RTL-2	Germany	ASTRA 1A	19.2°	East	Ku1-Band	11.214	Hor	1	Multi
RTL-2	Germany	Eutelsat II f1 (Hotbird 1/2)	13.0°	East	Ku1-Band	11.095	Hor	21B	Super
RTL-9	France	Télécom 2B	5.0°	West	Ku3-Band	12.543	Hor	K7	Spot
RTL-TVi	France	Télécom 2B	5.0°	West	Ku1-Band	12.627	Hor	K9	Spot
RTM 1 (TV Marocaine)	Morocco	Eutelsat II f3	16.0°	East	Ku1-Band	10.972	Ver	25	Wide
RTM-1 Morocco	Morocco	Intelsat 706	53.0°	West	C-Band	3.798	RHCP	12	East Hemisph.
RTP Internacional	Portugal	Gorizont 12/22	40.0°	East	C-Band	3.930	RHCP	11	Global Beam
RTP Internacional	Portugal	Express 2	14.0°	West	C-Band	4.025	RHCP	15	Semi Global
RTP Internacional	Portugal	Eutelsat II f1 (Hotbird 1/2)	13.0°	East	Ku1-Band	11.727	Ver	50	Wide
RTS Beograd (RTV Srbija)	Yugoslavia	Eutelsat II f4	7.0°	East	Ku1-Band	11.178	Hor	22B	Wide
RTT TV-7	Tunisia	Eutelsat II f3	16.0°	East	Ku1-Band	11.658	Ver	39	Wide
SABC 2	AFS	PanAmSat-4	68.5°	East	Ku3-Band	12.697	Ver	4KL	AFS Spot
SABC 3	AFS	PanAmSat-4	68.5°	East	Ku3-Band	12.664	Ver	15U	AFS Spot
SABC CCV	AFS	PanAmSat-4	68.5°	East	Ku3-Band	12.698	Ver	16L	AFS Spot
SABC NNTV	AFS	PanAmSat-4	68.5°	East	Ku3-Band	12.724	Ver	15L	AFS Spot
SABC TV1	AFS	PanAmSat-4	68.5°	East	Ku3-Band	12.665	Ver	4	AFS Spot
SABC-TV	South Africa	Intelsat 602	63.0°	East	C-Band	3.807	LHCP	7B	South West Zone
Samanloyu TV	Turkey	Türksat 1B	42°	East	Ku1-Band	11.121	RHCP	15	East Spot
Sara Vision		Arabsat 2A	26.0°	East	C-Band	4.043	Hor	A1	Zone
Sat-1	Germany	DFS-3 Kopernikus	23.5°	East	Ku1-Band	11.475	Ver	6	Spot
Sat-1	Germany	ASTRA 1A	19.2°	East	Ku1-Band	11.288	Hor	2	Multi
Saudi Arabia	Saudi Arabia	Arabsat 2A	26.0°	East	Ku3-Band	12.536	RHCP	11	Zone
Saudi Arabia TV 1	Saudi Arabia	Arabsat 2A	26.0°	East	C-Band	3.925	RHCP	13	Zone
Saudi Arabia TV 2	Saudi Arabia	Arabsat 2A	26.0°	East	C-Band	3.978	RHCP	8	Zone
Saudia	Saudi Arabia	Arabsat 2A	26.0°	East	Ku1-Band	12.661	Ver	27	Multi
Sci-Fi Channel	UK	ASTRA 1B	19.2°	East	Ku2-Band	11.611	LHCP	18	Spot
SciFi Channel Nordic	Denmark	Telenor	0.8°	West	C-Band	12.054	Hor	5	Multi
Sell-a-Vision	Germany	ASTRA 1A	19.2°	East	Ku1-Band	11.273	Hor	1	Zone
Sharjah TV	Mauritania	Arabsat 2A	26.0°	East	C-Band	3.720	Ver	16	Multi
Sky Movies	UK	ASTRA 1A	19.2°	East	Ku1-Band	11.436	Ver	60	Multi
Sky Movies Gold	UK	ASTRA 1D	19.2°	East	Ku1-Band	10.876	Hor	26	Multi
Sky Movies Gold	UK	ASTRA 1B	19.2°	East	Ku1-Band	11.597	Ver	12	Multi
Sky News	UK	ASTRA 1A	19.2°	East	Ku1-Band	11.377	Hor	36	East Hemisph.
Sky One	UK	Intelsat 512	21.5°	West	C-Band	4.055	RHCP	8	Multi
Sky Scottish	UK	ASTRA 1D	19.2°	East	Ku1-Band	1.318	Ver	59	Multi
Sky Soap	UK	ASTRA 1C	19.2°	East	Ku1-Band	10.861	Hor	47	Multi
Sky Sports	UK	ASTRA 1B	19.2°	East	Ku1-Band	11.170	Ver	20	Multi
Sky Sports 2	UK	ASTRA 1C	19.2°	East	Ku1-Band	11.508	Hor	47	Multi
Sky Sports 3	UK	ASTRA 1B	19.2°	East	Ku1-Band	11.670	Ver	31	Multi
Sky Sports Gold	UK	ASTRA 1C	19.2°	East	Ku1-Band	11.170	Hor	47	Multi
Sky Travel	UK	ASTRA 1C	19.2°	East	Ku1-Band	11.170	Hor	47	Multi
Sky-2	UK	ASTRA 1A	19.2°	East	Ku1-Band	11.303	Hor	7	Multi
Sony Entertainment TV	India	PanAmSat-4	68.5°	East	C-Band	3.910	Hor	C7	East Hemisph.
Sudan TV	Sudan	Arabsat 2A	26.0°	East	C-Band	3.905	LHCP	10	Zone

ITU REGION 1 TV IN ALPHABETICAL ORDER

Station	Country	Satellite	Position	Band	Frequency	Pol.	Tr	Beam
Sudan TV	Sudan	Intelsat 703	57.0° East	C-Band	4.020	LHCP	13	East Hemisph.
Super RTL	Germany	ASTRA 1A	19.2° East	Ku1-Band	11.391	Hor	7	Multi
Super Television Channel (USA)	USA	Orion 1	37.5° West	Ku1-Band	11.617	Ver	48	Multi
SWF/SDR (S-3)	Germany	ASTRA 1C	19.2° East	Ku1-Band	11.185	Ver	14	Zone
Syrian TV	Syria	Arabsat 2A	26.0° East	C-Band	3.998	LHCP	35	Multi
TCC (The Children's Channel)	UK	ASTRA 1C	19.2° East	Ku2-Band	10.993	Hor	9	Spot
TCC The Children's Channel	UK	Thor	0.8° West	Ku3-Band	11.938	RHCP	R4	Spot
Tele Monte-Carlo (TMC)	Monte-Carlo	Télécom 2B	5.0° West	Ku3-Band	12.648	Ver	3	East Spot
Tele-5	Spain	Hispasat	31.0° West	Ku1-Band	12.631	Ver	55	Multi
Teleclub	Switzerland	ASTRA 1D	19.2° East	Ku1-Band	10.802	Hor	23	East Spot
TeleDeporte	Spain	Hispasat	31.0° West	Ku2-Band	12.149	LHCP	37	Wide
TelePace/Vatican TV	Italy/Vatican State	Eutelsat II f3	16.0° East	Ku1-Band	11.575	Ver	35	East Spot
Telesat 5	Spain	Hispasat	31.0° West	Ku2-Band	12.379	LHCP	R5	Spot
TF-1	France	Télécom 2B	5.0° West	Ku3-Band	12.690	Ver	26B	Wide
TGRT	Turkey	Eutelsat II f2	10.0° East	Ku1-Band	11.095	Ver	63	Multi
The Adult Channel	UK	ASTRA 1D	19.2° East	Ku1-Band	10.920	Ver	58	Multi
The Computer Channel	UK	ASTRA 1B	19.2° East	Ku1-Band	10.847	Ver	26	Wide
The Disney Channel	UK	ASTRA 1B	19.2° East	Ku1-Band	11.597	Ver	27B	Multi
The Egyptian Space Channel	Egypt	Eutelsat II f3	16.0° East	Ku1-Band	11.178	Hor	35	Multi
The Family Channel	UK	ASTRA 1C	19.2° East	Ku2-Band	10.993	Hor	47	Multi
The History Channel	UK	ASTRA 1C	19.2° East	Ku1-Band	11.170	Ver	41	Multi
The Learning Channel (TLC)	UK	ASTRA 1C	19.2° East	Ku1-Band	11.082	Hor	18	Multi
The Movie Channel	UK	ASTRA 1B	19.2° East	Ku1-Band	11.479	Ver	48	Multi
The Playboy Channel	UK	ASTRA 1C	19.2° East	Ku1-Band	11.185	Hor	60	Multi
The Racing Channel	UK	ASTRA 1D	19.2° East	Ku1-Band	10.876	Ver	47	Multi
The Sci-Fi Channel	UK	ASTRA 1D	19.2° East	Ku1-Band	11.170	Hor	60	Multi
The Weather Channel	UK	ASTRA 1D	19.2° East	Ku1-Band	10.876	Hor	60	Multi
The Weather CHannel	UK	ASTRA 1C	19.2° East	Ku1-Band	11.283	Ver	64	Multi
TM 3	Germany	ASTRA 1C	19.2° East	Ku1-Band	11.023	Hor	7	Spot
TM3	USA	Eutelsat II f1(Hotbird 1/2)	13.0° East	C-Band	4.115	Hor	37	Multi
TNT	USA	PanAmSat-4	68.5° East	Ku1-Band	11.181	Hor	C13U	East Hemisph.
TNT/Cartoon Network	Turkey	Eutelsat II f1(Hotbird 1/2)	13.0° East	Ku1-Band	11.555	Hor	22B	Super
TRT Int.	Turkey	Türksat 1B	42° East	Ku1-Band	11.173	Ver	10A	East Spot
TRT Int.	Turkey	Türksat 1B	42° East	Ku1-Band	11.462	Ver	16	East Spot
TRT-1	Turkey	Türksat 1B	42° East	Ku1-Band	11.490	Ver	9A	East Spot
TRT-3	Turkey	Türksat 1B	42° East	Ku1-Band	11.490	Ver	9B	East Spot
TRT-4	India	Intelsat 703	57.0° East	C-Band	4.133	LHCP	7	East Hemisph.
TV 1-India	Denmark	Intelsat 707	1.0° West	Ku1-Band	11.472	Ver	75	West Spot
TV 3 Danmark	Spain	Hispasat	31.0° West	Ku3-Band	12.671	Ver	13	East Spot
TV 3 de Catalunya	Sweden	Intelsat 707	1.0° West	Ku1-Band	11.597	Ver	75L	West Spot
TV 3 Sweden	France	Eutelsat II f1(Hotbird 1/2)	13.0° East	Ku1-Band	11.325	Ver	6	Wide
TV 5 Internationale	Sweden	Sirius	5.2° East	Ku2-Band	12.015	RHCP	16	Spot
TV 6 Sweden	Angola	Intelsat 601	27.5° West	C-Band	3.795	LHCP	102L	South East Zone
TV Angola	UK	Eutelsat II f3	16.0° East	Ku1-Band	11.163	Hor	22B	Super
TV Eurotica	Norway	Intelsat 707	1.0° West	Ku1-Band	11.016	Hor	61	West Spot
TV Norge								

237

ITU REGION 1 TV IN ALPHABETICAL ORDER

Station	Country	Satellite	Position		Band	Frequency	Pol.	Tr	Beam
TV Polonia	Poland	Eutelsat II f1(Hotbird 1/2)	13.0°	East	Ku1-Band	11.472	Hor	13	Spot
TV România int.	Romania	Eutelsat II f3	16.0°	East	Ku1-Band	11.575	Ver	37	Wide
TV Rossija 2	CIS	Gorizont 12/22	40.0°	East	C-Band	3.675	RHCP	6	Global
TV Sport Eurosport	France	Télécom 2A	8.0°	West	Ku3-Band	12.711	Hor	R11	Spot
TV-1000	Sweden	Intelsat 707	1.0°	West	Ku1-Band	11.053	Ver	72	West Spot
TV-2	Norway	Intelsat 707	1.0°	West	Ku1-Band	11.553	Ver	65	West Spot
TV-3	Norway	Intelsat 707	1.0°	West	Ku1-Band	11.086	Hor	62	West Spot
TV-4 (Sweden)	Sweden	Sirius	5.2°	East	Ku2-Band	11.938	RHCP	12	Spot
TV-6 Moscow	C.I.S.	Gals 1/2	36.0°	East	Ku2-Band	11.840	LHCP		West Spot
TV3+	Denmark	Telenor	0.8°	West	Ku2-Band	11.977	LHCP	14	Spot
TV5 Internationale/Afrique	France	Intelsat 707	1.0°	West	C-Band	3.840	RHCP	2	East Hemisph.
TVE-Internacional	Spain	Eutelsat II f1(Hotbird 1/2)	13.0°	East	Ku1-Band	11.221	Hor	1	Spot
TVX The Fantasy Channel	UK	ASTRA 1C	19.2°	East	Ku1-Band	10.979	Ver	34	Multi
UK Gold	UK	ASTRA 1B	19.2°	East	Ku1-Band	11.553	Hor	23	Multi
UK Living	UK	ASTRA 1C	19.2°	East	Ku1-Band	10.979	Ver	34	Multi
Vesa TV - Cable TV	Slovakia	Eutelsat II f2	10.0°	East	Ku1-Band	10.972	Ver	25	Wide
VH-1	UK	ASTRA 1B	19.2°	East	Ku1-Band	11.538	Ver	22	Multi
VH-1	Germany	ASTRA 1B	19.2°	East	Ku2-Band	11.611	Hor	27	Multi
VH1 Nordic	Sweden	Sirius	5.2°	East	Ku2-Band	11.785	RHCP	4	Spot
VIVA	Germany	Eutelsat II f1(Hotbird 1/2)	13.0°	East	Ku1-Band	11.005	Hor	20B	Super
Viva 2	Germany	Eutelsat II f1(Hotbird 1/2)	13.0°	East	Ku1-Band	10.972	Hor	20A	Super
VOX	Germany	Eutelsat II f1(Hotbird 1/2)	13.0°	East	Ku1-Band	11.596	Hor	33	Spot
VOX	Germany	ASTRA 1A	19.2°	East	Ku1-Band	11.273	Hor	5	Multi
WBTV (The Warner Channel)	UK	ASTRA 1D	19.2°	East	Ku1-Band	10.876	Ver	60	Multi
WDR	Germany	ASTRA 1C	19.2°	East	Ku1-Band	11.052	Hor	39	Multi
What's In store	Netherlands	ASTRA 1C	19.2°	East	Ku1-Band	10.935	Ver	64	Multi
WorldNet	USA	Intelsat 601	27.5°	West	C-Band	3.743	RHCP	81	East Hemisph.
WorldNet USIA	USA	Intelsat 704	66.0°	East	C-Band	4.177	RHCP	38	Global
Yemen TV	Yemen	Arabsat 2A	26.0°	East	C-Band	4.180	LHCP	22	Zone
Yemeni Television	Rep of Yemen	Intelsat 707	1.0°	West	C-Band	4.185	LHCP	39	Global
ZDF	Germany	ASTRA 1C	19.2°	East	Ku1-Band	10.964	Hor	33	Multi
Zee TV	UK	ASTRA 1D	19.2°	East	Ku2-Band	10.788	Ver	54	Multi
ZTV (Sweden)	Sweden	Sirius	5.2°	East		11.861	RHCP	8	Spot

ITU REGION 1 RADIO IN ALPHABETICAL ORDER

12.2. ITU REGION 1 RADIO IN ALPHABETICAL ORDER

Station	Country	Satellite	Position	Band	Frequency	Pol.	Tr	Beam
A. France & L'Essentiel	France	Télécom 2B	5.0° West	Ku3-Band	12.564	Ver	R2	Spot
AKRA FM	Turkey	Eutelsat II f2	10.0° East	Ku1-Band	11.095	Ver	26B	Wide
Algerian Cultural Radio	Algeria	Eutelsat II f3	16.0° East	Ku1-Band	11.095	Ver	26B	Wide
Algerian Radio	Algeria	Intelsat 601	27.5° West	C-Band	4.008	RHCP	24	West Hemisph.
America One	USA	ASTRA 1B	19.2° East	Ku1-Band	11.538	Ver	22	Multi
Antenne Bayern	Germany	ASTRA 1A	19.2° East	Ku1-Band	11.214	Hor	1	Multi
ARD starpoint 1/2	Germany	ASTRA 1B	19.2° East	Ku1-Band	11.493	Hor	19	Multi
ART	Saudi Arabia	Arabsat 2A	26.0° East	Ku3-Band	12.720	Hor	11	Zone
ASDA FM	UK	ASTRA 1C	19.2° East	Ku1-Band	11.170	Hor	47	Multi
Bartok Radio	Hungary	Eutelsat II f3	16.0° East	Ku1-Band	11.595	Hor	33	Wide
Bayern 1	Germany	ASTRA 1C	19.2° East	Ku1-Band	11.141	Hor	45	Multi
Bayern 2	Germany	ASTRA 1C	19.2° East	Ku1-Band	11.141	Hor	45	Multi
Bayern 3	Germany	ASTRA 1C	19.2° East	Ku1-Band	11.141	Hor	45	Multi
Bayern 4 Klassik	Germany	ASTRA 1C	19.2° East	Ku1-Band	11.141	Hor	45	Multi
Bayern 5 Aktuel	Germany	DFS-3 Kopernikus	23.5° East	Ku3-Band	12.625	Hor	K4	Spot
Bayern-4 Klassik	Germany	ASTRA 1B	19.2° East	Ku1-Band	11.553	Hor	23	Multi
BBC 5 Live	UK	ASTRA 1B	19.2° East	Ku1-Band	11.553	Hor	23	Multi
BBC Radio 2	UK	ASTRA 1C	19.2° East	Ku1-Band	10.979	Ver	34	Multi
BBC Radio 3	UK	ASTRA 1B	19.2° East	Ku1-Band	11.553	Hor	23	Multi
BBC Radio 4	UK	ASTRA 1C	19.2° East	Ku1-Band	10.979	Ver	34	Multi
BBC Radio One FM	UK	ASTRA 1B	19.2° East	Ku1-Band	11.553	Hor	23	Multi
BBC World Service	UK	Eutelsat II f1(Hotbird 1/2)	13.0° East	Ku1-Band	11.616	Ver	38	Wide
BBC World World Service	Turkey	Türksat 1B	42° East	Ku1-Band	11.030	Hor	13	East Spot
Best FM	Turkey	Türksat 1B	42° East	Ku1-Band	11.144	Ver	8A	West Spot
Best FM	Turkey	Türksat 1B	42° East	Ku1-Band	11.030	Hor	13	East Spot
Beur FM	France	Télécom 2B	5.0° West	Ku3-Band	12.564	Ver	R2	Spot
Cadena Cero	Spain	ASTRA 1B	19.2° East	Ku1-Band	11.656	Ver	30	Multi
Cadena Cope Convencional	Spain	Hispasat	31.0° West	Ku3-Band	12.711	Hor	14	East Spot
Cadena Dial	Spain	ASTRA 1B	19.2° East	Ku1-Band	11.656	Ver	30	Multi
Cadena Principales	Spain	Hispasat	31.0° West	Ku2-Band	12.302	LHCP	31	East Spot
Catalonia Principales Com	Spain	Hispasat	31.0° West	Ku3-Band	12.631	Ver	3	East Spot
Classique FM	France	Télécom 2B	5.0° West	Ku3-Band	12.606	Ver	R3	Spot
CMT Radio	UK	ASTRA 1D	19.2° East	Ku1-Band	10.729	Ver	50	Multi
CNN Radio	USA & UK	ASTRA 1B	19.2° East	Ku1-Band	11.626	Hor	28	Multi
CNN Radio	USA	PanAmSat-4	68.5° East	C-Band	4.090	Ver	C13L	East Hemisph.
D-Radio (Deutschlandradio Berlin)	Germany	ASTRA 1A	19.2° East	Ku1-Band	11.347	Ver	10	Multi
Der Oldiesender	Germany	Eutelsat II f1(Hotbird 1/2)	13.0° East	Ku1-Band	11.596	Hor	33	Spot
Der Oldiesender (RTL)	Germany	DFS-3 Kopernikus	23.5° East	Ku3-Band	12.625	Hor	K4	Spot
Deutsche Welle	Germany	ASTRA 1A	19.2° East	Ku1-Band	11.229	Ver	2	Multi

ITU REGION 1 RADIO IN ALPHABETICAL ORDER

Station	Country	Satellite	Position	Band	Frequency	Pol.	Tr	Beam
Deutsche Welle	Germany	ASTRA 1A	19.2° East	Ku1-Band	11.229	Ver	2	Multi
Deutsche Welle [Africa]	Germany	Eutelsat II f1|Hotbird 1/2)	13.0° East	Ku1-Band	11.162	Ver	27B	Wide
Deutsche Welle (Eur. prgr1/2)	Germany	ASTRA 1B	19.2° East	Ku1-Band	11.493	Hor	19	Multi
Deutsche Welle (German)	Germany	ASTRA 1B	19.2° East	Ku1-Band	11.493	Hor	19	Multi
Deutsche Welle Radio	Germany	Intelsat 704	66.0° East	C-Band	4.177	RHCP	38	Global
Deutschlandfunk	Germany	DFS-3 Kopernikus	23.5° East	Ku3-Band	12.625	Hor	K4	Spot
Deutschlandradio Berlin	Germany	DFS-3 Kopernikus	23.5° East	Ku3-Band	12.625	Hor	K4	Spot
Deutschlandradio Köln	Germany	ASTRA 1A	19.2° East	Ku1-Band	11.347	Ver	10	Multi
Deutschlandradio Köln	Germany	Eutelsat II f1|Hotbird 1/2)	13.0° East	Ku1-Band	11.596	Hor	33	Spot
DLF	Germany	ASTRA 1B	19.2° East	Ku1-Band	11.493	Hor	19	Multi
DLR Berlin	Germany	ASTRA 1B	19.2° East	Ku1-Band	11.493	Hor	19	Multi
Eins Live	Germany	ASTRA 1C	19.2° East	Ku1-Band	11.052	Hor	39	Multi
Energy 93.3	Germany	DFS-3 Kopernikus	23.5° East	Ku3-Band	12.625	Hor	K4	Spot
ENTV La Chaîne 3	Algeria	Eutelsat II f3	16.0° East	Ku1-Band	11.095	Ver	26B	Wide
ENTV Radio 1 Arabic	Algeria	Eutelsat II f3	16.0° East	Ku1-Band	11.095	Ver	26B	Wide
ENTV Radio 2 Kabyle	Algeria	Eutelsat II f3	16.0° East	Ku1-Band	11.095	Ver	26B	Wide
ERF 2	USA / Germany	ASTRA 1C	19.2° East	Ku1-Band	11.038	Ver	38	Multi
ERTU (cultural)	Egypt	Eutelsat II f3	16.0° East	Ku1-Band	11.178	Ver	27	Wide
ERTU (general prgr)	Egypt	Eutelsat II f3	16.0° East	Ku1-Band	11.178	Ver	27	Wide
Europe 2	Monte Carlo	Télécom 2B	5.0° West	Ku3-Band	12.648	Ver	R4	Spot
Europe-1	France	Télécom 2B	5.0° West	Ku3-Band	12.522	Ver	R1	Spot
Euroradio Channel R	various	Eutelsat II f4	7.0° East	Ku1-Band	11.0266	Ver	25	Wide
Euroradio Channel T	various	Eutelsat II f4	7.0° East	Ku1-Band	11.0399	Ver	26	Wide
FM Super Station (FM 99.7)	Kuwait	Arabsat 1C	30.5° East	C-Band	4.181	RHCP	26	Zone
Fourviere FM	France	Télécom 2B	5.0° West	Ku3-Band	12.522	Ver	R1	Spot
France Culture Europe	France	Eutelsat II f1|Hotbird 1/2)	13.0° East	Ku1-Band	11.325	Ver	6	Wide
France Info	France	Eutelsat II f1|Hotbird 1/2)	13.0° East	Ku1-Band	11.325	Ver	6	Wide
France Inter	France	Eutelsat II f1|Hotbird 1/2)	13.0° East	Ku1-Band	11.325	Ver	6	Wide
French Digital Satellite Radio	France	Télécom 2C	3.0° East	Ku3-Band	12.522	Ver	R1	Spot
Frequence Mousquetaire	France	Télécom 2B	5.0° West	Ku3-Band	12.522	Ver	R1	Spot
G-FM	UK	ASTRA 1D	19.2° East	Ku1-Band	10.788	Ver	54	Multi
Granada FM	UK	Eutelsat II f1|Hotbird 1/2)	13.0° East	Ku1-Band	10.987	Ver	25	Wide
Grand Magazines	Monte Carlo	Télécom 2B	5.0° West	Ku3-Band	12.648	Ver	R4	Spot
GWR Classic Gold	UK	ASTRA 1C	19.2° East	Ku1-Band	11.170	Hor	47	Multi
HBB FM	Turkey	Türksat 1B	42° East	Ku1-Band	11.100	Ver	12	West Spot
Hit Radio FFH	Germany	ASTRA 1C	19.2° East	Ku1-Band	10.964	Ver	33	Multi
HR2 Kultur	Germany	DFS-3 Kopernikus	23.5° East	Ku3-Band	12.625	Hor	K4	Spot
Institute for Radio Techniques	Germany	ASTRA 1C	19.2° East	Ku1-Band	11.141	Hor	45	Multi
IRIB Radio-1	Iran	Intelsat 602	63.0° East	Ku1-Band	11.155	Ver	71	East Spot
IRIB Radio-1	Iran	Intelsat 602	63.0° East	Ku1-Band	11.002	Ver	71	East Spot
IRIB Radio-2 (int.)	Iran	Intelsat 602	63.0° East	Ku1-Band	11.002	Ver	71	East Spot
Irish Satellite Radio Netw.	Ireland	ASTRA 1B	19.2° East	Ku1-Band	11.538	Ver	22	Multi
Israel Radio	Israel	Intelsat 707	1.0° West	Ku1-Band	11.178	Ver	73	East Spot
Kiss-FM, Mexico	Mexico	PanAmSat f1	45.0° West	Ku1-Band	11.515	Hor	19B	Spot
Klassik Radio	Germany	DFS-3 Kopernikus	23.5° East	Ku3-Band	12.625	Hor	K4	Spot
Kol Israel Network B	Israel	Intelsat 707	1.0° West	Ku1-Band	11.017	Ver	71	East Spot

ITU REGION 1 RADIO IN ALPHABETICAL ORDER

Station	Country	Satellite	Position		Band	Frequency	Pol.	Tr	Beam
Kol Israel Network C	Israel	Intelsat 707	1.0°	West	Ku1-Band	11.017	Ver	71	East Spot
KRAL FM	Turkey	Eutelsat II f2	10.0°	East	Ku1-Band	11.617	Ver	38	Wide
Kussuth Radio	Hungary	Eutelsat II f3	16.0°	East	Ku1-Band	11.595	Hor	33	Wide
Libya TV	Libya	Arabsat 2A	26.0°	East	Ku3-Band	12.700	Ver	10	Zone
Lotus	AFS	PanAmSat-4	68.5°	East	Ku3-Band	12.664	Ver		AFS Spot
Mamara FM	Turkey	Türksat 1B	42°		Ku1-Band	11.144	Ver	8A	West Spot
MBC	UK	Arabsat 2A	26.0°	East	Ku3-Band	12.735	Ver	12	Zone
MBC FM	UK	Arabsat 2A	26.0°	East	C-Band	4.098	Ver	18	Zone
MBC Radio	UK	Arabsat 1C	30.5°	East	C-Band	4.107	LHCP	22	Super
MBC Sputnik	UK	Eutelsat II f1(Hotbird 1/2)	13.0°	East	Ku1-Band	11.554	Hor	32	Spot
MDR Sputnik	Germany	DFS-3 Kopernikus	23.5°	East	Ku3-Band	12.625	Hor	K4	Multi
MDR-Info	Germany	ASTRA 1C	19.2°	East	Ku1-Band	11.112	Hor	43	Multi
MDR-Kultur	Germany	ASTRA 1C	19.2°	East	Ku1-Band	11.112	Hor	43	Multi
MDR-Live	Germany	ASTRA 1C	19.2°	East	Ku1-Band	11.112	Hor	43	Multi
MDR-Sputnik	Germany	ASTRA 1C	19.2°	East	Ku1-Band	11.112	Hor	43	Multi
Metro FM	Turkey	Eutelsat II f2	10.0°	East	Ku1-Band	11.617	Ver	38	Wide
Middle East Radio	Egypt	Eutelsat II f3	16.0°	East	Ku1-Band	11.178	LHCP	27	Wide
Mmbabatho Radio	Senegal	Intelsat 601	27.5°	West	C-Band	3.884	Hor	103A	East Hemisph.
Morales FM	Turkey	Türksat 1B	42°		Ku1-Band	11.144	Ver	8A	West Spot
N-Joy Radio	Germany	ASTRA 1B	19.2°	East	Ku1-Band	11.464	Hor	17	Multi
N-Joy Radio	Germany	ASTRA 1B	19.2°	East	Ku1-Band	11.582	Hor	25	Multi
NDR 2	Germany	ASTRA 1B	19.2°	East	Ku1-Band	11.582	Hor	25	Multi
NDR 3	Germany	ASTRA 1B	19.2°	East	Ku1-Band	11.582	Hor	25	Multi
NDR 4	Germany	ASTRA 1B	19.2°	East	Ku1-Band	11.582	Hor	25	Multi
NHK Radio Japan Nippon	Japan	Intelsat 601	27.5°	West	C-Band	3.915	RHCP	23B	East Hemisph.
Oman Radio	Oman	Arabsat 1C	30.5°	East	C-Band	4.078	LHCP	20	Zone
ORF Wien	Austria	ASTRA 1C	19.2°	East	Ku1-Band	11.141	Hor	45	Multi
Petőfi Radio	Hungary	Eutelsat II f3	16.0°	East	Ku1-Band	11.595	Hor	33	Wide
PIK Radio 3	Cyprus	Eutelsat II f4	7.0°	East	Ku1-Band	11.146	Hor	22A	Wide
PRT 1	Poland	Eutelsat II f1(Hotbird 1/2)	13.0°	East	Ku1-Band	11.472	LHCP	13	Spot
PRT 2	Poland	Eutelsat II f1(Hotbird 1/2)	13.0°	East	Ku1-Band	11.472	LHCP	13	Spot
PRT 3	Poland	Eutelsat II f1(Hotbird 1/2)	13.0°	East	Ku1-Band	11.472	LHCP	13	Spot
PRT 5	Poland	Eutelsat II f1(Hotbird 1/2)	13.0°	East	Ku1-Band	11.472	LHCP	13	Spot
Radio 2000	AFS	PanAmSat-4	68.5°	East	Ku3-Band	12.697	Ver	4KL	AFS Spot
Radio Africa Nr.-1	Gabon	Intelsat 601	27.5°	West	C-Band	3.915	RHCP	23B	East Hemisph.
Radio Almaty International	Kasachstan	Intelsat 703	57.0°	East	Ku1-Band	11.543	Hor		East Hemisph.
Radio Beograd	Yugoslavia	Eutelsat II f4	7.0°	East	Ku1-Band	11.178	Hor		Wide
Radio Bop	Senegal	Intelsat 601	27.5°	West	C-Band	3.884	Hor	22B	East Hemisph.
Radio Campanik	Germany	ASTRA 1B	19.2°	East	Ku1-Band	11.523	LHCP	103A	Multi
Radio Canada Int.	Canada	Eutelsat II f1(Hotbird 1/2)	13.0°	East	Ku1-Band	11.265	LHCP	21	Spot
Radio Damascus	Syria	Arabsat 2A	26.0°	East	C-Band	3.998	LHCP	3	Zone
Radio Diffusion Nat. Tchadienne (RNT)	Chad	Intelsat 707	1.0°	West	C-Band	3.860	RHCP	14	Zone
Radio Dubai	UAE	Arabsat 1C	30.5°	East	C-Band	3.955	RHCP	53	Zone
Radio Dubai	UAE	Arabsat 1C	30.5°	East	C-Band	3.955	RHCP	14	Zone
Radio Dubai	UAE	Arabsat 2A	26.0°	East	C-Band	4.057	LHCP	16	Zone

241

ITU REGION 1 RADIO IN ALPHABETICAL ORDER

Station	Country	Satellite	Position		Band	Frequency	Pol.	Tr	Beam
Radio Dubai	UAE	Arabsat 2A	26.0°	East	C-Band	4.057	LHCP	16	Zone
Radio Due	Italy	Eutelsat II f1(Hotbird 1/2)	13.0°	East	Ku-Band	11.449	Ver	12	Spot
Radio Due	Italy	Eutelsat II f1(Hotbird 1/2)	13.0°	East	Ku-Band	11.366	Ver	8	Spot
Radio Eiva!	Switzerland	ASTRA 1A	19.2°	East	Ku-Band	11.332	Hor	9	Multi
Radio Eiva!	Switzerland	ASTRA 1D	19.2°	East	Ku-Band	10.802	Hor	55	Multi
Radio Exterior de España	Spain	Eutelsat II f1(Hotbird 1/2)	13.0°	East	Ku-Band	11.221	Hor	1	Spot
Radio Finland Int.	Finland	Eutelsat II f1(Hotbird 1/2)	13.0°	East	Ku-Band	11.162	Ver	27B	Wide
Radio France Int.	France	Eutelsat II f3	16.0°	East	Ku-Band	11.095	Ver	26B	Wide
Radio France Internationale	France	ASTRA 1C	19.2°	East	Ku-Band	11.156	Ver	46	Multi
Radio France Internationale 2	France	ASTRA 1C	19.2°	East	Ku-Band	11.156	Ver	46	Multi
Radio Golos Rossija	CIS	Gorizont 12/22	40.0°	East	C-Band	3.675	RHCP	6	Global
Radio Horeb	Germany	ASTRA 1A	19.2°	East	Ku-Band	11.406	Ver	14	Multi
Radio Italia	Italy	Eutelsat II f1(Hotbird 1/2)	13.0°	East	Ku-Band	11.408	Ver	10	Spot
Radio Kuwait	Kuwait	Arabsat 2A	26.0°	East	C-Band	4.166	RHCP	21	Zone
Radio Kuwait (FM 89.5 & 98.9)	Kuwait	Arabsat 1C	30.5°	East	C-Band	4.181	RHCP	26	Zone
Radio Maria	Italy	Eutelsat II f1(Hotbird 1/2)	13.0°	East	Ku-Band	11.408	Ver	10	Spot
Radio Mayak	CIS	Gorizont 11/17 (Stationar 5)	53.0°	East	C-Band	3.675	RHCP	6	Global
Radio Mediterranée Int. (Medi-1)	Morocco	Intelsat 706	53.0°	West	C-Band	3.990	RHCP	14	East Hemisph.
Radio Meldodie	Germany	Eutelsat II f1(Hotbird 1/2)	13.0°	East	Ku-Band	11.055	Hor	21A	Super
Radio Metro	South Africa	Intelsat 602	63.0°	East	C-Band	3.807	LHCP	4	South West Zone
Radio MF-702	South Africa	Intelsat 602	63.0°	East	C-Band	3.897	LHCP	4	South West Zone
Radio Moscow Int / Radio Mayak	CIS	Gorizont 26	11.0°	West	C-Band	3.675	RHCP	6	Spot
Radio Nostalgij	CIS	Gorizont 12/22	40.0°	East	C-Band	3.675	RHCP	6	Global
Radio Portugal Int.	Portugal	Eutelsat II f1(Hotbird 1/2)	13.0°	East	Ku-Band	11.727	Ver	50	Wide
Radio RAI Int.	Italy	Eutelsat II f1(Hotbird 1/2)	13.0°	East	Ku-Band	11.449	Ver	12	Spot
Radio Renascença 1	Portugal	Eutelsat II f1(Hotbird 1/2)	13.0°	East	Ku-Band	11.727	Ver	50	Wide
Radio REDA	Israel	Intelsat 707	1.0°	West	Ku-Band	11.017	Ver	71	East Spot
Radio Rossija 4 Kanal	CIS	Express 2	14.0°	West	C-Band	3.675	RHCP	6	Spot
Radio Shqiptar	Albania	Eutelsat II f3	16.0°	East	Ku-Band	11.575	Ver	37	Wide
Radio Sonder Grense	AFS	PanAmSat-4	68.5°	East	Ku3-Band	12.697	Ver	4KL	AFS Spot
Radio SRI	Switzerland	Intelsat 601	27.5°	West	C-Band	3.915	RHCP	23B	East Hemisph.
Radio Sudan	Sudan	Arabsat 2A	26.0°	East	C-Band	3.905	LHCP	10	Zone
Radio Suisse Int.	Switzerland	Eutelsat II f1(Hotbird 1/2)	13.0°	East	Ku-Band	11.325	Ver	6	Spot
Radio Sweden	Sweden	Eutelsat II f1(Hotbird 1/2)	13.0°	East	Ku-Band	10.987	Ver	25	Wide
Radio Sweden Int.	Sweden	ASTRA 1C	19.2°	East	Ku-Band	10.964	Hor	33	Multi
Radio Sweden Int.	Sweden	ASTRA 1C	19.2°	East	Ku-Band	10.964	Hor	33	Multi
Radio Sweden/Radio Arlando	Sweden	Tele-X	5.0°	East	Ku2-Band	12.475	LHCP	40	Spot
Radio Top 40	Spain	Hispasat	31.0°	West	Ku3-Band	12.711	Hor	14	East Spot
Radio Tunisia	Tunisia	Eutelsat II f3	16.0°	East	Ku-Band	11.658	Ver	39	Wide
Radio UM	Portugal	Gorizont 12/22	40.0°	East	C-Band	3.930	RHCP	11	Global Beam
Radio Uno	Italy	Eutelsat II f1(Hotbird 1/2)	13.0°	East	Ku-Band	11.366	Ver	8	Spot
Radio Vlaanderen/BRT Nachtradio	Belgium	ASTRA 1D	19.2°	East	Ku-Band	10.920	Hor	63	Multi
Radio Voz Galicia	Spain	Hispasat	26.0°	West	C-Band	12.631	Ver	3	East Spot
Radio Yemen	Yemen	Arabsat 2A	26.0°	East	C-Band	4.180	LHCP	22	Zone
Radio Zagreb	Croatia	Eutelsat II f3	16.0°	East	Ku-Band	10.987	Hor	20	Wide
Radio-5	South Africa	Intelsat 602	63.0°	East	C-Band	3.807	LHCP	4	South West Zone

ITU REGION 1 RADIO IN ALPHABETICAL ORDER

Station	Country	Satellite	Position		Band	Frequency	Pol.	Tr	Beam
Radioropa Info.	Germany	DFS-3 Kopernikus	23.5°	East	Ku3-Band	12.625	Hor	K4	Spot
Radiostancija Mayak	CIS	Gorizont 12/22	40.0°	East	C-Band	3.675	RHCP	6	Global
Radyo D	Turkey	Türksat 1B	42°	East	Ku1-Band	11.569	Ver	2A	West Spot
Radyo Forex	Turkey	Türksat 1B	42°	East	Ku1-Band	11.030	Hor	13	East Spot
Radyo Kulup	Turkey	Türksat 1B	42°	East	Ku1-Band	11.030	Hor	13	East Spot
RB2 Kulturell	Germany	DFS-3 Kopernikus	23.5°	East	Ku3-Band	12.625	Hor	K4	Spot
RDP Antena 1	Portugal	Express 2	14.0°	West	C-Band	4.025	RHCP	15	Semi Global
RDP Antena 1	Portugal	Eutelsat II f1(Hotbird 1/2)	13.0°	East	Ku1-Band	11.727	Ver	50	Wide
RDP Commercial Lisboa	Portugal	Eutelsat II f1(Hotbird 1/2)	13.0°	East	Ku1-Band	11.727	Ver	50	Wide
RDP Internacional	Portugal	Express 2	14.0°	West	C-Band	4.025	RHCP	15	Semi Global
Rep. of Yemen Radio	Rep. of Yemen	Intelsat 707	1.0°	West	C-Band	4.185	LHCP	39	Global
RFE/VOA	Germany/USA	Eutelsat II f1(Hotbird 1/2)	13.0°	East	Ku1-Band	11.095	Hor	21B	Super
RFI En Langues Européennes	France	Intelsat 601	27.5°	West	C-Band	3.915	RHCP	23B	East Hemisph.
RFI Mondial en France	France	Intelsat 601	27.5°	West	C-Band	3.915	RHCP	23B	East Hemisph.
RFM/Radio Renascença 2	Portugal	Eutelsat II f1(Hotbird 1/2)	13.0°	East	Ku1-Band	11.727	Ver	50	Wide
Rire et Chansons	France	Télécom 2B	5.0°	West	Ku3-Band	12.564	Ver	R2	Spot
RNE Radio 1	Spain	Eutelsat II f1(Hotbird 1/2)	13.0°	East	Ku1-Band	11.221	Hor	1	Multi
RNE Radio 2	Spain	Hispasat	31.0°	West	Ku2-Band	12.149	LHCP	23	Spot
RNE Radio Cinco	Spain	Hispasat	31.0°	West	Ku2-Band	12.149	LHCP	23	Multi
RNE Radio Exterior	Spain	Hispasat	31.0°	West	Ku2-Band	12.149	LHCP	23	Multi
RNE Radio Uno	Spain	Hispasat	31.0°	West	Ku2-Band	12.149	LHCP	23	Multi
RNW (Hola Holanda)	Dutch	ASTRA 1D	19.2°	East	Ku1-Band	10.847	Ver	58	Wide
RNW (Network Europa)	Dutch	ASTRA 1D	19.2°	East	Ku1-Band	10.847	Ver	58	Wide
RPR-2	Germany	DFS-3 Kopernikus	23.5°	East	Ku3-Band	12.625	Hor	K4	Spot
RTE Radio One	Ireland	ASTRA 1C	19.2°	East	Ku1-Band	11.538	Ver	22	Multi
RTL OldieSender	Germany	ASTRA 1B	19.2°	East	Ku1-Band	11.391	Hor	13	Multi
RTM Berber	Morocco	Eutelsat II f3	16.0°	East	Ku1-Band	10.972	Ver	25	Wide
RTM Int.	Morocco	Eutelsat II f3	16.0°	East	Ku1-Band	10.972	Ver	25	Wide
S2 Kultur	Germany	DFS-3 Kopernikus	23.5°	East	Ku3-Band	12.625	Hor	K4	Spot
S2 Kultur	Germany	ASTRA 1C	19.2°	East	Ku1-Band	11.185	Ver	K4	Multi
SAFM	AFS	PanAmSat-4	68.5°	East	Ku3-Band	11.185	Ver	18	AFS Spot
SDR-1	Germany	ASTRA 1C	19.2°	East	Ku1-Band	11.493	Ver	38	Multi
SDR-3	Germany	ASTRA 1B	19.2°	East	Ku1-Band	11.617	RHCP	21	Multi
Sky Radio	Netherlands	Arabsat 2A	26.0°	East	C-Band	4.166	Ver	48	Zone
SR1 Eurowelle Saar	Germany	ASTRA 1C	19.2°	East	Ku1-Band	11.185	Ver	48	Multi
SR1 Saarland	Germany	ASTRA 1B	19.2°	East	Ku1-Band	11.493	Ver	8	Multi
Star*Sat Radio	Germany	ASTRA 1C	19.2°	East	Ku1-Band	11.185	Ver	19	Multi
Sunrise Radio	UK	ASTRA 1A	19.2°	East	Ku1-Band	11.332	Hor	K4	Spot
Süper FM	Turkey	Eutelsat II f2	10.0°	East	Ku1-Band	11.479	Ver	K4	Multi
Superstation Kuwait	Kuwait	ASTRA 1D	19.2°	East	Ku1-Band	11.479	Hor	9	Wide
SWF-1	Germany	ASTRA 1C	19.2°	East	Ku1-Band	11.185	Ver	48	Multi
SWF-3	Germany	ASTRA 1B	19.2°	East	Ku1-Band	11.493	Ver	19	Multi
SWF-3	Germany	ASTRA 1C	19.2°	East	Ku1-Band	11.185	Ver	48	Multi
Swiss Radio International	Switzerland	ASTRA 1A	19.2°	East	Ku1-Band	11.332	Hor	9	Spot
Swiss Radio International	Switzerland	ASTRA 1D	19.2°	East	Ku1-Band	10.802	Ver	55	Wide
Tatlises Radio	Turkey	Türksat 1B	42°	East	Ku1-Band	11.144	Hor	8A	West Spot

243

ITU REGION 1 RADIO IN ALPHABETICAL ORDER

Station	Country	Satellite	Position		Band	Frequency	Pol.	Tr	Beam
TGRT FM (Huzur Radio)	Turkey	Eutelsat II f2	10.0°	East	Ku1-Band	11.095	Ver	26B	Wide
The Voice (of Scandinavia)	Sweden	Tele-X	5.0°	East	Ku2-Band	12.475	LHCP	40	Spot
Tunis Radio Int.	Tunisia	Eutelsat II f3	16.0°	East	Ku1-Band	11.658	Ver	39	Wide
Turkish Radio	Turkey	Eutelsat II f1 (Hotbird 1/2)	13.0°	East	Ku1-Band	11.181	Hor	22B	Super
TWR / ERF	USA / Germany	ASTRA 1C	19.2°	East	Ku1-Band	11.038	Ver	38	Multi
UNICO Supermarché	France	Télécom 2B	5.0°	West	Ku3-Band	12.522	Ver	R1	Spot
United Christian Broadcasters	UK	ASTRA 1B	19.2°	East	Ku1-Band	11.508	Ver	20	Multi
unknown radio	Iran	Intelsat 602	63.0°	East	Ku1-Band	11.100	Ver	73	East Spot
Virgin 1215	UK	ASTRA 1A	19.2°	East	Ku1-Band	11.377	Ver	12	Multi
Virgin Megastore Music	UK	Eutelsat II f1 (Hotbird 1/2)	13.0°	East	Ku1-Band	10.987	Ver	25	Wide
VOA	USA	Intelsat 601	27.5°	West	C-Band	3.743	RHCP	81	East Hemisph.
VOA Europe [English]	USA	Eutelsat II f1 (Hotbird 1/2)	13.0°	East	Ku1-Band	11.162	Ver	27B	Wide
VOA Europe [foreign]	USA	Eutelsat II f1 (Hotbird 1/2)	13.0°	East	Ku1-Band	11.162	Ver	27B	Wide
VOA Europe Network	USA	Intelsat 704	66.0°	East	C-Band	4.177	RHCP	38	Global
VOA Ext. Sce.	USA	Intelsat 704	66.0°	East	C-Band	4.177	RHCP	38	Global
VOA [foreign 1]	USA	Intelsat 601	27.5°	West	C-Band	3.743	RHCP	81	East Hemisph.
VOA [foreign 2]	USA	Intelsat 601	27.5°	West	C-Band	3.743	RHCP	81	East Hemisph.
VOA [foreign 3]	USA	Intelsat 601	27.5°	West	C-Band	3.743	RHCP	81	East Hemisph.
VOA [foreign 4]	USA	Intelsat 601	27.5°	West	C-Band	3.743	RHCP	81	East Hemisph.
Voice of Music	Israel	Intelsat 707	1.0°	West	Ku1-Band	11.178	Ver	73	East Spot
Voice of the Arabs	Egypt	Eutelsat II f3	16.0°	East	Ku1-Band	11.178	Ver	27	Wide
WDR 2	Germany	ASTRA 1C	19.2°	East	Ku1-Band	11.052	Hor	39	Multi
WDR 3	Germany	ASTRA 1C	19.2°	East	Ku1-Band	11.052	Hor	39	Multi
WDR 4	Germany	ASTRA 1C	19.2°	East	Ku1-Band	11.052	Hor	39	Multi
WDR Radio 5	Germany	ASTRA 1C	19.2°	East	Ku1-Band	11.052	Hor	39	Multi
WDR-3 Köln	Germany	DFS-3 Kopernikus	23.5°	East	Ku3-Band	12.625	Ver	K4	Spot
World Radio Network 1	UK	Eutelsat II f1 (Hotbird 1/2)	13.0°	East	Ku1-Band	10.987	Ver	25	Wide
World Radio Network (WRN 1)	UK	ASTRA 1B	19.2°	East	Ku1-Band	11.538	Ver	22	Multi
World Radio Network (WRN 1)	UK	Eutelsat II f1 (Hotbird 1/2)	13.0°	East	Ku1-Band	11.265	Hor	3	Spot
World Radio Network (WRN 2)	UK	Eutelsat II f1 (Hotbird 1/2)	13.0°	East	Ku1-Band	11.554	Ver	32	Super
Youth	AFS	PanAmSat-4	68.5°	East	Ku3-Band	12.664	Ver		AFS Spot

ITU REGION 2 TV IN ALPHABETICAL ORDER

12.3. ITU REGION 2 TV IN ALPHABETICAL ORDER

Station	Country	Satellite	Position		Band	Frequency	Pol.	Tr	Beam
Action Pay-per-View	USA	Galaxy 7	91.0°	West	C-Band	3.760	Ver	2	Conus
AFRTS	USA	Satcom SN-2	69.0°	West	C-Band	4.100	Ver	17U	Conus
AFRTS [U.S.]	USA	Intelsat 707	1.0°	West	C-Band	4.175	RHCP	24	Global
Airport News and Transportation Netw.	USA	Galaxy 7	91.0°	West	Ku2-Band	12.065	Hor	17K	Conus
Alaska Satellite TV Project	USA	Satcom C5	139.0°	West	C-Band	4.180	Hor	24	Conus
America's Collectibles Netw.	USA	Galaxy 3	93.5°	West	C-Band	3.920	Hor	11	Conus
America's Talking	USA	Galaxy 1	133.0°	West	C-Band	3.900	Hor	10	Spot
America-1	USA	TDRSS	41.0°	West	C-Band	3.920	Hor	11	Spot
American Collectibles Netw.	USA	Telstar 402	89.0°	West	C-Band	3.940	Hor	12	Conus
American Movie Classics	USA	Satcom C4	135.0°	West	C-Band	3.720	Ver	1	Conus
American One	USA	Satcom SN3R	87.0°	West	C-Band	4.180	Hor	18U	Conus
American TV Netw.	USA	Satcom C3	131.0°	West	C-Band	4.140	Hor	22	Conus
APNA TV	India	Ekspress 2	14.0°	West	C-Band	3.825	RHCP	9	Global
Arab Netw. America (ANA)	USA	Galaxy 6	74.0°	West	C-Band	3.900	Ver	10	Conus
Argentina TV a Color	UK	Intelsat 706	53.0°	West	C-Band	4.160	RHCP	12	Zone
Around the World after Dark	USA	Telstar 302	85.0°	West	C-Band	4.060	Hor	18	Conus
Arts & Entertainment TV	USA	Galaxy 5	125.0°	West	C-Band	4.180	Hor	23	Conus
Atlantic Satellite Netw. Halifax	Canada	Anik E2	107.3°	West	Ku2-Band	12.096	Hor	29	Spot
Automotive Sat. Training Netw.	USA	Galaxy 7	91.0°	West	Ku2-Band	12.065	Hor	17K	Conus
BBC Breakfast News	UK	Galaxy 7	91.0°	West	C-Band	3.840	Hor	7	Conus
BBC Breakfast News	UK	Galaxy 9	123.0°	West	C-Band	3.720	Ver	1	West Hemisph.
BBC World	UK	Intelsat 706	53.0°	West	C-Band	3.985	RHCP	7	Zone
Black Entertainment TV (BET)	USA	Galaxy 5	125.0°	West	C-Band	4.100	Ver	20	Conus
Bravo	USA	Satcom C4	135.0°	West	C-Band	3.840	Ver	7	Conus
C-Span	USA	TDRSS	41.0°	West	C-Band	3.880	Ver	9	Spot
C-SPAN 1	USA	Satcom C3	131.0°	West	C-Band	3.840	Ver	7	Conus
C-Span 2 ASAP	USA	Satcom C4	135.0°	West	C-Band	4.080	Ver	19	Conus
Cable Health Club	USA	Satcom C4	135.0°	West	C-Band	3.980	Ver	14	Conus
Cal-Span	USA	Satcom C1	137.0°	West	C-Band	4.040	Hor	17	Conus
Canadian Exxxtacy	Canada	Telstar 401	97.0°	West	C-Band	3.720	Ver	1	Conus
Canal 13 (XHDF-TV)	Mexico	Morelos F2	116.7°	West	C-Band	4.180	Ver	12N	Spot
Canal France Int.	France	Ekspress 2	14.0°	West	C-Band	3.825	RHCP	9	Global
Canal Sur	Peru	PanAmSat f1	45.0°	West	C-Band	4.140	Hor	18a	CA Spot
Cartoon Network	USA	Intelsat 601	27.5°	West	C-Band	3.900	RHCP	10	West Hemisph.
CBC Newsworld	Canada	Anik E2	107.3°	West	C-Band	3.820	Ver	7	Spot
CBC North	Canada	Anik E2	107.3°	West	C-Band	3.880	Hor	9	Spot
CBCM	Canada	Anik E2	107.3°	West	C-Band	3.960	Hor	7	Spot
CBMT, Montreal/CBC	Canada	Americom GE 1	103.0°	West	Ku2-Band	12.140	Hor	22	Spot
CCC (Chinese Comm. Channel)	USA	Ekspress 2	14.0°	West	C-Band	4.125	RHCP	17	Global
Channel 1 Moscow	Russia								

245

ITU REGION 2 TV IN ALPHABETICAL ORDER

Station	Country	Satellite	Position	Band	Frequency	Pol.	Tr	Beam
Cinemax (East)	USA	Galaxy 1	133.0° West	C-Band	4.080	Hor	19	Conus
Cinemax East 2	USA	Galaxy 3	93.5° West	C-Band	4.040	Hor	17	Conus
Cinemax II (East)	USA	Galaxy 1	133.0° West	C-Band	4.160	Hor	23	Conus
Cinemax (West)	USA	Galaxy 5	125.0° West	C-Band	4.020	Ver	16	Conus
Classic Sports Network	USA	Galaxy 7	91.0° West	Ku2-Band	11.810	Ver	6	Conus
CNN	USA	Satcom SN3R	87.0° West	Ku2-Band	3.980	Hor	16L	Conus
CNN Airport Channel	USA	GStar 2	125.0° West	Ku2-Band	11.988	Ver	13	Spot
CNN Headline News	USA	Galaxy 5	125.0° West	C-Band	3.800	Ver	5	Conus
CNN Headline News	USA	Galaxy 5	125.0° West	C-Band	4.140	Hor	22	Spot
CNN Int.	USA	PanAmSat f1	45.0° West	C-Band	3.720	Hor	1	LA Spot
CNN Int.	USA	Galaxy 1	133.0° West	C-Band	4.000	Hor	15	Conus
CNN International	USA	Satcom SN3R	87.0° West	C-Band	4.020	Ver	16U	Conus
CNN International	USA	GStar 2	125.0° West	Ku2-Band	12.035	Hor	6	Spot
CNN Newsource	USA	GStar 4	105.0° West	Ku2-Band	11.852	Hor	3	Conus
Comedy Central	USA	Galaxy 1	133.0° West	C-Band	3.720	Ver	1	Conus
Comedy Central (East)	USA	Satcom C3	131.0° West	C-Band	4.120	Ver	21	Conus
Computer Television Netw.	USA	Galaxy 1	123.0° West	C-Band	4.160	Ver	23	West Hemisph.
Consumer News & Business Ch. (CNBC)	USA	Galaxy 5	125.0° West	C-Band	3.960	Hor	13	Conus
Cornerstone TV	USA	Galaxy 6	74.0° West	C-Band	3.980	Ver	14	Conus
Cornerstone	USA	Americom GE 1	103.0° West	C-Band	3.840	Hor	6	Conus
Country Music TV	USA	Satcom C4	135.0° West	C-Band	4.180	Ver	24	Spot
CTV Television Netw.	Canada	Anik E2	107.3° West	C-Band	4.180	Ver	24	Spot
CycleSat commercials	USA	Americom GE 1	103.0° West	Ku2-Band	12.100	Ver	20	Conus
Deutsche Welle TV	Germany	Satcom C4	135.0° West	C-Band	3.800	Hor	5	Conus
Deutsche Welle TV	Germany	Satcom SN-2	69.0° West	C-Band	3.760	Hor	2	N & Lat. Am. Spot
Deutsche Welle TV (foreign)	Germany	Intelsat K	21.5° West	Ku1-Band	11.605	Hor	7	N & Lat. Am. Spot
Deutsche Welle TV (German)	Germany	Intelsat K	21.5° West	Ku1-Band	11.605	Hor	7	Conus
Discovery Channel	USA	Galaxy 5	125.0° West	C-Band	3.940	Ver	12	Conus
Discovery (East)	USA	Satcom C4	135.0° West	C-Band	4.120	Ver	21	Conus
Disney Channel (East)	USA	Galaxy 5	125.0° West	C-Band	3.720	Hor	1	Conus
Disney Channel (West)	USA	Galaxy 1	133.0° West	C-Band	3.840	Ver	7	Conus
El Entertainment	USA	Satcom C3	131.0° West	C-Band	4.160	Ver	23	Conus
ECO-Televisa	Mexico	PanAmSat f1	45.0° West	C-Band	3.883	Ver	5	LA Spot
Encore	USA	Galaxy 1	101.0° West	C-Band	3.760	Hor	3	Conus
Encore 2	USA	Spacenet 4	101.0° West	C-Band	3.720	Ver	1	Conus
Encore 8	USA	Spacenet 4	101.0° West	C-Band	4.060	Ver	18	Conus
Encore 8	USA	Spacenet 4	101.0° West	C-Band	3.920	Hor	11	Conus
Encore 8	USA	Spacenet 4	101.0° West	C-Band	3.920	Hor	11	Conus
ESPN	USA	Galaxy 5	125.0° West	C-Band	3.880	Ver	9	Conus
ESPN	USA	Galaxy 9	123.0° West	C-Band	4.120	Ver	21	West Hemisph.
ESPN 2	USA	Galaxy 1	125.0° West	C-Band	3.980	Ver	14	Conus
ESPN Blackout Channel	USA	Galaxy 1	133.0° West	C-Band	3.980	Ver	14	Conus
ESPN International	USA	PanAmSat f1	45.0° West	C-Band	4.119	Ver	15a	LA Spot
Estacion Montello	USA	Galaxy 7	91.0° West	C-Band	3.920	Hor	11	Conus
Eternal Word TV	USA	Galaxy 1	133.0° West	C-Band	3.920	Hor	11	Conus
Europlus/Teleplus	Italy	Intelsat K	21.5° West	Ku1-Band	11.735	Hor	1	NA Spot

ITU REGION 2 TV IN ALPHABETICAL ORDER

Station	Country	Satellite	Position	Band	Frequency	Pol.	Tr	Beam
Eurotica	USA	Telstar 402	89.0° West	C-Band	4.020	Hor	16	Spot
Exxxtasy 2	Canada	Telstar 401	97.0° West	C-Band	4.000	Ver	15	Conus
Exxxtreme/Climaxxx Promo Ch.	USA	Telstar 302	85.0° West	C-Band	4.000	Ver	15	Conus
Fantasy Cafe TV	USA	Telstar 402	89.0° West	C-Band	3.860	Hor	8	Spot
FLIX movie services	USA	Telstar 302	85.0° West	C-Band	3.960	Ver	13	Conus
FOXNet - PrimeTime	USA	Spacenet 4	101.0° West	C-Band	3.900	Ver	10	Conus
fX Movies	USA	Galaxy 7	91.0° West	Ku2-Band	4.160	Ver	23	Conus
G.O.P. TV	USA	Galaxy 7	91.0° West	C-Band	3.820	Ver	1K	Conus
Game Show Network	UK	Satcom SN 2	69.0° West	C-Band	3.740	Hor	6	Conus
GEMS TV	USA	Telstar 401	97.0° West	Ku2-Band	12.092	Hor	13L	Conus
Georgia Public TV (GPTV)	USA	Telstar 401	97.0° West	Ku2-Band	12.123	Hor	14A	Conus
Global Shopping Network	USA	Galaxy 1	133.0° West	C-Band	4.180	Ver	14B	Conus
Global TV	Canada	Anik E2	107.3° West	C-Band	38.60	Ver	24	Conus
GOP TV	USA	Telstar 402	89.0° West	C-Band	4.100	Hor	8	Spot
Gospel Music TV	USA	Galaxy 3	93.5° West	C-Band	4.000	Ver	20	Spot
HBO East 2	USA	Galaxy 3	93.5° West	C-Band	4.020	Ver	15	Conus
HBO East 3	USA	Galaxy 3	93.5° West	C-Band	4.080	Hor	16	Conus
HBO III (East)	USA	Galaxy 7	91.0° West	C-Band	3.860	Ver	19	Conus
HBO II (East)	USA	Galaxy 7	91.0° West	C-Band	3.980	Ver	8	Conus
HBO III (West)	USA	Galaxy 7	91.0° West	C-Band	4.180	Ver	14	Conus
HBO West 2	USA	Galaxy 3	93.5° West	C-Band	4.100	Ver	24	Conus
HBO/Cinemax	USA	PanAmSat f1	45.0° West	C-Band	3.840	Ver	20	SA Spot
HBO/Cinemax	USA	PanAmSat f1	45.0° West	C-Band	3.900	Ver	4	NA Spot
Hispavision	Spain	Hispasat	30.0° West	Ku2-Band	12.015	RHCP	11	Spot
Home and Garden TV Netw.	USA	Galaxy 1	133.0° West	C-Band	4.100	Ver	15	Conus
Home Box Office (East)	USA	Galaxy 5	125.0° West	C-Band	4.000	Hor	20	Conus
Home Box Office (West)	USA	Galaxy 5	125.0° West	C-Band	3.860	Ver	15	Conus
Home Shopping Club 2	USA	Satcom C3	131.0° West	C-Band	3.900	Hor	8	Conus
Home Shopping Network 1	USA	Satcom C4	135.0° West	C-Band	3.900	Ver	10	Conus
Home Sports Entertainment	USA	Satcom C1	137.0° West	C-Band	3.990	Ver	10	Conus
Home Team sports, Baltimore	USA	Satcom SN3R	87.0° West	C-Band	4.160	Ver	10	Conus
Hong Kong TVB Jade Channel	Hong Kong	Galaxy 4	99.0° West	Ku2-Band	12.050	Hor	12	Conus
Hospitality TV	USA	Galaxy 7	91.0° West	Ku2-Band	11.900	Ver	18	Conus
HRT Croatia	Croatia	Telstar 402	89.0° West	C-Band	4.060	Hor	10K	Spot
HSE2	USA	Satcom C1	137.0° West	C-Band	4.040	Hor	18	Conus
HTV Hispanic music videos	USA	Spacenet 4	101.0° West	C-Band	3.940	Ver	17	Conus
Independent Film Channel	USA	Galaxy 6	74.0° West	C-Band	3.960	Hor	12	Conus
Infomercial Channel	USA	Satcom C3	131.0° West	C-Band	4.100	Hor	13	Conus
Infomerica TV	USA	Galaxy 3	93.5° West	C-Band	4.120	Hor	20	Conus
Infomerica TV	USA	Satcom C1	137.0° West	C-Band	4.060	Ver	21	Conus
KCNC, Denver	USA	Satcom C1	137.0° West	C-Band	3.980	Hor	18	Conus
KDVR, Denver (FOX afflate)	USA	Satcom C1	137.0° West	C-Band	3.800	Hor	14	Conus
KMGH, Denver	USA	Satcom C1	137.0° West	C-Band	3.740	Hor	5	Conus
KMGH, Denver	USA	Satcom C1	137.0° West	C-Band	3.820	Ver	2	Conus
KNBC-TV, LA	USA	Spacenet 4	101.0° West	C-Band	3.820	Hor	6	Conus
							9	Conus

ITU REGION 2 TV IN ALPHABETICAL ORDER

Station	Country	Satellite	Position	Band	Frequency	Pol.	Tr	Beam
Knowledge TV	USA	Galaxy 5	125.0° West	C-Band	4.120	Hor	21	Conus
KOMO-TV, Seattle	USA	Spacenet 4	101.0° West	C-Band	3.860	Hor	10	Conus
KOMO-TV Seattle	USA	Spacenet 4	101.0° West	C-Band	3.860	Hor	8	Conus
KPIX TV, San Francisco	USA	Spacenet 4	101.0° West	C-Band	4.180	Hor	24	Conus
KRMA, Denver	USA	Satcom C1	137.0° West	C-Band	3.760	Hor	3	Conus
KTLA, Los Angeles	USA	Satcom SN3R	87.0° West	C-Band	4.000	Hor	8	Conus
KWGN, Denver	USA	Satcom C1	137.0° West	C-Band	4.160	Hor	23	Conus
La Cadena de Milagro	USA	Telstar 302	85.0° West	C-Band	4.100	Hor	20	Spot
La Cadena de Milagro	USA	Telstar 402	89.0° West	C-Band	4.160	Hor	23	Conus
La Cadena di Milagro	USA	Galaxy 7	91.0° West	C-Band	4.120	Hor	21	Spot
La Chaine French	Canada	Anik E2	107.3° West	Ku2-Band	11.974	Hor	25	Spot
Lifetime (East)	USA	Satcom C4	135.0° West	C-Band	3.780	Hor	4	Conus
Lifetime West	USA	Satcom C3	131.0° West	C-Band	3.780	Hor	4	Conus
Madison Square Garden	USA	Satcom C4	135.0° West	C-Band	3.820	Hor	6	Conus
MCET Educational Netw.	USA	SBS 6	74.0° West	Ku2-Band	12.092	Ver	16	Conus
Merchandise and Entertainm. TV (MET)	USA	Galaxy 6	74.0° West	C-Band	4.060	Ver	18	Conus
Midwest Sports Channel	USA	Galaxy 6	74.0° West	C-Band	4.000	Ver	15	Conus
MOR Music Television	USA	Galaxy 5	125.0° West	C-Band	3.900	Ver	10	Conus
Movie Channel (West)	USA	Galaxy 9	123.0° West	C-Band	4.060	Ver	18	West Hemisph.
MTV (West)	USA	Galaxy 9	123.0° West	C-Band	4.080	Ver	19	West Hemisph.
MuchMusic	Canada	Galaxy 6	74.0° West	C-Band	3.880	Hor	9	Conus
Music Television (West)	USA	Satcom C3	131.0° West	C-Band	4.020	Hor	16	Conus
Music TV(East)	USA	Satcom C4	135.0° West	C-Band	4.040	Ver	17	Conus
Muslim Television	USA	Galaxy 7	91.0° West	Ku2-Band	11.990	Ver	15K	Global
MuslimTV	Russia	Ekspress 2	14.0° West	C-Band	3.825	RHCP	9	Conus
NASA Select Channel	USA	Satcom SN-2	69.0° West	C-Band	3.880	Hor	5	Spot
Nat. Enpowerment TV Netw.	USA	Americom GE 1	103.0° West	C-Band	4.040	Hor	19	Spot
Nat. Jewish TV Network	USA	Telstar 402	89.0° West	C-Band	4.080	Hor	19	Conus
Nat. Prgr. Sce. Preview channel	USA	Spacenet 4	101.0° West	C-Band	3.980	Ver	14	Conus
National Empowerment TV Netw.	USA	Galaxy 7	91.0° West	C-Band	4.100	Ver	20	Conus
National Weather Netw.	USA	Satcom C1	137.0° West	Ku2-Band	12.065	Hor	17K	Conus
NBC	USA	Satcom SN3R	87.0° West	C-Band	3.860	Ver	8	Conus
Nebraska Educational TV	USA	Satcom SN3R	87.0° West	C-Band	3.740	Ver	13L	Conus
Nebraska Educational TV	USA	Satcom C1	137.0° West	C-Band	3.780	Ver	13U	West Hemisph.
Network One (N1)	USA	Satcom C3	131.0° West	C-Band	3.920	Ver	11	Conus
New England Sports Channel	USA	Satcom SN-2	69.0° West	C-Band	3.980	Ver	14	Conus
Newsport	USA	Satcom C1	137.0° West	C-Band	4.140	Ver	18L	West Hemisph.
Newsport	USA	Satcom C1	137.0° West	C-Band	4.020	Ver	16	Conus
NewsTalk Television	USA	Galaxy 7	91.0° West	C-Band	4.140	Ver	22	Conus
NHK Tokyo	Japan	Galaxy 9	123.0° West	C-Band	3.760	Ver	3	West Hemisph.
Nickelodeon (East)	USA	Satcom C4	135.0° West	C-Band	4.060	Ver	18	Conus
Nickelodeon (West)	USA	Satcom C3	131.0° West	C-Band	4.040	Ver	17	Conus
Nickelodeon (West)	USA	Galaxy 9	123.0° West	C-Band	4.140	Ver	22	West Hemisph.
Nostalgia Channel	USA	Galaxy 1	133.0° West	C-Band	3.940	Hor	12	Conus
Nustar	USA	Satcom C4	135.0° West	C-Band	3.800	Ver	5	Conus
Odyssey Network	USA	Satcom C3	131.0° West	C-Band				

ITU REGION 2 TV IN ALPHABETICAL ORDER

Station	Country	Satellite	Position	Band	Frequency	Pol.	Tr	Beam
Ontario Legislature	Canada	Anik E2	107.3° West	Ku2-Band	11.939	Hor	24	Spot
Oragon Ed Net	USA	Satcom SN3R	87.0° West	Ku2-Band	12.060	Hor	23	West Spot
ORT TV-1	Russia	Ekspress 2	14.0° West	C-Band	3.675	RHCP	1	Global
ORT TV-1	Russia	Gorizont 26	11.0° West	C-Band	3.675	RHCP	6	Global
Panda America (Home Shopping)	USA	Telstar 402	89.0° West	C-Band	4.180	Ver	24	Spot
PBS (schedule C)	USA	Telstar 401	97.0° West	Ku2-Band	11.902	Hor	7	Conus
PBS (schedule D)	USA	Telstar 401	97.0° West	Ku2-Band	11.914	Hor	8	Conus
Peru TV	Argentina	PanAmSat f1	45.0° West	C-Band	3.940	Hor	12	CA Spot
Peru TV Ch. 13	Peru	PanAmSat f1	45.0° West	C-Band	3.851	Hor	10a	CA Spot
Peru TV Ch. 2	Peru	PanAmSat f1	45.0° West	C-Band	3.871	Hor	10b	CA Spot
Peru TV Ch. 4	Peru	PanAmSat f1	45.0° West	C-Band	3.789	Hor	8b	CA Spot
Peru TV Ch. 5	Peru	PanAmSat f1	45.0° West	C-Band	3.771	Hor	8a	CA Spot
Playboy at Night	USA	Galaxy 5	125.0° West	C-Band	3.740	Ver	2	Conus
PrevuGuide	USA	Satcom C4	135.0° West	C-Band	3.860	Hor	8	Conus
Prime Network	USA	Satcom C3	131.0° West	C-Band	3.920	Ver	11	Conus
Prime Sports Intermountain (West)	USA	Satcom C1	137.0° West	C-Band	4.040	Hor	17	Conus
Prime Sports Northwest	USA	Satcom C1	137.0° West	C-Band	4.140	Ver	22	Conus
Prime Sports Showcase	USA	Satcom C1	137.0° West	C-Band	4.060	Ver	18	Conus
Prime Ticket, California	USA	Satcom C1	137.0° West	C-Band	3.840	Hor	7	Conus
Pro-Am sports, Detroit	USA	Satcom SN3R	87.0° West	C-Band	4.120	Hor	11	Conus
Public Broadcasting Sce. (PBS)	USA	Telstar 401	97.0° West	C-Band	3.860	Hor	8	Spot
Q-CVC	Mexico	Morelos F2	116.7° West	C-Band	3.860	Ver	4N	Conus
Quorum multi-level marketing	USA	Galaxy 7	91.0° West	Ku2-Band	12.020	Ver	16	Conus
QVC Fashion Channel	USA	Satcom C3	131.0° West	C-Band	3.860	Hor	8	Conus
QVC, Home Shopping Netw.	USA	Satcom C4	135.0° West	C-Band	3.880	Ver	9	Conus
RAI	Italy	Galaxy 7	91.0° West	C-Band	3.880	Ver	9	Global
RAI	Italy	Galaxy 3	93.5° West	C-Band	3.940	Hor	12	NA Conus
RCTV (Venezuela 7)	Venezuela	Intelsat 709	50.0° West	C-Band	3.962	RHCP	7	Zone
Real Estate TV Network	USA	Galaxy 7	91.0° West	Ku2-Band	12.140	Ver	22	Conus
Request TV 1	USA	Satcom C4	135.0° West	C-Band	4.020	Hor	16	Conus
RTP Internacional	Portugal	Ekspress 2	14.0° West	C-Band	4.025	RHCP	15	Global
RTV Beograd	Croatia	PanAmSat f1	45.0° West	Ku2-Band	11.760	Hor	22b	NA Conus
Russian TV Network	USA	Telstar 402	89.0° West	C-Band	4.080	Ver	19	Spot
S. Carolina Educational TV	USA	Telstar 401	97.0° West	Ku2-Band	11.790	Ver	3A	Conus
Satellite City TV	USA	Galaxy 5	125.0° West	C-Band	3.740	Hor	2	Conus
Sci-Fi Channel	USA	Galaxy 5	125.0° West	C-Band	3.780	Ver	4	Conus
Sellevision	USA	Satcom SN3R	87.0° West	C-Band	4.060	Ver	17L	Conus
Shepherd's Chapel	USA	Galaxy 4	99.0° West	C-Band	3.820	Ver	5	Conus
Shop at Home	USA	Telstar 402	89.0° West	C-Band	3.780	Hor	4	Spot
Shop-at-Home	USA	Galaxy 1	133.0° West	C-Band	3.980	Ver	14	Conus
Shop-at-Home Network	USA	Satcom C3	131.0° West	C-Band	3.880	Ver	9	Conus
Showtime East	USA	Satcom C4	135.0° West	C-Band	4.100	Hor	15	Conus
Showtime (West)	USA	Galaxy 9	123.0° West	C-Band	4.000	Ver	20	Conus
Showtime West	USA	Telstar 302	85.0° West	C-Band	4.120	Ver	15	West Hemisph.
SkyVision Home Shopping Ch.	USA	Galaxy 7	91.0° West	C-Band	4.060	Ver	21	Conus
SkyVision Promo Channel	USA			C-Band		Ver	18	Conus

ITU REGION 2 TV IN ALPHABETICAL ORDER

Station	Country	Satellite	Position	Band	Frequency	Pol.	Tr	Beam
Space	Argentina	PanAmSat f1	45.0° West	C-Band	3.929	Ver	6b	SA Spot
Sport South	USA	Satcom SN3R	87.0° West	C-Band	4.080	Hor	10	Conus
Sportschannel alternatives	USA	Americom GE 1	103.0° West	C-Band	4.040	Hor	17	Spot
SportsChannel Chicago	USA	Satcom C1	137.0° West	C-Band	3.880	Hor	9	Conus
SportsChannel Chicago	USA	Americom GE 1	103.0° West	C-Band	3.960	Hor	13	Conus
SportsChannel Chicago	USA	Satcom C1	103.0° West	C-Band	3.860	Ver	8	Spot
SportsChannel Chicago Plus	USA	Satcom C1	137.0° West	C-Band	3.720	Hor	1	Conus
SportsChannel Cincinnati	USA	Satcom C1	137.0° West	C-Band	3.720	Hor	1	Conus
SportsChannel Florida	USA	Satcom C1	137.0° West	C-Band	3.880	Hor	9	Conus
SportsChannel Hawaii	USA	Satcom C1	137.0° West	C-Band	3.720	Hor	1	Conus
SportsChannel New England	USA	Satcom SN-2	69.0° West	C-Band	4.120	Hor	11	Conus
SportsChannel New England	USA	Americom GE 1	103.0° West	C-Band	3.980	Hor	14	Spot
SportsChannel New York	USA	Satcom SN 2	69.0° West	C-Band	3.720	Ver	1	Conus
SportsChannel New York Plus	US	Satcom SN-2	69.0° West	C-Band	4.180	Ver	18U	Conus
SportsChannel Ohio	USA	Americom GE 1	103.0° West	C-Band	3.780	Ver	3	Spot
SportsChannel Ohio, Flori, Cinci.	USA	Satcom C1	137.0° West	C-Band	4.000	Ver	15	Conus
SportsChannel Pacific	USA	Satcom C1	137.0° West	C-Band	3.780	Ver	4	Conus
SportsChannel Pacific	USA	Americom GE 1	103.0° West	C-Band	4.020	Ver	16	Spot
SportsChannel Philadelphia	US	Satcom SN-2	69.0° West	C-Band	3.920	Hor	6	Conus
Sportsouth (regional)	USA	Americom GE 1	103.0° West	C-Band	3.880	Hor	9	Conus
SRC Educational Netw.	USA	Telstar 401	97.0° West	Ku2-Band	11.855	Hor	6	Spot
SSVC	UK	TDRSS	41.0° West	C-Band	3.720	Hor	1	West Hemisph.
Sundance Channel	USA	Galaxy 9	123.0° West	C-Band	3.980	Hor	14	Conus
Sunshine Balckout Channel	USA	Satcom C1	137.0° West	C-Band	3.880	Ver	9	Conus
Sunshine Network	USA	Satcom C1	137.0° West	C-Band	4.180	Ver	24	Conus
Super Television Channel (USA)	USA	Orion F1	37.5° West	Ku1-Band	11.617	Ver	7	Spot
System United for Retransm.	Latin America	Satcom SN 2	69.0° West	C-Band	3.780	Ver	13U	Conus
TCI Preview Channel UA	USA	Galaxy 7	91.0° West	C-Band	4.000	Hor	15	Conus
TCI TV	USA	Galaxy 5	125.0° West	Ku2-Band	12.110	Ver	21	Conus
Telecasa	Mexico	Morelos F2	116.7° West	C-Band	4.120	Ver	6W	Spot
Telefe	Argentina	PanAmSat f1	45.0° West	C-Band	3.910	Ver	6a	SA Spot
Television Boliviana	Bolivia	Intelsat 709	50.0° West	C-Band	4.143	RHCP	11	Zone
TF1	France	Intelsat 601	27.5° West	C-Band	3.761	LHCP	3	NW Zone
The Babe Network	USA	Telstar 402	89.0° West	C-Band	3.760	Ver	3	Spot
The Baseball Netw. (TBN)	USA	Telstar 302	85.0° West	C-Band	3.800	Ver	3	Conus
The Baseball Network	USA	Telstar 401	97.0° West	C-Band	4.080	Ver	19	Conus
The Cartoon Network	USA	Galaxy 1	133.0° West	C-Band	3.860	Ver	8	Conus
The Family Channel	USA	Galaxy 5	125.0° West	C-Band	3.920	Hor	11	Conus
The Family Channel (West)	USA	Satcom C3	131.0° West	C-Band	3.720	Ver	1	Conus
The Filipino Channel	Philippines	Galaxy 4	99.0° West	Ku2-Band	12.170	Hor	24	Conus
The Golf Channel	UK	Galaxy 7	91.0° West	C-Band	3.840	Ver	7	Conus
The International Channel	USA	Satcom C1	137.0° West	C-Band	4.100	Ver	20	Conus
The International Channel	USA	Galaxy 7	91.0° West	C-Band	3.940	Ver	12	Conus
The Kentucky Netw.	USA	Satcom SN-2	69.0° West	Ku2-Band	12.060	Ver	23	Conus
The Learning Channel	USA	Satcom C3	131.0° West	C-Band	3.740	Hor	2	Conus
The Movie Channel	USA	Satcom C3	131.0° West	C-Band	4.040	Ver	17	Conus

ITU REGION 2 TV IN ALPHABETICAL ORDER

Station	Country	Satellite	Position	Band	Frequency	Pol.	Tr	Beam
The Movie Channel (West)	USA	Satcom C4	135.0° West	C-Band	4.140	Hor	22	Conus
The Nashville Network	USA	Galaxy 5	125.0° West	C-Band	4.060	Ver	18	Conus
The New Inspirational Netw.	USA	Galaxy 1	133.0° West	C-Band	4.040	Hor	17	Conus
The Outdoor Channel	USA	Telstar 402	89.0° West	C-Band	3.920	Ver	11	Spot
The People's Network	USA	Galaxy 7	91.0° West	Ku2-Band	11.900	Ver	10K	Conus
The People's Network	USA	Galaxy 7	91.0° West	Ku2-Band	12.080	Ver	19K	Conus
The Travel Channel	USA	Satcom C4	135.0° West	C-Band	3.960	Ver	13	Conus
The University Netw.	USA	Galaxy 6	74.0° West	C-Band	4.080	Ver	19	Conus
The X Channel	USA	Telstar 402	89.0° West	C-Band	3.820	Hor	6	Spot
Three Angels Broadcasting	USA	Galaxy 3	93.5° West	C-Band	4.160	Hor	23	Conus
TNT Internacional	Turkey	PanAmSat f1	45.0° West	C-Band	3.798	Ver	5	LA Spot
Trinity Broadcasting Network	USA	Galaxy 5	125.0° West	C-Band	3.760	Hor	3	Conus
Turner Classic Movies	USA	Galaxy 1	133.0° West	C-Band	4.020	Ver	16	Conus
Turner Network Television (TNT)	USA	Galaxy 5	125.0° West	C-Band	4.040	Hor	17	Conus
Turner Vision Promo Channel	USA	Galaxy 7	91.0° West	Ku2-Band	4.060	Ver	18	Conus
Turner Vision Promo Sce.	USA	Telstar 302	85.0° West	C-Band	3.960	Ver	7	Conus
TV69	USA	Galaxy 6	74.0° West	C-Band	3.860	Hor	8	Spot
TV Asia	USA	Telstar 302	85.0° West	C-Band	3.940	Ver	12	Conus
TV Erotica	Chile	PanAmSat f1	45.0° West	C-Band	4.040	Hor	17	Conus
TV Naçionale de Chile (ch 10)	Chile	PanAmSat f1	45.0° West	C-Band	3.769	Ver	2b	SA Spot
TV Naçionale de Chile (ch 7)	Chile	PanAmSat f1	45.0° West	C-Band	3.750	Ver	2a	SA Spot
TV Ontario	Canada	Anik E2	107.3° West	Ku2-Band	12.000	Hor	26	Spot
Tv-Japan	Japan	Galaxy 6	74.0° West	C-Band	3.820	Ver	6	Conus
Tv-Japan	Japan	Galaxy 6	74.0° West	C-Band	3.820	Ver	6	Conus
TVE internaçional	Spain	Hispasat	30.0° West	Ku2-Band	12.078	RHCP	19	Spot
TVN Theatre 1	USA	Galaxy 3	93.5° West	C-Band	3.720	Hor	1	Conus
TVN Theatre 2	USA	Galaxy 3	93.5° West	C-Band	3.740	Ver	2	Conus
TVN Theatre 3	USA	Galaxy 3	93.5° West	C-Band	3.760	Hor	3	Conus
TVN Theatre 4	USA	Galaxy 3	93.5° West	C-Band	3.780	Ver	4	Conus
TVN Theatre 5	USA	Galaxy 3	93.5° West	C-Band	3.800	Hor	5	Conus
TVN Theatre 6	USA	Galaxy 3	93.5° West	C-Band	3.820	Ver	6	Conus
TVN Theatre 7	USA	Galaxy 3	93.5° West	C-Band	3.840	Hor	7	Conus
TVN Theatre 8	USA	Galaxy 3	93.5° West	C-Band	3.860	Ver	8	Conus
TVN Theatre 9	USA	Galaxy 3	93.5° West	C-Band	3.880	Hor	9	Conus
VN Theatre 10	USA	Galaxy 3	93.5° West	C-Band	3.900	Ver	10	Conus
UAE TV Dubai	UAE	Intelsat K	21.5° West	Ku1-Band	11.915	Hor	4	NA Spot
United Arab Emirates TV (UAE)	UAE	Galaxy 7	91.0° West	C-Band	3.900	Ver	10	Conus
United Paramount Network (UPN)	USA	Telstar 401	97.0° West	C-Band	4.080	Ver	19	Conus
United States Info. Agency	USA	Satcom SN-2	69.0° West	C-Band	3.760	Hor	2	Conus
Univision	USA	Satcom SN3R	87.0° West	C-Band	3.800	Hor	3	Conus
USA Network East	USA	Galaxy 5	125.0° West	C-Band	4.080	Hor	19	Conus
USA Network West	USA	Galaxy 1	133.0° West	C-Band	4.120	Hor	21	Conus
USA WorldNet	USA	Intelsat 601	27.5° West	C-Band	3.995	RHCP	15	West Hemisph.
Valuevision	USA	Galaxy 1	133.0° West	C-Band	3.940	Ver	12	Conus
Venus Adult	Canada	Anik E2	107.3° West	C-Band	4.140	Ver	22	Spot
VH-1	USA	Satcom C4	135.0° West	C-Band	4.160	Ver	23	Conus

251

ITU REGION 2 TV IN ALPHABETICAL ORDER

Station	Country	Satellite	Position	Band	Frequency	Pol.	Tr	Beam
Via TV (interactive TV)	UAE	Galaxy 7	91.0° West	C-Band	4.040	Hor	17	Conus
Video Catalog Channel	Canada	Anik E2	107.3° West	C-Band	4.060	Ver	18	Spot
Viewer's Choice	USA	Satcom C3	131.0° West	C-Band	3.760	Ver	3	Conus
VTC Satelite Network	Canada	Telstar 401	97.0° West	C-Band	3.720	Ver	1	Conus
WABC, New York	USA	Galaxy 4	99.0° West	C-Band	3.900	Ver	10	Conus
Weather Channel	USA	Satcom C3	131.0° West	C-Band	3.960	Ver	13	Conus
Westerns Encore 3	USA	Spacenet 4	101.0° West	C-Band	3.780	Ver	4	Conus
WFLD-TV, Chicago	USA	Spacenet 4	101.0° West	C-Band	3.900	Hor	11	Conus
WGN, Chicago Superstation	USA	Galaxy 5	125.0° West	C-Band	3.840	Hor	7	Conus
WHDH-TV, Boston, MA	USA	Spacenet 4	101.0° West	C-Band	3.740	Hor	7	Conus
Wholesale Shopping Netw.	USA	Telstar 302	85.0° West	C-Band	3.760	Ver	2	Conus
WJLA, Washington	USA	Americom GE 1	103.0° West	C-Band	3.900	Ver	10	Spot
WMNB	USA	SBS 5	123.0° West	Ku2-Band	11.898	Ver	12	Conus
WMNB Russian Language Station	USA	Galaxy 4	99.0° West	Ku2-Band	11.930	Hor	12	Conus
WNBC New York	USA	Americom GE 1	103.0° West	C-Band	3.820	Ver	6	Spot
World Harvest TV	USA	Galaxy 4	99.0° West	C-Band	4.000	Hor	15	Conus
WorldNet	USA	Satcom SN-2	69.0° West	C-Band	3.760	Ver	2	Conus
Worship TV	USA	Galaxy 6	74.0° West	C-Band	4.160	Hor	23	Conus
WPIX, New York	USA	Satcom SN3R	87.0° West	C-Band	3.880	Ver	5	Conus
WRAL, Raleigh, NC	USA	Galaxy 4	99.0° West	C-Band	3.980	Ver	14	Conus
WRAL, Raleigh, NC	USA	Americom GE 1	103.0° West	C-Band	4.180	Hor	24	Spot
WSBK, Boston	USA	Satcom SN3R	87.0° West	C-Band	3.760	Ver	2	Conus
WTBS, Atlanta	USA	Galaxy 5	125.0° West	C-Band	3.820	Ver	6	Conus
WUSA-TV, Washington, DC	USA	Spacenet 4	101.0° West	C-Band	3.780	Hor	8	Conus
WWOR-TV	USA	Satcom C4	135.0° West	C-Band	4.000	Ver	15	Conus
WXIA, Atlanta	USA	Galaxy 4	99.0° West	C-Band	4.140	Ver	22	Conus
XEIPN-TV Canal 11, Mexico City	Mexico	Morelos F2	116.7° West	C-Band	3.920	Hor	3W/U	Spot
XEQ-TV 9	Mexico	Solidaridad F1	109.2° West	C-Band	3.840	Ver	4N	Spot
XEW-TV, Mexico City	Mexico	Morelos F2	116.7° West	C-Band	3.980	Ver	7N	Spot
XHGC-TV (Canal 5)	Mexico	Morelos F2	116.7° West	C-Band	3.860	Ver	4N	Spot
XHIMT, Canal 22	Mexico	Morelos F2	116.7° West	C-Band	4.020	Ver	8N	Spot
XHIMT-TV, TV7	Mexico	Morelos F2	116.7° West	C-Band	4.120	Ver	6W	Spot
XXXPlore	USA	Telstar 402	89.0° West	C-Band	3.900	Hor	10	Spot
Z-Music	USA	Galaxy 1	133.0° West	C-Band	3.820	Ver	6	Conus

ITU REGION 2 RADIO IN ALPHABETICAL ORDER

12.4. ITU REGION 2 RADIO IN ALPHABETICAL ORDER

Station	Country	Satellite	Position	Band	Frequency	Pol.	Tr	Beam
AFRTS Radio	USA	Intelsat 707	1.0° West	C-Band	4.175	RHCP	24	Global
AFRTS Radio Sce.	USA	Satcom SN-2	69.0° West	C-Band	4.100	Ver	17U	Conus
Ambassador Insp. Radio	USA	Satcom SN3R	87.0° West	C-Band	4.000	Hor	8	Conus
Ambassador Insp. Radio	USA	Satcom SN3R	87.0° West	C-Band	4.000	Hor	8	Conus
American Freedom Netw.	USA	Americom GE 1	103.0° West	C-Band	3.840	Hor	6	Spot
American Urban Radio Netw.	USA	Satcom SN3R	87.0° West	C-Band	3.880	Hor	5	Conus
ANA Radio Netw.	USA	Galaxy 6	74.0° West	C-Band	3.900	Ver	10	Conus
Antenna Greece	Greece	Telstar 402	89.0° West	C-Band	4.060	Hor	18	Spot
BBC World Radio	USA	Satcom C1	137.0° West	C-Band	3.760	Hor	3	Conus
BBC World Service	UK	PanAmSat f1	45.0° West	Ku2-Band	11.808	Hor	23a	NA Conus
BBC WS (Europe)	UK	TDRSS	41.0° West	C-Band	3.720	Hor	1	Spot
BBC WS (foreign)	UK	TDRSS	41.0° West	C-Band	3.720	Hor	1	Spot
BBC WS (Ukrainian)	UK	TDRSS	41.0° West	C-Band	3.720	Hor	1	Spot
BFBS 1 Radio	UK	TDRSS	41.0° West	C-Band	3.720	Hor	1	Spot
BFBS 2 Radio	UK	TDRSS	41.0° West	C-Band	3.720	Hor	1	Spot
BFBS 3 Radio	UK	TDRSS	41.0° West	C-Band	3.720	Hor	1	Spot
Brother Stair	USA	Galaxy 5	125.0° West	C-Band	3.820	Ver	6	Conus
Business Radio Network	USA	Satcom C4	135.0° West	C-Band	3.900	Hor	10	Conus
C&W	USA	Satcom SN3R	87.0° West	C-Band	4.060	Ver	17L	Conus
C-SPAN 1 ASAP	USA	Satcom C3	131.0° West	C-Band	3.840	Ver	7	Spot
C-SPAN audio sce 1 (a.o. VOA)	USA	Satcom C3	131.0° West	C-Band	3.840	Ver	7	Spot
C-SPAN audio sce 2 (a.o. BBC)	USA	Satcom C3	131.0° West	C-Band	3.840	Ver	7	Spot
CBC FM	Canada	Anik E2	107.3° West	C-Band	3.720	Ver	7	Spot
CBC Radio	Canada	Anik E2	107.3° West	C-Band	3.720	Hor	1	Spot
CBC Radio affiliates	Canada	Anik E2	107.3° West	C-Band	3.720	Ver	7	Spot
CBC Radio (Atlantic)	Canada	Anik E2	107.3° West	C-Band	3.820	Ver	7	Spot
CBC Radio (Eastern)	Canada	Anik E2	107.3° West	C-Band	3.820	Ver	7	Spot
CBC Stereo Radio (Atlantic)	Canada	Anik E2	107.3° West	C-Band	3.720	Ver	7	Spot
CBM AM	USA	Satcom C3	131.0° West	C-Band	4.160	Hor	1	Spot
CBN News	USA	Satcom C3	131.0° West	C-Band	3.720	Ver	23	Conus
CBN Radio	USA	Galaxy 5	125.0° West	C-Band	3.920	Hor	1	Conus
CBN Radio Netw.	Canada	Satcom SN-2	69.0° West	C-Band	4.120	Hor	11	Conus
Christian Music Sce.	Canada	Anik E2	107.3° West	Ku2-Band	12.000	Hor	11	Spot
CJRT FM, Toronto	USA	Galaxy 5	125.0° West	C-Band	3.800	Ver	26	Conus
CNN Radio	USA	Galaxy 5	125.0° West	C-Band	4.140	Hor	5	Conus
CNN Radio	USA	GStar 2	125.0° West	Ku2-Band	12.035	Ver	22	Spot
CNN Radio Netw.	USA	Satcom SN3R	87.0° West	C-Band	3.880	Hor	6	Conus
Colorado Talking Book	USA	Satcom C1	137.0° West	C-Band	3.760	Hor	3	Conus
Deutsche Welle	Germany	Satcom C4	135.0° West	C-Band	3.800	Ver	5	Conus

253

ITU REGION 2 RADIO IN ALPHABETICAL ORDER

Station	Country	Satellite	Position	Band	Frequency	Pol.	Tr	Beam
Deutsche Welle	Germany	Intelsat K	21.5° West	Ku-Band	11.605	Hor	7	N & Lat. Am. Spot
Digital Music Express	USA	Satcom C3	131.0° West	C-Band	4.180	Hor	24	Conus
Easy Listening	USA	Galaxy 4	99.0° West	C-Band	3.820	Ver	6	Conus
EZ Listening	USA	Satcom SN3R	87.0° West	C-Band	4.060	Ver	17L	Conus
EZ Listening	USA	Satcom SN3R	87.0° West	C-Band	4.060	Ver	17L	Conus
EZ Listening	USA	Satcom SN3R	87.0° West	C-Band	4.060	Ver	17L	Conus
For The People Netw.	USA	Satcom C1	137.0° West	C-Band	3.740	Ver	2	Conus
Georgia Radio Reading Sce. (GRRS)	USA	Telstar 401	97.0° West	Ku2-Band	12.092	Hor	14A	Conus
In Touch	USA	Satcom C4	135.0° West	C-Band	3.900	Ver	10	Conus
In Touch	USA	Satcom C5	139.0° West	C-Band	4.180	Hor	24	Conus
Italian Radio	Italy	Intelsat K	21.5° West	Ku1-Band	11.915	Hor	4	NA Spot
KBVA-FM, Bella Vista	Puerto Rico	Galaxy 4	99.0° West	C-Band	3.820	Ver	6	Conus
KHNC-AM	USA	Galaxy 6	74.0° West	C-Band	3.980	Ver	14	Conus
KILA, Las Vegas	USA	Satcom C4	135.0° West	C-Band	3.860	Ver	8	Conus
KJAZ	USA	Galaxy 5	125.0° West	C-Band	3.780	Ver	4	Conus
KLON FM, Longbeach, CA	USA	Satcom C1	137.0° West	C-Band	3.740	Ver	2	Conus
KSKA-FM, Anchorage	USA	Satcom C5	139.0° West	C-Band	4.180	Hor	24	Conus
KSL-AM, Salt Lake City	USA	Satcom C1	137.0° West	C-Band	3.820	Ver	6	Conus
KUCV-FM	USA	Satcom SN3R	87.0° West	C-Band	3.740	Ver	13L	Conus
KUCV-FM	USA	Satcom SN3R	87.0° West	C-Band	3.780	Ver	13U	Conus
La Voz de la Resistancia	US	Satcom SN 2	69.0° West	C-Band	3.780	Ver	13U	Conus
Nebraska Talking Book Netw.	USA	Satcom SN3R	87.0° West	C-Band	3.740	Ver	13U	Conus
Northern Native Radio	Canada	Anik E2	107.3° West	Ku2-Band	12.000	Hor	26	Spot
Peach State Public Radio	USA	Telstar 401	97.0° West	Ku2-Band	12.092	Hor	14A	Conus
Radio Beograd 1	Croatia	PanAmSat f1	45.0° West	C-Band	11.760	Ver	22b	NA Conus
Radio Dubai	UAE	Galaxy 7	91.0° West	C-Band	3.900	Ver	10	Conus
Radio Dubai	UAE	Intelsat K	21.5° West	Ku1-Band	11.915	RHCP	4	NA Spot
Radio Exterior	Spain	Hispasat	30.0° West	Ku2-Band	12.015	RHCP	15	Spot
Radio France Int. (RFI)	France	Satcom SN 2	69.0° West	C-Band	3.780	Ver	13U	Conus
Radio Maria	UAE	Galaxy 7	91.0° West	C-Band	3.900	Ver	10	Conus
Radio Naçional de España	Spain	Hispasat	30.0° West	Ku2-Band	12.015	RHCP	15	Spot
Radio Otto	Italy	Intelsat K	21.5° West	Ku1-Band	11.915	Hor	4	NA Spot
Radio Portugal	Portugal	Ekspress 2	14.0° West	C-Band	4.025	RHCP	15	Global
Radio Rossija	Russia	Ekspress 2	14.0° West	C-Band	3.675	RHCP	6	Global
Radio Sedeye Iran	USA	Satcom SN3R	87.0° West	C-Band	4.000	Hor	8	Conus
Radio Tropical	Haiti	Satcom SN-2	69.0° West	C-Band	3.920	Ver	6	Conus
RAI Italy Radio Netw.	Italy	Satcom C1	137.0° West	C-Band	4.000	Ver	15	Conus
Religious Music	UAE	Galaxy 7	91.0° West	C-Band	3.900	Ver	10	Conus
Rete Otto	Italy	Telstar 402	89.0° West	C-Band	4.060	Hor	18	Spot
RTP foreign sce.	Portugal	Ekspress 2	14.0° West	C-Band	4.025	RHCP	15	Global
Safeway Markets	USA	Satcom SN3R	87.0° West	C-Band	4.060	Ver	17L	Conus
Soft music	USA	Satcom SN3R	87.0° West	C-Band	4.060	Ver	17L	Conus
Soft music	USA	Satcom SN3R	87.0° West	C-Band	4.060	Ver	17L	Conus
Spanish sce.	USA	Satcom C1	137.0° West	C-Band	4.000	Ver	15	Conus
Sun Radio Network	USA	Galaxy 5	125.0° West	C-Band	4.120	Hor	21	Conus
SuperAudio (American Country Favorites)	USA							

254

ITU REGION 2 RADIO IN ALPHABETICAL ORDER

Station	Country	Satellite	Position	Band	Frequency	Pol.	Tr	Beam
SuperAudio (Classic Hits)	USA	Galaxy 5	125.0° West	C-Band	4.120	Hor	21	Conus
SuperAudio (Classical Collections)	USA	Galaxy 5	125.0° West	C-Band	4.120	Hor	21	Conus
SuperAudio (Light Rock)	USA	Galaxy 5	125.0° West	C-Band	4.120	Hor	21	Conus
SuperAudio (New Age of Jazz)	USA	Galaxy 5	125.0° West	C-Band	4.120	Hor	21	Conus
SuperAudio (Prime Demo)	USA	Galaxy 5	125.0° West	C-Band	4.120	Hor	21	Conus
SuperAudio (Soft Sounds)	USA	Galaxy 5	125.0° West	C-Band	4.120	Hor	21	Conus
Superstation WLIR 1300 AM NY	USA	Satcom SN 2	69.0° West	C-Band	3.720	Ver	1	Conus
Swiss Radio International	Switzerland	Intelsat K	21.5° West	Ku1-Band	11.605	Hor	7	N & Lat. Am. Spot
Talk Radio feeds	USA	Galaxy 6	74.0° West	C-Band	3.980	Ver	14	Conus
Talk Radio Network	USA	Satcom C1	137.0° West	C-Band	3.800	Hor	5	Conus
Tech Talk Radio	USA	Telstar 302	85.0° West	C-Band	4.120	Ver	21	Conus
Trinity Broadcasting (Spanish)	USA	Galaxy 5	125.0° West	C-Band	3.760	Hor	3	Conus
Unistar	USA	Satcom SN3R	87.0° West	C-Band	3.880	Hor	5	Conus
United Video Background Music	USA	Satcom C4	135.0° West	C-Band	3.860	Hor	8	Conus
USA Radio Netw.	USA	Satcom SN3R	87.0° West	C-Band	3.800	Hor	3	Conus
VOA	USA	Intelsat 601	27.5° West	C-Band	3.995	RHCP	15	West Hemisph.
VOA	USA	Intelsat 601	27.5° West	C-Band	3.995	RHCP	15	West Hemisph.
VOA	USA	Satcom SN-2	69.0° West	C-Band	3.760	Hor	2	Conus
Voice of Russia World Service	C.I.S.	Ekspress 2	14.0° West	C-Band	4.125	RHCP	17	Global
Voice Print	Canada	Anik E2	107.3° West	C-Band	3.820	Ver	7	Spot
WCBS-AM, Newsradio 88, NY	USA	Galaxy 7	91.0° West	C-Band	4.080	Hor	19	Conus
WCBS-FM New York	USA	Galaxy 4	99.0° West	C-Band	4.100	Hor	20	Conus
WCCO-AM, Minneapolis	USA	Galaxy 6	74.0° West	C-Band	4.000	Hor	15	Conus
WCMQ-FM Hiaeah, FL	US	Satcom SN 2	69.0° West	C-Band	3.780	Ver	13U	Conus
WCNJ FM, Hazlet, NY	USA	Americom GE 1	103.0° West	C-Band	3.820	Ver	6	Spot
WCRP, Guayama, Puerto Rico	Puerto Rico	Galaxy 4	99.0° West	C-Band	3.820	Ver	6	Conus
WFMT, Chicago	USA	Galaxy 5	125.0° West	C-Band	3.840	Ver	7	Conus
WFMT, Chicago	USA	Galaxy 5	125.0° West	C-Band	3.840	Ver	7	Conus
WHME	USA	Galaxy 4	99.0° West	C-Band	4.000	Ver	15	Conus
WNTL-AM Indian Head	USA	Galaxy 6	74.0° West	C-Band	3.900	Ver	10	Conus
WNVK Newark, NY	USA	Satcom SN-2	69.0° West	C-Band	3.920	Hor	6	Conus
World Harvest Radio	US	Galaxy 4	99.0° West	C-Band	4.000	Hor	15	Conus
WOCD-FM (CD 101.9), NY	USA	Satcom C4	135.0° West	C-Band	3.820	Ver	6	Spot
WQXR-FM	USA	Satcom C4	135.0° West	C-Band	4.000	Hor	15	Spot
WRN (World Radio Network 1)	USA	Galaxy 5	125.0° West	C-Band	3.820	Ver	6	Conus
WRN (World Radio Network 2)	USA	Galaxy 5	125.0° West	C-Band	3.820	Ver	6	Conus
WROL-AM	USA	Satcom SN3R	87.0° West	C-Band	3.760	Ver	2	Conus
WSM-AM, Nashville	USA	Galaxy 5	125.0° West	C-Band	4.060	Ver	18	Conus
WUSF-FM	USA	Satcom C4	135.0° West	C-Band	3.900	Ver	10	Spot
WW Freedom Radio Network.	USA	Americom GE 1	103.0° West	C-Band	3.840	Ver	6	Conus
XEW-FM, WFM, Mexico	Mexico	Solidaridad F1	109.2 West	C-Band	3.840	Ver	4N	Spot
XEWA-FM, Mexico City	Mexico	Morelos F2	116.7° West	C-Band	3.860	Ver	4N	Spot
XEX-FM 101.7, Mexico City	Mexico	Morelos F2	116.7° West	C-Band	3.980	Ver	7N	Spot
Yesterday USA	USA	Satcom SN3R	87.0° West	C-Band	3.800	Ver	3	Conus
Yesterday USA	USA	Telstar 402	89.0° West	C-Band	3.920	Ver	11	Spot
Yesterday USA	USA	Galaxy 5	125.0° West	C-Band	3.840	Hor	7	Conus

ITU REGION 3 TV IN ALPHABETICAL ORDER

12.5. ITU REGION 3 TV IN ALPHABETICAL ORDER

Station	Country	Satellite	Position	Band	Frequency	Pol.	Tr	Beam
ABC Australia	Australia	Intelsat 511	180.0° East	C-Band	3.894	RHCP	10	West Hemisph.
ABC HACBSS	Australia	Optus B1	160.0° East	Ku2-Band	12.661	Ver	7	SE
ABC HACBSS	Australia	Optus A3	156.0° East	Ku2-Band	12.725	Ver	8	WA
ABC HACBSS	Australia	Optus A3	156.0° East	Ku2-Band	12.501	Hor	12	CA
ABC HACBSS	Australia	Optus A3	156.0° East	Ku2-Band	12.693	Hor	15	CA
ABC TV HACBSS	Australia	Optus B1	160.0° East	Ku2-Band	12.626	Hor	14	NE
ABC TV (interchange)	Australia	Optus B1	160.0° East	Ku2-Band	12.548	Hor	13L	NA
ABS-CBN Philippines	Philippines	Palapa B2P	113.0° East	C-Band	3.960	Hor	13	Spot
ABS/CBN	AUS/CHN	PanAmSat-2	169.0° East	C-Band	4.090	Hor		
AFRTS	USA	Intelsat 702	177.0° East	C-Band	4.177	LHCP	38	Global
AN-TEVE	Indonesia	Palapa B2P	113.0° East	C-Band	4.020	Ver	16	Spot
Army TV (ch. 5)	Thailand	Thaicom-1	78.5° East	C-Band	3.760	Ver		Spot
ART	Russia	Gorizont 23	103.0° East	C-Band	3.876	RHCP	8	Global
Asahi New Star	Japan	Superbird B1	162.0° East	Ku3-Band	12.610	Ver	13	Spot
Asia Business News	China	PanAmSat-2	169.0° East	C-Band	3.723	Hor		
Asia Business News	USA	PanAmSat-2	169.0° East	C-Band	4.150	Ver		
Asia Business TV	Japan	PanAmSat-4	68.5° East	C-Band	3.785	Ver		Asian
Asia TV Network	India	PanAmSat-4	68.5° East	C-Band	4.179	Ver		Asian
Asianet	India	Intelsat 703	57.0° East	C-Band	3.980	RHCP	51	East Zone
ATVi Australia	Australia	Palapa B2P	113.0° East	C-Band	3.880	Hor	9	Spot
Azerbaijan Radio TV	Azerbaijan	Gorizont 19	96.5° East	C-Band	3.875	RHCP	10	North Hemisph.
BBC Asia	UK	AsiaSat 1(Hong Kong)	105.5° East	C-Band	3.940	Ver	5	SB
BBC Asia (mandarin)	UK	AsiaSat 1(Hong Kong)	105.5° East	C-Band	3.940	Ver	5	SB
BBC World Service TV	USA	PanAmSat-2	169.0° East	C-Band	3.901	Ver		
BBC World Service TV	Japan	AsiaSat 2(Hong Kong)	100.5° East	C-Band	3.900	Ver		SB
BBC World Service TV	UK	PanAmSat-4	68.5° East	C-Band	3.800	Ver		Asian
BBC WSTV	UK	TDRS-F5	185.7° East	C-Band	3.987	Hor		
BBC WSTV	Japan	JCSat 2	154.0° East	Ku2-Band	12.673	Hor	24	Spot
BGV Channel	Japan	JCSat-3	128.0° East	Ku2-Band	12.568	Ver		Spot
Bloomberg Info. TV	Japan	JCSat-3	128.0° East	Ku2-Band	12.568	Ver		Spot
Business News Network	Japan	JCSat 2	154.0° East	Ku2-Band	12.298	Ver	3	Spot
Cable Soft Netw.	Japan	Superbird B1	162.0° East	Ku2-Band	12.410	Ver	6A	Spot
Cable Soft Network	Japan	Superbird B1	162.0° East	Ku2-Band	12.430	Hor	5B	Spot
Cable TV Access Channel	Japan	JCSat 2	154.0° East	Ku2-Band	12.283	Hor	2	Spot
Canal France Int.	France	Palapa B2P	113.0° East	C-Band	4.160	Hor	22	Spot
Canal France International	France	Palapa B2P	113.0° East	C-Band	4.160	Ver	23	Spot
Canal France Internationale	France	Intelsat 704	66.0° East	C-Band	4.055	RHCP	15	East Hemisph.
Car Information TV	Japan	JCSat-3	128.0° East	Ku2-Band	12.523	Hor		Spot
CBHS Hour	China	Apstar-1	138.0° East	C-Band	3.636	Ver		
CCTV-4	China	AsiaSat 1(Hong Kong)	105.5° East	C-Band	4.120	Hor	11	NB

ITU REGION 3 TV IN ALPHABETICAL ORDER

Station	Country	Satellite	Position	Band	Frequency	Pol.	Tr	Beam
CCTV-4	China	Gorizont 19	96.5° East	C-Band	3.825	RHCP	9	North Hemisph.
CETV Shandong	China	Apstar-1	138.0° East	C-Band	3.800	Hor		Spot
Channel 11 Thailand	Thailand	Thaicom-1	78.5° East	C-Band	3.870	Ver		Spot
Channel 3 Thailand	Thailand	Thaicom-2	78.5° East	C-Band	3.950	Hor		West Hemisph.
Channel 7 Australia	Australia	Intelsat 511	180.0° East	C-Band	3.876	RHCP	13	Spot
Channel 7 Thailand	Thailand	Thaicom-2	78.5° East	C-Band	3.750	Hor		West Hemisph.
Channel 9 Australia	Australia	Intelsat 511	180.0° East	C-Band	3.930	RHCP	13	Spot
Channel 9 Thailand	Thailand	Thaicom-2	78.5° East	C-Band	3.970	Hor		
Channel KTV	Ciina	PanAmSat-2	169.0° East	Ku3-Band	12.730	Ver		
China Central TV 2	China	Apstar-1a	134.0° East	C-Band	4.180	Ver		
China Central TV 4	China	AsiaSat 2(Hong Kong)	100.5° East	C-Band	3.960	Hor		SB
China Central TV 4	China	PanAmSat-4	68.5° East	C-Band	3.836	Ver		Asian
China Central TV-1	China	Chinasat 5	115.5° East	C-Band	3.883	Hor		Spot
China Central TV-2	China	Chinasat 5	115.5° East	C-Band	3.803	Hor		Spot
China Educational TV-1	China	Apstar-1	138.0° East	C-Band	3.840	Ver		
China Educational TV-2	China	Apstar-1	138.0° East	C-Band	3.880	Hor		
China Entertainment TV	China	Apstar-1	138.0° East	C-Band	4.160	Hor		
Chinese Channel	Hong Kong	AsiaSat 1(Hong Kong)	105.5° East	C-Band	3.920	Hor	6	NB
Chinese Channel (Mandarin)	Hong Kong	AsiaSat 1(Hong Kong)	105.5° East	C-Band	3.920	Ver	6	NB
Chinese Satellite TV (CSTV)	China	Apstar-1	138.0° East	C-Band	3.666	Ver		
Chinese TV Network	China	PanAmSat-2	169.0° East	C-Band	3.723	Hor		
Chinese TV Network	USA	PanAmSat-2	169.0° East	C-Band	4.150	Ver		Spot
Chinese TV Network	Ciina	PanAmSat-2	169.0° East	Ku3-Band	12.524	Ver		
Chinese TV Network	China	JCSat-3	128.0° East	Ku2-Band	12.643	Hor		
Cna Central TV 1	China	Apstar-1a	134.0° East	C-Band	3.860	Ver		
Cna Central TV 2	China	Apstar-1a	134.0° East	C-Band	3.820	Hor		
Cinefil Imagica	Japan	JCSat-3	128.0° East	Ku2-Band	12.688	Ver	4	Spot
CNBC Asia	USA	Palapa B2P	113.0° East	C-Band	3.620	Hor		Spot
CNN Int	USA	PanAmSat-2	169.0° East	C-Band	3.967	Ver		
CNN International	USA	Palapa B2P	113.0° East	C-Band	3.980	Ver	14	Spot
CNN International	China	Apstar-1	138.0° East	C-Band	3.980	Hor		
CNN International	USA	PanAmSat-4	68.5° East	S-Band	4.085	Hor		Asian
Community TV	India	Insat 1D	82.9° East	C-Band	2.615	LHCP	S1	Spot
Country Music TV	USA	PanAmSat-2	169.0° East	C-Band	3.901	Ver		
CSTV Music Channel	China	Apstar-1	138.0° East	C-Band	3.666	Ver		Spot
CSTV News Channel	China	Apstar-1	138.0° East	C-Band	3.950	Hor		Spot
Dai Truyen Hinh Vietnam	Vietnam	Thaicom-1	78.5° East	C-Band	3.710	Ver		Spot
Dai Truyen Hinh Vietnam	Vietnam	Measat-1	91.5° East	C-Band	4.120	Ver	C10	Spot
DD Channel 1	India	Insat 2B	94.0° East	C-Band	4.140	Ver	C11	Spot
DD Channel 10	India	Insat 2B	94.0° East	C-Band	4.160	Ver	C12	Spot
DD Channel 2	India	Insat 2B	94.0° East	C-Band	3.865	Ver	C3	SB
DD Channel 7	India	Insat 2B	94.0° East	C-Band	4.000	Hor		Spot
Deutsche Welle TV	Germany	AsiaSat 2(Hong Kong)	100.5° East	Ku2-Band	12.343	Ver	6	Spot
Diamond Channel	Japan	JCSat 2	154.0° East	Ku2-Band	12.568	Ver		Spot
Digital Tampa 501	Japan	JCSat-3	128.0° East	Ku2-Band	12.568	Ver		Spot
Digital Tampa 502	Japan	JCSat-3	128.0° East	Ku2-Band	12.568	Ver		Spot

ITU REGION 3 TV IN ALPHABETICAL ORDER

Station	Country	Satellite	Position	Band	Frequency	Pol.	Tr	Beam
Discovery Channel	USA	Palapa B2P	113.0° East	C-Band	3.720	Hor	4	Spot
Discovery Channel	USA	PanAmSat-2	169.0° East	C-Band	3.780	Hor		
Discovery Channel	USA	Apstar-1	138.0° East	C-Band	4.140	Ver		
Discovery Channel	India	PanAmSat-4	68.5° East	C-Band	3.785	Ver		Asian
Doordarshan TV	India	Insat 1D	82.9° East	S-Band	2.575	LHCP	S1	Spot
Doordarshan TV	India	Insat 2A	74.0° East	C-Band	4.115	Hor	21	Spot
Doordarshan TV	India	Insat 2B	94.0° East	S-Band	2.615	LHCP	S1	Spot
Doordarshan TV	India	PanAmSat-4	68.5° East	C-Band	4.030	Ver		Asian
Doordarshan TV	India	Insat 2B	94.0° East	S-Band	2.575	Ver	S1	Spot
Dub'l II	CIS	Gorizont 20	90.0° East	C-Band	3.675	RHCP	6	Spot
Dub'l IV	CIS	Gorizont 23	103.0° East	C-Band	3.675	RHCP	6	Global
Dub'l IV	CIS	Gorizont 16/25	80.0° East	C-Band	3.675	RHCP	6	Spot
Dub'l-1	CIS	Statsionar 16	145.0° East	C-Band	3.875	RHCP	6	Spot
Egyptian Satellite Channel	Egypt	AsiaSat 2(Hong Kong)	100.5° East	C-Band	3.640	Hor		SB
EM-TV	Papua New Guinea	Rimsat-2 (Gorizont 30)	142.5 East	C-Band	3.875	LHCP		SB
Enterprise Channel	Australia	Optus A3	156.0° East	Ku2-Band	12.661	Ver	7	WA
ESPN	USA	PanAmSat-4	68.5° East	C-Band	3.865	Hor		Asian
ESPN International	USA	Palapa B2P	113.0° East	C-Band	4.100	Ver	20	Spot
ESPN International	USA	PanAmSat-2	169.0° East	C-Band	3.862	Ver		
ESPN International	USA	Apstar-1	138.0° East	C-Band	4.100	Ver		
Family Theatre	Japan	Superbird B1	162.0° East	Ku2-Band	12.370	Ver	2B	Spot
Feisuo Satellite TV	Japan	Superbird A	158.0° East	Ku3-Band	12.290	Ver	2A	Spot
Friendly TV	Japan	Superbird A	158.0° East	Ku3-Band	12.290	Ver	2A	Spot
Fuji TV Network	Japan	TDRS-F5	185.7 East	C-Band	3.845	Hor		
Gemini TV	Sri Lanka	Intelsat 703	57.0° East	C-Band	3.750	RHCP	51	East Zone
GMA Philippines	Philippines	Palapa B2P	113.0° East	C-Band	3.920	Ver	11	Spot
Golden West Network	Australia	Optus A3	156.0° East	Ku2-Band	12.341	Ver	2	WA
Green Channel	Japan	Superbird B1	162.0° East	Ku3-Band	12.550	Hor	19	Spot
Green Channel	Japan	JCSat 2	154.0° East	Ku2-Band	12.373	Hor	8	Spot
Guangdong Satellite TV	China	AsiaSat 2(Hong Kong)	100.5° East	C-Band	3.840	Ver		SB
Guizhou TV Station	Mongolia	AsiaSat 1(Hong Kong)	105.5° East	C-Band	3.760	Hor	2	SB
Guizhou TV Station-1	Hong Kong	AsiaSat 1(Hong Kong)	105.5° East	C-Band	4.040	Hor	9	NB
HBO Asia	USA	Palapa B2P	113.0° East	C-Band	4.000	Hor		NB
HBO Asia	USA	Apstar-1	138.0° East	C-Band	4.060	Hor	15	Spot
Henan Satellite TV		AsiaSat 2(Hong Kong)	100.5° East	C-Band	3.720	Hor		SB
IBC TV Network	Japan	JCSat-3	128.0° East	Ku2-Band	12.703	Hor		Spot
Japan Cable Television	Japan	Superbird B1	162.0° East	Ku3-Band	12.650	Ver	15	Spot
Japan Leisure Channel	Japan	JCSat 2	154.0° East	Ku2-Band	12.268	Ver	1	Spot
Japan Religious Channel	Japan	JCSat 2	154.0° East	Ku2-Band	12.283	Hor	2	Spot
Japan Satellite Broadcasting Co.	Japan	BS-3b	110.0° East	Ku2-Band	11.804	RHCP	5	Spot
Japan Sports Channel	Japan	JCSat 2	154.0° East	Ku2-Band	12.583	Hor	22	Spot
Karaoke Channel	Japan	JCSat-3	128.0° East	Ku2-Band	12.568	Ver		Spot
KBP Peoples Network	Philippines	Palapa B2P	113.0° East	C-Band	4.050	Hor	17	Spot
KBS Satellite TV 1	Korea	Koreasat 1	116.0° East	Ku2-Band	11.823	LHCP		Spot
KBS Satellite TV 2	Korea	Koreasat 1	116.0° East	Ku2-Band	11.823	LHCP		Spot
Keirin Channel	Japan	JCSat 2	154.0° East	Ku2-Band	12.538	Ver	21	Spot

ITU REGION 3 TV IN ALPHABETICAL ORDER

Station	Country	Satellite	Position	Band	Frequency	Pol.	Tr	Beam
Kids Station	Japan	JCSat 2	154.0° East	Ku2-Band	12.553	Hor	21	Spot
Kids Station	Japan	JCSat-3	128.0° East	Ku2-Band	12.658	Hor		Spot
Kikkei Satellite News	Japan	JCSat-3	128.0° East	Ku2-Band	12.688	Ver		Spot
KN Television	Japan	JCSat-3	128.0° East	Ku2-Band	12.703	Hor		Spot
Korea Vision	Korea	JCSat-3	128.0° East	Ku2-Band	12.643	Ver		Spot
Kuoshin Satellite TV	Japan	Superbird A	158.0° East	Ku3-Band	12.290	Ver	2A	Zone
Lao National TV	Laos	Gorizont-G41	130.0° East	C-Band	3.775	LHCP		Zone
Life Design Channel	Japan	Superbird B1	162.0° East	Ku3-Band	12.530	Ver	8	Spot
M Channel (JMTV)	Japan	JCSat 2	154.0° East	Ku2-Band	12.403	Hor	10	Spot
M Channel (JMTV)	Japan	JCSat 2	154.0° East	Ku2-Band	12.358	Ver	6	SB
MCM	France	AsiaSat 2(Hong Kong)	100.5° East	C-Band	4.000	Ver		
Meishi Entertainment TV	Taiwan	PanAmSat-2	169.0° East	Ku3-Band	12.704	Ver	13	Spot
Midnight Blue	Japan	JCSat 2	154.0° East	Ku2-Band	12.448	Ver		Spot
Mondo 21	Japan	JCSat-3	128.0° East	Ku2-Band	12.538	Ver		East Zone
Money TV	USA	Intelsat 703	57.0° East	C-Band	3.750	RHCP	51	Spot
MTV Asia	UK	Palapa B2P	113.0° East	C-Band	4.120	Hor	21	NB
MTV Asia	Hong Kong	AsiaSat 1(Hong Kong)	105.5° East	C-Band	3.840	Ver	4	SB
MTV Asia	Hong Kong	AsiaSat 1(Hong Kong)	105.5° East	C-Band	3.900	Ver	5	Asian
MTV Asia	India	PanAmSat-4	68.5° East	C-Band	4.185	Ver		Spot
MTV Japan (Music Channel)	Japan	Superbird B1	162.0° East	Ku3-Band	12.690	Ver	15	
MTV Mandarin	UK	PanAmSat-2	169.0° East	C-Band	3.804	Ver		
MTV Mandarin	China	Apstar-1	138.0° East	C-Band	3.860	Ver		Spot
Music Asia	China	Rimsat-2 (Gorizont 30)	142.5 East	C-Band	3.675	LHCP	51	East Zone
Muslim TV Ahmadiyya Int.	India	Intelsat 703	57.0° East	C-Band	4.178	LHCP	11	SB
Myanmar TV	India	AsiaSat 1(Hong Kong)	105.5° East	C-Band	4.140	Ver		
NBC Asia	Burma	PanAmSat-2	169.0° East	C-Band	4.093	Ver		East Zone
NEPC TV	USA	Intelsat 703	57.0° East	C-Band	4.065	LHCP	51	West Hemisph.
Network 10 Austraia	India	Intelsat 511	180.0° East	C-Band	3.765	RHCP	11	
NHK Int. TV	Australia	PanAmSat-2	169.0° East	C-Band	4.030	Hor		West Hemisph.
NHK Tokyo	Japan	Intelsat 511	180.0° East	Ku3-Band	3.925	LHCP	13	Asian
NHK TV-Japan	Japan	PanAmSat-4	68.5° East	Ku3-Band	12.600	Ver	8	Spot
Nihon Cable TV Netw.	Japan	Superbird B1	162.0° East	Ku2-Band	12.490	LHCP		Spot
Nihon TV Network	Japan	JCSat-3	128.0° East	Ku2-Band	12.598	Ver		Spot
Nikkei Satellite News	Japan	JCSat 2	154.0° East	Ku2-Band	12.313	Hor	4	Spot
Nine Netw. Australia	Australia	Intelsat 511	180.0° East	C-Band	4.135	RHCP	37	Global
NTV	CIS	Gorizont 18	140.0° East	C-Band	3.725	RHCP	10	Global
Orbita II	CIS	Gorizont 19	96.5° East	C-Band	3.675	RHCP	6	Spot
Orbita-1	CIS	Gorizont 18	140.0° East	C-Band	3.675	RHCP	6	Spot
P-Sat	Japan	JCSat 1	150.0° East	Ku2-Band	12.463	Ver	15	SB
Pakistan TV	Pakistan	AsiaSat 1(Hong Kong)	105.5° East	C-Band	4.100	Ver	10	
PHTV-Information Channel	China	Apstar-1	138.0° East	C-Band	4.180	Ver		
PHTV-Sanlih Channel	China	Apstar-1	138.0° East	C-Band	4.180	Ver		
PHTV-Toei Channel	China	Apstar-1	138.0° East	C-Band	4.180	Ver		
Pioneer Music Satellite	Japan	JCSat-3	128.0° East	Ku2-Band	12.538	Hor		Spot
Playboy Channel	Japan	Superbird B1	162.0° East	Ku3-Band	12.710	Ver	19	Spot
Prefec Mulch	Japan	JCSat-3	128.0° East	Ku2-Band	12.658	Ver		Spot

ITU REGION 3 TV IN ALPHABETICAL ORDER

Station	Country	Satellite	Position	Band	Frequency	Pol.	Tr	Beam
Prefec.Today	Japan	JCSat-3	128.0° East	Ku2-Band	12.658	Ver		Spot
Prime International	USA	PanAmSat-2	169.0° East	C-Band	3.990	Ver		
Prime Sports	Hong Kong	AsiaSat 1(Hong Kong)	105.5° East	C-Band	3.800	Hor	3	NB
Prime Sports	Hong Kong	AsiaSat 1(Hong Kong)	105.5° East	C-Band	3.860	Ver	11	SB
Prime Sports (Mandarin)	Hong Kong	AsiaSat 1(Hong Kong)	105.5° East	C-Band	3.800	Hor	3	NB
Prime Sports (Mandarin)	Hong Kong	AsiaSat 1(Hong Kong)	105.5° East	C-Band	3.860	Ver	11	SB
Queensland Television	Australia	Optus B1	160.0° East	Ku2-Band	12.688	Ver	15	NE
Radio TV Brunei	Brunei	Palapa B2P	113.0° East	C-Band	4.140	Hor	22	Spot
Rainbow Channel	Japan	JCSat 2	154.0° East	Ku2-Band	12.403	Hor	10	Spot
Rainbow Channel	China	Apstar-1	138.0° East	C-Band	3.636	Ver		
Rajawari Citra Televisi Indonesia (RCTI)	Indonesia	Palapa B2P	113.0° East	C-Band	3.740	Hor	5	Spot
Ray TV	Shri Lanka	Rimsat-2 (Gorizont 30)	142.5° East	C-Band	3.725	LHCP		Spot
RCTS (Imparja)	Australia	Optus A3	156.0° East	Ku2-Band	12.629	Hor	14	CA
RFO Tahiti	France	Intelsat 511	180.0° East	C-Band	4.045	RHCP	35	Global
RTE Int.	Spain	AsiaSat 2(Hong Kong)	100.5° East	C-Band	4.000	Ver		SB
RTP International	Portugal	AsiaSat 2(Hong Kong)	100.5° East	C-Band	3.980	Ver		SB
Satellite ABC	Japan	JCSat 2	154.0° East	Ku2-Band	12.613	Hor	24	Spot
Satellite ABC	Japan	JCSat-3	128.0° East	Ku2-Band	12.523	Hor		Spot
Satellite Culture Japan	Japan	JCSat 2	154.0° East	Ku2-Band	12.493	Hor	16	Spot
Satellite Culture Japan	Japan	JCSat-3	128.0° East	Ku2-Band	12.568	Hor		Spot
Satellite News	Japan	JCSat-3	128.0° East	Ku2-Band	12.643	Hor		Spot
Satellite Theatre	Japan	JCSat 2	154.0° East	Ku2-Band	12.643	Hor	23	Spot
Shandong TV Staton 1	China	Apstar-1a	134.0° East	C-Band	4.100	Ver		
Shopping Channel	Japan	JCSat-3	128.0° East	Ku2-Band	12.718	Hor		Spot
Sichuan TV Station	China	Apstar-1a	134.0° East	C-Band	4.080	Hor		
Singapore Int. TV	Singapore	Palapa B2P	113.0° East	C-Band	4.140	Ver	24	Spot
Sky News	Japan	AsiaSat 2(Hong Kong)	100.5° East	C-Band	3.900	Ver	6	SB
Sky TV	Australia	Optus B1	160.0° East	Ku2-Band	12.597	Ver		NA
Sony Entertainment Network	USA	PanAmSat-4	68.5° East	C-Band	3.910	Hor		Asian
Sony Entertainment TV	Japan	PanAmSat-4	68.5° East	C-Band	3.905	Ver		Asian
Sound with Radio	Japan	JCSat-3	128.0° East	Ku2-Band	12.568	Hor	18	Spot
Space Shower TV	Japan	JCSat 2	154.0° East	Ku2-Band	12.523	Hor	11	Spot
Space Vision Network	Japan	Superbird B1	162.0° East	Ku3-Band	12.570	Ver		Spot
Space Vision Network	Japan	JCSat-3	128.0° East	Ku2-Band	12.598	Ver		Spot
Star Channel	Japan	Superbird B1	162.0° East	Ku3-Band	12.730	Ver	19	Spot
Star Movies	China	AsiaSat 1(Hong Kong)	105.5° East	C-Band	3.880	Ver	5	NB
Star Plus	Hong Kong	AsiaSat 1(Hong Kong)	105.5° East	C-Band	3.960	Ver	7	NB
Star Plus Japan	Japan	AsiaSat 2(Hong Kong)	100.5° East	C-Band	3.900	Ver		SB
Star TV	Australia	Palapa B2P	113.0° East	C-Band	4.180	Hor	23	Spot
Star TV	Burma	AsiaSat 1(Hong Kong)	105.5° East	C-Band	4.180	Ver	11	SB
Star TV	China	AsiaSat 2(Hong Kong)	100.5° East	C-Band	3.760	Hor		SB
Star TV Chinese	Hong Kong	AsiaSat 1(Hong Kong)	105.5° East	C-Band	4.060	Ver	9	SB
Star TV Plus	Hong Kong	AsiaSat 1(Hong Kong)	105.5° East	C-Band	4.020	Ver	8	SB
Sun Music TV (Tamil Soc.)	Sri Lanka	Intelsat 703	57.0° East	C-Band	3.750	RHCP	51	East Zone
Sun TV	Sri Lanka	Intelsat 703	57.0° East	C-Band	3.808	RHCP	51	East Zone
Super Channel	USA	Superbird B1	162.0° East	Ku2-Band	12.450	Ver	7	Spot

ITU REGION 3 TV IN ALPHABETICAL ORDER

Station	Country	Satellite	Position	Band	Frequency	Pol.	Tr	Beam
Taiwan Satellite TV	Taiwan	Apstar-1	138.0° East	C-Band	3.666	Ver		Spot
Televisi Pendidikan Indonesia (TPI)	Indonesia	Palapa B2P	113.0° East	C-Band	4.080	Hor	19	Spot
Theatre Television	Japan	JCSat-3	128.0° East	Ku2-Band	12.688	Ver		
TNT/Cartoon Netw.	USA	PanAmSat-2	169.0° East	C-Band	4.150	Ver		
TNT/Cartoon Network	USA	Apstar-1	138.0° East	C-Band	4.020	Ver		Asian
TNT/Cartoon Network	USA	PanAmSat-4	68.5° East	Ku2-Band	4.115	Ver		Spot
Travel Channel	Japan	JCSat-3	128.0° East	Ku2-Band	12.538	Ver		
TV 4 Channel	China	Apstar-1	138.0° East	C-Band	3.636	Ver		
TV 5	France	AsiaSat 2 (Hong Kong)	100.5° East	C-Band	4.000	Ver		SB
TV 6 Mockba	C.I.S.	Gorizont 16/25	80.0° East	C-Band	3.875	RHCP	10	North Hemisph.
TV Asahi	Japan	Intelsat 511	180.0° East	Ku1-Band	11.508	Hor	79	East Hemisph.
TV Mongol	Mongolia	AsiaSat 2 (Hong Kong)	100.5° East	C-Band	3.680	Ver		SB
TV New Zealand	New Zealand	Intelsat 511	180.0° East	C-Band	4.188	RHCP	38	Global
TV Oceania	Japan	Optus B1	160.0° East	Ku2-Band	12.438	Hor	11	NA
TV Oceania	Japan	Optus A3	156.0° East	Ku2-Band	12.661	Hor	7	WA
TV Oceania	Japan	Optus A3	156.0° East	Ku2-Band	12.436	Hor	11	NE
TV Shopping Network	China	AsiaSat 2 (Hong Kong)	100.5° East	C-Band	3.660	Ver		SB
TV-1 (Malaysia)	Malaysia	Palapa B2P	113.0° East	C-Band	3.820	Ver	6	Spot
TV-3 Malaysia	Malaysia	Palapa B2P	113.0° East	C-Band	3.900	Ver	10	Spot
TVBS		PanAmSat-2	169.0° East	Ku2-Band	12.410	Ver		
TVBS	China	Apstar-1	138.0° East	C-Band	3.900	Ver		
TVI	India	Intelsat 703	57.0° East	C-Band	4.135	LHCP	51	East Zone
TVRI	Indonesia	Palapa B2P	113.0° East	C-Band	3.840	Ver	7	Spot
TVRI Indonesia	Indonesia	Palapa B2R	108.0° East	C-Band	4.000	Hor	15	Spot
Unique Business Channel	China	Apstar-1	138.0° East	C-Band	3.636	Ver		
Vi Jay TV	Shri Lanka	Rimsat-2 (Gorizont 30)	142.5 East	C-Band	3.825	LHCP		Spot
Viva Channel	Japan	AsiaSat 2 (Hong Kong)	100.5° East	C-Band	3.900	Ver		SB
Voice of the Earth	Japan	JCSat-3	128.0° East	Ku2-Band	12.538	Ver		Spot
VTV-4 (Vietnam)	Vietnam	Gorizont 16/25	80.0° East	C-Band	3.875	RHCP	7	North Hemisph.
Walt Disney TV	USA	Apstar-1	138.0° East	C-Band	4.040	Ver		
Weather Channel	USA	Superbird B1	162.0° East	Ku2-Band	12.450	Ver		Spot
World Entertainment	Japan	JCSat-3	128.0° East	Ku2-Band	12.538	Ver		Spot
World Net	India	Intelsat 703	57.0° East	C-Band	4.050	RHCP	51	East Zone
Worldnet	USA	PanAmSat-2	169.0° East	C-Band	3.925	Hor		
WorldNet	Mongolia	AsiaSat 2 (Hong Kong)	100.5° East	C-Band	3.680	Ver		SB
WorldNet/C-Span/Deutsche Welle TV	USA	Intelsat 511	180.0° East	C-Band	3.975	RHCP	14	West Hemisph.
Xinjiang TV Staton 1	China	Apstar-1a	134.0° East	C-Band	4.120	Hor		NB
Xizang TV Station	China	Apstar-1a	134.0° East	C-Band	4.040	Hor		SB
Yunnan TV Station-1	Hong Kong	AsiaSat 1 (Hong Kong)	105.5° East	C-Band	4.040	Hor	9	
Zee TV (English)	Hong Kong	AsiaSat 1 (Hong Kong)	105.5° East	C-Band	3.980	Hor	7	SB
Zee TV (mandarin)	Hong Kong	AsiaSat 1 (Hong Kong)	105.5° East	C-Band	3.980	Hor	7	SB
Zhejiang TV Station	China	Chinasat 5	115.5° East	C-Band	3.760	Hor		
Zhejiang TV Station	China	Apstar-1a	134.0° East	C-Band	4.020	Ver		Spot

12.6. ITU REGION 3 RADIO IN ALPHABETICAL ORDER

Station	Country	Satellite	Position	Band	Frequency	Pol.	Tr	Beam
AFRTS Radio	USA	Intelsat 702	177.0° East	C-Band	4.177	LHCP	38	Global
BBC World Service	UK	AsiaSat 1(Hong Kong)	105.5° East	C-Band	3.840	Hor	4	NB
BBC World Service	UK	AsiaSat 1(Hong Kong)	105.5° East	C-Band	3.900	Ver	5	SB
BBC World Service	UK	AsiaSat 1(Hong Kong)	105.5° East	C-Band	3.980	Ver	7	SB
BBC World Service	UK	AsiaSat 1(Hong Kong)	105.5° East	C-Band	4.020	Ver	8	SB
BBC World Service (Arabic)	UK	AsiaSat 1(Hong Kong)	105.5° East	C-Band	3.960	Hor	7	NB
BBC World Service (Chinese)	UK	AsiaSat 1(Hong Kong)	105.5° East	C-Band	3.920	Hor	6	NB
CNN Radio	USA	PanAmSat-2	169.0° East	C-Band	3.967	Hor		
CNN Radio	USA	Apstar-1	138.0° East	C-Band	3.980	Ver		
Guizhou PBS	Hong Kong	AsiaSat 1(Hong Kong)	105.5° East	C-Band	4.040	Hor	9	NB
Music Bird / Satellite Music	Japan	JCSat 2	154.0° East	Ku2-Band	12.628	Ver	25	Spot
PCM Zipang	Japan	JCSat 2	154.0° East	Ku2-Band	12.598	Ver	23	Spot
Radio Australia	Australia	Palapa B2P	113.0° East	C-Band	3.880	Hor	9	Spot
Radio Hanoi	Vietnam	Gorizont 16/25	80.0° East	C-Band	3.875	RHCP	10	North Hemisph.
Radio Mayak	CIS	Gorizont 18	140.0° East	C-Band	3.675	RHCP	6	Spot
Radio Mayak	CIS	Gorizont 19	96.5° East	C-Band	3.675	RHCP	6	Spot
Radio Mayak	CIS	Gorizont 20	90.0° East	C-Band	3.875	RHCP	10	Spot
Radio Rossia	CIS	Gorizont 20	90.0° East	C-Band	3.675	RHCP	6	Spot
Radio Rossia	CIS	Statsionar 16	145.0° East	C-Band	3.875	RHCP	6	Spot
Radio Sky	Japan	JCSat 2	154.0° East	Ku2-Band	12.598	Ver	23	Spot
Radiostancija Mayak (RSM)	CIS	Gorizont 16/25	80.0° East	C-Band	3.675	RHCP	6	Spot
RTP Antena 1	Portugal	AsiaSat 2(Hong Kong)	100.5° East	C-Band	3.980	Ver		SB
Sakha TV	CIS	Statsionar 16	145.0° East	Ku1-Band	11.525	RHCP	6	Spot
St. Giga	Japan	BS-3b	110.0° East	Ku2-Band	11.804	RHCP	5	Spot
VOA (Chinese)	USA	Intelsat 511	180.0° East	C-Band	3.975	RHCP	14	West Hemisph.
VOA (English)	USA	Intelsat 511	180.0° East	C-Band	3.975	RHCP	14	West Hemisph.
VOA (English/Spanish)	USA	Intelsat 511	180.0° East	C-Band	3.975	RHCP	14	West Hemisph.
VOA (French/Spanish)	USA	Intelsat 511	180.0° East	C-Band	3.975	RHCP	14	West Hemisph.
VOA (Lao/Korean)	USA	Intelsat 511	180.0° East	C-Band	3.975	RHCP	14	West Hemisph.
VOA (Vietnamese/Tibetan etc.)	USA	Intelsat 511	180.0° East	C-Band	3.975	RHCP	14	West Hemisph.
World Radio Network	UK	AsiaSat 2(Hong Kong)	100.5° East	C-Band	4.000	Ver		SB
Yunnan PBS	Hong Kong	AsiaSat 1(Hong Kong)	105.5° East	C-Band	4.040	Hor	9	NB

ITU REGION 1 (ALBANIA-FRANCE)

CHAPTER 13
NAMES & ADDRESSES

13.1. ITU REGION 1

Radio Tirana, Rruga Ismail Quemali, Tirana, Albania, **Tel:** +355 42 27512, **Fax:** +355 42 23650
RTA-TV (Algerian TV), 21 Blvd.des Martyrs, 16000 Algiers, Algeria, **Tel:** +213-2-60 2010, **Fax:** +213-2-60 3753
TV Angola, Avenida H-Chi-Miin, P.O. Box 2604, Luanda, Angola, **Tel:** +244-2-320 025, **Fax:** +244-2-391 091
Euratel Brussels, 12 avenue Ariane, 1200 Brussels, Belgium, **Tel:** +32-2-736 1795, **Fax:** +32-2-773 4895
FilmNet Belgium, Tollaan 97, B-1932 St. Stevens-Woluwe, Belgium, **Tel:** +32-2-716 5311, **Fax:** +32-2-725 8854
Bop-Tv Mmabatho, P.O. Bag X 2150, Mmabatho, Bophuthatswana, **Tel:** +27-140-289 111, **Fax:** +27-140-2600 9
TV Bosnia-Hercegovina (TVBiH), Bulevar M. Selimovica 4, 71000 Sarajevo, Bosnia-Hercegovina
HRT (Hvratska Televisija), Prisavlje 3, 10000, Zagreb, Croatia, **Tel:** +385 (1) 616 3366/368 3962/516 475, **Fax:** +385 (1) 616 3392, **WWW:** http://www.hrt.hr
PIK CYBC, P.O. Box 4824, Nicosia, Cyprus, **Tel:** +357-2-422 231, **Fax:** +357-2-314 055
Egyptian Radio, Corniche El-Nil, Maspero, 1186 Cairo, Egypt, **Tel:** +20-2-747158, **Fax:** +20-2-746 189
ESC Egyptian Space Channel, P.O. Box 1186, Cairo, Egypt, **Tel:** +20-2-574 6881, **Fax:** +20-2-749 310
The Egyptian Satellite Channel, Corniche El-Nil, Maspero, 1186 Cairo, Egypt, **Tel:** +20-2-747158, **Fax:** +20-2-746 189
ETV Ethiopian Television, P.O. Box 5544, 5544 Addis Abeba, Ethiopia, **Tel:** +251-1-1167 015
ARTE, 2a, rue de la Fonderie, 67080 Strasbourg Cedex, France, **Tel:** +33-88 522 222, **Fax:** +33-88-522 200, **WWW:** http://www.arte-tv.com/
Canal Courses, 83 rue de la Boutie, 75008 Paris, France, **Tel:** +33-1-4953 3900, **Fax:** +33-1-4289 0895
Canal France Internationale (CFI), 59 Boulevard Exelmans, 75016 Paris, France, **Tel:** +33-1-40711171, **Fax:** +33-1-40711172, **WWW:** http://www.radio-france.fr/
Canal J(eunesse), 91 bis, rue du Cherche-Midi, 75286 Paris Cedex 06, France, **Tel:** +33-1-4954 54 54, **Fax:** +33-1-4222 8717
Canal Jimmy, Immeuble Quai Quest, 42 quai du Point-du-Jour, 92659 Boulogne-Billancourt Cedex, France, **Tel:** +33-1-4610 1066, **Fax:** +33-1-4761 9400, **WWW:** www.ellipse.fr/Web/HTML/Jimmy_home/html
Canal Plus, 85/89 quai André Citroên, 75711 Paris Cedex 15, France, **Tel:** +33-1-4425 1000, **Fax:** +33-1-4425 1234, **WWW:** http://www.cplus.fr/
Ciné Cinéfil, Immeuble Quai Quest, 42 quai du Point-du-Jour, 92659 Boulogne-Billancourt Cedex, France, **Tel:** +Tel:33-1-4610 1099, **Fax:** +Fax:33-1-4761 9400
Ciné Cinémas, Immeuble Quai Quest, 42 quai du Point-du-Jour, 92659 Boulogne-Billancourt Cedex, France, **Tel:** +33-1-4610 1099, **Fax:** +33-1-4761 9400
Eurosport, 54 avenue de la Voie Lactée, 92100 Boulogne, France, **Tel:** +33-1-4141 1234, **Fax:** +33-1-4141 2485
France Supervision, 22 Avenue Montagne, F-75008 Paris, France, **Tel:** +33 1 4421 4242, **Fax:** +33 14421 5673
France-2, 22 Avenue Montagne, 75387 Paris Cedex 08, France, **Tel:** +33-1-4421 4242, **Fax:** +33-1-4421 5673, **WWW:** http://www.france2.fr/
La Cinquième, 10-14, rue Horace Vernet, 92130 Issy-Les-Moulineaux, France, **Tel:** +33-1 4146 5555, **Fax:** +33-1-4108 0222, **WWW:** http://www.lacinquieme.fr/
M-6 Metropole 6, 16 Cours Albert 1er, 75008 Paris, France, **Tel:** +33-1-4421 6666, **Fax:** +33-1-4563 7852, **WWW:** http://www.m6.fr/
MCM, 109 rue du Faubourg St-Honoré, 75008 Paris, France, **Tel:** +33-1-44 9591 00, **Fax:** +33-1-44 9591 25, **WWW:** http://www.netbeat.com/mcm
Planète, Immeuble Quai Quest, 42 quai du Point-du-Jour, 92659 Boulogne-Billancourt Cedex, France, **Tel:** +33-1-4610 1055, **Fax:** +33-1-4761 9400
Radio Hector, 116 av. du Presedent Kennedy, 75786 Paris cedex 16, France, **Tel:** +33-1-4230 2222/3183, **Fax:** +33-1-4230 4070
RFO, Centre Bourdan, 5 Avenue de Recteur-Poincaré, 75016 Paris, France, **Tel:** +33-1-4524 7100, **Fax:** +33-1-4224 9596
RTL Radio, 22, rue Bayard, 75008 Paris, France, **Tel:** +33-1-4070 4070, **Fax:** +33-1-4070 4411
RTL9, 3, Alleé St. Symphorien, F-5700 Metz,

ITU REGION 1 (FRANCE-GERMANY)

France, **Tel:** +33 (87) 565758, **Fax:** +33 (87) 565759, **WWW:** http://www.rtl9.com
TF-1, 1 quai Point-du Jour, 92656 Boulogne Cedex, France, **Tel:** +33-1-4141 1234, **Fax:** +33-1-4141 2840, **WWW:** http://www.tf1.fr/
TV Sport, 305 Avenue Le Jour se Lève, 92657 Boulogne-Billancourt, France, **Tel:** +33-1-4694 6100, **Fax:** +33-1-4761 0030
TV5, 15 rue Cognacq-Jay, 7th arrondissement, F-75007 Paris, France, **Tel:** +33-1-4418 5555, **Fax:** +33-1-4418 0655, **WWW:** http://www.tv5.org/
RTG TV Gabon, P.O. Box 150, Libreville, Gabon, **Tel:** +241-732 152, **Fax:** +241-732 153
3-Sat, Essenheimer Straße, 55128 Mainz, Germany, **Tel:** +49-6131-901, **Fax:** +49-6131-7054, **WWW:** http://www.3sat.com/
AFN-Television Europe, Bertramstraße 6, 60320 Frankfurt am Main, Germany, **Tel:** +49-69-1568 8262, **Fax:** +49-69-1568 8300
ARD, Arnulfstraße 42, 80335 München, Germany, **Tel:** +49-89-5900 01, **Fax:** +49-89-5900 3249, **WWW:** http://www.ard.de
ARD Eins, Hans-Bredow-Straße, 76530 Baden-Baden, Germany, **Tel:** +49-7221-2914, **Fax:** +49-7221-9226
BR-3, Rundfunkplatz 1, 80335 München, Germany, **Tel:** +49-89-5900 01, **Fax:** +49-89-5900 2375, **WWW:** http://www.br-online.de/
Deutsche Bundespost Telekom, Heinrich-von-Stephan Straße, 53175 Bonn, Germany, **Tel:** +49-228-140, **Fax:** +49-228-148 872
Deutsche Welle, Raderberggürtel 50, 50968 Köln, Germany, **Tel:** +49-221-3890, **Fax:** +49-221-3893 000, **WWW:** http://www.dw.gmd.de
Deutsche Welle TV, 50588 Cologne, Germany, **Tel:** +49-221-389 0, **Fax:** +49-221-389 4155, **WWW:** http://www.dw.gmd.de
Deutschlandradio Berlin, Hans Rosenthal Platz, 10825 Berlin, Germany, **Tel:** +49-30-8503 0, **Fax:** +49-30-8503 390, **WWW:** http://www.dfn.de/~drnl/
Deutschlandradio Köln, Raderberggürtel 40, 50968 Köln, Germany, **Tel:** +49-221-345 0, **Fax:** +49-221-3807 66, **WWW:** http://www.dfn.de/~drnl/
DS Kultur, Nalepastraße 10-50, 1160 Berlin, Germany, **Tel:** +49-30-6384 0
DSF Deutsches Sportfernsehen, Bahnhofstraße 27, 85774 Unterföhring, Germany, **Tel:** +49-89-95002-0, **Fax:** +49-89-9500 2329
Evangeliums RF, Postfach 14 44, 35573 Wetzlar, Germany, **Tel:** +49-6441-957 0, **Fax:** +49-6441-957 120
HR2 Kultur, Bertramstraße 8, 60320 Frankfurt am Main, Germany, **Tel:** +49-69-155 1, **Fax:** +49-69-155 2900, **WWW:** http://www.hr-online.de
Kabel-1, Bahnhofstraße 28, 85774 Unterföhring, Germany, **Tel:** +49-89-9508 07-0, **Fax:** +49-89-9508 0714, **WWW:** http://www.kabel1.de/

Klassik Radio Hamburg, Brandswiete 4, 20457 Hamburg, Germany, **Tel:** +49-40-4141 090
MDR Fernsehen, Kantstraße 71-73, 04275 Leipzig, Germany, **Tel:** +49-341-22760, **Fax:** +49-341-5663 544, **WWW:** http://www.mdr.de
MDR-Sputnik, Waisenhausring 8-10, 06108 Halle, Germany, **Tel:** +49-345-3796 1, **Fax:** +, **WWW:** http://www.mdr.de
n-tv Der Nachrichtensender, Taubenstraße 1, 10117 Berlin, Germany, **Tel:** +49-30-201 90 0, **Fax:** +49-30-201 90 505
NDR, Gazellenkamp 57, 22529 Hamburg, Germany, **Tel:** +49-40-41560, **Fax:** +49-40-447602, **WWW:** http://www.dfn.de/presse/ndr/home.html
NDR2, Gazellenkamp 57, 22529 Hamburg, Germany, **Tel:** +49-40-41560, **Fax:** +49-40-447602, **WWW:** http://www.ndr2.de
Nickelodeon (Germany), Couvenstraße 8, D-40211 Düsseldorf, Germany
Premiere, Am Stadtrand 52, 22047 Hamburg, Germany, **Tel:** +49-40-6944 50, **Fax:** +49-40-6944 51, **WWW:** http://www.premiere.de/
Pro-7, Bahnhofstraße 27a, 85774 Unterföhring, Germany, **Tel:** +49-89-95001 0, **Fax:** +49-89-95001 230, **WWW:** http://www.pro-sieben.de/
Radio ffn, Dorfstraße 2, 30916 Isernhagen KB, Germany, **Tel:** +49-5139 8080 0
Radio Xanadu, Pestalozzistraße 23, 80469 München, Germany, **Tel:** +49-89-231 9070, **Fax:** +49-89-260 4417
Radioropa Info, TechnicPark, D-54550 Daun/Eifel, Germany, **Tel:** +49-6592-203 0, **Fax:** +49-6592-203 238
Radioropa Info, Sudio Leipzig, Reichpietschstraße 23, D-04317 Leipzig, Germany, **Tel:** +49-341-69606 0, **Fax:** +49-341-69606 4
RTL Fernsehen, Aachener Straße 1036, 50858 Köln, Germany, **Tel:** +49-221-456 0, **Fax:** +49-221-456 4290, **WWW:** http://www.RTL.de/
RTL Nord, Jenfelder Allee 80, 22045 Hamburg, Germany, **Tel:** +49-40-6688 4350, **Fax:** +49-40-6688 4400
RTL-2 Fernsehen GmbH, Bavariafilmplatz 7, 82031 Grünwald, Germany, **Tel:** +49-89-641 85 0, **Fax:** +49-89-641 85 999, **WWW:** http://www.rtl2.de/
Sat-1, Otto-Schott-Straße 13, 55127 Mainz, Germany, **Tel:** +49-6131-900 0, **Fax:** +49-6131-900 100, **WWW:** http://www.sat1.de/
SDR/SWF 2 Kultur, Neckar Straße 230, 70190 Stuttgart, Germany, **Tel:** +49-711-288 0, **Fax:** +49-711-288 3366, **WWW:** http://www.sdr.de
SR-1/2/3, Postfach 1050, 66111 Saarbrücken, Germany, **Tel:** +49-681-602 0, **Fax:** +49-681-602 3874
Star*Sat Radio, Postfach 506, 54541 Daun, Germany, **Tel:** +49-6592-712 674, **Fax:** +49-6592-712 231
Super RTL, Richard-Byrd-Straße 6, 55829

ITU REGION 1 (GERMANY-NETHERLANDS)

Cologne, Germany, **Tel:** +49-221-91550
SWF-3, Hans-Bredow-Straße, 76530 Baden-Baden, Germany, **Tel:** +49-7221-276 1, **Fax:** +49-7221-32915, **WWW:** http://www.swf3.de
TM-3 (Tele München-3), Bavariafilmplatz 7, D-82031 Grünwald, Germany, **Tel:** +49 (89) 6410 104, **Fax:** +49 (89) 6410 268, **Contact:** Herr Jochem Kröhne (Man. Dir.)
VH-1, Bramfelder Straße 117, 22305 Hamburg, Germany, **Tel:** +49-40-611500, **Fax:** +, **WWW:** http://www.vh1.de/
VIVA II, Caudius-Dornier-Straße 5b, 50829 Cologne, Germany, **Tel:** +49-221-956820
VOX, Richard-Byrd-Straße 6, 55829 Cologne, Germany, **Tel:** +49-221-9534 0, **Fax:** +49-221-9534 440, **WWW:** http://www.vox.de/
WDR-3 Köln, Appellhofplatz 1, 50667 Köln, Germany, **Tel:** +49-221-1602-132, **Fax:** +49-221-1602 135, **WWW:** http://www.wdr.de
World Tamil Television, Urbanstraße 31, 10967 Berlin, Germany, **Tel:** +49-30-6913 858, **Fax:** +49-30-6917 256
ZDF, Postfach 4040, 55100 Mainz, Germany, **Tel:** +49-6131-702 060, **Fax:** +49-6131-702 052, **WWW:** http://www.zdf.de/
ET-1, Mesogion 432, 15310 Athens, Greece, **Tel:** +301-639 5970, **Fax:** +301-638 1075
Duna TV, Mefzarosu 48-56, 1140 Budapest, Hungary, **Tel:** +36-1-1560122, **Fax:** +36-1-1566772, **WWW:** www.hungary.com/dunatv/
IRIB TV/Radio, P.O. Box 19395, 1774 Tehran, Iran, **Tel:** +98-21-2942 84, **Fax:** +98-21-2940 24
IRIB TV/Radio (Islamic Rep. of Iran TV), P.O. Box 19395, 1774 Tehran, Iran, **Tel:** +98-21-2942 84, **Fax:** +98-21-2940 24
Irish Satellite Radio, 100 O' Connell Street, Limerick, Ireland, **Tel:** +353-61-31 9595, **Fax:** +353-61-41 9890
Radio Telefis Éireann (RTE Radio-1), Donnybrook, Dublin 4, Ireland, **Tel:** +353-1-208 3111, **Fax:** +353-1-208 3080, **WWW:** http://www.bess.tcd.ie/ireland/rte.htm
Israel ch2/3, AT Communications, 8 Shef-tal Street, 67013 Tel Aviv, Israel, **Tel:** +972-3-562 6761, **Fax:** +972-3-562 6438
Canale-5, Viale Europa 48, Cologna Monzese, 20093 Milan, Italy, **Tel:** +39-2-2514 1, **Fax:** +39-2-2514 7715
Cinquestelle, Via del Garivaggio 4, 20144 Milano, Italy, **Tel:** +39-2-433 889, **Fax:** +39-2-498 5514
Italia-1, Viale Europa 48, Cologna Monzese, 20093 Milan, Italy, **Tel:** +39-2-2514 1, **Fax:** +39-2-2514 7715
RAI, Via Mazzini 14, 00195 Rome, Italy, **Tel:** +39-6-38781, **Fax:** +39-6-3226070, **WWW:** http://www.rai.it
RAISat, via del Babuino 9, 00187 Rome, Italy, **Tel:** +39-6-368 62276, **Fax:** +39-6-322 0390, **WWW:** http://www.rai.it
Rete-4, Viale Europa 48, Cologna Monzese, 20093 Milan, Italy, **Tel:** +39-2-2514 1, **Fax:** +39-2-2514 7715
Telespacio / Italsat, Via Alberto Bergamini 50, 00159 Rome, Italy, **Tel:** +39-6-406 931
JRTV Jordan, P.O. Box 1041, Aman, Jordan, **Tel:** +962-6-638 760, **Fax:** +962-6-788 115
Kasachstan Radio/TV, Mira Ul. 173, 480013 Alma Ata, Kasachstan, **Tel:** +7-3272-633 716
Med TV (Kurdish Satellite Television), The Linen Hall, 162-168 Regent Street, London W1R 5AT, Kurdistan, **Tel:** +44 (71) 494 2523, **Fax:** +44 (71) 494 2528, **E-mail:** med@med-tv.be, **WWW:** www.ib.be/med/
The Kuwait Satellite Channel, P.O. Box 621, 13007 Safat, Kuwait, **Tel:** +965-242 3774, **Fax:** +965-245 6660
Libyan TV, P.O. Box 3731, Tripol1, Libya, **Tel:** +218-21-32 451, **Fax:** +218-21-33 470
RTL Luxembourg, Villa Louvigny, Allée Marconi, 2850 Luxembourg, Luxembourg, **Tel:** +352-47661, **Fax:** +352-4766 2730, **WWW:** http://www.rtl.lu/
RTL Radio, R.C. Luxembourg B. 6139, 2850 Luxembourg, Luxembourg, **Tel:** +352-452 9901, **Fax:** +352-452 99021
Société Européenne des Satellites (SES), Chateau Betzdorf, 6815 Betzdorf, Luxembourg, **Tel:** +352-71072 51, **Fax:** +352-71072 5433, **Contact:** Bill Wijdeveld
RTM Malagasy, B.P 1202, Antananarivo, Madagascar, **Tel:** +261-223 81
TV Nationale Mauritaine, B.P 2000, Nouakchott, Mauritania, **Tel:** +222-53266, **Fax:** +222-52164
Radio Monte Carlo, P.O. Box 128, Monte Carlo, Monaco, **Fax:** +33-93-159 448
Radiodiffusion Télévision Marocaine, 1 rue al Brihi, Rabat, Morocco, **Tel:** +212-7-764871, **Fax:** +212-7-762010
Oman TV, Ministry of Information, P.O. Box 600, Muskat Oman, **Tel:** +968-290 831, **Fax:** +968-290 831
Euro-7, Postbus 1594, 1200 BN Hilversum, Netherlands, **Tel:** +31 35 210 799
Eurostep, P.O. Box 11112, 2301 EC Leiden, The Netherlands, **Tel:** +31-71-120 863, **Fax:** +31-71-134 545
Holland FM, Eendrachtsweg 36, 3012 LC Rotterdam, The Netherlands, **Tel:** +31-10-413 9013, **Fax:** +31-10-413 9744
KinderNet, Joan Muysksensweg 42, 1099 CK Amsterdam, The Netherlands, **Tel:** +31-20-665 9576, **Fax:** +31-20-692 8772
MultiChoice (RTL 4/5, Veronica), Archimedesbaan 21, 3439 ME Nieuwegein, The Netherlands, **Tel:** +31-30-6086611, **Fax:** +31-30-6050005, **Contact:** Judith Huisman
RTL Rock, Postbus 15000, 1200 TV Hilversum, The Netherlands, **Tel:** +31-35-718 718, **Fax:** +31-35-236 892
RTL-4/5, Franciscusweg 219, 1219 SE Hilversum,

ITU REGION 1 (NETHERLANDS-SWEDEN)

The Netherlands, **Tel:** +31-35-718 718, **Fax:** +31-35-236 892
SBS6, Plantage Middenlaan 14, 1018 DD Amsterdam, The Netherlands, **Tel:** +31-20-6228183, **Fax:** +31-20-6224146, **WWW:** http://www.cameo.nl/sbs6/default.html
Sky Radio, Naarderpoort 2, 1411 MA Naarden, Netherlands, **Tel:** +31-35-6991007, **Fax:** +31-35-6991008, **Contact:** Mr. Ton Lathouwers (GM), **WWW:** http://www.skyradio.nl
The Music Factory, Graaf Wichmanlaan 46, 1405 HB Bussum, Netherlands, **Tel:** +31-2159-96666, **Fax:** +31-2159-49959
TV 10, Mediapark, Postvak F44, Sumatralaan 45, 1217 GP Hilversum, Netherlands, **Tel:** +31-35-772700, **Fax:** +, **WWW:** http://www.tv10.nl/
Veronica, Laapersveld 75, 1213 VB Hilversum, Netherlands, **Tel:** +31-35-716716, **Fax:** +31-35-249771, **WWW:** http://veronica.nl
ORTV Niger, P.O. Box 309, Niamey, Niger, **Tel:** +227-723 155, **Fax:** +227-723 548
NRK1, Bjornstein Bjørnssons Place 1, N-0340 Oslo, Norway, **Tel:** +47-23-047000, **Fax:** +47-23-047440, **WWW:** http://www.nrk.no/
TV-2 Norge, Verftsgatan 2C, 5011 Bergen, Norway, **Tel:** +47-5-239 900, **Fax:** +47-5-239 905, **WWW:** http://www.tvnorge.no/
TVN Norge, Sagveien 17, 0458 Oslo 4, Norway, **Tel:** +47-2-350 350, **Fax:** +47-2-376 560, **WWW:** http://www.tvnorge.no/
Z-TV, Maredalsveijen 17, 0458 Oslo, Norway, **Tel:** +47-22-71 4000, **Fax:** +47-22-71 6550, **WWW:** http://www.ztv.se/
FilmNet Poland, Domaniewska 39a, 02 672 Warsaw, Poland, **Tel:** +48-2-640 2905, **Fax:** +48-2-640 2907
Polsat, Marchakovska 83, P-00517 Warsaw 84, Poland, **Tel:** +48-22-295 684
TV Polonia, ul. J.P. Woronicza 17,, P00950 Warsaw, Poland, **Tel:** +48-22-478191, **Fax:** +48-22-447419, **WWW:** http://linx.tvp.com.pl/welcome.htm
RTP Internaçional, av 5 de Outubro 197, 1000 Lisbon, Portugal, **Tel:** +351-1-793 1774, **Fax:** +351-1-793 3054, **WWW:** www.rtp.pt/rtpinternacional
2x2 Telekanal, Meschdunarodnaja Associacja Radio i Televisidienia, Moscow, Russia, **Tel:** +7-095-2926 121, **Fax:** +7-095-2925 918
Channel-6 St. Petersburg, Parachutnaya ulitsa 6, 197341 St. Petersburg, Russia, **Tel:** +7-812-3013878, **Fax:** +7-812-3010887
ORT-1, Piatnickaia 25, 113326 Moscow, Russia, **Tel:** +7-095-217 7260, **Fax:** +7-095-215 1324
Russia TV-2, 12 Korolyov Street, 127000 Moscow, Russia, **Tel:** +7-095-2177 898, **Fax:** +7-095-2889 508
Telekanal Rossija, 5 Ulitsa Yamskogo Polya 19/21, 125124 Moscow, Russia, **Tel:** +7 095 2500 511, **Fax:** +7 095 2500 105

Arabsat, P.O. Box 1038, 11431 Riyadh, Saudi Arabia, **Tel:** +966-1-464 6666, **Fax:** +966-1-465 6983
BSKSA Saudi Arabia, P.O. Box 57137, 11574 Riyadh, Saudi Arabia, **Tel:** +966-1-4010 40, **Fax:** +966-1-4044 192
Canal Horizons, P.O. Box 1765, Dakar, Senegal, **Tel:** +221-211 472, **Fax:** +221-223 490
RTV Beograd, Takovska 10, 11000 Belgrade, Serbia, **Tel:** +381-11-342 9000, **Fax:** +381-11-322 9766
RTV Srbije (Beograd), Takovska 10, 11000 Belgrade, Serbia, **Tel:** +381-11-342 9000, **Fax:** +381-11-322 9766
VTV (Vasa Televizia), Stara Prievozska 2, 821 09 Bratislava 2, Slovakia, **Tel:** +42 (7) 4521 3336 (ex. 253)
M-Net, Hendrik Verwoerd Drive, 2125 Randburg, South Africa, **Tel:** +27-11-889 1911, **Fax:** +27-11-789 4447
SABC-TV, Broadcasting Centre Auckland Park, 2092 Johannesburg, South Africa, **Tel:** +27-11-714 9111, **Fax:** +27-11-714 5050, **WWW:** www.sabc.co.za/TV/index.htm
Antena -1 Internaçional, Carretera San Sebastian de los Reyes, 28700 Madrid, Spain, **Tel:** +34-1-632 0500, **Fax:** +34-1-632 7144
Antena Tres TV/Radio, Carretera San Sebastian de los Reyes, 28700 Madrid, Spain, **Tel:** +34-1-632 0500, **Fax:** +34-1-632 7144
Canal Plus España, Gran Via 32, 3rd. floor, 28013 Madrid, Spain, **Tel:** +34-1-396 5500, **Fax:** +34-1-396 5600
Documania, Sogecable, Gran Vía 32 3 Piso, 28013 Madrid, Spain, **Tel:** +34-1-396 5687, **Fax:** +34-1-396 5484
Hispasat, P.O. Box 95000, 28080 Madrid, Spain, **Tel:** +34-1-372 9000, **Fax:** +34-1-307 6683
REE Radio Exterior España, Apartado 156.202, 28080 Madrid, Spain, **Tel:** +34-1-346 1160, **Fax:** +34-1-261 6388
Sogecable, Torre Picasso, Plaza Ruiz Picasso, E-28020 Madrid, Spain, **Tel:** +34-1-3965 514
Tele Cinco, Piazza Pablo Picasso, Torre Picasso 36, Madrid, Spain, **Tel:** +34-1-5550 302, **Fax:** +34-1-5550 146, **WWW:** http://www.telecinco.es
TVE-Internaçional, Torespaña O'Donnell 77, 28007 Madrid, Spain, **Tel:** +34-1-4095 773, **Fax:** +34-1-574 2015, **WWW:** http://www.tve.es/
Sudanese TV Khartoum, P.O. Box 1094, Omdurman, Sudan, **Tel:** +249-55022
FilmNet Scandinavia, Hornsgatan166, Box 9006, 10271 Stockholm, Sweden, **Tel:** +46-8-772 2500, **Fax:** +46-8-669 90 98, **WWW:** http://www.filmnet.se/
Kanal 5, P.O. Box 26205, S-100 41 Stockholm, Sweden, **Tel:** +46-8-674 1500, **Fax:** +46-8-612 0595, **WWW:** http://www.kanal5.se/

ITU REGION 1 (SWEDEN-UK)

Radio Sweden, 10510 Stockholm, Sweden, **Tel:** +46-8-7847 281, **Fax:** +46-8-6676 283
SVT 1/2, Oxenstiernsgatan 26-34, 10510 Stockholm, Sweden, **Tel:** +46 (8) 784 0000, **Fax:** +46 (8) 784 1500, **WWW:** http://www.svt.se/
SVT Swedish Television, Oxenstiernsgatan 26-34, 105 10 Stockholm, Sweden, **Tel:** +46-8-784 0000, **Fax:** +46-8-660 3263
TV-1000 Succé Kanalen(Scansat), P.O. Box 2133, 10314 Stockholm, Sweden, **Tel:** +46-8-100 019, **Fax:** +46-8-243 840, **WWW:** http://www.tv1000.se
TV-3 (Scansat), P.O. Box 2094, 10313 Stockholm, Sweden, **Tel:** +46-8-403 333, **Fax:** +46-8-216 564
TV-3 Sverige, Karlbergsvägen 77, P.O. Box 21052, 10031 Stockholm, Sweden, **Tel:** +46 (8) 610 3300, **Fax:** +46 (8) 610 3330, **WWW:** http://www.tv3.se
TV-4 Nordisk Television, Storängskroken 10, 11479 Stockholm, Sweden, **Tel:** +46-8-644 4400, **Fax:** +46-8-644 4440
TV-4 Sverige, Tegeludds Vagen 3, 11579 Stockholm, Sweden, **Tel:** +46 (8) 644 4400, **Fax:** +46 (8) 644 4440, **WWW:** http://www.tv4.se/
Radio E Viva!, Kreuzstraße 26, 8032 Zürich, Switzerland, **Tel:** +41-1-262 3636, **Fax:** +41-1-262 4990
Swiss Radio International, Giacomettistraße 1, 3000 Bern 15, Switzerland, **Tel:** +41-31-350 92 22, **Fax:** +41-31-350 95 44
Teleclub, Löwenstraße 11, Postfach, 8021 Zürich, Switzerland, **Tel:** +41-1-225 2525, **Fax:** +41-1-225 2500, **WWW:** http://www.teleclub.ch/
Syrian Television, Ommayyad Square, Damascus, Syria, **Tel:** +963-11-720 700
Arab States Broadcasting Union (ASBU), 17 rue el-Mansoura, el-Mensah 4, 1014 Tunis, Tunisia, **Tel:** +216-1-23 8044, **Fax:** +216-1-766 551
RTT Tunisia, 71 Avenue de la Liberté, 1002 Tunis, Tunisia, **Tel:** +216-1-287 300, **Fax:** +216-1-781 058
ATV, Medya Plaza Basin Express Yolu, 34540 Günesli, Istanbul, Turkey, **Tel:** +90-212-502 8005, **Fax:** +90-212-502 8802
Cine-5, Halaskargazi Cad 180, Osmanbey Istanbul, Turkey, **Tel:** +90-1-2461 260, **Fax:** +90-1-2475 778
HBB TV, Büyükdere Cad 187, Levent, Istanbul, Turkey, **Tel:** +90-212-281 4800, **Fax:** +90-212-281 4808
interSTAR (AKK), Divan Yolu Turbedar Sok 2, Iketelli Günesli, Istanbul, Turkey, **Tel:** +90-212-698 4901, **Fax:** +90-212-698 4970, **WWW:** www.medyatext.com/star/index.html
Kanal D, G 48 Sok 12, 806604 Istanbul, Turkey, **Tel:** +90-1-285 1250, **Fax:** +90-1-285 1263
Kanal-6, G48 Sok 12, 806604 Istanbul, Turkey, **Tel:** +90-212-285 1250, **Fax:** +90-212-285 1263/229 3173, **WWW:** www.medyatext.com/kanal6
Radio Kulüp, Istanbul, Turkey, **Tel:** +90-212-2827 610, **Fax:** +90-212-2827 613
Show TV, Halaskargazi Cad 180, Osmanbey Istanbul, Turkey, **Tel:** +90-1-246 1260, **Fax:** +90-1-247 5778, **WWW:** www.medyatext.com/shotext
TeleON, Divanyolu, Türbedar Sok 22, Cagaloglu, 34410 Istambul, Turkey
TRT Radio-TV Corporation, TRT Sitesi Oran, City Ankara, 06450 Ankara, Turkey, **Tel:** +90-312-490 4983, **Fax:** +90-312-490 4985
Voice of Turkey, P.O. Box 333, 06443 Ankara, Turkey, **Tel:** +90-41-189 453, **Fax:** +90-441-353 816
Abu-Dhabi Satellite Channel, P.O. Box 637, Abu Dhabi, UAE, **Tel:** +971-2-452 000, **Fax:** +971-2-451 470
EDTV, P.O. Box 1695, Dubai, UAE, **Tel:** +971-4-370 255, **Fax:** +971-4-3710 79
ASDA FM, The Studio, Spring Bank, Astley, M29 7BR Manchester, UK, **Tel:** +44-942-89 6111, **Fax:** +44-942-88 4397
BBC Prime/World, Woodlands 80, W12 0TT London, UK, **Tel:** +44-181-576 2884, **Fax:** +44-181-576 2782, **WWW:** http://www.bbc.co.uk/worldwide/television/channel/html
BFBS Radio, P.O. Box 1234, W2 London, UK, **Tel:** +44-171-724 1234, **Fax:** +44-171-706 1582
Bravo, Twyman House, 16 Bonny Street, NW1 9PG London, UK, **Tel:** +44-171-482 4824, **Fax:** +44-171-284 2042, **WWW:** http://www.uaep.co.uk/BRAVO.html
British Sky Broadcasting, 6 Centaurs Business Park, Grant Way, Off Syon Lane, TW7 5QD Isleworth, Mddx, UK, **Tel:** +44-171-782 3000, **Fax:** +44-171 782 3030, **WWW:** http://www.sky.co.uk/
Cartoon Network, 37 Kentish Town Road, NW1 8NX London, UK, **Tel:** +44-171-284 7000, **Fax:** +44-171-284 7060
CMT Country Music TV Europe, Twyman House, 16 Bonny Street, NW1 9PG London, UK, **Tel:** +44-171-813 5000, **Fax:** +44-171-284 2042
CNE (Chinese News Entertainment), Marvic House, Bishops Road, Fulham, SW6 7AD London, UK, **Tel:** +44-171-610 3880, **Fax:** +44-171-610 3118
CNN International, CNN-House, 19-22 Rathbone Place, W1P 1DF London, UK, **Tel:** +44-171-637 7100, **Fax:** +44-171-637 6768, **WWW:** http://www.cnn.com/
Discovery Channel, Twyman House, 16 Bonny Street, NW1 9PG London, UK, **Tel:** +44-171-482 4824, **Fax:** +44-171-284 2042, **WWW:** http://www.discovery.co.uk
Euronet, P.O. Box 1234, N1 6XH London, UK, **Tel:** +44-171-613 1101
European Business News (EBN), P.O. Box 900,

ITU REGION 1 (UK-ZAIRE)

London EC4M 7RX, UK, **Tel:** +44 (171) 653 9300, **Fax:** +44 (171) 653 9393, **Contact:** Mrs. Theresa Dear (Human Resources Co-ordinator), **E-mail:** feedback@ebn.co.uk, **WWW:** http://www.ebn.co

Granada/BSkyB, 6 Centaurs Business Park, Grant Way, Off Syon Lane, TW7 5QD Isleworth, Mddx, UK, **Tel:** +44-171-782 3000, **Fax:** +44-171 782 3030, **WWW:** http://www.granada.co.uk/

JSTV Japan Satellite TV, 17 King's Exchange, Tileyard Road, N7 9AH London, UK, **Tel:** +44-171-607 7677, **Fax:** +44-171-607 7442/7720

Middle East Broadcasting Centre (MBC), 10 Heathmans Road, Parson's Green, Fulham, SW6 4TJ London, UK, **Tel:** +44-171-3719 597, **Fax:** +44-171-3719 601

MTV, Centro House, 20-23 Mandela Street, Camden Town, MW1 0DU London, UK, **Tel:** +44-171-383 4250, **Fax:** +44-171-388 2064

MTV Muslim TV, Al Shirkatul Islamiyyah, SW18 5QL London, UK, **Tel:** +44- 181- 870 0922, **Fax:** +44- 181- 870 0684

NBC Super Channel, Melrose House, 14 Lanark Square, Limeharbour, E14 9QD London, UK, **Tel:** +44-171-418 9418, **Fax:** +44-171-418 9419/20/21, **WWW:** www.nbc.com/nbc-europe

NHK Tokyo, Japan Satellite TV (Europe) Ltd., Unit 17 King's Exchange, Tileyard Road, N7 9AH London, UK, **Tel:** +44-171-607 7677, **Fax:** +44-171-607 7442, **WWW:** http://www.nrk.org.jp/

Nickelodeon, 6 Centaurs Business Park, Grant Way, TW7 5QD Isleworth Middlesex, UK, **Tel:** +44-171-705 3000, **Fax:** +44-171-705 3030, **WWW:** http://www.ee.ac.uk/Contrib/Entertainment/nickelodeon.html

Reuters Television, 40 Cumberland Avenue, NW10 7EH London, UK, **Tel:** +44-181-965 7733, **Fax:** +44-181-965 0620

Sci-Fi Channel Europe, UIP House, 45 Beadon Road, Hammersmith, W6 0EG London, UK, **Tel:** +44-171-563 4048, **Fax:** +44-171-741 3293, **WWW:** http://www.scifi.com/

Screensport, ETN, The Quadrangle, 180 Wardour Street, W1V 4AE London, UK, **Tel:** +44-171-439 1177, **Fax:** +44-171-439 1415

SSVC, Chalfont Grove, Gerrards Cross, SL9 8TN Buckinghamshire, UK, **Tel:** +44-1494-874 461, **Fax:** +44-1494-872 982

StepUp (University of Plymouth), The Hoe Centre, Notte Street, PL1 2AR Plymouth, Devon, UK, **Tel:** +44-1752-233 639, **Fax:** +44-1752-233 638

Sunrise Radio, Cross Lances Road, TW3 2AD Hounslow, Mddx, UK

SuperGold, Chiltern Road, Dunstable, LU6 1HQ Bedfordshire, UK, **Tel:** +44-1582-666 001, **Fax:** +44-1582-661 725

TCC The Children's Channel, 9-13 Grape Street, WC2 8DR London, UK, **Tel:** +44-171-2403 422, **Fax:** +, **WWW:** http://www.tcc.co.uk/

Television Direct Ltd., Prince House, 43-51 Prince Street, BS1 4PS London, UK, **Tel:** +44-1272-252627

The Adult Channel, Unit 023, Canalot Prod. Studios, 222 Kensal Rd., W10 5BN London, UK, **Tel:** +44-181-964 1411, **Fax:** +44-181-964 5934

The Chinese Channel, 30-31 Newman Street, W1P 3PE London, UK, **Tel:** +44-171-636 6818, **Fax:** +44-171-636 6848

The Landscape Channel, Crowhurst, TN33 9BX East Sussex, UK, **Tel:** +44-1424-830 688, **Fax:** +44-1424-830 680

The Parliamentary Channel, Twyman House, 16 Bonny Street, NW1 9PG London, UK, **Tel:** +44-171-482 4824, **Fax:** +44-181-284 2042

TLC The Learning Channel, Twyman House, 16 Bonny Street, NW1 9PG London, UK, **Tel:** +44-171-284 1570, **Fax:** +44-171-284 2042

Travel, 66 Newman Street, London W1P 3LA, UK, **Tel:** +44 (171) 636 5401, **Fax:** +44 (171) 636 6424, **WWW:** http://www.travelchannel.com/traveluk/uk.html

Turner Broadcasting (TNT), 37 Kentish Road Town, NW1 8NX London, UK, **Tel:** +44-171-284 7000, **Fax:** +44-171-284 7060

UK Gold, The Quadrangle, 180 Wardour Street, W1V 4AE London, UK, **Tel:** +44-171-306 6100, **Fax:** +44-171-306 6101

UK Living, BBC Enterprises Ltd, Woodlands, 80 Wood Lane, W12 0TT London, UK, **Tel:** +44-181-5762 510

United Christian Broadcasters Ltd., P.O. Box 255, ST4 2UE Stoke on Trent, UK, **Tel:** +44-1782-64 2000, **Fax:** +44-1782-64 1121

Virgin 1215, 1 Golden Square, W1R 4DJ, London, UK, **Tel:** +44-171-434 1215, **Fax:** +44-171-434 1197, **WWW:** http://www.virginradio.co.uk/

World Radio Network Ltd, Wyvil Court, 10 Wyvil Road, SW8 2TY London, UK, **Tel:** +44-171-896 9000, **Fax:** +44-171-896 9007, **Contact:** Carl Miosga, **WWW:** www.wrn.org

World Radio Network Ltd, Wyvil Court, 10 Wyvil Road, SW8 2TY London, UK, **Tel:** +44-171-896 9000, **Fax:** +44-171-896 9007, **Contact:** Carl Miosga

Worldwide Television News, The Interchange, Oval Road, Camden Lock, NW1 7EP London, UK, **Tel:** +44-171-410 5200, **Fax:** +44-171-413 8302

Zee TV, Unit 3, Springvilla Park, Springvilla Road, HA8 7EB Edgware, Middlesex, UK, **Tel:** +44-181-381 2233, **Fax:** +44-181-381 2443

La Voix du Zaïre, P.O. Box 3171, Kinshasa-Gombe, Zaïre, **Tel:** +243-12-23171

ORTZ Zaïre TV, P.O. Box 3171, Kinshasa-Gombe, Zaïre, **Tel:** +243-12-23171

ITU REGION 2 (ARGENTINA-USA)

13.2 ITU REGION 2

Asociación de Teleradiodifusoras Argentinas (ATA, Av. Córdoba 323, 6to., 1094 Buenos Aires, Argentina, **Tel:** +54-1-312 4208/4219/ 4533, **Fax:** +54-1-312 4208
Empresa Nacional deTelevision Bolivia (ENTB Can. 7), Avenida Camacha 1485, Edificio la Urbana 6to. piso, Casilla, 900 La Paz, Bolivia
Amazon Sat TV, Avenida Carvalho Leal, 1.270, 69065-000 Manaus, Brazil
Bandeirantes, Radiantes, 13 Morumbi, Caixa Postal 372 CEP, 05699-900 Sao Paulo, Brazil
Rede Globo TV, 303 Lopez Quintas, 22463-900 Rio de Janeiro, Brazil
Rede Manchete, Rua do Russel 766-804, 22210-010 Rio de Janeiro, Brazil
Alberta Access Network, 295 Midpark Way S.E, T2X 2A8 Calgary, Alberta, Canada
Atlantic Satellite Network, 2885 Robie Street, B3J 2Z4 Halifax, Nova Scotia, Canada
Canadian Broadcasting Corp., P.O. Box 8478, Ottawa, Ont K1G 3J5, Canada, **Tel:** +1 (613) 724 1200, **Fax:** +1 (613) 738 6749, **Contact:** Mr. Perin Beatty (Pres/CEO), **WWW:** http://www.cbc.ca/
CBC Newsworld, P.O. Box 500, Station A, M5W 1E6 Toronto, Ontario, Canada, **Fax:** +, **WWW:** http://www.cbc.ca/
CTV Television, 42 St. Charles East, M4Y 1T4 Toronto, Ontario, Canada
First Choice, 98 St. Queens East, Suite 300, M4S 2C1 Toronto, Ontario, Canada, **Tel:** +1-416-364 9115, **Fax:** +1-416-365 1572
Global Television Network, 81 Barber Green Road, M3C 2A2 Don Mills, Ontario, Canada
La Sept, 1680 Boulevard Provence, Suite 310, Place du Commerce, G4W 2Z7 Brossard, Québec, Canada
Le Reseau des Sports, 1755 Boulevard René Levesque, H2K 4P6 Montreal, Québec, Canada
Much Music, 299 Queens West, M5V 2Z5 Toronto, Ontario, Canada, **Tel:** +1-416-591 5757, **Fax:** +1-416-591 7791
Musique Plus, 209 St. Catherine Avenue, H2X 1L2 Montreal, Québec, Canada
Première Choix, 2100 St. Catherine Avenue, H2H 2T3 Montreal, Québec, Canada
Société de Radio-Television du Québec, 800 rue Fullum, H2K 3L7 Montreal, Québec, Canada
StoryVision Netw. Inc., 191 Lombard Avenue, Suite 1400, R3B 0X1 Winnipeg, Manitoba, Canada, **Tel:** +1-204-942 1005, **Fax:** +1-204-957 7647
Super Channel, 5324 Calgary Trail, Suite 200, T6H 4J8 Edmonton, Alberta, Canada, **Tel:** +1-403-437 7744, **Fax:** +1-403-437 3181
Television Quatre Saisons, 405 Ogilvy Avenue, H3M 1N4 Montreal, Québec, Canada

The Sports Network, 1155 Leslie Street, M3C 2J6 Don Millls, Ontario, Canada
The Weather Network, 1755 Boulevard René Levesque, H2K 4P6 Montreal, Québec, Canada
TV Cinq, 1755 Boulevard René Levesque, H2K 4P6 Montreal, Québec, Canada, **Fax:** +, **WWW:** http://www.tv5.ca/
TV Ontario, P.O. Box 200, Station Q, M4T 2T1 Toronto, Ontario, Canada
Vision TV, 315 Queen Street East, M5A 1S7 Toronto, Ontario, Canada
YTV, 64 Jefferson Avenue, M6K 3H3 Toronto, Ontario, Canada
Television Nacional de Chile, Bellavista 0990, P.O. Box 16104, Correo 9, Santiago, Chile, **Tel:** +56-774 552, **Fax:** +56-353 000
Instituto Cubano de Radio y Televion (Cubavision), Calle 23 No. 258, Vedado, 10400 Habana, Cuba, **Tel:** +53-7-322305/327511/ 325501
Cablevision, Dr. Rio de Loza 190, Colonia Doctores, 06720 Mexico DF, Mexico
GalaVision, Av. Chapultepec 28, DF 06724 Col Doctores, Mexico, **Tel:** +52-5-709 2314, **Fax:** +52-5-709 1222
Galavisión ECO, Av. Chapultepec 28, Col Dotores, DF 06742 Mexico, Mexico, **Tel:** +52-5-709 2314, **Fax:** +52-5-709 1222
Televisa S.A. (Canales 2, 5/Galavisiòn), Edificio Televicentro, Av. Chapultepec 28, 06724 Mexico DF, Mexico
XEIPN (Canal 11), Prolongiaçion Carpio 475, Colonio Santo Toomas, 11340 Mexico DF, Mexico
XHDF (Canal 13), Av. Periferico Sur 4121, 12 DF Mexico City, Mexico
Canal 11, Avenida Manco Capac 333, La Victoria, Lima, Peru
Canal 13, Jiron Huaraz 2098, Pueblo Libre, Lima, Peru
Canal 2, Avenida San Felipe 968, Jesus Maria, Lima, Peru
Canal 4, Montero Rosas 1099, Santa Beatriz, Lima 14, Peru
Canal 5, Avenida Arequipa 1110, Lima, Peru
Canal 7, José Galves 1040, Santa Beatriz, Lima, Peru
Canal 9, Avenida Arequipa 3570, San Isdro, Lima, Peru
ABS-CBN Broadcasting Corp. (Ch. 12), Mother Ignacia Street, Queson City, Philippines
ABC Television, 77 West 66th street, NY 10023 New York, USA, **Tel:** +1-212-456 7777, **Fax:** +, **WWW:** http://www.abctv.com/
ABC World News, 77 West 6th. Street, 10023 New York, USA, **Tel:** +1-212-456 7777, **Fax:** +1-212-465 6384, **WWW:** http://www.abctv.com/
Action Pay-Per-View, 2425 West Olympic Boule-

ITU REGION 2 (USA)

vard, Suite 4050 W., CA 90404 Santa Monica, USA, **Tel:** +1-310-453 450

ACTS Satellite Network, 6350 West Freeway, TX 76150 Fort Worth, USA, **Tel:** +1-817-737 3241

AEI Music Network, 900 East Pine Street, WA 98122 Seattle, USA, **Tel:** +1-206-329 1400, **Fax:** +1-206-329 9952

AFRTS, 1016 North McCadden Place, 90038 Los Angeles, USA

AFRTS, 10888 La Tuna Canyon Road, CA 91362 Sun Valley, USA, **Tel:** +1-818-504 1200, **Fax:** +1-818-504 1234

Agridata Resources Inc., 330E Kilbourn Avenue, WI 53202 Milwaukee, USA, **Tel:** +1-414-273 0873

Agrisat, P.O. Box 314, WI 53510 Belmont, USA

All News Channel, 3415 University of Minneapolis, MN 55414 Minneapolis, USA, **Tel:** +1-612-642 4645, **Fax:** +, **WWW:** http://www.all-news.com/

Ambassador Inspirational Radio, 515 East Commonwealth, CA 92632 Fullerton, USA, **Tel:** +1-714-738 1501, **Fax:** +1-714-738 4625

AMCEE, 500 Technic Parkway NW, GA 30313 Atlanta, Georgia, USA, **Tel:** +1-404- 894 3362

America's Disability Channel/The Silent Network, 1777 North East Loop 410, Suite 1401, TX 78217 San Antonio, USA, **Tel:** +1-512-820 2680

American Law Network, 4025 Chestnut Street, PA 19104 Philadelphia, USA, **Tel:** +1-215-243 1600

American Movie Classics, 150 Crossways Park West, NY 11797 Woodbury, USA, **Tel:** +1-516-364 2222, **Fax:** +1-516-364 8924

AP News Plus, 1825 K Street, NW, DC 20006 Washington, USA

Arizona Sports Progamming Netw. (ASPN), 17602 N. Black Canyon Highway, Suite 1, AZ 85069 Phoenix, USA, **Tel:** +1-602-866 0072

Arts & Entertainment, 235 East 45th Street, NY 10017 New York, USA, **Tel:** +1-212-661 4500, **Fax:** +1-212-210 0640, **WWW:** http://www.aetv.com

Automotive Satellite TV Netw., 1303 Marsh Lane, TX 76205 Carrollton, USA, **Tel:** +1-214-417 4100, **Fax:** +1-214-416 8892

Bankers TV Netw., 1346 Oakbrook Drive, Suite 170, GA 30093-2229 Norcross, USA, **Tel:** +1-404-446 3900, **Fax:** +1-404-448 6600

Baptist Telecom Netw., 127 Ninth Avenue North, TN 37234 Nashville, USA, **Tel:** +1-615-251 2283

Biznet / US Chamber of Commerce, 1615 High Street N.W., DC 20062 Washington DC, USA, **Tel:** +1-202-463 5921, **Fax:** +1-202-463 5835

Black Entertainment TV, 1700 North Moore Street, Suite 22, VA 22209 Rosslyn, USA, **Tel:** +1-703-516 6436, **Fax:** +1-703-875 0441

Black Radio Netw. Inc., 166 Madison Avenue, NY 10016 New York, USA, **Tel:** +1-212-686 6850, **Fax:** +1-212-686 7308

Bravo, 150 Crossway Park West, NY 11797 Woodbury, USA, **Tel:** +1-516-364 2222, **Fax:** +1-516-364 2297

BrightStar, International Building, Rockefeller Center, 630 5th. Avenue, 22nd. floor, NY 10111 New York, USA, **Tel:** +1-212-582 8578

C-Span, 400 N. Capitol Street NW, 20001 Washington DC, USA, **Tel:** +1-202-737-3220, **Fax:** +1-202-737 3323, **WWW:** http://www.c-span.org

C-Span, 444 N. Capitol Street N.W., Suite 412, DC 20001 Washington DC, USA, **Tel:** +1-202-737 3220, **Fax:** +1-202-737 3323

Caribbean SuperStation (CSS), 333 Roosevelt Boulevard, FL 33021 Hollywood Hills, USA, **Tel:** +1-305-985 2641

CBN, CBN Center, VA 23463 Virginia Beach, USA, **Tel:** +1-804-424 3511

CBS News, 51 West 2nd. Street, 10019 New York, USA, **Tel:** +1-212-975 4321, **Fax:** +1-212-975 9130, **WWW:** http://www.cbs.com/

CBS Television, 51 West 52nd. Street, NY 10019 New York, USA, **Tel:** +1-212-975 4321, **Fax:** +, **WWW:** http://www.cbs.com/

Channel America, 19 West 21st. Street, Suite 204, NY 10010 New York, USA, **Tel:** +1-212-366 9880

Christian Broadcasting Netw. (CBN), 700 CBN Center, VA 23463 Virginia Beach, USA, **Tel:** +1-804-523 7307, **Fax:** +1-804-424 3005

Cinemax, HBO Building, 1100 Avenue of the Americas, NY 10036 New York, USA, **Tel:** +1-212-512 1000, **Fax:** +1-212-512 5517

CMT (Country Music TV), 2806 Opryland Drive, TN 37214-1209 South Nashville, USA, **Tel:** +1-615-255 8836

CNBC, 2200 Fletcher Avenue, NJ 07024 Fort Lee, USA, **Tel:** +1-201-585 6464, **Fax:** +, **WWW:** http://nbceurope.com/cnbc.html/

CNN, One CNN Center, Box 105366, GA 30303 Atlanta, USA, **Tel:** +1-404-827 1500, **Fax:** +1-404-827 1593, **WWW:** http://www.cnn.com/

Comedy Central, 1775 Broadway, 20th. floor, NY 10036 New York, USA, **Tel:** +1-212-767 8600

Conus Communications, 3415 University Avenue, MN 55114 St. Paul, USA, **Tel:** +1-612-642 4645

Corporate Satellite Television Netw., P.O. Box 550265, GA 30355-2765 Atlanta, USA, **Tel:** +1-404-364 2300, **Fax:** +1-404-364 2322

Court TV, 600 3rd. Avenue, NY 10016 New York, USA, **Tel:** +1-212-973 2868, **Fax:** +, **WWW:** http://www.courttv.com

Cycle Sat Inc., 119 John K. Hanson Drive, IA 50436 Forest City, USA, **Tel:** +1-515-582 6999, **Fax:** +1-515-582 6998

Deutsche Welle (Washington), 2800 South Shirlington Rd Suite 901, Arlington VA 22206-3601, USA, **Tel:** +1 (703) 9316644

ITU REGION 2 (USA)

Discovery Channel, 7700 Wisconsin Avenue, Suite 700, MD 20814-3578 Bethesda, USA, **Tel:** +1-301-577 1999, **Fax:** +1-301-577 0733, **WWW:** http://www.discovery.com/
Disney Television, 3800 West Alameda Avenue, CA 91505 Burbank, USA, **Tel:** +1-818-569 7700, **Fax:** +, **WWW:** http://www.disney.com/
Drive-In Cinema, 532 Broadway, 6th. floor, NY10012 New York, USA, **Tel:** +1-212-941 1434, **Fax:** +1-212-941 4746
E! Entertainment, 5670 Wilshire Boulevard, CA 90036-5679 Los Angeles, USA, **Tel:** +1-213-954 2400, **Fax:** +1-213-954 2662
ECO (Galavisiòn), 2121 Avenue of the Stars, Suite 2300, CA 90067 Los Angeles, USA, **Tel:** +1-213-286 0122, **Fax:** +1-213-286 0126
ESPN International, 605 3rd. Avenue, NY 10158 New York, USA, **Tel:** +1-212-916 9200, **Fax:** +1-212-916 9325, **WWW:** http://www.espenet.SportZone.com
Eternal Word TV Network, 5817 Old Leeds Road, AL 35210 Birmingham, USA, **Tel:** +1-205-956 9537
Family Channel, 1000 Centerville Turnpike, VA 23463 Virginia Beach, USA, **Tel:** +1-804-523 7301, **Fax:** +1-804-424 9223
Family Radio Network, 618 S. Sheridan Road, IA 51501 Shenandoah, USA, **Tel:** +1-712-246 5151
FamilyNet, 6350 West Freeway, TX 76150 Ft. Worth, USA
FilSat, 548 S. Kingsley Drive, CA 90020 Los Angeles, USA, **Tel:** +1-213-265 2929/4290, **Fax:** +1-818-753 8518
Financial News Netw., 2200 Fletcher Ave., NY 07024 Fort Lee, USA, **Tel:** +1-201-585 2622, **Fax:** +1-201-585 6205
Florida News Network, 1851 Southhampton, 32207 Jacksonville, USA, **Tel:** +1-904-393 9847
HBO / Cinemax, 1100 Avenue of the Americas, NY 10036 New York, USA, **Tel:** +1-212-512 1000, **Fax:** +1-212-512 5517, **WWW:** http://www.hbo.com/
Home Shopping Network, P.O. Box 9090, FL 34618 Clearwater, USA, **Tel:** +1-813-530 9455
Home Sports Entertainment (HTS), 600 E. Los Colinas Boulevard, Suite 2200, TX 75039 Irving, USA, **Tel:** +1-214-401 0099, **Fax:** +1-214-869 2999
Home Team Sports, 7700 Wisconsin Avenue, MD 20814 Bethesda, USA, **Tel:** +1-301-718 2300, **Fax:** +1-301-718 3300
In Store Satellite Netw., 7050 Union Park Circle, Suite 1050, UT 84047 Midvale, USA, **Tel:** +1-801-562 2252, **Fax:** +1-801-562 1773
JC Penney TV Network, 1845 Empire Avenue, CA 91504 Los Angeles, USA, **Tel:** +1-213-871 8170
Jewish Television Network (JTN), 9021 Melrose, Suite 309, CA 90069 Los Angeles, USA, **Tel:** +1-213-273 6841
Jukebox Network, 1200 Biscayne Boulevard, FL 33181-2742 Miami, USA, **Tel:** +1-305-899 3600
KBL Sports, 1301 Grandview Avenue, Suite 300, PA 15211 Pittsburg, USA, **Tel:** +1-412-381 9500, **Fax:** +1-412-381 9528
KCNC, 1044 Lincoln Street, CO 80203 Denver, USA, **Tel:** +1-303-861 4444
KMGH, 123 Speer Boulevard, CO 80203 Denver, USA, **Tel:** +1-303-832 7777
KRMA, 1261 Glenarm Place, CO 80204 Denver, USA, **Tel:** +1-303-892 6666
KTLA, P.O. Box 500, CA 90078 Los Angeles, USA, **Tel:** +1-213-460 5500
KTVT, P.O. Box 2495, TX 76103 Fort Worth, USA
KUBD, 9805 East Cliff Avenue, CO 80231 Denver, USA
KUSA, 1089 Bannock Street, CO 80204 Denver, USA, **Tel:** +1-303-893 9000
KWGN, P.O. Box 5222, CO 80155 Denver, USA, **Tel:** +1-303-740 2844
Liberty Broadcasting Network, 1150 West King Street, FL 32922 Cocoa, USA, **Tel:** +1-305-632 1000
Lifetime Television, Worldwide PLaza, 309 West 49th. Street, NY 10019 New York, USA, **Tel:** +1-212-424 7000
Midwest Sports Channel, 90 South 11th. Street, MN 55403 Minneapolis, USA, **Tel:** +1-612-330 2575
Mind Extension University, 9697 East Mineral Avenue, CO 80112 Englewood, USA, **Tel:** +1-303-792 311
Modern Satellite Network, 5000 Park Street North, FL 33709 St. Petersburg, USA, **Tel:** +1-813-541 7571
Moody Broadcasting Network, 820 North LaSalle Drive, IL 60610 Chicago, USA, **Tel:** +1-312-329 4433, **Fax:** +1-312-329 4468
MovieTime, 5670 Wilshire Boulevard, CA 90036-5679 Los Angeles, USA, **Tel:** +1-213-461 5511
MSGN (Madison Square Garden Network), 2 Pennsylvania PLaza, NY 10121-0091 New York, USA, **Tel:** +1-212-465 6000, **Fax:** +1-212-465 6021
MTV, 1515 Broadway, NY 10036 New York, USA, **Tel:** +1-212-258 8000, **Fax:** +, **WWW:** http://www.mtv.com/
NASA Select Channel, Kennedy Space Center, FL 32899 Cape Canaveral, USA, **Tel:** +1-407-867 2468
National Black Network, 463 7th. Avenue, NY 10018 New York, USA, **Tel:** +1-212-714 1000, **Fax:** +1-212-714 1563
National Satellite Programming Network, 1909 Avenue G., P.O. Box 1489, TX 77471 Rosenberg, USA, **Tel:** +1-713-342 9655, **Fax:** +1-713-342 7016
NBC, 30 Rockefeller Plaza, NY 10112 New York, USA, **Tel:** +1 (212) 644 7572, **Fax:** +1 (212) 315

ITU REGION 2 (USA)

4037, WWW: http://www.nbc.com/
NESN (New England Sports Network), 70 Brooklyn Avenue, MA 02215 Boston, USA, **Tel:** +1-617-536 9233
Newspaper Satellite Network, 18333 North Preston Road, LB6, TX 75232 Dallas, USA, **Tel:** +1-214-931 9858, **Fax:** +1-214-931 0069
NHK, Japan Network Group, Inc, 1325 Avenue of the Americas, 8th floor, NY 10019 New York, USA, **Tel:** +1-212-262 3377, **Fax:** +1-212-262 5577, **WWW:** http://www.nrk.org.jp/
Nick at Nite, 1515 Broadway, 21st. floor, NY 10036 New York, USA, **Tel:** +1-212-258 8000, **Fax:** +1-212-258 7938, **WWW:** http://nick-at.nite.com/
Nickelodeon, 1515 Broadway, 21st. floor, NY 10036 New York, USA, **Tel:** +1-212-258 8000, **Fax:** +1-212-258 7938, **WWW:** http://www.nickelodeon.com/
PBS (Public Broadcasting Service), 1320 Braddock PLace, VA 22206 Alexandria, USA, **Tel:** +1-703- 739 5000, **Fax:** +1-703-739 8938/0775, **WWW:** http://www.pbs.org/
Phoenix Communications Group-Sports News Satellite, 3 Empire Boulevard, NY 07606-180 South Hackensack, USA, **Tel:** +1-212-921 8100, **Fax:** +1-212-719 0614
Playboy Channel, 8560 Sunset Boulevard, CA 90069 Los Angeles, USA, **Tel:** +1-213-659 4080, **WWW:** http://www.playboy.com/pb-tv/pb-tv.html
Prevue Inc., 3801 South Sheridan Ave., OK 74136 Tulsa, USA, **Tel:** +1-918-664 5566, **Fax:** +1-918-663 6228
Prime Network, 660 East Los Colinas, Suite 2200, TX 75039 Irving, USA, **Tel:** +1-214-401 0099, **Fax:** +1-214-869 2999
Prime Sports Network, 44 Cook Street, Suite 600, CO 80206 Denver, USA, **Tel:** +1-303-355 7777
Primetime 24, 342 Madison Avenue, Suite 1520, NY 10173 New York, USA, **Tel:** +1-212-599 4440, **Fax:** +1-212-599 2402
Pro Am Sports System, P.O. Box 3812, MI 48106-3812 Ann Arbor, USA, **Tel:** +1-313-930 7277, **Fax:** +1-313-930 0723
QVC Network, 8101 East Prentice, Suite 610, CO 80111 Englewood, USA, **Tel:** +1-303-721 7718, **Fax:** +1-303-721 7058, **WWW:** http://www.qvc.com/
RAI, 1350 Avenue of the Americas, NY 10019 New York, USA, **Tel:** +1-212-975 0200, **WWW:** http://www.rai.it
Rainbow Programming, 150 Crossways Park West, NY 11797 Woodbury, USA, **Tel:** +1-516-364 2222
Reuters News View, 1700 Broadway, NY 10019 New York, USA, **Tel:** +1-212-603 3575
Satellite Music Network, 12655 North Central Expressway, Suite 600, TX 75243 Dallas, USA, **Tel:** +1-214-991 9200, **Fax:** +1-214 991 1071

Satellite Sports Network, 2080 N. 360, Suite 200, TX 75050 Grand Prairie, USA, **Tel:** +1-214-988 1088
Seeburg Satellite Systems, 3717 National Drive, Suite103, NC 27612-4877 Raleigh, USA, **Tel:** +1-919-851 5823
Shop at Home, Highway 32, Village Square Center, TN 37821 Newport, USA
Showtime, 1633 Broadway, NY 10019 New York, USA, **Tel:** +1-212-708 1600
Skylight Satellite Network, 3003 Snelling Avenue, North, MN 55113 St. Paul, USA, **Tel:** +1-612-631 5000
Spice Pay-Per-View, 532 Broadway, 6th. floor, NY10012 New York, USA, **Tel:** +1-212-9411434, **Fax:** +1-212-941 4746
Sports Channel, 600 East Market Street, Suite102, TX 78205 San Antonio, USA, **Tel:** +1-512-224 2067
Sports Channel Chicago, 820 W. Madison, IL 60302 Oak Park, USA, **Tel:** +1-708-524 9444
Sports Channel L.A., 1545 26th. Street, CA 90404 Santa Monica, USA, **Tel:** +1-310-453 1985
Sports Channel New England, 10 Tower Office Park, Suite 600, MA 01801 Woburn, USA, **Tel:** +1-617-933 9300, **Fax:** +1-617-933 3877
Sports Channel New York, 150 Crossways Park W., NY 11797 Woodbury, USA, **Tel:** +1-516-364 2222
Sports Channel Ohio, 6500 Rockside Road, Suite 340, OH 44131 Independence, USA, **Tel:** +1-216-328 0333
Sports Channel Philadelphia, 225 City Line Avenue, Bala Cynwyk Plaza, PA 19004 Philadelphia, USA, **Tel:** +1-215-668 2210, **Fax:** +1-215-668 9499
Sports News Satellite, 250 Harbor Drive, Box 10210, CT 06904 Stamford, USA, **Tel:** +1-203-965 6789
SportsChannel America, 3 Crossways Park West, NY 11797 Woodbury, USA, **Tel:** +1-516-921 3764
SportsChannel Florida, 2295 Corporate Boulevard N.W., Suite 140, FL 33431 Boca Raton, USA, **Tel:** +1-407-994 0250
Starion Entertainment, 11766 Wilshire Boulevard, Suite 710, CA 90025-6538 Los Angeles, USA, **Tel:** +1-213-393 3746
Sunshine Network, 390 North Orange Avenue, Suite 1075, FL 32801 Orlando, USA, **Tel:** +1-407- 648 1150, **Fax:** +1-407-648 1679
TBS, P.O. Box 105 366, GA 30348 Atlanta, USA, **Tel:** +1-404-827 3013, **WWW:** http://www.turner.com/
Telemundo Productions Inc., 2470 West 8th. Avenue, FL 33010 Hialeah, USA, **Tel:** +1-305-883 7951, **Fax:** +1-305-884 3542
Tempo Sound, 4643 South Ulster Street, Suite 400, CO 80237-2864 Denver, USA, **Tel:** +1-918-495 3200

ITU REGION 2/3 (USA-JAPAN)

The Comedy Channel, 120 East 23rd Street, NY 10010 New York, USA, **Tel**: +1-212-512 8900
The Family Channel, 1000 Centerville Turnpike, VA 23463 Virginia Beach, USA, **Tel**: +1-804-523 7301, **Fax**: +1-804-424 9223
The Fox Television Network, 205 East 67th. Street, NY 10021 New York, USA, **Tel**: +1-212-452 5555
The Inspirational Network, P.O. Box 410 325, NC 28421 Charlotte, USA, **Tel**: +1-213-826 2429
The Movie Channel, 1633 Broadway, NY 10019 New York, USA, **Tel**: +1-212-708 1600
The Weather Channel, 2840 MT. Wilkinson Parkway, Suite 200, GA 30339 Atlanta, USA, **Tel**: +1-404-433 6800, **Fax**: +1-404-433 5130
TLC (The Learning Channel), 7700 Wisconsin Ave, MD 20814 Bethesda, USA, **Tel**: +1-301-986 1999, **Fax**: +1-301-986 4826
TNT Communications, 575 Madison Avenue, NY 10022 New York, USA, **Tel**: +1-212-644 0200
Turner Network Television, One CNN Center, GA 30348 New York, USA, **Tel**: +1-404-827 1136
Univision, 605 3rd. Avenue, 12th. floor, NY 10158-0180 New York, USA, **Tel**: +1-212-455 5200
USA Network, 1230 Avenue of the Americas, NY 10020 New York, USA, **Tel**: +1-212-408 9100, **Fax**: +1-212-408 2750
VH-1 (Video Hits 1), 1515 Broadway, 22nd. floor, NY 10036 New York, USA, **Tel**: +1-212-258 7800
Viewer's Choice Pay-Per-View, 909 3rd. Avenue, 21st. floor, NY 10022 New York, USA, **Tel**: +1-212-486 6600, **Fax**: +1-212-688 9497
Virginia Satellite Educational Network (VSEN), P.O. Box 2120, VA 23216 Richmond, USA, **Tel**: +1-804-225 2833, **Fax**: +1-804-371 2456
VOA, US Information Agency, 20547 Washington DC, USA
WorldNet (US Information Agency), 601 D Street. N.W., Rm. 5000, DC 20547 Washington DC, USA, **Tel**: +1-202-501 7806, **Fax**: +1-202-501 664
WorldNet USIA, Patrick Henry Building, 601 D Street NW, 20547 Washington DC, USA, **Tel**: +1-202-501 7806, **Fax**: +1-202-501 6664
Worldwide Television News (WTN), 1995 Broadway, NY 10023 New York, USA, **Tel**: +1-212-362 4440, **Fax**: +1-212-496 1269
Worldwide Television News (WTN), 1995, NY 10023 New York, USA, **Tel**: +1-212-362 4440, **Fax**: +1-212-496 1269

13.3. ITU REGION 3

Afghanistan TV, Ansari Watt, Kabul, Afghanistan
Peoples Radio & TV Afghanistan, P.O. Box 544, Kabul, Afghanistan, **Tel**: +93-25241
Australian Broadcasting Corporation (ABC), Ultimore Centre, 700 Harris Street, NSW 2007 Ultimo, Austalia, **Tel**: +61-2-333 1500, **Fax**: +61-2-333 5305
Imparja Television Pty Ltd, 14 Leichhardt Terrace, NT 0871 Alice Springs, Australia, **Tel**: +61-89-530 300, **Fax**: +61-89-530 322
Network Ten Australia, P.O. Box 10, NSW 2066 Lane Cove, Australia, **Tel**: +61-2-887 0222
The Nine Network, P.O. Box 27, NSW 2068 Willoughby, Australia, **Tel**: +61-2-430 0444, **Fax**: +61-2-436 9193
The Seven Network, Television Centre, Mobbs Lane, NSW 2121 Epping, Australia, **Tel**: +61-2-858 7777, **Fax**: +61-2-858 7888
Azerbaidzhani Television, Mehdi Huseyn küçäsi 1, 370000 Baki, Azerbaidzhan, **Tel**: +7-8922-398 585, **Fax**: +7-8922-395 452
Bangladesh Television, Rampura, 17 Dhaka, Bangladesh, **Tel**: +880-2-400139/9
Cambodian Television, 19th. Street 242, Chaktomuk, Daun Penh, Phnom Penh, Cambodia, **Tel**: +855-2-2983-223 49/4449-241 49
China Central Television (CCTV), 11 Fuxinglu Road, 100859 Beijing, China, **Tel**: +86-1-801 1144, **Fax**: +86-1-801 1149
Star TV (Satellite Televisiion Asian Region Ltd.), 8th Floor, One Harbourfront, 18 Tak Fung Street, Hong Kong, **Tel**: +852-2621 8888, **Fax**: +852-2621 8000, **Contact**: Mr. Don Atyeo (GM), **WWW**: http://www.startv.com
Asianet Communications Ltd., 1-C, Apex Plaza, 3, Nungambakkam High Road, 600034 Madras, India, **Tel**: +91-44-825 1513/826 0731, **Fax**: +91-44-827 2604
Doordarshan India, Directorate General of Doordarshan, Mandi House, Copernicus Marg., 110 001 New Delhi, India, **Tel**: +91-11-3820 94/99, **Fax**: +91-11-386 507
Jalan Kelapa Puan Timur TV, NB5, 4 Kelapa Gading Permai, Jakarta, Indonesia, **Tel**: +61-21-450 1208, **Fax**: +61-21-450 1208
PT Indonesia Satellite Corp. (PT Indosat), Jalan Medan Merdeka Barat No. 21, P.O. Box 2905, Jakarta 10110, Indonesia, **Tel**: +62 (21) 380 2614, **Fax**: +62 (21) 3450155/3809633
Televisi Republik Indonesia (TVRI), J1 Jerbang Pemuda-Senayan, Jakarta, Indonesia, **Tel**: +62-21-5733135/2279, **Fax**: +62-21-5732408/3122
Israel Broadcasting Authority (IBA), P.O. Box 28080, 91280 Jerusalem, Israel, **Tel**: +972-2-252 905, **Fax**: +972-2-257 034, **Contact**: Mr. Mordechai Krishenbaum (DG)
BBC World Service TV (subscriptions), Satellite News Corp., JBP Oval Building,, 52-2 Jingumae 5-chome Shibuya-ku, Tokyo, Japan, **Tel**: +81-3 -3406 9777, **WWW**: http://www.bbcnc.org.uk/
Japan Satellite Broadcasting Inc. (JSB), No. 6 Central Building, 1-19-10 Toranomon, Mina-

ITU REGION 3 (JAPAN-THAILAND)

to-ku, Tokyo, Japan
MTV Japan, 5F Shirokanedai Crest Building, 4-2-11, Shirokanedai, Minato-ku, 108 Tokyo, Japan, **Tel:** +81-3-5448 1103, **Fax:** +81-3-5448 9131
NHK Tokyo, 2-2-1 Jinnan, Shibuya-ku, 150-01 Tokyo, Japan, **Tel:** +81-3-3465 5813, **Fax:** +81-3-3465 5814, **WWW:** http://www.nrk.org.jp/
Malaysian Television System Berhad (TV3), 7-9th. floor, KUB Building, No. 1 Lorong Kapar, P.O. Box 11124, 50736 Kuala Lumpur, Malaysia
Radio Television Malaysia (RTM), Department of Broadcasting, Angkasapuri, 50614 Kuala Lumpur, Malaysia, **Tel:** +60-3-2745 333, **Fax:** +60-3-2744 290
Mongol Televiz, Mongolian TV Centre, Hasbataar Street, Ulaanbataar, Mongolia

TV Myanmar, G P.O. Box 1432, Yangon, Myanmar, **Tel:** +95-1-31 355
Television New Zealand, P.O. Box 3819, Auckland, New Zealand, **Tel:** +64-9-770 630, **Fax:** +64-9-750 979
Pakistan Television Corp. (PTV), P.O. Box 1221, Constitution Avenue, Islamabad, Pakistan
EMTV, Media-Niugini Pty. Ltd., P.O. Box 443, Boroko, Papua New Guinea, **Tel:** +675-257 322, **Fax:** +675-254 450
HBO Asia, 151 Lorong Chuan, 4th floor #04-B/5, Singapore, Singapore, **Tel:** +65-288-6303, **Fax:** +65-287-2210
Bangkok Entertainment Co. Ltd., 1126/1 New Petchbury Road, 10400 Bangkok, Thailand, **Tel:** +66-2-253 9970-3, **Fax:** +66-2-253 9978
Television of Thailand, Ratchadamoen Road, 10200 Bangkok, Thailand, **Tel:** +66-2-222 8821

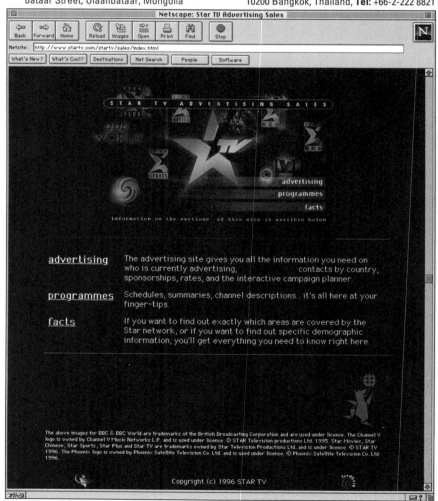

CHAPTER 14
USEFUL WORLD WIDE WEB ADDRESSES

URL	Description	Country
http://AHMADIYYA.ORG	Muslim TV Ahmadiyya	UK
http://area51.upsu.plym.ac.uk/~sat/	Robin Clark's satellite web page	UK
http://area51.upsu.plym.ac.uk/~sat/telesatellite/	Telesatellite French satellite magazine	France
http://chaparral.net/	Chaparral	Netherlands
http://europe2.fr/	Europe2	France
http://globalshopping.com	Global Shopping Network	USA
http://home.sprynet.com/sprynet/gapsat/	GAP Mobile Broadcasting Satellite Services	USA
http://homepage.midusa.net/~litz/index.html	Brian Litzenberger's web page	USA
http://l-channel.com	International Channel web page	USA
http://i-star.com/dmorgandb/index.htm	Dan Morgan's SWN web page	USA
http://irdeto.nethold.nl	Irdeto - maker of Multichoice's Conditional Access Module	Netherlands
http://ireland.iol.ie/~megatek/	Megatek	Ireland
http://itre.ncsu.edu/misc/sat.html	Jay Novello's WWW Site	USA
http://itre.ncsu.edu/radio/	SW & radio catalog of hypertext links	USA
http://libertysports.com	Liberty Sports web page	USA
http://linx.tvp.com.pl/welcome.htm	TVpolonia	Poland
http://members.aol.com/cbsfield/index.html	CBS TV Field Operatons	USA
http://net.foref.org/	NET political TV	USA
http://nick-at.nite.com/	Nick at Nite	USA
http://nitro.nettel.com/net-1/	N1	USA
http://nssdc.gsfc.nasa.gov/spacewarn/spacewarn.html	Spacewarn - informative newsletter	USA
http://ourworld.compuserve.com/homepages/S_fridenberg1/	British Entertainment Network	UK
http://pathfinder.com/Aahs/aahome.html	Radio AAHS	USA
http://tripled.com	Triple D Publishing - publishes ONSAT	USA
http://ttn.nai.net/	The Talk Network	Canada
http://web.idirect.com/~klg/sat.html	K.L.G. Microcomputer Software's Links to Satellite Sites	USA
http://wtn.com/wtn/wtn.html	Washington Telecomm Newswire	USA
http://www.skyradio.nl	SKY-Radio	Netherlands
http://www.3sat.com/	3sat	Germany
http://www.abctv.com/	ABC	USA
http://www.abweb.com/sat/index.html	AB Sat	France
http://www.aetv.com	A&E Cable Network	USA
http://www.aini.com	American Independent Network	USA
http://www.allnews.com/	All News Channel/Conus Communications	USA
http://www.americaone.com/	America One Television Network	USA
http://www.ameritalk.com/	America's Talking	USA

275

USEFUL WORLD WIDE WEB ADDRESSES

URL	Description	Country
http://www.ard.de	Arbeitsgemeinschaft der Oeffentlichrechtlichen Rundfunkanstalten der Bundesrepublik Deutschland	Germany
http://www.ari.net/satellite.html	Satellite links	USA
http://www.arte-tv.com/	arte	Germany
http://www.asb.com/usr/keith/	Keith Knipschild's home page	USA
http://www.asiaonline.net/spacerep/	Asia'Pacific Space Report	
http://www.asn.ca	Atlantic Satellite Network	
http://www.astra.lu	SES/ASTRA	Luxembourg
http://www.att.com/skynet/	AT&T Skynet Services web page	USA
http://www.baton.com/	Baton Broadcasting Inc	Canada
http://www.bbcnc.org.uk/	BBC Media home page	UK
http://www.bbcnc.org.uk/worldservice/	BBC World Service	UK
http://www.bdt.com/home/capjcruz/pinoytv.html	Filipino Sat TV Online (The Filipino Channel)	Philippines
http://www.bertelsmann.de/	BMG Bertelsmann	Germany
http://www.bess.tcd.ie/ireland/rte.htm	RTE Eire	Ireland
http://www.betnetworks.com	Black Entertainment TV	USA
http://www.betnetworks.com/actionppv/index.html	Action PPV	USA
http://www.betnetworks.com/jazz/index.html	BET on Jazz	USA
http://www.bnn.nl/	BNN - Bart Nieuws Network	Netherlands
http://www.booktv.com	BookTV - a satellite shopping channel for books	USA
http://www.br-online.de/	Bayrischer Rundfunk	Germany
http://www.broadcast.net/	Broadcast Net	USA
http://www.brtn.be	BRTN - the public channel in Dutch	Belgium
http://www.brtn.be/	BRTN	Belgium
http://www.c-span.org	C-Span	USA
http://www.cable-online.com/	Cable Online	USA
http://www.cablelabs.com	Cable Television Laboratories Inc.	USA
http://www.cameo.nl/sbs6/default.html	SBS-6	Netherlands
http://www.capitolnet.com/	Capitol Radio Networks web page	UK
http://www.capitolnet.com/ncnn/	North Carolina News Network	USA
http://www.casema.nl/	CASEMA	Netherlands
http://www.cbc.ca/	CBC Canada	Canada
http://www.cbs.com/	CBS	USA
http://www.channel-onetv.co.uk/	Channel One	UK
http://www.chaparral.net	Chaparral Communications	USA
http://www.chch.com/	CHCH-TV home page	USA
http://www.cityscape.co.uk/channel4/	Channel 4	UK
http://www.ctmulti.com/	CLT MultiMedia	Luxembourg
http://www.cnn.com/	CNN Online	USA
http://www.courttv.com/	CourtTV	USA
http://www.cplus.be	Canal+ Belgique - French language pay-tv	Belgium
http://www.cplus.fr/	CANAL+	France
http://www.crs4.it/~luigi/MPEG/mpegfaq.html	MPEG information	Italy
http://www.crtc.gc.ca	CRTC Gopher site	USA
http://www.ctvnet.com	Computer TV Network	USA
http://www.cyberspice.com	Spice PPV	USA
http://www.df1.de/	DF-1 Das digitale Fernsehen	Germany

276

USEFUL WORLD WIDE WEB ADDRESSES

URL	Description	Country
http://www.dfn.de/	Deutschlandfunk	Germany
http://www.dfn.de/d-radio	DeutschlandRadio Berlin	Germany
http://www.dfn.de/presse/ndr/home.html	Norddeutscher Rundfunk	Germany
http://www.direcpc.com	DirecPC	USA
http://www.directv.com/	DirecTV	USA
http://www.discovery.ca	Discovery Canada schedule	Canada
http://www.disney.com/	Discovery Channel	USA
http://www.disneychannel.co.uk/	Disney Channel	USA
http://www.dmxmusic.com/	Digital Music Express DMX	UK
http://www.dmxmusic.com/	DMX official web page	USA
http://www.dot.gov/dotinfo/faa/cst/bulletin.html	Commercial Space Launch Bulletin	USA
http://www.dr.dk/	DR- Danmarks Radio	Denmark
http://www.drgenescott.org/	Dr. Gene Scott web page	USA
http://www.dtag.de/	Deutsche Telekom AG	Germany
http://www.dw.gmde.de/	DeutscheWelle	Germany
http://www.ebn.co.com/	EBN	UK
http://www.echostar.com/	Echostar	Netherlands
http://www.ee.ac.uk/Contrib/Entertainment/nickelodeon.html	Nickelodeon	UK
http://www.eeng.dcu.ie/~dtobin/home2.htm	Derek Tobin's page	
http://www.endo.nl/	CasTel	Netherlands
http://www.espenet.SportZone.com/	ESPN	USA
http://www.esrin.esa.it/	European Space Agency	Italy
http://www.esrin.esa.it/htdocs/esa/ariane/	Ariane 5 launcher web page	Italy
http://www.eurosat.com	Eurosat - the European sat-hack page	UK
http://www.eurotv.com	Euro TV - Belgium based European TV-guide	Belgium
http://www.eutelsat.de/	Eutelsat	Germany
http://www.eutelsat.org	Eutelsat's web pages	France
http://www.everyday.de/MTG	Kinevik	Denmark
http://www.fcc.gov	FCC WWW Server	USA
http://www.filmnet.se/	FilmNet	Sweden
http://www.flash.net/~nmtp/	New Mexico Teleport	USA
http://www.france2.fr/	France 2	France
http://www.france3.fr/	France 3	France
http://www.fritz.de/	Radio fritz	Germany
http://www.funet.fi/index/esi	Henrik Berle's European Satellite TV web Site	
http://www.ge.com/capital/americom/	GE Americom web site	USA
http://www.ge.com/capital/spacenet/	GE Spacenet web site	USA
http://www.gi.com	General Instruments	USA
http://www.gblaccess.com	Global Access Telecommunication Services	USA
http://www.gpl.net/paulmax/index.html	Paul-Maxwell King	UK
http://www.granada.co.uk/	Granada Sky Broadcasting	UK
http://www.grove.net	Bob Grove's Web Site	US
http://www.grundig.com/	Grundig	Netherlands
http://www.hackwatch.com/	Hackwatch	UK
http://www.hbo.com/	Home Box Office	USA

USEFUL WORLD WIDE WEB ADDRESSES

URL	Description	Country
http://www.hcisat.com	Hughes Communications Inc.	USA
http://www.historychannel.com	The History Channel	USA
http://www.hitsathome.com	Request TV	USA
http://www.homeboxoffice.com/	HomeBoxOffice	USA
http://www.hq.nasa.gov/office/pao/ntv.html	NASA TV	USA
http://www.hr-online.gmd.de/	Hessischer Rundfunk	Germany
http://www.hrt.com.hr/	HRT Croatia	Croatia
http://www.htssports.com	Home Team Sports regional sports network	USA
http://www.hughespace.com/	Hughes Space and Communications web page	USA
http://www.infosat-lu/	infosat	Luxembourg
http://www.inmarsat.org/inmarsat/	Inmarsat WWW http://www.mit.edu:8001/activities/wmbr/otherstations.html Radio Page - links	USA
http://www.innovision1.com/wf/	H. Steele Price IV's Wildfeed Submission Form for the Internet Wild Feeds List	
http://www.inrete.it/sat/eurotica.html	EuroTica	Italy
http://www.intelfax.co.uk/	Intelfax	UK
http://www.intelsat.int	Intelsat	USA
http://www.ionet.net/~michaelr/	WOKIE Network web page	
http://www.isso.org/publications/LaunchReport/LaunchReport.html	Keith Stein's "The Community Air & Space Report"	
http://www.itu.ch	ITU	Switzerland
http://www.itv.ca/	ITV now digital	Canada
http://www.kabel1.de/	Kabel-1	Germany
http://www.kanal5.se/	Kanal 5	Sweden
http://www.ket.org/	Kentucky Educational TV	USA
http://www.keystonecom.com	Keystone Communications WWW page	Netherlands
http://www.kinkfm.com/	KinkFM	Germany
http://www.KirchGruppe.de/	Kirch Gruppe	USA
http://www.klon.org/~jazzave	KLON-FM jazz	USA
http://www.knbc4la.com/home.html	KNBC-TV Los Angeles	USA
http://www.komotv.com	KOMO-TV Seattle - WA	USA
http://www.kpix.com	KPIX-TV San Francisco - CA	France
http://www.lacinquieme.fr/	La Cinquième	USA
http://www.leitch.com	LEITCH	UK
http://www.livetv.co.uk/	Live-TV	USA
http://www.lockheed.com/	Lockheed Martin	Sri Lanka
http://www.lsi.usp.br/~rbianchi/clarke/ACC.Homepage.html	Arthur C. Clarke unofficial homepage	France
http://www.m6.fr/	M6	Israel
http://www.mandy.com/rrc001.html	RR Communication Israel	Germany
http://www.mdr.de/	Mitteldeutscher Rundfunk	USA
http://www.medialinkvideonews.com/index.html	Medialink Video Press Release	USA
http://www.msnbc.com	MSNBC web site	USA
http://www.msnbc.com/	MSNBC	USA
http://www.mtv.com/	MTV Online	USA
http://www.mtv.com	MTV Networks	USA
http://www.muchmusic.com/bravo.html	Bravo Canada	Canada
http://www.muchmusic.com/muchmusic.html	MuchMusic	USA
http://www.multichannel.com/	Multichannel News	
http://www.multitasking.com/ana/	Arab Network of America web page	USA

USEFUL WORLD WIDE WEB ADDRESSES

URL	Name	Country
http://www.nbc.com/	NBC	USA
http://www.nbc4ny.com	WNBC-TV New York	USA
http://www.nbceurope.com/	NBC	UK
http://www.nbctonightshow.com	NBC Tonight Show	USA
http://www.nesn.com	New England Sports Network	USA
http://www.netbeat.com/mcm	MCM Euromusique	France
http://www.nethold.nl/	Nethold	Netherlands
http://www.netservers.com/~dandrew/sat_tv.html	Dan Andrew's Satellite TV and TVRO Info	USA
http://www.netten.net/adults/secrets.html	Secrets	USA
http://www.newstalk.com	NewsTalk TV	USA
http://www.nickelodeon.com/	Nickelodeon	USA
http://www.nmis.org/NewsInteractive/CNN/Newsroom/contentts.html	CNN Newsroom	USA
http://www.nokia.com/	Nokia	Netherlands
http://www.norsat.com/norsat/	Norsat International page	Norway
http://www.nos.nl/	De Nederlande publieke omroepen	Netherlands
http://www.nrk.no/	nrk- Norks rikskringkasting	Norway
http://www.nrk.org.jp/	NHK Nippon	Japan
http://www.ntt.jp/japan/NHK/TV/	NHK International	Japan
http://www.ntv.newcomm.net/	CJON-TV Newfoundland TV	Canada
http://www.ola.bc.ca/	Knowledge Network/Open Learning Agency	USA
http://www.omen.com.au/~pmerrett	TVRO info for Perth	Australia
http://www.orbit.net/	Orbit Networks	USA
http://www.orf.at	ORF	Germany
http://www.orn.com	SPECTRUM radio show page	USA
http://www.outdoor-channel.com	Outdoor Channel	USA
http://www.ozfm.newcomm.net/	CHOZ-FM	Canada
http://www.panamsat.com	PanAmSat	USA
http://www.paramount.com/	Paramount TV	USA
http://www.pbs.org/	PBS	USA
http://www.philips.com/	Philips Consumer Eletronics	Netherlands
http://www.ping.at/users/staytuned/program.html	Test slides and other neat stuff	USA
http://www.playboy.com/pb-tv/pb-tv.html	Playboy Channel	USA
http://www.ppv.com.	Viewers Choice	USA
http://www.premiere.de/	premiere	Germany
http://www.primetime24.com	PrimeTime 24	USA
http://www.pro-sieben.de/	Pro Sieben AG	Germany
http://www.qvc.com	QVC	USA
http://www.qvc.com	QVC	USA
http://www.radio-france.fr/	Radio France Int.	France
http://www.radiobremen.de/	Radio Bremen	Germany
http://www.rai.it/	RAI Italia	Italy
http://www.rldrake.com/	R.L. Drake Company radios and satellite receivers	USA
http://www.rnw.nl/	Radio Nederland Wereldomroep	Netherlands
http://www.rogers.com	Rogers Communications Inc	USA
http://www.rtbf.be/	RTBF	Belgium
http://www.rtl-westlive.de/	WestLive	Germany

USEFUL WORLD WIDE WEB ADDRESSES

URL	Description	Country
http://www.RTL.de/	RTL-Television	Germany
http://www.rtl.lu/	RTL Television Letzebuerg	Luxembourg
http://www.rtl2.de/	RTL2	Germany
http://www.rtlradio.com/	RTL Radio	Germany
http://www.sat-city.com	Sat City - a couple of other links	USA
http://www.sat-net.com/	Sat-Net various satellite tv info	UK
http://www.sat-net.com/drdish	Dr.Dish	Netherlands
http://www.sat-net.com/srtol/satellite.html	Satellite Channels Lists	UK
http://www.sat1.de/	German private satellite network	Germany
http://www.sat1.de/	Sat1	Germany
http://www.satcity.com/	SatCity	UK
http://www.satcodx.com	SATCO-DX - Worldwide satellite frequency list	UK
http://www.satguide.com	Satellite Entertainment Guide/DirectGuide	USA
http://www.sbc-online.de/	SBC online	Germany
http://www.sciatl.com/	Scientific Atlanta Inc web page	USA
http://www.scifi.com/	SciFi Channel	USA
http://www.screen.com/CPAC	Canadian Parliamentary Access Channel (CPAC)	Canada
http://www.scsn.net/~serc/	SERC educational TV	USA
http://www.sdr.de/	S_deutscher Rundfunk	Germany
http://www.searcher.com/links-tv.html	INFOSEARCH Broadcasting Links	USA
http://www.showtimeonline.com	Showtime Online	USA
http://www.siemens.de/	Siemens AG	Germany
http://www.sky.co.uk/	British Sky Broadcasting	UK
http://www.skylink.it/rete8/frame.htm	Reteotto Network	Italy
http://www.skyscottish.co.uk/	Sky Scottish	UK
http://www.smw.se/	Swedish Microwave	Sweden
http://www.spacecom.com	Spacecom Systems	USA
http://www.sr-online.de/	Saarl%ndischer Rundfunk	Germany
http://www.src.ca/	Radio Canada French Services	Canada
http://www.src.ca/	SRC	Canada
http://www.starsight.com/	Starsight Telecast Inc. (interactive programing info)	USA
http://www.sundancechannel.com/	Sundance Channel	USA
http://www.sundancefilm.com/	Sundance	USA
http://www.svt.se/	SVT- Svensk Television	Sweden
http://www.swf3.de/	SWF3 Radio	Germany
http://www.swmicronet.com	Southwest MicroNet/Northeast MicroNet web page	USA
http://www.tais.com	Toshiba	USA
http://www.tcc.co.uk/	Talk America Radio Network	UK
http://www.tcel.com/~save_on/satinks.html	TCC	USA
http://www.tcel.com/~save_on/satinks.html	Save-On Satellite Service	USA
http://www.tele-satellit.com/	Tele-Satellit - Europe's largest satellite magazine	Germany
http://www.teleclub.ch/	TeleClub	Switzerland
http://www.telenor.no/	TeleNor	Norway
http://www.teleportmn.com/	Teleport Minnesota web page	USA
http://www.telesat.ca	Telesat Canada web page	Canada
http://www.televar.com/~calhoun/satellite.html	Dave Calhoun's Satellite/Ham Radio links	

280

USEFUL WORLD WIDE WEB ADDRESSES

URL	Description	Country
http://www.terrapublishing.com/tr/	The Transponder Electronic Retail Industry Magazine	
http://www.tf1.fr/	TF1	France
http://www.tmf.nl/	TMF9	Netherlands
http://www.travelchannel.com/	Travel Channel	USA
http://www.travelchannel.com/traveluk/uk.html	Travel Channel Europe	UK
http://www.travelchannel.com/tv/tv.htm	Travel Channel	USA
http://www.tsn.ca	The Sports Network	Canada
http://www.tufts.edu/~cswineha	Curt Swinehart's home page	USA
http://www.turner.com	Turner Broadcasting Co.	USA
http://www.turner.com/	Turner Broadcasting	USA
http://www.turner.com/tcm	Turner Classic Movies	USA
http://www.tv.warnerbros.com/	Warner Brothers TV Network	USA
http://www.tv10.nl	TV10	Netherlands
http://www.tv1000.se	TV1000	Sweden
http://www.tv3.se	TV3 Sverige	Sweden
http://www.tv4.se/	TV4 Sverige	Sweden
http://www.tv5.ca/	TV5 Quebec	Canada
http://www.tv5.org/	TV5	France
http://www.tve.es/	TVE	Spain
http://www.tvnorge.no/	TVNorge	Norway
http://www.tvo.org/	TV Ontario	Canada
http://www.u-net.com/~arrowe/	Satmaster Satellite Page	UK
http://www.uaep.co.uk/BRAVO.html	BRAVO	UK
http://www.unik.no/~robert/hifi/tests/dmx/dmx.html	Digital Music Express	USA
http://www.upn.com/	United Paramount Network web page	USA
http://www.usanetwork.com/	USA Network web page	USA
http://www.uvsg.com	United Video Satellite Group	USA
http://www.uvsg.com/sse/sse.htm	Superstar Satellite Entertainment	USA
http://www.vbs.bt.co.uk/	British Telecom	UK
http://www.vecai.nl/	VECAI Dutch cable company	Netherlands
http://www.veronica.nl/	Veronica	Netherlands
http://www.vh-1.com/	VH-1	USA
http://www.vh1.de/	VH-1	Germany
http://www.vh1.de/	VH-1derland	Germany
http://www.viacom.com/	Viacom web page	USA
http://www.virginradio.co.uk/	Virgin Radio	UK
http://www.vnet.net/users/gingell/sat/sat.html	Mike Gingell's TVRO page	USA
http://www.vox.de/	VOX	Germany
http://www.vtm.be/	VTM	Belgium
http://www.vyvx.com	VYVX	USA
http://www.wdr.de/	Westdeutscher Rundfunk	Germany
http://www.weather.com/	The Weather Channel	USA
http://www.wgntv.com/	WGN-TV 9	USA
http://www.wintermute.co.uk/users/orrock/	The Den at Orrock - Satellite TV and Radio!	UK
http://www.winternet.com/~conus/	Conus Communications	USA
http://www.wmbakerassociates.com/tv.html	WEBtown's Television Home Page	USA

USEFUL WORLD WIDE WEB ADDRESSES

URL	Description	Country
http://www.wral-tv.com	WRAL-TV Raleigh - NC	USA
http://www.wrn.org/	WRN	UK
http://www.wrn.org/tesug	TESUG	UK
http://www.xs4all.nl/eurojazz	EuroJazz	Netherlands
http://www.xs4all.nl/~ceylon/	CompuSat BBS	Netherlands
http://www.xtc-com.com/channel.htm	XXXtasy Premiere	USA
http://www.xxxotica.com/	Exxxotica	USA
http://www.yle.fi/	YLE- Yleis Radio	Finland
http://www.zdf.de/	ZDF	Germany
http://www2.best.com/~olivert	Oliver Tse's home page - great for soccer on TV	USA
http://www2.helix.net/~lekei/	Don Lekei's home page	USA
http://xan.esrin.it:2602/satellite.html	European Satellite Information	Italy

US TVRO/DSS DEALER LINKS

URL	Description	Country
http://hometech.com/hometech/	Hometech Enterprisespage	USA
http://ourworld.compuserve.com/homepages/Iremigio	Communications Research Group	USA
http://satscan.com	Satscan Electronics Corp.	USA
http://usat2.nsnnet.com	NSN Network Services	USA
http://www.all-pla.net/rainsat	Rainbow Sat Communications	USA
http://www.atcon.com/stores/hobby/	Hobby Corner Satellite page	USA
http://www.biddeford.com/~gradys/	Grady's Radio and Satellite TV	USA
http://www.cadvision.com/nolimits/satinfo.html	PME Satellite TV	USA
http://www.castles.com/sse	Superior Satellite Engineers	USA
http://www.comstream.com/	Comstream broadcast equipment manufacturer	USA
http://www.datasat.com/	DataSat Communications	USA
http://www.dcserv.com/	Diversified Communications Services	USA
http://www.digigo.com/satcat.htm	DIR Programming Service	USA
http://www.gev.com/	G.E.V. C-band and Authorized PRIMESTAR	USA
http://www.gourmet-ent.com	Gourmet Entertaining! Home Page./Arc-set/Sat-set	USA
http://www.grove.net	Grove Enterprises web site	USA
http://www.InstantWeb.com/S/satellite2000/	Great TV Action	USA
http://www.microspace.com/	Microspace Communications	USA
http://www.minidishtv.com/	Continental Satellite Company	USA
http://www.miralite.com/Index.html	Miralite Communications	USA
http://www.nauticom.net/users/drdon/dishdr/dishdr.html	Dish Doctor Home Electronics	USA
http://www.nethomes.com/allsat/	All Sat	USA
http://www.ocala.com/afc	Antennas for Communications	USA
http://www.prodelin.com	Prodelin Corporation	USA
http://www.rwt.co.uk/	Real World Technology	USA
http://www.satnews.com/	Satfinder	USA
http://www.sigweb.com/sepatriot.html	Satellite Export and Engineering	USA
http://www.skyvision.com	Skyvision Fergus Falls	USA
http://www.tedsat.com/Welcome.html	Tedsat Canadian satellite dealer	USA
http://www.wcs-online.com/jones/	Jones Satellite Programming	USA
http://www.wcs-online.com/pt24/	PrimeTime 24 Programming	USA
http://www.xcity.com/nhe/nhehome.htm	Nelson Hill Electronics	USA

CHAPTER 15

WORLD TELEVISION

EUROPE

ALBANIA

TV-sets: 300.000 — **Systems:** B & G — **Colour:** PAL.
RADIOTELEVISIONE SHQIPTAR (Gov.)
✉ Rruga "Ismail Qemali" 11, Tirana. ☎ +355 (42) 27512. 🖷 23650. **Cable:** RTSH Tirana.
L.P: DG: Bardhyl Pollo. Dir. of Int. Rel: Enver Lekaj.

Stations:	ch	kW/Pol	Stations:	ch	kW/Pol
Tirana	5	60H	Butrint	12	5H
Elbasan	6	10	Kükes	12	100H
Gjirokaster	7	10H	Vlora	12	10H
Peshkopi	8	2H	Peskopi	32	10H
Berat	9	10H	Letaj	39	25H
Tirana	10	0.2V	Tirana	57	800H
Pogradec	11	100H			

D.Prgr: 0700-2200. **F.PI:** 2nd Channel.

AUSTRIA

TV-sets: 2.706.000 — **Colour:** PAL — **Systems:** B & G.
ÖSTERREICHISCHER RUNDFUNK
✉ ORF-Zentrum Wien, A-1136 Wien, Würzburggasse 30. ☎ 43 (222) 87 8780. 🖷 +43 (222) 87 878 2250. ① 133601.
L.P: DG: Gerhard Zeiler. Dir. of TV: Joh. Kunz, E. Wolfram Marboe. Tech.Dir: Dr. Wolfgang Pasewald. Head of PR: Thomas Prantner.
Stations: System B ch1-12, System G ch21-68.

Location	Prgr 1	kW	Prgr 2	kW	Pol
St. Polten	2A	60	21	500	H
Innsbruck	4	80	23	800	H
Neumarkt	4	1	47	10	H
Bad Ischl	5	2	38	5	H
Bregenz	5	100	24	350	H
Kufstein	5	3	24	50	H
Wien	5	100	24	1000	H
			34*	50	H
Linz	6	100	43	500	H
Rechnitz	6	3			V
			43	55	H
Graz	7	100	23	800	H
Weitra	7	5	58	90	H/V
Salzburg	8	100	32	800	H
			36	300	H
Wofsberg	8	2	28	20	H
Bruck/Mur	9	20	41	200	H
Bludenz	9	2	33	30	H
Klagenfurt	10	150	24	1250	H
Semmering	10	10	36	65	H/V
Schladming	11	10	40	80	H

Spittal/Drau	12	2	54	20	H
Mattersburg	38	30	52	30	V
St. Pölten	38	150	21	600	H
Poysdorf	51	10	43	10	H
Lienz			41	15	H
Gmunden			59	10	H

+ 867 low power tr's — *) also carries local prgr.
N.B: ORF 1&2 partly stereo.
D.Prgr: Prgr. 1: W 0800-2345, Sun 0800-2330 (all times approx). **Prgr. 2:** Mon-Fri 1530-2330, Sat 1245-2400, Sun 0800-2345 (all times approx).

AZORES (Portuguese)

TV-sets: 3380 — **Colour:** PAL — **System:** B.
RADIOTELEVISÃO PORTUGUESA (RTP)
✉ Ponta Delgada, S. Miguel.
Stations: Pico da Barrosa chE7 150kW H, Santa Barbara chE9 100kW H, Lages (Isle of Terceira) chE4 1 kW + 8 repeaters (only relay of RTP's 1st. prgr.).
D. Prgrs: 2000-2400.

AFRTS (US Air Force)
✉ Detachment 3, Air Force European Broadc. Squadron, APO New York, NY. 09406-5000, USA
Station: (System M): Lajes Field chA8 1kW H.
D. Prgr: 0900-0200 (Fri/Sat 0400).

BELARUS

TV-sets: 3.600.000 — **Colour:** SECAM — **Systems:** D & K

BELARUSKAE TELEBACHANNE (Nat. State TV and Radio Company)
✉ Makaenka 9, Miensk, 220807. ☎ +375 (172) 649286, 🖷 +375 (172) 648182, ① 252267 TV SU.
L.P: President: A.R. Stljarou.

Stations	ch	Pol	Stations	ch	Pol
Miensk	1	H	Bragin	27	H
Mjadzel	8	H	Grodna	3	H
Brest	7	H	Slonim	10	H
Pinsk	4	H	Heraneny	7	H
Viciebsk	2	H	Smargon	36	H
Ushachy	9	H	Magileu	4	H
Homel	10	H	Babrujsk	12	H
Smjatanichy	5	H	Kastjukovichy	9	V
Zlobin	22	H			

+ 3 low power repeaters.
D.Prgr: 0600-2200

TELEVISION BROADCASTING NETWORK (TBN)
✉ 15a F. Skariny Street, Miensk 22072. ☎ 🖷 +375 (172) 394171, 394536. **E-mail:** mmc@glas.apc.org
TBN unites 12 private TV companies in the biggest cities of Belarus:

Belarus — Bulgaria

City	TV Company	Ch.
Mogilev	2nd Channel	2
Piensk	Varyag	7
Miensk	NTRC Bel TV	8
Bobruysk	Tele-Vesta	9
Baranovici	Intex	23
Soligorsk	Soltec	34
Kobrin	Inform TV	25
Zhlobin	Nuans	29
Orsha	Skif	34
Gomel	Nireya	35
Svetlogorsk	Ranak	36
Vitebsk	Delta TV	48

BELGIUM

TV sets: 4.200.000 — **Colour:** PAL — **Systems:** B & H.

BELGISCHE RADIO EN TELEVISIE (BRTN)
⌕ 1043 Brussels. **Cable:** BRT-TV. ☎ +32 (2) 741 3111. 🖷 +32 (2) 734 9351.
L.P: DG: J. Ceuleers. Dir. Prgr. Planning: J. Bauwens.
Stations: ch2-11 System B, ch21-68 system H. NICAM stereo audio.

Location	Prgr 1	kW	Prgr 2	kW	Pol
Antwerpen	2			0.1	V
Waver/Overijse	10	100	25	10	H
Egem	43	1000	46	1000	H
Genk	44	200	47	200	H
Schoten			62	200	H

D.Prgr:
BRT1: 1400-2200.
BRT2: 1800 (Sun 1300)-2230.

RADIO TELEVISION BELGE DE LA COMMUNAUTE CULTURELLE FRANCAISE (RTBF)
⌕ 1044 Brussels. ☎ +32 (2) 737 2111. 🖷 +32 (2) 737 4357. **Cable:** RTBF-TV — **L.P:** PD: G. Konen.
Stations: ch2-11 System B, ch21-68 system H. NICAM stereo audio.

Location	Prgr 1	kW	Télé 21	kW	Pol
Liège	3	100	42	1000	H
Wavre	8	100	28	500	H
Couvin			30		H
Léglise	57	10	60	0.5	H
Bruxelles			45	0.5	H
Profondeville	52	200	49	200	H
Tournai	57	20	63	20	V
Anderlues			61	200	H

+ 8 low power relay st's.
D.Prgr: RTBF1: 1545-2230. **Télé 21:** 1800 (Sat 1600, Sun 1300)-2200.

VTM (Vlaamse Televisie Maatschappij)
⌕ 1818 Vilvoorde. ☎ +32 (2) 254 5611. 🖷 +32 (2) 252 5016
L.P: MD: J. Merckx. Commercial TV Service in Dutch on cable only.

CANAL PLUS (Comm.)
⌕ Chaussee de Lauvain 656, 1050 Brussels. ☎ +32 (2) 7300 211. 🖷 +32 (2) 732 1848

Location	ch	kW	Pol
Liège	39	200	H
Wavre	50	500	H
Anderlues	58	200	H
Leglise	63	?	H

SATELLITE AND TV HANDBOOK

RTL-TVi (Comm.)
⌕ 1051 Brussels. ☎ +32 (2) 640 51 50. 🖷 +32 (2) 640 9307. ① 64430.
Station: ch27 (tr located in Dudelange, Luxembourg).
D.Prgr in French: 12h daily.

TV-5 - Europe
Station: Bruxelles ch56 1kW H.
Rebroadcasts the TV-5 satellite sce.

AMERICAN FORCES NETWORK - SHAPE
Colour: NTSC — **System:** M.
Station: chE33 1kW V, ch34 4.5kW (rel. AFN Germany).

BOSNIA/HERCEGOVINA

TV-sets: 1.012.094 — **Colour:** PAL — **Systems:** B&H.

RADIO TELEVIZIJA BOSNE I HERCEGOVINE
⌕ VI Proleterske brigade 4, 71000 Sarajevo. ☎ +38 (71) 522333.
Stations: System B=ch2-12, System H=ch21-68. ChE. audio powers 1/10 of vision powers indicated.

TV Sarajevo
1st Prgr.

	kW		kW
ch	(ERP)	ch	(ERP)
Majevica 5	10	Hum 8	6
Bjelašnica 5	6	Trovrh 9	10
Kozara 6	18	Leotar 10	6
Tusnica 6	10	Plješvica-B 10	2
Vele 7	10	Vlašič 11	100

+ 152 low power sts.
D.Prgr: Mon-Fri 0500-1130, 1345-0030. Sat 1100-0015, Sun 0700-0015.
TV Sarajevo
2nd Prgr.

Trovrh	21	100	Tusnica	43	7
Velez	26	1000	Majevica	46	750
Kozara	27	1000	Bjelašnica	47	60
Vlašić	29	1000	Plješvica	47	125
Leotar	37	100	Hum	52	22

+ 140 repeaters.
D.Prgr: Mon-Fri 1355-1600, 1730-2230. Sat 0900-2230, Sun 0635-2200.
TV Sarajevo
3rd Prgr.
Hum   37   100   (D.Prgr: 1700-2400.)

BULGARIA

TV-sets: 3.127.000 — **Colour:** SECAM — **Systems:** D & K.

BALGARSKA TELEVIZIJA (Gov.)
⌕ Ul. San Stefano 29, 1504 Sofia. ☎ +359 (2) 446329. 🖷 +359 (2) 662388. **Cable:** BT SOF BG. ① 22581 bt sof bg.
L.P: DG: Ivan Granitsky.
Stations: (System D ch1-12, System K ch21-60):

Prgr. 1

	ch	kW		ch	kW
Shumen	R5	100	Kjustendil	R10	50
Smoljan	R6	5	Botev Vrâh	R11	250

EUROPE — Bulgaria — Croatia

Burgas	R7	100	Sliven	R12	20
Sofija	R7	10	Belogradčik	R12	50
Silistra	R8	1	Dobrich (ex-		
Zelena Glava	R9	50	Tolbuhin tr)	R12	5
Kârdjali	R9	10	Ruše	R9	20
Varna	R9	5			

D.Prgr: approx 70h per week.

Prgr. 2	ch	kW		ch	kW
Botev Vrâh	24	1000	Sliven	31	200
Gotze Delchev	25	?	Varna	33	100
Burgas	26	1000	Kârdjali	34	100
Mihajlovgrad	26	1000	Kjustendil	34	100
Ruše	27	250	Silistra	35	100
Dobrich (ex-			Stara Zaora	37	?
Tolbuhin tr)	28	100	Smoljan	38	100
Plovdiv	28	?	Shumen	39	1000
Sofija	29	?	Belogradčik	46*	1000
Belogradčik	30	?			

+ 250 low power tr's (Prgr. 1 + 2)
D.Prgr: approx 70h per week.

NOVA TV (Comm.)
16 Sveta Nedelja Sq, 1000 Sofia. ☎ +359 (2) 805025. 🖷 +359 (2) 870298 — **L.P:** Exec. Mgr: Rumen Kovachev.
Station: Kopitoto ch48 1kW (covers greater Sofia area).
D.Prgr: 80h per week.

RUSSIAN TELEVISION RELAY
Station: Sofia ch36 2kW, Ruse ch32 100kW.
D.Prgr: Mon-Fri 1230-2300; Sat/Sun 0630-2330. Relays of OK-1 via satellite from Moscow.

TV 5 EUROPE
Station: Plovdiv chR6, Sofija ch41 0.2kW, Burgas ch32.

CROATIA

TV-sets: 950.000 — **Colour:** PAL — **System:** B&H

HRVATSKA TELEVIZIJA (HTV)
Prisavlje 3, Zagreb, Croatia. ☎ + 385 (1) 616 3366. 🖷 + 385 (1) 616 3392. ① 21477 HTV RH. **WWW:** http://www.hrt.hr
L.P: GM: Ivan Parac. Head of Prgrs: Hloverka Srzic-Novak. Head of Int. Rel. Dept: Marija Nemcic.
Stations (main st's in bold)

Transmitter	HTV1	HTV2	HTV3
Babino Polje	11	38	-
Bakar	7	45	41
Biokovo	9	41	45
Blato na Korculi	38	55	-
Bolfan	29	21	52
Brac	11	29	53
Brezje	58	34	62
Brezovica	44	47	54
Brinje	35	38	30
Brodski Stupnik	51	28	-
Buje	21	24	56
Cabar	10	33	36
Cavtat	56	59	-
Celavac	8	31	25
Cres	8	23	37
Crni Lug	23	47	5
Cucerje	30	36	42
Delnice	37	49	53
Dinjiska	24	27	-
Doljani	-	-	-
Donji Lapac	6	32	35
Drenovci	39	51	27
Dvor na Uni	22	42	60
Fara	44	51	-
Fuzine	8	60	49
Fuzine-Jezero	10	-	-
Gerovo	8	47	50
Gospic	6	-	-
Govedjari	11	36	-
Grobnik	5	47	53
Gruda	38	41	-
Gunjavci	40	43	52
HRT Bldg.	49	52	59
Hvar	6	39	31
Imotski	12	27	35
Ist	26	41	44
Ivanec	27	-	-
Ivanscica	36	40	63
Jablanac	37	45	50
Jelsa	27	32	57
Kalnik	5	43	-
Kasina	31	38	44
Klis	55	58	-
Knin	6	44	53
Komiza	7	55	59
Komolac	12	36	46
Koprivnica	33	37	48
Koromacno	7	36	22
Korcula	11	21	33
Kostajnica	38	43	31
Krapinske Toplice	21	55	45
Kriz	49	53	56
Kuna	11	43	-
Kupjacki Vrh	40	55	58
Kutjevo	26	29	-
Labinstica	4	23	34
Lastovo	53	59	-
Lepoglava	51	58	47
Lic	10	31	24
Licka Pljesivica	5	53	57
Lokve	6	21	32
Lopud	10	47	43
Majkovi	7	55	-
Mali Losinj	7	53	36
Mandicevac	31	40	-
Maranovici	36	43	-
Markusevec	55	44	41
Martinscica	6	52	-
Metkovic	47	53	31
Milna	31	43	39
Mirkovica	7	43	46
Mokosica	10	33	39
Molunat	11	21	-
Moslavacka Gora	67	21	34
Mrkopalj	8	36	33
Murter	25	28	43
Nova Gradiska	31	42	48
Novalja	41	49	23
Novigrad	42	45	53
Novigrad (Zadar)	6	25	28
Obrovac	10	34	-
Omis	7	37	40
Oriovac	36	45	-
Osijek	6	23	33
Osijek Donji Grad	27	29	49

Croatia — Czech Republic

Transmitter	HTV1	HTV2	HTV3
Ostra	56	58	50
Otes	30	38	-
Otok	34	59	-
Pag	7	35	32
Pakrac	44	47	31
Papuk	53	56	21
Peljesac	5	38	58
Planina	41	48	60
Plocice	43	46	-
Plomin	55	58	46
Podvinje	54	60	-
Porec	26	31	59
Prezid	9	23	26
Primosten	29	21	-
Promina	38	59	25
Psunj	4	50	58
Pucisca	36	39	28
Pula	35	26	48
Rabac	7	22	53
Razromir	9	55	58
Rasa	8	47	50
Resetari	21	38	25
Rovinj	43	49	31
Ruda	21	24	37
Sibenik-Martinska	5	52	58
Sibenik-Subicevac	11	49	55
Sibinj	21	24	32
Sinj	5	29	26
Skradin	51	54	-
Slano	49	46	-
Slatina	29	44	47
Slavonski Brod	9	42	48
Srb	10	-	-
Sljeme	9	28	25
Srinjine	51	59	-
Srdj	6	28	22
Starigrad Paklenica	5	41	-
Stipanov Gric	12	24	27
Ston	11	53	56
Straza	12	30	60
Strigova	44	48	30
Suvaja	-	-	-
Sv.Gera	50	58	63
Sv.Martin	45	47	-
Sv.Nedjelja	12	39	54
Svilno	26	42	34
Trstenik	8	40	-
Trsce	52	59	49
Ugljan	51	57	37
Uljenje	12	35	48
Umag	25	32	52
Unije	24	27	32
Ucka	11	29	39
Vela Luka	7	29	43
Veleb. Pljesivica	34	44	51
Velika	52	38	-
Velika Petka	8	45	48
Vinkovci	12	44	36
Virovitica	42	45	37
Vis	51	55	59
Visovac	36	50	53
Vrgorac-Gomila	21	31	10
Vrgorac-Polje	57	60	52
Vrlika	11	26	29
Vrsar	23	36	44
Vrucica	6	12	-
Vucinici	52	22	32
Zadar	-	-	-
Zaton	36	40	-
Zlarin	32	45	-
Zlatarevac	25	28	34
Zupa	50	54	40
Zupanja	49	58	65

D.Prgr: HTV1: 0700-2300; HTV2: 1000-2300; HTV3: 0800-2400.

OTV (Open TV)
✉ Teslina 7, Zagreb, 10000. ☎ +385 (1) 424 124. 🖷 +385 (1) 455 1386.

Regional Stations

SLAVONSKA TELEVIZIJA OSIJEK
✉ Hrvatske Republike 20, Osijek 31000. ☎ +385 (31) 124 666. 🖷 +385 31 124 111.

TV MARJAN
✉ Savska bb, Split 21000. ☎ +385 (21) 364 525. 🖷 +385 (21) 523 455.

VINKOVACKA TELEVIZIJA
✉ Genschera 2, Vinkovci, 32000 Croatia. ☎ +385 (32) 331 990. 🖷 +385 (32) 331 985.

ZADARSKA TELEVIZIJA
✉ Molotska bb, Zadar, 23000 Croatia. ☎ +385 (23) 311 791. 🖷 +385 (23) 314 749.

CZECH REPUBLIC

TV-sets: 3.800.000 (est.) — **Colour:** SECAM — **System:** D & K

CZECH TELEVISION (Public Sce.)
✉ Kavcíhory, 140 70 Praha 4. ☎ +42 (2) 61131111. 🖷 +42 (2) 6927202. **WWW:** http://www.czech-tv.cz
L.P: DG: Ivo Mathé. PD: Jirí Pittermann. TD: Jan Horsk˝. Head of PR: Jiri Moc.

Location	CT1	kW	CT2	kW
Brno	29	20	46	20
Brno-mesto	35	2	52	2
Ceske Budejovice	39	20	49	20
Domazlice	41	2	12	0.2
Frydek-Mistek	37	17		
Hodonin			33	1
Hradec Kralove	22	20	57	20
Cheb	36	3.5	53	2
Chomutov	52	6	35	5
Jachymov	38	5		
Jesenik	36	20	50	20
Jihlava	25	10	42	10
Klatovy	22	5	58	0.1
Liberec	31	5	43	0.01
Mikulov	26	10		
Novy Jicin	34	5		
Olomouc	33	2		
Ostrava	31	20	51	2
Pacov	36	5		
Plzen	31	20	48	20
Plzen-mesto	34	5	27	1.5
Praha	26	50	53	1
			29	1
Praha-mesto	51	10	41	1
Rychnov nad Kneznou	28	5		

EUROPE — Czech Republic — Denmark

Location	CT1	kW	CT2	kW
Susice	35	5	52	0.08
Svitavy	24	10	58	0.1
Tabor	27	1		
Trutnov	23	16	40	20
T	28	10	45	10
Uhersky Brod			47	20
Usti nad Labem	33	20	50	20
Valasske Klobouky	24	2	42	2
Vimperk	32	5	47	0.08
Votice	30	5	56	5
Zilina	22	5	51	0.2
Z	32	5	49	0.1

D.Prgr: ET1: 0630-2330 (approx). **ET2:** 0600-0030 (approx).

Private Stations:

NOVA (Comm.)
☏ Vladislavova 20, 11313 Praha 1. ☏ +42 (2) 2110 0111. 🗎 +42 (2) 2110 0565 — **L.P:** Gen Dir: Vladimir Zelezny. PD: Jan Vit. Head of PR: Karel Soukup.

Stations	ch	kW	Pol.
Domazlice	12	1.6	H
Klatovy	22	100	H
Hradec Kralove	22	600	H
Trutnov	23	1000	H
Praha (mesto)	24	60	H
Svitavy	24	100	H
Jihlava	25	1	H
V. Klobouky	25	10	H
Praha	26	1000	H
Tabor	27	7.8	H
Trebec	28	100	H
Rychnov	28	100	H
Brno	29	600	H
Blatna	29	600	H
Ostrava	31	600	H
Plzen	31	600	H
Liberec	31	100	H/V
Vimperk	32	100	H
Usti n. Labem	33	600	H
Plzen (city)	34	100	H
Chomutov-Jedl	35	100	H
Brno (city)	35	20	H
Cheb	36	100	H
Pacov	36	1.6	H
Jesenik	36	600	H
Bardejov	37	10-0	H
Jáchymov	38	300	H
Jihlava-Javorice	42	300	H
Brno-Kojal	46	600	H
Plzen-Krasow	48	600	H
Jesenik-Praded	50	300	H
Hradec Kralove	57	600	H

DENMARK

TV-sets: 2.700.200 — **Colour:** PAL — **System:** B.

TELECOM A/S
☏ Telegade 2, DK-2630 Taastrup, Denmark. ☏ +45 (42) 529111. 🗎 +45 (42) 529331. **L.P:** Man. Dir: Jens Kiil.

DANMARKS RADIO (Gov.)
☏ Danmarks Radio, TV-Byen, DK-2860 Søborg. ☏ +45 (35) 203040. 🗎 +45 (35) 202644. ☻ 22695.

WWW: http://www.dr.dk
L.P: Dir TV: Finn Rowold; Dir of Prgrs: Thomas Dahlberg; TV Fakta: Bo Lynnerup; TV Int: Mogens Vemmer.
Stations: Pol H exc. Ølgod: V — Stereo: NICAM digital system, only on Copenhagen-Hove ch 31.

Location	ch	kW	Location	ch	kW
Sdr. Højrup (Fyn)	3	10	Sønderjylland	7	60
Copenhagen-	4	50/5	Århus	8	60/6
Bornholm	5	10	Vestjylland	10	60
Sydvestjylland	5	5	Copenhagen-	31	600/60
Sydsjælland	6	60	Tolne, Vendsys.	57	40

+ 26 low-powered repeaters not mentioned.
D.PRGR: MF 1400-2230, SS 1300-2300.

TV2 (Comm.)
☏ Rugaardsvej 25, DK-5100 Odense C. ☏ +45 (65) 91 12 44. 🗎 +45 (65) 91 33 22.
L.P: Dir. Gen: Tøger Seidenfaden. Head of News: Svenning Dalgaard. Head of Prgr's: Jørgen Steen Nielsen. Head of Facts Dept: Lally Hoffmann. Head of Fiction Dept: Lone Bastholm. Head of Tech. Dept: Lars Esben Hansen.
Stations: Pol H, exc. Thisted: V — Stereo: NICAM digital system.

Location	ch	kW		ch	kW
Odense	22	500	Ringkøbing	40	600
(Tommerup)			(Videbæk)		
Hadsten	26	600	Jyderup	48	600
Åbenrå	27	600	Nakskov	52	100
Thisted	28	250	Copenhagen-	53	600
Vejle (Hedensted)	30	500	Hove		
Svendborg	32	150	Bornholm (Rø)	56	600
Varde	33	500	Viborg	56	500
Nibe	35	600	Vordingborg	68	600
Tolne	37	50			

+25 low-power translators.
D.PRGR: 1600-2300 MF, 1300-2300 SS (all colour) —
N: 1800-1830 & 2100-2115 MF. **Comm:** between prgr's.
Regional Prgr's: 1830-1900 (exc. Sat).
Regional Addr's:
TV2/Bornholm, Ravnsgade 5, 3720 Aakirkeby. ☏ +45 (53) 97 57 10. 🗎 +45 (53) 97 45 85. **Dir:** Bent Nørby Bonde. **D.Prgr.** on ch. 56.
TV/Fyn, Kongensgade 68 — I, 5000 Odense C. ☏ +45 (66) 14 18 18. 🗎 +45 (66) 12 44 24. **Dir:** Ebbe Larsen. **D.Prgr.** on ch. 22 & 32.
TV2/Lorry, Allegade 7-9, 2000 Frederiksberg. ☏ +45 (38) 88 00 11. 🗎 +45 (38) 88 31 11. **Dir:** Dan Tschernia. **D.Prgr.** on ch. 53.
TV/Midt-Vest, Søvej 2, P.O. Box 1460, 7500 Holstebro. ☏ +45 (97) 40 33 00. 🗎 +45 (97) 40 14 44. **Dir:** Ivar BrÆndgaard. **D.Prgr.** on 28, 40 & 56.
TV/Nord, Søparken 4, 9440 Åbybro. ☏ +45 (98) 24 46 00. 🗎 +45 (98) 24 46 98. **Dir:** Bent Bjørn. **D.Prgr.** on ch. 35 & 37.
TV/Syd, Laurids Skausgade 12, 6100 Haderslev. ☏ +45 (74) 53 05 11. 🗎 +45 (74) 53 23 16. **Dir:** Helge Lorenzen. **D.Prgr.** on ch. 27, 30, 33 & 42.
TV/Øst, Kildemarksvej 7, 4760 Vordingborg. ☏ +45 (55) 34 02 00. 🗎 +45 (55) 34 01 34. **Dir:** Ole Dalgaard. **D.Prgr.** on ch. 48, 52 & 58.
TV2/Østjylland, Niels Brocks Gade 16, 8900 Randers. ☏ +45 (86) 40 40 00. 🗎 +45 (86) 40 56 86. **Dir:** Erling Bundgaard. **D.Prgr.** on ch. 26.
ALF: DKK 195 of the Radio Denmark license income, plus commercial income.

Denmark — Finland

TV3 (Comm.)
✎ Overgaden oven Vandet 10, DK-1415 København K. — ☏ +45 (31) 57 30 80. 🖷 +45 (31) 57 80 81.
L.P: Dir: Klaus Fog. Head of prgr's: Susanne Teilmann.
Station: see satellite section.
D.PRGR: 1600-2400. Commercials interrupt prgr's.
Local TV Stations: Approx. 35 organizations are on the air on UHF with 0.2-3 kW ERP.

ESTONIA

TV-sets: 615.000 — **Colour:** PAL/SECAM — **Systems:** B/D&K

TALLINN EESTI TELEVISIOON (ETV)
✎ 12 Faehlmanni Street, 0100 Tallinn. ☏ +372 (2) 434102. 🖷 +372 (2) 434155. ③ 173869 ETV EE. **E-mail:** etv@etv.ee **WWW:** http://www.etv.ee
L.P: DG: Hagi Shein. Dep. DG & Editor in Chief: Raul Rebane.

EVTV (Comm.)
✎ Peterburi 81, 0014 Tallinn. ☏ +372 (6) 328228. 🖷 +372 (6) 323650 — **L.P:** Chmn: Victor Siilats. MD: Enn Eesmaa. Mktg. Mgr: Antero Laanela. PR: Kristina Haiba.

KANAL KAKS (Channel 2) (Comm.)
✎ Harju 9, 0001 Tallinn. ☏ +372 (2) 442356. 🖷 +372 (2) 446862 — **L.P:** Chmn: Ilmar Taska. Sen. VP: Eva Banhidi. PD: Liina Kirt. Int. Rel: Vanessa Vogel.

REKLAAMITELEVISIOON (Comm.)
✎ Endla 3, 0106 Tallinn. ☏ +372 (2) 666743. 🖷 +372 (6) 311077 — **L.P:** DG: Tomas Lepp.

TIPP TV (Comm.)
✎ Regati Pst-1-6, 0019 Tallinn. ☏ +372 (2) 238535. 🖷 +372 (2) 238555 — **L.P:** Pres: Juri Makarov.

Transmitting stations (composite listing for all networks):

Site	ch.	prgr.
Kohtla-Nômme	R1	EVT/RTV
Kunda	R1	ORT/BFD/ORS
Pôltsamaa	R1	ORT
Tallinn Kl.	R2	ETV
Narva-Keskus	R3	ETV
Vôru-Keskus	R3	?
Rakvere-Tabani	R3	ETV
Tallinn Kl.	R3	KK
Haapsula-Uuemôisa	R3	ORT/BFD
Kallaste-Torila	R3	ORT
Sillamäe	R3	?
Pärnu-Tammiste	R4	ETV
Tartu-Soinaste	R4	EVT/RTV/TAT
Narva-Keskus	R4	ORT/BFD
Kohtla-Nômme	R5	KK
Jôgeva-Eristvere	R5	ORT/BFD
Ruhnu-Majakas	R5	ETV
Valgjärve	R6	ETV
Paide-Keskus	R8	ETV
Sôrve-Lôopôllu	R8	ETV
Pärnu-Tammiste	R9	ORT/BSD
Järvakandi	R10	ORT/BFD/ORS
Kohtla-Nômme	R11	ETV
Orissaare	R11	ETV
Vôhma	R11	ORT
Tallinn-Kl.	R12	ORT/BFD/ORS
Tartu-Puiestee	R12	ETV
Viljandi-Viiratsi	22	ORT
Orissaare	22	ORT/BFD
Tallinn-Ülemiste	22	ALO
Narva-Keskus	23	KK
Rakvere-Tabani	27	ORT/BFD
Kihelkonna	27	ORT
Koeru	27	?
Tartu-Tamme	28	ALO/RTV
Tallinn-Kl.	28	EVT/RTV
Kohtla-Nômme	29	ORT/BFD
Valgjärve	30	ORT/BFD
Kuressaare-Keskus	31	ORT/BFD
Kärdla-Rehemäe	34	ORT/BFD
Paide-Keskus	39	ORT/BFD
Koeru	39	?
Sôrve-Lôopôllu	41	ORT
Tallinn-Kl.	42	KK
Tallinn-Kl.	45	SEE

ETV=Eesti Televisioon; EVT=EVTV; ORT=Ostankino TV; RTV=Reklaamitelevisioon; ALO=Alo TV Tartu (comm.); KK=Channel 2 (comm.); BFD=BFD Reklaamiklubi (comm.); ORS=TV Orsent (loc. comm.)

FAROE ISLANDS (Danish)

TV-sets: 14.000 — **System:** B&G — **Colour:** PAL.

SJÓNVARP FØROYA
✎ Sjónvarp Føroya, M.A. Winthersgøta 2, Postboks 21, 110 Tórshavn. ☏ +298 17780.
L.P: Gen. Mgr: J.A. Skaale. Dir. of Adm: F. Lómstein. Head of Technical Dept: R.C. Joensen.
Stations: Tórshavn ch6 145kW H, Suduroy ch9 6.4kW H, Eysturoy ch10 1kW V + 29 low power repeaters.
Sat. relays: ch22/28 MTV Europe, ch30/37 BBC World, ch34/44 Scansat TV3, Eurosport.
D.Prgrs: Tues 1930-2300; Thurs 2100-2200; Fri 1930-2330; Sat 1930-2400; Sun 1500-1700 & 1930-2300.

FINLAND

TV-sets: 2.081.000 — **Colour:** PAL — **Systems:** B & G.

YLEISRADIO OY (Public Broadc. Sce.)
✎ TV-1, Box 96, FIN-00024, YLEISRADIO, Finland. ☏ +358 (0) 14801. **Cable:** YLEtv. ③ 121270. **Tele**🖷 +358 (0) 1480 5148 **FST (Swedish Language TV)**, Box 83, FIN-00024 YLEISRADIO. ☏ +358 (0) 14801. 🖷 +358 (0) 1481256. **TV-2**, Box 196, FIN-33101 Tampere. ☏ +358 (31) 345 6111.③ 22749. 🖷 +358 (31) 345 6892.
L.P: Dir. TV: H. Lehmusto; Dir. Prgr TV-1: A. Gartz; Dir Prgr TV-2: A. Hoffren; Dir. Swedish lge Radio & TV: A. Sandelrin; Dir Prgrs: J. Harms.

MTV Oy (Comm.)
✎ 00033 MTV3, Finland ☏ +358 (0) 15001. 🖷 +358 (0) 1500707. **Cable:** Comtele. ③ 124319 COMTV SF.
L.P: Pres: E. Pilkama. Exec. Vice Pres: J. Paavela. Prgr. Dir: T. Äijälä. News Dir. (Editor-in-chief): P. Hyvârinen. Tec. Dir: H. Marsalo. Marketing Dir: Eero Aalto. Sales Dir: Heikki Rotko. Vice Pres (communications): J. Mietinen. Comm. Mgr: M. Paaso. Proj. Mgr: M. Rainbird (FinnImage)

EUROPE — Finland — France

CHANNEL THREE FINLAND/OY KOL-MOSTELEVISIO AB (Comm., subsidiary of MTV Oy)
✉ 00033 MTV3, Finland. ☎ +358 (0) 15001. ℻ 126068 CHANT SF. 📠 +358 (0) 150 0677.
L.P: Man. Dir: J-P. Louhelainen. Sales Dir: H. Vahala.
V.P (aquisitions): J. Sairanen. Hd. of Sports: T. Lehmuskallio. Prgr Mgr: S. Kievari. Producer: R. Haavisto.
Prgrs: Sport and other subcontracted programming on 3rd network (MTV3 channel).

Prgr. I	ch	kW	Pol		ch	kW	Pol
Tervola	3	20	H	Vuotso	8	0.25	H
Vuokatti	4	40	H	Haapavesi	9	60	H
Kuttanen	5	3	H	Kerimäki	9	10	H
Lapua	5	80	H	Lahti	9	80	H
Pyhätunturi	5	60	H	Joutseno	10	10	H
Anjalankoski	5	30	H	Pihtipudas	10	60	H
Ahvenanmaa	5	10	H	Posio	10	10	H
Espoo	6	180	H	Tammela	10	10	H
Iisalmi	6	1	H	Inari	10	30	H
Mikkeli	6	10	H	Eurajoki	11	20	H
Pyhävuori	6	20	H	Jyväskylä	11	10	H
Taivalkoski	6	60	H	Vaasa	11	1	H
Utsjoki	6	1	H	Ylläs	11	60	H
Karigasniemi	7	1	H	Ähtäri	26	100	H
Kruunupyy	7	10	H	Fiskars	31	100	H
Kuopio	7	80	H	Kiihtelysvaara	35	50	H
Oulu	7	60	H	Ruka	36	300	V
Turku	7	60	H	Pernaja	39	100	H
Koli	8	20	H	Rovaniemi	53	100	H
Tampere	8	80	H				

+ 142 low power repeaters.
YLE-prgrs: approx. 82h weekly.

Prgr. II	ch	kW	Pol		ch	kW	Pol
Espoo	8	60	H	Utsjoki	34	20	H
Turku	9	30	H	Vuotso	35	1	H
Vaasa	9	1	H	Kuopio	36	600	H
Pyhävuori	10	20	H	Pernaja	36	100	H
Oulu	11	20	H	Tammela	37	600	H
Iisalmi	22	20	H	Mikkeli	38	600	H
Tervola	22	1000	H	Ähtäri	39	100	H
Taivalkoski	23	600	H	Lahti	40	600	H
Karigasniemi	23	20	H	Inari	40	500	H
Lapua	24	1000	H	Fiskars	41	100	H
Jyväskylä	25	600	H	Kiihtelysv.	41	50	H
Haapavesi	28	1000	H	Anjalankoski	49	600	H
Ruka	28	300	H	Koli	51	600	H
Ahvenanmaa	28	300	H	Tampere	53	1000	H
Kruunupyy	30	200	H	Ylläs	55	900	H
Posio	30	500	H	Kerimäki	55	600	H
Joutseno	32	600	H	Rovaniemi	56	100	H
Pihtipudas	32	1000	H	Vuokatti	56	1000	H
Pyhätunturi	32	650	H	Kuttanen	60	20	H
Eurajoki	33	600	H				

+ 142 low power repeaters.
YLE-prgrs: approx. 65h weekly.
Relay st. FINLAND'S TV in Stockholm: see under "Sweden".

Prgr. III	ch	kW	Pol		ch	kW	Pol
Pyhävuori	12	20	H	Vuotso	28	1	H
Espoo	24	600	H	Mikkeli	28	600	H
Pyhätunturi	24	650	H	Ähtäri	29	100	H
Tervola	25	1000	H	Iisalmi	29	20	H
Posio	27	500	H	Tammela	30	600	H

	ch	kW	Pol		ch	kW	Pol
Haapavesi	31	1000	H	Vaasa	51	20	H
Utsjoki	31	20	H	Ylläs	52	900	H
Oulu	33	600	H	Lahti	51	600	H
Kruunupyy	33	200	H	Ruka	51	300	H
Jyväskylä	35	600	H	Inari	54	500	H
Taivalkoski	35	600	H	Turku	54	1000	H
Eurajoki	36	600	H	Pernaja	55	100	H
Karigasni.	37	20	H	Anjalankoski	56	600	H
Kiihtelysva.	38	50	H	Koli	57	600	H
Lapua	40	1000	H	Kerimäki	58	600	H
Pihtipudas	43	1000	H	Fiskars	58	100	H
Kuopio	49	600	H	Vuokatti	59	1000	H
Rovaniemi	49	100	H	Tampere	59	1000	H
Kuttanen	50	20	H	Joutseno	60	600	H

+ 141 low power repeaters.
MTV-3 prgrs: approx. 105h weekly.

TV-4	ch	kW	Pol	TV-4	ch	kW	Pol
Fiskars	23	100	H	Espoo	35	600	H
Pernaja	26	100	H	Turku	57	1000	H

+6 low power repeaters.
Prgrs: (produced by Swedish television, SVT) approx. 60h weekly.

FRANCE

TV-sets: 29.300.000 — **Colour:** SECAM — **System:** L.

TÉLÉVISION FRANÇAISE 1 (TF1) (Priv, Comm)
✉ Société TF1, 1 quai du Point du Jour 92100 Boulogne-Billancourt. ☎ +33 (1) 4141 1234.
L.P: Pres. & Dir. Gen: P. Le Lay. Dir. Tec: F. Hericourt.
N.B: Using NICAM stereo on main transmitters

FRANCE 2 (Public Television)
✉ Société Nationale Antenne 2, 22 Ave Montaigne, F-75008 Paris. ☎ +33 (1) 4421 4242. 📠 +33 (1) 4421 5145. ℻ 204 068. **L.P:** Pres: Xavier Gouyou-Beauchamp.
N.B: Using NICAM stereo on main transmitters

FRANCE 3 (FR3) (Public Television)
✉ 116 Avenue du President Kennedy, 75016 Paris. ☎ +33 (1) 4230 1313. 📠 +33 (1) 4289 0327.
L.P: Pres: Xavier Gouyou-Beauchamp.

CANAL PLUS (Private)
✉ 85/89 Quai Andre Citroën, 75015 Paris. ☎ +33 (1) 4425 1000. 📠 +33 (1) 4425 1234. ℻ CPLUS 201 141 F.
L.P: Man. Dir: Pierre Lescure, Tec. Dir: Lucien Banton, FilmBuyer: Rene-Bonnell.
Canal Plus is a subscription network. The signals armostly coded and subscribers need a decoder.Times for uncoded sig-nals: W 0600-0630, 1130-1230, 1715-1930.
Stations (Pol=H, exc. where indicated)
*) Lower power applies to Canal Plus only.

Location	TF1	F2	F3	C Plus	kW(ERP)
Abbeville	63	57	60	—	250
Ajaccio	31	21	24	4	500/8
Albertville	45	39	42	7	5/0.1
Alençon	48	51	54	—	100
Ales	27	21	24	65	100/3
Amiens	41	47	44	10	500/0.01
Angers	47	44	41	10	20/0.04

Location	TF1	F2	F3	C Plus	kW(ERP)
Annemasse	—	—	66	—	1.75
Argenton	46	40	43	—	80
Aurillac	59	65	62	9V	500/4
Aurillac	—	—	54	—	350
Autun	48	51	54	10	500/0.1
Auxerre	37	31	34	6	300/0.03
Avignon	42	45	39a	—	125
Avignon	—	—	33	—	55
Bar-le-Duc	54	51	48	6	200/0.05
Bastia	41	47	—	2V	500/17
Bayonne	64	58	61	7V	500/2.5
Bergerac	37	34	31	—	250
Besançon	47	41	44	3V	500/60
Besançon	29	23	26	10	250/0.02
Bordeaux	63	57	60	8	1000/50
Boulogne	29	34	37	10	100/0.5
Bourges	23	26	29	8	1000/211
	—	—	43	—	34
Brest	27	21	24	10	1000/230
Brive	23	29	26	6	150/0.5
Caen	22	25	28	9	1000/198
Carcassonne	64	58	66/61	3V	1000/100
Chambery	29	23	26	8	100/1
Chambery	—	—	49	—	4
Chamonix	25b	28	22	7	50/0.02
Champagnole	58	61	64	—	80
Chartres	55	50	53	9	250/0.18
Chartres	—	—	34	—	30
Chaumont	52/62	49	55	—	80
Cherbourg	65	59	62	6	100/8
Clermont-Fd.	22	28	25	5	1000/200
Cluses	56	50	53	6	8/0.1
Corte	51	61	54	7	100/1.5
Dijon	—	—	56	—	30
Dijon	59	62	65	9	1000/30
Dunkerque	42	39	45	10	200
Epinal	65	60	63	10	100/0.1
Forbach	47	22	25	28	20/2
Gap	27	21	24	9	8/0.3
Gex	27	21	24	5V	1000/30
Grenoble	56	50	53	6	50/0.6
Gueret	64	58	61	9	80/0.3
Hirson	54	48c	51	32	200/0.25
Hyères	65	59	62	6	35/2
Laval	63	57d	60	8	80/0.02
Le Creusot	35	33	30	67	20/10
Le Havre	46	43	40	5	100/2
Le Havre	—	—	35	—	2
Le Mans	24	27	21	5V	1000/200
LePuy	63	57	60	6	100/0.13
Lesparre	39	45	42	—	10
Le Vilhain	—	—	36	—	3.6
Lille	27	21	24	5	1000/200
Limoges	—	—	32	—	450
Limoges	56	50	53	10	1000/260
Longwy	52	47	44	8	100/0.1
Lyon (Mt. Pilat)	46	40	43	10	1000/400
Lyon	61	58	64	66	10/10
Macon	57	55	49	—	5
Mantes	64	58	61	9	80/0.02
Marseille	29	23	26	5	1000/200
Marseille	40	46	43	57	22/4.5
Maubeuge	39	42	45	29	5/2.5
Mende	37	31	34	68	80/1
Menton	62	50	56	68	50/17
Metz	37	34	31	5V	1000/45
Mezières	29	23	26	36	500/?
Millau	47	44	41	10	20/0.15
Montmelian	64	58	61	9	3/?
Montpellier	56	50	53	9	1000/2.5
Mortain	50	52	55	—	10
Morteau	48	54	51	10	20/0.01
Mulhouse	27	21	24	5	1000/300
Nancy	23	29	26	8	500/30
Nantes	23	29	26	9V	1000/300
Neufchatel	51	48	54	65	80/8
Nice	64	58	61	66	10/6
Niort	28	22	25/58e	6V	1000/405
Orleans	42	39	45	—	100
Paris	25	22	28	6	700/104
Paris Est	43	46	40	53	5/5
Paris Nord	45	39	56	59	10/4
Paris Sud	49	52	62	65	3/3
Parthenay	52	49	55	—	80
Perpignan	22	25	28	7	10/3
Pignans	46	43	40	56	2/3
Porto-Vecchio	40	37	34	6	25/?
Privas	64	58	61	6	50/0.02
Reims	43	46	40	9	1000/80
Rennes	39	45	42	7	1000/500
Rouen	23	33	26	7	500/65
Sarrebourg	50	40	53	—	250
Sens	57	63	50/605		100/1.3
Serres	50	53	56	4	2/1
St. Etienne	35	30	33	38	10/12
St. Flour	52	49	55	7	6/0.4
St. Martin	48	51f	54	8	5/0.03
St. Raphael	25	28	22	10V	1000/70
Strasbourg	62	56	43	10V	1000/20
Tarascon	52	55	49	8	5/0.5
Toulon	51	48	54	9	100/1.6
Toulouse	27g	21	24	5	500/100
Toulouse	—	—	47	—	600
Toulouse (town)	45	39	42	7	10/1
Tours	65	59	62	10	200/0.1
Troyes	27	24	38/217		1000/100
Ussel	42	45	39	—	50
Utelle	47	44	41	—	10
Vannes	50	56	53	5	500/16
Verdun	65	59	62	8	500/0.08
Villers-Cotterets	65	59	62	—	4
Vittel	30	35	32	—	80
Wissembourg	54	48	51	—	50

+ low power st's not mentioned under 1kW.
a=300kW; b=50kW; c=500kW; d=100kW; e=310kW; f=3kW; g=1000kW.
D.Prgrs: TF1: 24h — **F2:** 24h — **FR3:** 0500-0000 — **CanalPlus:** 0600-0200, Sat/Sun 24h.

LA CINQUIEME (Educational, Public Television)
Addr: 10 rue Horace Vernet, 92130 Issy les Moulineaux. ☎ +33 (1) 4146 5555. 📠 +33 (1) 4108 0222
L.P: Jean-Marie Cavada
D.Prgrs: 0515-1800 (using ARTE txs)

ARTE (Cultural, Public Television)
2A rue de la Fonderie, 67080 Strasbourg Cedex. ☎ +33 (3) 8814 2222. 📠 +33 (3) 8814 2200 — **L.P:** Jérôme Clément.
D.Prgrs: 1800-0100.

M6 Metropole TV (Priv. Comm.)
16 Cours Albert 1er, 75008 Paris. ☎ +33 (1) 4421 6666. 📠 +33 (1) 4563 7852.
L.P: Dir. Gen: Jean Drucker. **D.Prgrs:** 24h.

EUROPE — France

Stations:	ARTE kW (ERP)	M6	kW (ERP)
Abbeville	45	80	63	80
Ajaccio	41	110	44	110
Alencon	42	12	45	12
Amiens	49	5	52	(5)
Angers	50	20	53	(15)
Angouleme	31	5	34	(5)
Annemasse	54	3.8	51	3.8
Argenton	38	18	51	18
Aurillac	57	80	67	80
Autun	45	300	42	(300)
Auxerre	55	50	49	(100)
Avignon	47	40	36	40
Bar-Le-Duc	—	—	38	70
Bastia	49	5	35	5
Bayeux	52	1.6	49	1.6
Bayonne	56	40	45	(40)
Beauvais	49	100	52	(100)
Bergerac	66	90	58	90
Bordeaux	65	130	43	10
Boulogne	59	3.2	62	(3.2)
Bourges	21	130	56	(130)
Brest	34	100	60	(100)
Brive	21	2	48	(2)
Caen	64	100	61	100
Carcassonne	46	56	43	56
Chalon-Sur-Saone	44V	2	41V	(2)
Chambery	55	4	52	(4)
Charleville	44	5	41	(5)
Chartres	47	50	44	(50)
Chaumont	57	24	39	20
Cherbourg	35	10		
Cluses	32	1.6	63	1.6
Clermont-Ferrand	30	100	33	(100)
Dijon	46V	1		
Dunkerque	59	1	62	1.8
Gueret	30	10	33	8
Haguenau	—	—	36	1
Hyeres	67	34	38	10
La Rochelle	48	1	51	1
Laval	33	20	30	(20)
Le Creusot	38	2.5	60	(2.5)
Le Havre	53	20	56	(20)
Le Mans	32	150	35	(150)
Lens	51	9.5	54	9.5
Les Sables D'Olonne	61	1	63	1
Le Touquet	55	2.1	32	2.1
Lille	65	6	53	(6)
Limoges	38	2	35	2
Lorient	62	1	65	1
Lyon	28	10	22	(10)
Lyon	59	250	62	(250)
Mantes	55	10	53	(10)
Marseille	32	150	38	(100)
Marseille	54	22	49	(22)
Maubeuge	32	3	37	10
Mende	50	1.5	53	(1.5)
Metz	39	200		
Montpellier	48	250	40	100
Nancy	55	7	43	(5)
Nantes	21	50	65	(50)
Neufchatel	34	10	31	10
Nice	51	2	45	2
Niort	38	200	64	(200)
Orleans	52	20	36	(20)
Paris	30	100	33	(100)
Paris-Est	48	5	58	(5)
Paris-Nord	65	3	62	(3)
Paris-Sud	59	3	42	(3)
Parthenay	60	30	67	(30)
Pau	60	2	63	(2)
Perpignan	38	2	35	(2)
Pignans	58	3	61	3
Poitiers	41	1	44	1
Reims	53	25	56	(25)
Rennes	34	75	31	(75)
Rouen	59	100	62	(100)
Royan	—	—	33	2.1
Saint-Etienne	65	10	55	10
Saint-Flour	—	—	58	1
Saint-Nazaire	55	4	52	(4)
Saint-Raphael	36	80	39	(4)
Selestat	67	2.1		
Sens	47	40	44	(40)
Toulon	57	50	60	(50)
Toulouse	32	5	34	(5)
Toulouse	29	80	38	(80)
Tours	57	80	54	(80)
Troyes	29	250	50	(250)
Ussel	63	50	60	50
Valence	56	1.5	53	1.5
Valenciennes	49	6	34	(6)
Vannes	58	75	48	(75)
VillersCotterets	39V	25	57V	(25)

+1345 stations under 1kW.

private stations

TELE MONTE CARLO (relays in France)
✉ 16 boulevard Princesse Charlotte, Monte Carlo 98090, France. ☎ +377 9315 1617. 📠 +377 9325 0109.
Stations: Toulonch 33 25kW, Marseille ch35 150kW, Marseille(town) ch51 4.5kW all using L/SECAM.
D.Prgr: 1030-2400.

TELE TOULOUSE (Private, Comm.)
✉ 3 Place Alphonse Jourdain, 31069 Toulouse Cedex. ☎ +33 (5) 6123 6565. 📠 +33 (5) 612471.
L.P: Etienne Mallet.
Station: Toulouse ch37 2kW ERP, Muret ch63 40W.
D.Prgr: 0630-0000.

TELE BLEUE (Private, Comm.)
✉ Rue Brousan, 30128 Garons. ☎ +33 (66) 700123. 📠 +33 (4) 66700701 — **Station:** Nimes ch61 400W ERP, Ch66 500W. — **D.Prgr:** MF 1700-2230, Sat 0900-1130/1800-2200, Sun 1800-2230.

TELE LYON METROPOLE (Priv., Comm.)
✉ 15 bd Yves Farge, 69007 Lyon. ☎ +33 (4) 7271 1090. 📠 +33 (4) 7271 1095 — **L.P:** Jerome Bellay.
Station: Lyon ch25 2kW ERP, Lyon (south) ch56 1kW.
D.Prgr: 1700-2305.

8 Mont Blanc (Priv., Comm.)
✉ Route Pontets, 74320 Sevrier. ☎ +33 (4) 5052 6969. 📠 +33 (4) 5052 4991 — **L.P:** André Campana.
Stations: Cluses ch47 1.8kW, Annemasse ch57 3.8kW +29 low power stations under 1kW — **D.Prgr:** 1800-2300.

AQUI TV (Priv., Comm.)
✉ 19 rue Campniac, 24000 Perigueux. ☎ +33 (5) 5335 3000.
L.P: François Carrier.
Stations: Bergerac ch25 3.5kW + 10 stations under 1 kW— **D. Prgr:** 1130-2200

GERMANY

TV-sets: 30.500.000 — **Colour:** PAL — **System:** B.

ARD (PROGRAMMDIREKTION DEUTSCHES FERNSEHEN)
Arnulfstrasse 42, 80335 München. ☎ +49 (89) 59 00 01 +49 (89) 5900 3249
L.P: PD: Dr. Günter Struve. TD. Chrmn: Ingo Dahrendorf. Film Buyer: Klaus Lackschewitz
NB: The ARD is an umbrella organisation representing regionalized German public radio- and tv broadcasters. The ARD is responsible for the first public tv network (ARD Eins) and third public tv programs.
ARD members:
-**Bayerischer Rundfunk Fernsehen,** Rundfunkplatz 1, 80335 München. ☎ +49 (89) 59 00 01 +49 (89) 5900 2375
-**Hessischer Rundfunk Fernsehen,** Bertramstrasse 8, 60320 Frankfurt. (+ 49 (69) 1551 2 +49 (69) 1552 900
-**MDR Fernsehen,** Kantstrasse 71-73, 04275 Leipzig. ☎ +49 (341) 22760 +49 (341) 5663 544
-**NDR Fernsehen,** Rothenbaumchaussee 132, 20149 Hamburg. ☎ +49 (40) 4131 +49 (40) 4476 02
-**ORB Fernsehen,** August-Bebel-Strasse 25-53, 14482 Potsdam-Babelsberg. ☎ +49 (331) 72 36 00 +49 (331) 77395
-**Radio Bremen Fernsehen,** Hans-Bredow-Strasse 10, 28307 Bremen. ☎ +49 (421) 2460 +49 (421) 246 2010/1010
-**Saarländischer Rundfunk Fernsehen,** Funkhaus Halberg, 66100 Saarbrücken. ☎ +49 (681) 6020 +49 (681) 6023 874
-**SDR Fernsehen,** Neckarstrasse 230, 70190 Stuttgart. ☎ +49 (711) 929 1 +49 (711) 929 2600
-**SFB Fernsehen/B 1,** Masurenallee 8-14, W-14057 Berlin. ☎ +49 (30) 3031 0 +49 (30) 301 50 62
-**SWF Fernsehen,** Hans-Bredow-Strasse, 76530 Baden-Baden. P.O. Box 820, 76485 Baden-Baden. ☎ +49 (7221) 92 0 +49 (7221) 92 20 13
-WDR Fernsehen, Appellhoffplatz 1, D-50667 Köln. ☎ +49 (221) 22 01 2 +49 (221) 2204 300
D.Prgr: ARD Eins, Germany's first public tv network, broadcasts nationwide 24h (except for 2 fi hrs. in the evening when regional tv from 13 tv centers is relayed). In 1992 ARD together with Germany's second public tv network ZDF, started broadcasting a common breakfast tv program called "Morgenmagazin" (Mon-Fri 0500-0800). Germany's third program comprises a combination of seven separate channels: NDR/RB in "N 3"; MDR in "MDR 3"; ORB in "ORB 3"; WDR in "West 3"; HR in "HR 3"; SWF/SDR/Saarländischer Rundfunk in "Südwest 3"; Bayerischer Rundfunk in "BR 3".

ZWEITES DEUTSCHES FERNSEHEN (ZDF)
P.O. Box 4040, 55030 Mainz. ☎ +49 (6131) 70 1 +49 (6131) 7021 57
L.P: Gen.Dir: Prof.Dr. Dieter Stolte; TD: Dr. Albert Ziemer; Film Buyer: Dr. Hans-Jürgen Steimer
D.Prgr: The ZDF is Germany's second public tv network, broadcasting nationwide 24h.

COMMERCIAL TV STATIONS

SAT EINS
+ Martin Luther Strasse 1, 10777 Berlin, Germany. ☎ + 49 (30) 21241 0 2 +49 (30) 21241 140
L.P: Man Dir: Hans Grimm; Hd. of News: Heinz Klaus Mertes; Film Buyer: Akim Andorfer
D.Prgr: This commercial satellite tv station, set up by a number of German publishing houses, broadcasts 24h. Reg. prgr. from Hannover, Dortmund, Mainz, Stuttgart, München at 1630-1700 UTC.

RTL FERNSEHEN
Aachenerstrasse 1036, 50858 Cologne, Germany. ☎ +49 (221) 456 0 +49 (221) 456 4290
D.Prgr: Originally Luxembourg's German language tv channel, but due to German legal rules currently operating from Cologne. RTL Fernsehen broadcasts 24 hrs. with loc. prgrs. (Hamburg, Essen, Frankfurt, Mannheim, München, Berlin) from 1700-1730.
N.B: Also relayed via satellite

RTL-2
Bavariafilmplatz 7, 82031 Grünwald. ☎ +49 (89) 641850. +49 (89) 64185999.
Stations: Aschaffenburg ch21; Augsburg ch58; Bayreuth ch46; Deggendorf ch52; München ch27; Nürnberg ch53; Regensburg ch48; Rosenheim ch50; Weilheim ch47; Würzburg ch34/56 (ASTRA)

SUPER RTL
Richard-Byrd-Strasse 6, D-50829, Cologne. ☎ +49 (221) 9155 0
Stations: Hamburg ch34 (local prgr "Hamburg 1").

DEUTSCHES SPORTFERNSEHEN (DSF)
Bahnhofstrasse 27, 85774 Unterföhring. ☎ +49 (89) 95002 0 +49 (89) 9500 2392
L.P: Man.Dir: Dr. Dieter Hahn; Prgr. Dir: Rudolf Brückner; Marketing Dir: Kai Blasberg
D.Prgr: 24h. Sports & Leisure.

VOX
Richard-Byrd-Strasse 6, D-50829, Cologne. ☎ +49 (221) 9534 0 +49 (221) 9534 440
Stations: Bielefeld ch36 (1kW); Bochum ch50 (100W); Bonn ch34 (1.5kW); Bonn ch51 (200W); Bremen ch49 (63kW); Bremerhaven ch57 (63kW); Burscheid ch41 (4kW); Chemnitz ch47; Dortmund ch43 (500W); Düren ch22 (400W); Düsseldorf/Witzhelden 39 (100kW); Gevelsberg ch5 (2.5W); Herdecke ch22 (20W); Iserlohn ch7 (2W); Köln ch27 (200W); Köln/Bonn ch40 (3000W); Lippstadt ch24 (30W); Mülheim ch27 (250W); Münster ch34 (200W); Paderborn ch22 (100W); Rheine ch74 (4kW); Saarbrücken/ Winterberg 56 (1kW); Siegburg/Troisdorf ch51 (200W); Siegen ch36 (30W); Stolberg ch29 (1500W); Unna ch51 (200W); Wesel/ Büderich ch59 (200kW); Wuppertal ch8 (1kW).

PRO SIEBEN
Bahnhofstrasse 27, 85774 Unterföhring. ☎ +49 (89) 9507 1000.

Principal Transmitting stations

1)BAYERISCHER RUNDFUNK FERNSEHEN
BR. third program = "Bayerisches Fernsehen", at some moments split up in an "Altbayern" and "Frankenland" version.

Area	ARD	ZDF3	S.1	RTL	DSF	kW
Amberg	—	37	43	50	52	—	350/320
Ansbach	—	—	—	—	—	49	
Augsburg	—	23	44	38	30	58	310/430
Bamberg	52	24	56	45	48	54	50/85/90
Bayreuth	—	30	54	39	46	—	98/96
Brotjacklr./D.	7	33	40	22	35	52	100/380/270
Büttelberg	55	—	—	—	—	—	400
Coburg	—	22	41	—	—	—	190/41
Dillb./Nürnb.	6	34	59	40	36	53	100/400/492
Grünten-Allgäu	2	28	46	—	—	—	100/470/470
H.Linie/Rgnsb.	53	21	42	38	34	46	75/370/400
H.Peissenberg	25	22	53	42	33	—	10/350/350
Högl	—	42	50	—	—	—	40
Hoher Bogen	55	28	59	—	—	—	200
Hühnrb./Hsslb.	60	32	47	—	—	—	400/140/200
Kreuzberg/Röhn	3	29	49	44	—	21	100/350/270
Landshut	—	39	58	—	—	—	27/55
Ochsenk./Hof	4V	23	57	—	—	—	100/500/500
Passau	—	30	60	—	—	—	41/34
Pfaffenberg/S.	59	35	51	—	—	—	100/230/250
Pfaffenhofen	—	31	41	—	—	—	480/420
Pfarrkirchen	—	27	57	—	—	—	250
Regensburg	—	—	—	—	—	38	?
Schnaitsee	—	26	54	—	—	—	360/280
Wendelstein/	10	35	56	59	24	37	100/200/200
München	—	—	—	—	—	38	
Würzburg	10	25	45	21/	57/	—	5/410/360
	—	—	—	38	60	—	

2) HESSISCHER RUNDFUNK FERNSEHEN
HR. third program = "Hessen 3".

Area	ARD	ZDF3	S.1	RTL	DSF	kW
Biedenkopf/A.	2	24	52	—	—	—	100/500/500
Fulda/Rhön	47	29	37	60	41	—	10
Gr. Feldberg	8	34	54	—	—	—	100/500/500
Habichtswald	56	28	41	—	—	—	50/100/100
Hardb./Krehb.	5	33	43	—	—	—	10/100/100
Hoher Meissner	7	32	55	—	—	—	100/390/470
Hohes Lohr	—	—	32	45	—	—	63/63
Kassel	—	—	—	42	35	58/60	
Rimberg	57	25	39	—	—	—	400/400/332
Würzberg	56	—	—	—	—	—	100/350/350

3) MITTELDEUTSCHER RUNDFUNK FERNSEHEN
MDR. third program = "Mitteldeutsches Fernsehen"

Area	ARD	ZDF3	S.1	RTL	DSF	kW
Brocken	6H	49	34H	—	—	—	100/500/1000
Chemnitz	8H	49	32	—	—	—	100
Dequede	12V	—	31H	—	—	—	2/500
Dresden	10V	46	29H	—	—	—	100/570/1000
Görlitz	6	—	—	—	—	—	1
Halle	—	—	46	57	—	—	
Inselsberg	5H	—	31H	—	—	—	100/500
Kapaunberg	—	40	37	—	—	—	?
Keula	—	—	45	—	—	—	25
Kulpenberg	8	—	57	—	—	—	0.12/1
Leipzig	9V	42	22H	—	—	—	100/500/600
Leipzig-Zeitz.	35	—	52	—	—	—	85/100
Löbau	27H	56	39H	—	—	—	200/500/500
Naumburg	35	—	52	—	—	—	85/200
Remda	21	50	27	—	—	—	200/500/500
Ronneburg	59	—	25	—	—	—	20/2
Saalfeld-Remda	21	50	27	—	—	—	200/500/200
Sieglitzberg	7	—	29	—	—	—	0.8/5
Schöneck	28	—	—	—	—	—	5
Sonneberg	12H	—	33H	—	—	—	30/500

Weida 59 — 25 — — — 20/50
Wittenberg 30 38 55 — — — 150/5/210

D.Prgr: Mitteldeutsches Fernsehen: Mon-Fri 1300-2330 approx. (includes relays of local "Fenster" (Window) programmes. Sat/Sun: 1200-0100 approx.)

4) NDR FERNSEHEN/B1/RB FERNSEHEN
NDR/RB. third program = "N 3". NDR1 Hamburg on ch E56 (500kW), NDR1 Cuxhaven on ch E51 (250kW).

NB: On October 1st 1992 Sender Freies Berlin Fernsehen started up a new regional prgr. under the name "B 1". This prgr. replaces "N 3" in the Berlin area.

Area	ARD	ZDF3	S.1	RTL	DSF	kW
Aurich	53	33	43	—	—	—	230/410/400
Bannenb./Höhb.	43	21	45	—	—	—	250/500/440
Berlin	7	33	39	25	22	9V	100/430/200
Bremen	22	32	42	29	46/	—	100/500/500
	—	—	—	—	49	—	
Bremerh./Cuxh.	45	24	48	5	8/11	—	30/330/330
Bungsb./Eutin	50	21	47	31	44	—	260/235/280
Flensburg	4	39	57	28	24	—	50/160/150
Garz (Rügen)	40	21	29	—	—	—	250/250/250
Hamburg	9/26	30	40	48	46	34	20/100/460/500
Hannover	8	24	44	40	36	—	5/500/500
Harz/Göttingen	10	21	59	39	29	—	100/160/250
Heide/Eiderst.	10V	31	44	—	—	—	25/500/380
Helpterberg	37	52	22	—	—	—	500/200/1000
Kiel	5	35	55	53	24	—	2/250/200
Lingen	41	24	59	30V	—	—	400/500/400
Lübeck	—	23	33	42	36	—	240
Marlow	8	43V	24V	—	—	—	100/100
Minden	—	26	54	—	—	—	500/20
Neumünster/S.	28	26	45	—	—	—	500/100/100
Niebüll	29	34	60	—	—	—	100/200/150
Osnabrück	50	39	56	44V	36	—	160/250/138
Schwerin	11	—	29	—	—	—	100
Stadthagen	47	—	—	—	—	—	100
Steinkimmen/C.	55	37	40	27V	35V	—	400/47/47
Torfhaus	—	23	53	—	—	—	500
Ülzen	—	27	58	—	—	—	500/400
Visslh./Verden	7V	25	60	—	—	—	20/66/50
Züssow	36	51	23	—	—	—	70/190/190

5) OSTDEUTSCHER RUNDFUNK BRANDENBURG FERNSEHEN
ORB. third program = "Fernsehen Brandenburg"

Area	ARD	ZDF3	S.1	RTL	DSF	kW
Belzig	—	—	54	—	58	—	50/5
Berlin	7	33	27	—	—	—	100/700
Casekow	—	—	54	—	—	—	100
Cottbus	53	57	23	—	—	—	355/100/577
Frankfurt/Oder	11	50	43	—	—	—	0.8/83
Höhbeck	51	21	35	—	—	—	200
Treplin	—	50	43	—	—	—	?
Woldegk	—	—	—	34	—	—	2.9

6) SWF FERNSEHEN/SDR FERNSEHEN/SR FERNSEHEN
SWF/SDR/SR. third program = "Südwest 3" (further regional split-ups as indicated under a-d).

a) SWF-Mainz:

Area	ARD	ZDF3	S.1	RTL	DSF	kW
Ahrweiler	—	33	56	—	—	—	170/210
Bad Marienberg	47	21	44	—	—	—	50/190/165
Boppard	—	28	41	—	—	—	160/120
Donnersberg	10	37	60	—	—	—	100/250/250
Eifel	23	30	40	—	—	—	20/195/180
Haardtkopf	25	35	55	—	—	—	400

Germany — Greece

Hohe Derst	23	44	46	—	—	1
Kaiserslautern	3	22	44	50	33 —	25/25
Koblenz	6	31	51	57	36 —	50/85/83
Mainz	11	—	—	36	— 44	1
Saarbr./Tflsk.	29	45	53	26	— —	20/50/50
Trier	5	22	48	56	— —	0.6/41/22
Weinbiet	6	—	—	—	— —	30

b)SR-Saarbrücken

Area	ARD	ZDF	3	S.1	RTL	DSF	kW
Göttelborner Höhe/							
Saarbrücken	2	45	42	26	—	—	100/470/500
Nelb./Piesb.	28	36	44	56	—	—	?

c)SWF-Baden-Baden

Area	ARD	ZDF	3	S.1	RTL	DSF	kW
Brandenkopf	48	28	45	—	—	—	2.5/25/50
Eggb./Hochrhn.	22	39	52	—	—	—	0.8/225/295
Feldberg/	8	22	57	50/	36/	—	100/100/100
Donausch.	—	—	—	60	38	—	
Freib.im Brsg.	7	33	58	—	38	—	0.9/500/500
Grünten	43	—	—	—	—	—	500
Hornisg./B-B	9	31	41	49	—	55	80/220/290
Raichenbach/R.	4	35	55	—	—	—	100/260/330
Ravensburg	—	37	40	27	— 30V	270/290
Waldshut	21	—	—	—	—	—	1
Wannenberg	30	—	—	—	—	—	5

d)SDR-Stuttgart

Area	ARD	ZDF	3	S.1	RTL	DSF	kW
Aalen/Heubach	8	29	52	—	—	—	50/209/135
Bd Mergentheim	48	—	—	—	—	—	10
Heidelberg	7/50	27	53	—	—	—	100/500/ 440/492
Heilbron/Eberb.	49	30	58	—	—	—	10/215/195
Langenbrand/P.	21	34	59	23	—	—	100/170/250
Stuttgart	11	26	39	—	—	—	100/404/234
Neu-Ulm	—	33	54	36	— 48/51	370/330/2
Waldenberg/L.	9	28	42	—	—	—	100/165/185

7)WESTDEUTSCHER RUNDFUNK FERNSEHEN

WDR. third program = "West 3".

Area	ARD	ZDF	3	S.1	RTL	DSF	kW
Aachen	24	37	58	27	26	26	200/320/400
Bonn	43	26	49	—	5/36	—	100/81/100
Dortmund	—	25	53/	47	58	58	500/470
Ederk./Hochs.	50	27	40	—	—	—	50/250/250
Eggeberge	—	31	48	—	—	—	250
Kleve/	46	35	48/	—	52	52	100/500/475/
Wesel	—	—	59	—	—	—	80
Köln	11	—	—	36	—	—	10
Langenberg/	9	29	39/	44	36	36	100/500/100/
Düsseldorf	—	—	55	—	—	—	450
Minden	—	26	57	—	—	—	500/200
Münster	32	21	45	51	38	—	500/170/250
Nordh./Lüdens.	30	37	60	—	—	—	240/200/250
T.Wald/Bielef.	11	33	46	38	59	59	100/310/450
Wuppertal	6	22	42	—	—	—	0.6/100/85

GIBRALTAR

TV-sets: 7.500 — Colour: PAL — System: B & G.

GBC TELEVISION (partly comm.)
📧 Broadcasting House, 18 So. Barrack Rd, Gibraltar.
☎ +350 79760. **Cable:** Broadcasts. 📠 +350 78673. ① 2229 GBCEE GK.
L.P: GM: George Valarino. Senior Eng: John Tewkesbury.
Station: chE6 0.2/0.4kW H (+ low power repeaters ch12, 53, 56).
D.Prgr: rel BBC World + local prgrs.

GREECE

TV-sets: 2.300.000 — Colour: PAL — Systems: B & G.

ELLINIKI TILEORASSI-1 (ET-1)
📧 Leophoros Messogeion 432, GR-153 42 Aghia Paraskevi Attikis. ☎ +30 (1) 63 95 970. 📠 +30 (1) 63 92 263. ① 216066 — **L.P:** DG: George Stamatelppoulos.

Stations	ch	kW	Pol
Akarnanika	3	1.5/	H
Rhodes Isl.	9	10/2	V
		0.3	H
Mytilini	9	10/2	H
Thessaloniki	5	30/6	H
Tripolis	10	30/6	H
Pilion (Volos)	6	30/6	H
Heraklion	10	10/2	H
Alexandroupolis	6	30/6	V
Ioannina	10	10/2	H
Kalamata	6	10/2	H
Parnis (Athens)	11	30/6	H
Chania	7	10/2	H
Hymettos Kavala (Athens)	21	15/3	H
Kastoria	7	30/6	V
Tholopotamos	22	30/6	H
Kefalinia Isl.	8	30/6	V
Alexandroupolis	33	15/3	H
Thira	8	30/6	H
Thassos Isl.	39	550/55	H
Gerania	9	10/2	H
Euros	50	18/7	H
Corfu Isl.	9	10/2	H

+ 700 low power repeaters.
D.Prgr: 0800-2400 (approx.).

ELLINIKI TILEORASSI-2 (ET-2)
📧 Leophoros Messogeion 136, GR-115 25 Athens.
☎ +30 (1) 7701911. 📠 +30 (1) 77 97 776. ① 210886.
L.P: DG: Panos Panayotu.

Stations	ch	kW	Pol
Hymettos Pilion (Athens)	41	750	H
(Athens)	5	30/7.5	V
Rhodes Isl.	42	500	H
Alexandroupolis	21	170	H
Tripolis	42	500	H
Thassos Isl.	23	1000	H
Akarnanika	43	1000	H
Ioannina	25	250	H
Mytilini	48	200	H
Thira Isl.	29	500	H
Vitsi (Kastoria)	49	250	H
Thessaloniki	30	1000/	
Corfu Isl.	50	530/200	H
Kefalinia Isl.	57	430	H
Parnis (Athens)	34	450	H
Kavala	59	650	H

+300 low power repeaters
D.Prgr: 0800-2400 (approx.).

ET-3 (regional channel for Macedonia)
📧 Aggelaki 2, GR-546 21 Thessaloniki. ☎ +30 (31) 27 87 84. 📠 +30 (3) 23 64 66 — **L.P:** DG: Hichalis Alexandridis.
Station: Thessaloniki ch23/27 H (local prgrs), Pilion

EUROPE

ch 44H, Pag-gaion ch 35H, Parnis ch 52H, Hymettos ch 31H, Thassos Isl. ch 26H, Florina ch 39H, Polygiros ch 21H.
D.Prgr: 0800-2400 (approx).

Private Stations

Antenna TV, 10-12 Kifissias Ave, Maroussi, 15125 Athens. ☎ +30 (1) 6842220. 🗎 +30 (1) 3890304.
Argo TV, Metamorphosseos 9, 55132 Kalamaria, Thessaloniki. ☎ +30 (31) 351733. 🗎 +30 (31) 351739.
Channel Seven-X, 64 Leoforos Kiffissoas, Athens. ☎ +30 (1) 68976042. 🗎 +30 (1) 6897608.
City Channel, 14 Leoforos Kastoni, 41223 Larissa. ☎ +30 (1) 232839. 🗎 +30 (1) 232013.
Tele City, 58 Praxitelous, 17674 Athens. ☎ +30 (1) 9429222. 🗎 +30 (1) 9413589.
Jeronimo Groovy TV, Ag konstantin 40, 15124 Athens. ☎ +30 (1) 6896360. 🗎 +30 (1) 6896950.
Mega Channel, 10 Alamanas Str, 15125 Amarousion, Athens. ☎ +30 (1) 689900014. 🗎 +30 (1) 6899016.
Neo Kanali SA, 9-11 Pireos, 10552 Athens. ☎ +30 (1) 5238230. 🗎 +30 (1) 5247325.
Serres TV, Nigritis 27, 62124 Serres.
Skai TV, 2 Phalereos & Ethnarchou, Macaroiu, N. Phaliro.
Star Channel, 37 Dimitras, 1178 Tayros, Athens. ☎ +30 (1) 3450626. 🗎 +30 (1) 3452190.
Teletora, 17 Lycabetous, 10672 Athens. ☎ +30 (1) 3617285. 🗎 +30 (1) 3638712.
Traki TV, Central Square, 67100 Xanthi. ☎ +30 (541) 20670. 🗎 +30 (541) 27368.
TRT, 5 Zachou Str, 38333 Volos. ☎ +30 (421) 30500. 🗎 +30 (421) 36888.
TV Macedonia, 222 Nea Egnatia, 54642 Thessaloniki. ☎ +30 (31) 850512. 🗎 +30 (31) 850513.
TV Plus, Syngrou Ave. 97, 11745 Athens. ☎ +30 (1) 9028707. 🗎 +30 (1) 9028310.
TV-100 (Thessaloniki municipal st.), 16 Aggelaki Str, 54621 Thessaloniki. ☎ +30 (31) 265828. 🗎 +30 (31) 267532.

AFN TV (U.S. Mil.)
Station: chA2 Iraklion, A6.

HUNGARY

TV-sets: 4.261.600 — **Colour:** PAL — **Systems:** D & K.

MAGYAR TELEVÍZIO (MTV)
✉ Szabadság ter 17, 1810 Budapest, 5. ☎ +36 1111 4059. 🗎 +36 1115 74979. ① 325558.
L.P: Pres: Gyula Berecky.
Stations: Prgr. 1 (System D), Prgr. 2 (System K)

loc.	Prgr 1 (kW)		Pol	Prgr 2 (kW)	Pol
Aggtelek	R28	200	H	R45	200	V
Budapest	R1	150/50	H	R24	600	V
Budapest	R41	600	H	—	—	—
Csávoly	R28	80/8	H	R7	9/0.9	V
Csengod	25	200	H	42	200	H
Fehergyarmat	R24	600	V	R41	600	H
Györ	R8	10/1	V	35	105/10H
Kabhegy	R12	150/15	H	22	760/76H
Kékes	R8	30/3	H	36	880/88H
Komádi	R7	50	V	R32	200/20H
Nagykanizsa	R1	50/5	V	R31	380	V

loc.	Prgr 1 (kW)		Pol	Prgr 2 (kW)	Pol
Ozd	35	200	H	—	—	—
Pécs	R2	60/6	V	32	400/40H
Sopron	R9	5/1	V	R32	200	V
Szeged	26	200	H	31	?	H
Szentes	R10	200/20	V	R23	480	H
Tokaj	R43	80/20	H	26	420/42H
Tokaj	R4	80	H	—	—	—
Vasvar	R33	600	V	46	600	V

low power st's not mentioned.
D.Prgr: Prgr .1: Tue/Thu 0655-1045, 1440-2245; Mon/Wed 0800-1045, 1440-2245; Fri 0800-1045, 1440-2345; Sat 0530-1140, 1415-2300; Sun 0700-1140, 1250-2210. **Prgr. 2:** 1600-2215; Sat 1300-2320; Sun 1300-2300.

Private Stations

NAP TV
✉ Angol utca 13, 1149 Budapest. ☎ +36 (1) 251 0490. 🗎 +36 (1) 251 3372 — **L.P:** Chmn: Tamas Gyarfas.
Nap TV provides morning news and inf. sce. on Prgr 1 of MTV.

A3 (Pest-Buda TV) (local st.)
✉ Rona utca 140, 1147 Budapest. ☎ +36 (1) 251 4749.
L.P: Gen. Dir: Adam Namenyi.

TV3 BUDAPEST (local st.)
✉ Budakeszi utca 51, 1021 Budapest. ☎ +36 (1) 275 1800. 🗎 +36 (1) 275 1801 — **L.P:** Dir: Peter Kolin.

SIO TELEVISIO (local st.)
✉ Fo ter 2, 8600 Siofok. ☎ +36 (84) 317111. 🗎 +36 (84) 310887 (provides prgrs for visitors to Lake Balaton area).

ICELAND

TV-sets: 93.595 — **Colour:** PAL — **Systems:** B&G.

RÍKISÚTVARPID — SJÓNVARP
ICELANDIC NATIONAL TELEVISION
✉ Laugavegur 176, 150 Reykjavik. ☎ +354 515 3900. 🗎 +354 515 3008. ① 2035 VISION IS. **E-mail:** istv@ruv.is
L.P: MD: Petur Gudfinnsson. CE: E. Valdimarsson.
Stations: Pol H; Audio power 10% of visual power indicated.

Location	chE	kW		chE	kW
Stykkishólmur	3	90	Háfell	7	1100
Skálafell	4	300	Girdisholt	7	2.5
Gagnheidi	4	80	Hegranes	8	12
Vestmannaeyjar	5	39	Hnjúkar	9	20
Heidarfjall	5	10	Vatnsendi	10	20
Vadlaheidi	6	490	Vidarfjall	11	8.4

+ 156 low power relay st's.
D.Prgr: Mon-Thurs 1800-2300; Fri 1800-2400; Sat 1600-0100; Sun 1600-2330.

OMEGA (Rlg.)
✉ Grensasvegur 8, 108 Reykjavik. ☎ +354 568 3131. 🗎 +354 568 3741 — **L.P:** General Dir: Erik Eriksson.

Iceland — Italy

STOD 2 (Channel 2) & STOD 3 (Channel 3) (Subscription sces.)
✏ Krokhals 6, 112 Reykjavik. ☎ +354 515 6000 (Stod 2), +354 533 5600 (Stod 3). 🖷 +354 515 6810.

ICE TV Channel 3
✏ c/o Laufey Gudjonsdottir, Kringlan 7, Reykjavik 103. ☎ +354 533 5633. 🖷 +354 533 5699.

IRELAND

TV-sets: 1.056.000 — **Colour:** PAL — **System:** I.

RADIO TELEFIS EIREANN (Statutory Corporation)
✏ Donnybrook, Dublin 4. ☎ +353 (1) 208 3111. 🖷 +353 (1) 208 3080. **WWW:** http://www.rte.ie/tv/
L.P: Contr. of TV. Prod: Liam Miller.

Stations	I	II	kW	Pol	Stations	I	II	kW	Pol
Maghera	B		100	H	Mt.Leinster	F	I	100	V
	E	H	100	V	Kippure	E	H	100	H
Mullaghanish	D	G	100	V	Truskmore	I	G	100	H
N.E.Donegal	23	26	20	H	Longford	40	43	800	H
Dublin	29	33	25	H	Co.Louth	52	56	250	V

+36 low power transposer st's.
D.Prgr: RTE One: Mon-Fri 1100-2345 (Fri 0100 appr.), Sat 0900-0100, Sun 1005-2345. **RTE Two:** Mon-Fri 1430-2330, Sat 1230-0030, Sun 1015-2345.

ITALY

TV-sets: 17m — **Colour:** PAL — **Systems:** B & G.

RADIOTELEVISIONE ITALIANA
✏ Direzione Centrale TV, Viale Mazzini 14, 00195 Roma.
☎ +39 (6) 38781. 🖷 +39 (6) 3226070. ℂ 614432
L.P: Chmn: Enzo Siciliano. GM: Franco Iseppi. Dir RAIUNO: Giovanni Tantillo. Dir RAIDUE: Carlo Freccero. Dir RAITRE: Giovanni Minoli. Dir TG1: Rodolfo Brancoli. Dir TG2: Clemente Mimun. Dir TG3: Lucia Annunziata. Dir. TGR: Nino Rizzo Nervo. Dir. TGS: Marino Bartoletti. Dir. Televideo: Marcello Del Bosca. Dir. RAI-International: Roberto Morrione. PR: Carlo Sartori.

Stations: System: B/PAL in VHF band — G/PAL in UHF band. Pol. H except where indicated differently.

Location	RAIUNO	RAIDUE	RAITRE	kW
Agricento Giache (AG) D	—	—	2	
Alcamo M. Bonifato (TP)Ev	25	44		18/12
Bari (BA)	Fv	—	—	10
Benevento (BE)	—	33	51	12
Bertinoro (FO)	Fv	30	40	40/8
Bologna C. Barbiano (BO)31		28	48	200/100
Bra (CN)		22v	—	20
Brescia Vedetta (BS)	—		33	20
Cagliari Capoterra		28v	—	10
Camerino (MC)	F	—	—	3
Campo dei Fiori (VA)	59	28	22	20
Canepina (VT)	—	—	68v	12
Capo Milazzo (ME)	47	25	22	15
Capo Spartivento (RC) H		23v	39v	30/6
Casteluono (PA)	F	22v	62v	3/2
Catanzaro M. Tiriolo (CZ)Fv		30	55	20/2
Cima Penegal (TN)	—	27	69	20/10
Col Visentin (BL)	H	34	49	50/4
Crotone (CZ)	Bv	27	40	20/8
Crotone (CZ)	58	—	—	20
Firenze M. Morello (FI)	—	29	—	15
Fiuggi (FR)	—	25	67	40
Gambarie (RC)	D	26v	44v	400/20
Gambarie (RC)		—	51v	400
Genova Portofino (GE)	H	29	45	1000/160
Golfo di Palicastro (SA)	—	33v	50v	12
Golfo di Salerno (SA)	En	30v	69V	20
L'Aquila M. Luco (AQ)		24	—	15
Martina Franca (TA)	D	32	49	1000/200
Messina M. Piselli (ME)	—	29	42	40/20
Milano (MI)	G	26	33	40/20
Mione (TN)		69v	—	51	10
Monte Argentario (GR)E		24v	46v	500/25
Monte Argentario (GR)	—		68v	300
Monte Ascensione (AP)H2		56	—	50/5
Monte Beigua (SV)	—	32	42	120/60
Monte Caccia (BA)	A	25	44	1000/200
Monte Caccia (BA)		54	—	—	1000
Monte Cammarata (AG)A		34	48	400/30
Monte Canata (PR)	42v	31v	37v	30/10
Monte Cavo (RM)	H2	35	39	230/40
Monte Cimarani (AQ)	—	22	—	10
Monte Conero (AN)	E	26	42	400/100
Monte Creo (BG)	—	27	41	10
Monte D'Elio (FG)	Bv	24	40	20/2
Monte D'Elio (FG)	63	—	—	20
Monte Faito (NA)	B	23	39	200/44
Monte Faito (NA)	—	23v	39v	1200
Monte Favone (FR)	—	29	37	40/3
Monte Lauro (SR)	F	24	41	400/200
Monte Limbara (SS)	H	32	52	40/3
Monte Luco (SI)	H1	23	39	40/3
Monte Luco (SI)	H1v	—	—	3
Monte Maddalena (BS)F		43	D	13/2
Monte Maddalena (BS)	—	43v	—	10
Monte Nerone (PS)	A	33	54	400/34
Monte Peglia (TR)	H	31	49	400/30
Monte Peglia (TR)	—	—	37	400
Monte Penice (PV)	B	23	35	200/100
Monte Penice (PV)	—	—	36	250
Monte Pierfaone (PZ)	—	21	42	20
Monte Sambucco (FG) H		27	47	400/35
Monte Sambucco (FG)	—	—	50	400
Monte Scuro (CS)	G	28	46	60/8
Monte Serpeddi (CA)	G	30	49	500/30
Monte Serpeddi (CA)	67	—	—	10
Monte Serra (PI)	D	27	43	400/200
Monte Soro (ME)	E	32	67	50/20
Monte Subasio (PG)	51	35	47	10
Monte Subasio (PG)	51v	35v	47v	10
Monte Turu (TO)	H1v	28v	51v	10/2
Monte Venda (PD)	D	25	32	1000/100
Monte Vergine (AV)	D	31	43	200/20
Paganella (TN)	—	21	47	30/15
Palermo M. Pallegrino (PA)F		30v	46v	80/40
Palermo M. Pallegrino (PA)	—	27V	40V	80/40
Pantelleria C. Glindo (TP)Gv		—	—	25
Pescara S. Silvestro (PE)F		30v	46v	60/30
Pescara S. Silvestro (PE)60v		—	—	110
Piane di Mocogno (MO)	—	30v	—	13
Pigazzano (PC)		—	37	30
Punto Badde Urbana (OR)D		27	47	400/160
Roma M. Mario (RM)	G	28	43	300/36
Roseto Capo Spulico (CS) Fv		—	68	40/12

EUROPE — Italy — Latvia

Saint Vincent (AO)	—	31	41	20/10
Salento Turrisi (LE)	—	34	47	60
San Cerbone (FI)	G	—	—	50
Sassari M. Oro (SS)	—	30v	—	20
Sezze (LT)	F	31v	68v	5/4
Stazzona (CO)	—	—	53	10
Terracina (LT)	—	28	—	10
Torino Eremo (TO)	C	30	40	400/15
Torino Eremo (TO)	5	—	—	400
Trapani Erice (TP)	Hv	31	43	50/20
Trapani Erica (TP)	—	31v	43v	8/4
Trieste M. Belvedere (TS)	G	31	44	50/3
Trieste Muggia (TS)	35v	28v	48v	20
Udine (UD	F	22	47	300/160
Val Venosta (BZ)	A	22	36	3/2
Velletri (RM)	—	—	26	50	400/200

+ over 5200 st's below 2kW (VHF)/10kW (UHF) not mentioned

RUNDFUNKANSTALT SÜDTIROL (RAS)
(Public Statutory Body of the Autonomous Province of Southern Tyrol)

✉ Europaalee 164A, I-39100 Bozen. ☎ +39 (471) 3317 4258.
L.P: Pres: Helmuth Hendrich. MD: Klaus Gruber.
Stations: RAS1 (relay ORF-FS1); RAS2 (relay ZDF); RAS3 (relay ORF-FS2+SRG).

Location	ch	kW	Netw.
Perdonig	23	2.5	III
Penegal	55	10	II
Penegal	48	10	III
Perdonig	59	7.5	I
Vinschgau	50	3	III
Vinschgau	63	3	II
Penegal	52	10	I
Perdonig	65	7.5	II
Vinschgau	53	3	I

+ 122 low power st's not mentioned.

PRIVATE TV STATION NETWORKS

There are about 900 privately operated tv sts in Italy, mostly on local service. Due to space limitations, only those sts with nationwide networks are mentioned. Readers requiring more information should write (with Rp pls) to: D. Monferini, Via Davanzati 8, I-20158 Milano, Italy.

Stations:
1) Amica 8 & Amica 924h (key st. Firenze)
2) Canale 5   24h (key st. Milano)
2) Italia 1   24h (key st. Milano)
2) Rete Quattro   24h (key st. Milano)
3) Circuito 5 Stelle 24h (key st. Milano)
4) Junior TV   24h (key st. Firenze)
5) Cons.Italia 9 Netw. 24h (key st. Firenze)
6) Video Music/TMC2 24h (Key st. Castelv. Pasceli)
7) Italia 7   24h (key st. Milano)
8) Odeon   24h (key st. Rho-Milano)
9) Rete A   24h (key st. Milano)
10) Rete Mia   24h (key st. Lucca)
11) Tivultalia   24h (key st. Rho-Milano)
12) Rete MTV   24h (key st. Milano)
13) Super Six   24h (key st. Milano)
14) Tele Montecarlo 0700-0130 (key st. Roma)
15) Tele Più   24h (key st. Milano)

Addresses and other information
1) Sation Group DAPS, Via Vecchi 15, I 200-94 Assagro (MI). ☎+39 (2) 48843984. ✉ +39 (2) 488 43990. 20 st's affiliated.
2) Reti Televisive Italiane, Via Europa 48, I-20093 Cologno Monzese (MI). ☎ +39 (2) 25125. ✉ +39 (2) 2138019. 20 st's affiliated.
3) Società Circuito Cinque Stelle, Via Plinio 44, I-00193 Roma. ☎ +39 (6) 688001. ✉ +39 (6) 68800400. 17 st's affiliated.
4) Junior TV, Corso Garibaldi 35, I-20121 Milano. ☎ +39 (2) 801376. ✉ +39 (2) 72001156. 20 st's affiliated
5) Società Consorzio Italia 9 Network Television, Via Settala 57, I-20124 Milano. ☎ +39 (2) 29524964. ✉ +39 (2) 29404892. 20 -st's affiliated.
6) TMC2, Piazza delle Balduina 49, I-00136 Roma. ☎ +39 (6) 35584271. ✉ +39 (6) 35584227. 18 st's affiliated.
7) Italia 7, Via Einstein 21, I-20094 Assago (MI). ☎ +39 (2) 45701747. ✉ +39 (2) 45701724. 19 st's affiliated.
8) Odeon, Via Tavecchia 43/45, I-20017 Rho (MI). ☎ +39 (2) 935151. ✉ +39 (2) 93504423. 16 st's affiliated.
9) Rete A, Viale E Marelli 165, I-20099 Sesto San Giovanni (MI). ☎ +39 (2) 22477241. ✉ +39 (2) 2401630. 7 st's affiliated.
10) Rete Mia, Vallan Italiana Promomarket s.r.l., Via Tempaegnano 40, I-55100 Lucca. ☎ +39 (583) 490555. ✉ +39 (583) 490459; 20 st's affiliated.
11) Tivuitalia, Via Tavecchia 43/45, I-20017 Rho (MI). ☎ +39 (2) 935151. ✉ +39 (2) 93504423; 20 st's affiliated
12) MTV, Corso Europa 7, I-20122 Milano. ☎ +39 (2) 7621171. ✉ +39 (2) 7621227. 9 st's affiliated.
13) Super Six, GGS International s.r.l., Via Bolzano 29, I-20127 Milano. ☎ +39 (2) 26144822. ✉ +39 (2) 26144832; 19 st's affiliated
14) Tele Montecarlo, Piazza della Balduina 49, I-00136 Roma. ☎ +39 (6) 35584271. ✉ +39 (6) 35584257. 18 st's affiliated.
15) Tele Più, Via Piranesi 44/a, I-20137 Milano. ☎ +39 (2) 700271. ✉ +39 (2) 70027201. 20 st's affiliated.

LATVIA

TV-sets: 1.230.000 — **Colour:** SECAM/PAL — **System:** D&K

LATVIJAS TELEVIZIJA (LTV) (Gov.)

✉ Zakusalas krastmala 3, Riga LV-1509 ☎ +371 (2) 200314. ✉ +371 (2) 200025 ⓓ 161188 video lv — **L.P:** Dir. of TV: Imants Rakins. Dir. of Prgs: Daina Ostrovska. Int. Rel: Inese Vitkus.

Stations (pol = H)

	Netw 1	Netw 2	Netw 3	
	LTV-1	LTV-2	LNT	L.
Location	ch/kW	ch/kW	ch/kW	ch/kW
Cesvaine	R8/30	R5/20	41/603	36/0.6
Dagda	36/0.06	24/0.06	—	—
Daugavpils	R7/30	R10/0.65	40/603	21/0.6
Dundaga	25/47	30/47	R8/0.3	—
Engure	—	—	—	R2/0.015
Jekabpils	—	—	29/0.6	—
Kandava	R12/0.0015	—	—	—
Kraslava	—	R4/0.02	—	—
Kuldiga	R6/30	R1/10	40/0.6	—
Liepaja	33/203	21/203	35/81	R5/0.165
Malpils	—	—	—	R2/0.015

Latvia — Malta

Location	Netw 1 LTV-1 ch/kW	Netw 2 LTV-2 ch/kW	Netw 3 LNT ch/kW	L. ch/kW
Preili	—	R4/0.015	R2/0.015	—
Rekava *				
Rezekne	39/603	27/603	6/0.475	R10/0.155
Riga	R3/150	R10/158	28/380	R7/2.3
	—	—	—	31/95
Roja	—	—	—	R8/0/015
Sabile	R8/0.015	R5/0.015	R11/0.015—	
Valmiera	33/603	21/540	R11/0.55	R9/0.2
Ventspils	R5/0.165	R9/0.38	R12/0.38	R7/0.055
				38/0.9
				43/1
				45/1
Viesite *				
Vilani	—	—	R11/0.015R2/0.015	

L = State transmitters, leased by private broadcasters
* = under construction

D.Prgr: Netw. 1 (LTV national ch): 1600 (SS 0700)-2200.
Netw. 2 (LTV multilingual ch.): 1420 (SS 1000)-2200 in Latvian & Russian incl. rel. of private, regional and int. broadcasters.
Netw. 3: rel. of Lavijas Neatkariga Televizija (LNT) in Russian.

PRIVATE, MUNICIPAL AND REGIONAL STATIONS

Location	ch	kW (ERP)	Station
Dagda	R4°	0.04	TV Ezerzeme
Dobele	R5	0.05	Dobeles TV
Dundaga	R11	0.01	Dundagas TV
Gulbene	R11	0.1	Gulbenes TV 11. kanals
Jelgava	33	0.04	Jelgavas TV
Kraslava	R2	0.01	Kraslavas TV
Kuldiga	R8	0.07	Kuldigas TV
Ledurga	R12	0.04	Ledurgas TV
Liepaja	R23°	0.1	Kurzemes TV
Livani	R12	0.02	Livanu TV
Malpils	R12	0.02	Malpils TV "Spektrs"
Ogre	R2	0.015	Ogres TV
Riga	33	0.05	TV Miraza
	43°	1	TV Riga
Rujiena	R7	0.02	Rujienas TV
Salaogriva	R8	0.04	Salaogrivas TV
Skrunda	R5	0.01	Skrundas TV
Smiltene	R7	0.02	Smiltenes TV
Talsi	R4	1	Talsu TV
Valmiera	R7	0.02	Valmieras TV 7. kanals
Viesite	R6	1	Viesites TV

° = PAL colour, others SECAM.

LUXEMBOURG

TV-sets: 100.500 — **Colour:** PAL & SECAM — **Systems:** B, L&G.

CLT MULTI MEDIA
⌐ 45, blvd Pierre Frieden, L-1543 Luxembourg-Kirchberg. ☎ +352 421421. 🗎 +352 42142-2760
WWW: http://www.cltmulti.com — **L.P:** MD: Rémy Sautter.
CLT Multi Media operates various terrestrial and satellite-delivered TV services serving Belgium, France, German, Italy, Luxembourg, The Netherlands and Poland. It is also a major shareholder in Channel 5 (UK).
Terrestrial tr's: Dudelange ch7 (System B) 100/25kW H, Dudelange ch21 (System L) 1000/100kW, Dudelange ch27 (System G) 1000/100kW (2 studios), ch24 100kW.
RTL 9 in French: W 1045-1210, 1600-2230 (Sat 1400-2240); Sun 1100-2200 on ch21 (SECAM) for French viewers and ch27 (PAL) for Belgium & Luxembourg.
RTL-4 in Dutch & Multilingual: Luxembourg ch. 41 (30kW).
Club RTL (French for Belgium): ch27.

RTL TÉLÉ LETZEBUERG (local sce. of CLT)
⌐ 177, rue de Luxembourg, L-8077 Bertrange. ☎ +352 252 72 51. 🗎 +352 252 725431. **WWW:** www.rtl.lu
D. Prgr: in Luxembourgish: MF 1700-1900, Sat 1730-1900, Sun 1800-1930.

MACEDONIA

TV-sets: N/A — **Colour:** PAL — **Systems:** B&H

TELEVIZIJA MAKEDONIJE
⌐ Dolno Nerezi bb, 91000 Skopje. ☎ +38 (91) 258 230.
Stations:
TV Makedonije 1st Prgr.

Loc.	ch	kW	loc.	ch	kW
Pelister	4	30	Mali Vlaj	9	10
Golak	5	10	Vodno	9	10
Cnri Vrv	6	100	Popova Sapka	10	10
Stogovo	6	10	Turtel	11	50
Belasica	7	6.6	Vodno	12	10
Boskija	8	10			

+58 low power sts.
D.Prgr: Mon-Thu 1110-2200, Fri 1445-2200, Sat 0820-2145, Sun 0830-2100.
TV Skopje 2nd Prgr.

Loc.	ch	kW	loc.	ch	kW
Ohrid	21	4	Oteseva	37	10
Turtel	22	1000	Golak	38	200
Pelister	29	600	Popova Sapka	38	10
Crn Vrv	30	600	Mali Vlaj	44	200
Stogovo	31	10	Boskija	57	20
Belasica	35	13			

+56 low power sts.
D.Prgr: Mon-Fri 1355-2230, Sat 0900-2230, Sun 0650-2200.

MALTA

TV-sets: 145.300 — **Colour:** PAL — **System:** B.

PUBLIC BROADCASTING SERVICES LTD.
⌐ P.O. Box 82, Valletta. ☎ +356 225051. 🗎 +356 244601.
L.P: Chmn: Philip Farrugia Randon. Chief Exec: Tony Mallia. Head of TV: Andrew Psaila. TV Prgr. Mgr: Sylvana Cristina.
Station: chE10 10/2.5kW H.
D. Prgr: 1700-2230 (Sun also 1200-1530).

SUPER ONE TV
(Operated by Maltese Labour Party)
Station: Ch 29 10kW
D.Prgr: 0500-2230.

MOLDOVA

TV-sets: N/A — **Colour:** SECAM — **System:** D & K.

TV MOLDOVA

Location	TV Moldova	Românâ-1	ORT 1
Balti	2*	41*	8**
Cahul	8*	1**	31*
Camenca	10	5	8
Causeni	40*	—	28*
Ciadîr-Lunga	—	—	12
Cimislia	33**	—	2**
Comrat	—	—	10
Dnestrovsk	6	—	10
Edinet	7	27	31*
Hîncesti	—	7	10
Leovo	4	—	39
Rezina	12	25**	27*
Soroca	9	4	28
Stefan-Voda	9	—	—
Straseni	3*	11*	30*
Tighina	—	1	8
Ungheni	32*	7	29*

*) high power — **) medium power — others low power
N.B: Very low power repeaters not mentioned

VTV (priv-comm)
str. Hinncesti 61, Chisinau 277028
L.P: Dir: F.D. Bulan
Station: Chisinau ch23 600W
D.Prgr: 1700-2100 rel TV6 Moscow and own ads.

MONACO

TV-sets: 20.000 — **Colour:** SECAM & PAL — **Systems:** L & G.

TELE MONTE CARLO (Comm.)
16 Blvd. Princess Charlotte, MC 98090 Monaco-Cedex.
Cable: Tele-Carlo. ☎ +3393 505940. 🖷 +3393 250109. ① 469823 Tele Carlo. **WWW:** http://www.cecchigori.com/tv/
L.P: Pres: Jean-Louis Medecin. Dir. Prgrs: George Giaveret.
Stations: chF8 (L/SECAM) 50kW(ERP), ch30(L/SECAM) 500kW, ch33(G/PAL) 0.05kW, ch35(G/PAL) 40kW, ch39(L/SECAM) 0.02kW.
D.Prgr: French 1800-2400 on ch 8, 30, 39. (See also listing of relay st's in France). **Italian: TMC1 & TMC 2:** 1000-2400 on ch 33, 35.

NETHERLANDS

TV-Sets: 6.5m — **Colour:** PAL — **System:** B & G.

NEDERLANDSE OMROEPPROGRAMMA STICHTING (NOS)
Sumatralaan 45, 1217 GP Hilversum, P.O. Box 26600, 1202 JT Hilversum. ☎ +31 (35) 6779222. 🖷 +31 (35) 6772649 ① 43287.
L.P: Dir. Radio and TV: Ruurd Bierman; Hd. Comm: F. de Vries

Board Members: drs. A. Grewel; mr A. Herstel (NCRV); F.C.H. Slangen (KRO): ds. A. van der Veer (EO); mr A.J. Heerma van Voss (VPRO); G.C. Wallis de Vries (AVRO); K. van Doodewaard (TROS); mr H.J.E. Bruins Slot: dr. S.J. Noorda; prof. dr. D.Th. Kuiper; prof mr A. Geers; mr B. Staal; G.H. Veringa.

NEDERLANDSE PROGRAMMA STICHTING (NPS)
P.O. Box 29000, 1202 MA Hilversum. ☎ +31 (35) 6779333. 🖷 +31 (35) 6774517
L.P: Dir: W.J.M. van Beusekom

NB: The Dutch prgrs. are provided by the NOS, NPS and seven broadcasting organizations:
- **Algemeen Omroepvereniging AVRO**, 's Gravelandseweg 52, 1217 ET Hilversum, Postbus 2 1200 JA Hilversum. ☎ +31 (35) 6717911 🖷 +31 (35) 6717439
- **Vereniging Evangelische Omroep EO**, Oude Amersfoortseweg 79a, 1213 AC Hilversum, Postbus 21000, 1202 BB Hilversum. ☎ +31 (35) 6474747. 🖷 +31 (35) 6474727
- **Katholieke Radio Omroep KRO**, Emmastraat 52, 1213 AL Hilversum, Postbus 23000, 1202 EA Hilversum. ☎ +31 (35) 6713911 🖷 +31 (35) 6237345
- **Nederlandse Christelijke Radio Vereniging NCRV**, Bergweg 30, 1217 SC Hilversum, Postbus 25000, 1202 HB Hilversum. ☎ +31 (35) 6719911 🖷 +31 (35) 6719285
- **TROS**, Lage Naarderweg 45-47, 1217 GN Hilversum, Postbus 28450, 1202 LL Hilversum. ☎ +31 (35) 6715715 🖷 +31 (35) 6715236
- **Omroepvereniging VARA**, Heuvellaan 50, 1217 JN Hilversum, Postbus 175, 1200 AD Hilversum. ☎ +31 (35) 6711911 🖷 +31 (35) 6711333
- **Omroepvereniging VPRO**, 's Gravelandseweg 63-73, 1217 EH Hilversum, Postbus 11, 1200 JC Hilversum. ☎ +31 (35) 6712911 🖷 +31 (35) 6712254

Stations:

Loc.	Ned. 1	Ned. 2	Ned. 3	kW
Arnhem	50	53	43	30
Eys	51	54	48	1
Goes	29	32	35	250
Hulsberg	57	60	43	0.1
Lopik	4	27	30	100/2 x 1000
Losser	—	26	34	3
Maastricht	53	56	59	1
Markelo	7	54	51	30/2 x 300
Noorbeek	46	49	52	0.01
Roermond	5	31	34	50/2 x 250
Slenaken	29	32	35	0.02
Smilde	6	47	44	40/2 x 1000
St.Pietersberg	26	33	23	0.25
Wieringermeer	39	45	42	300
Wijk aan Zee	33	49	21	0.004

+ 5 low power repeaters.
D.Prgr: Ned 1: AVRO/KRO/NCRV; Ned 2: EO/NOS/TROS; Ned 3: NPS/VARA/VPRO.

NORWAY

TV-Sets: 2.000.000 — **Colour:** PAL — **Systems:** B & G.

NORSK RIKSKRINGKASTING
N-0340 Oslo 3. ☎ +47 (22) 459050. 🖷 +47 (22) 457440. ① International Relations 76820. Eurovision: 71794. TV News: 18530. Prgr. Purchases: 19676. **WWW:** http://www.nrk.no
L.P: Dir. Gen (Radio & TV): Einar Førde. Dep. TV Dir:

Norway

Anne Torjusson Diesen. PD: Kent Nilssen. Tech Dir: Geir Sundal. Dir of Inf. and PR: Hanne Løchstøer. Head. of Int. Rel: Kjell Lokvam.

TV-2 (Priv. Comm.)
Postboks 2, 5002 Bergen. ☎ +47 (55) 908070. 📠 +47 (55) 908090.
L.P: MD: Arne Jensen. Dir. of Prgrs: Finn H. Andreassen.

Stations: all sts. NICAM stereo

Location	NRK	TV-2	kW
Alta	7*	—	5
Andenes	6*/35*	49	4/14/15.5
Arendal	—	25	1.95
Bagn	3	—	55
Ballstad	—	26	5.5
Bangsberget	35	29	3.3/10.5
Bergen	9**	12	80
Biri	—	45	1.1
Bjerkreim	6+	27	15/42
Bokn	8+	44	100/270
Bransøy	—	30	1.1
Bremanger	4**	—	80
Dokka	—	26	1.05
Fannrem	—	25	1.45
Fennefossknipa	—	28	1.423
Fister	11+	—	1.5
Flisa	32	22	7/13.5
Frekhaug	—	23	1.1
Fyresdal	—	26	1.323
Førde	5**	—	1.2
Gamlemsveten	3	24	60/49
Gausta	8	—	45
Gol	—	26	5.445
Gran	—	52	6
Greipstad	2	—	60
Grisvågøy	—	25	1.07
Grong	5	24	100/9.45
Grøtevær	42*	—	1.6
Gulen	2**	—	48
Hadsel	4*	—	60
Halden	11	32	100/600
Hammerfest	9*	—	90
Harstad	—	47	4.8
Hasvik	—	26	1.25
Hemnes	3*	39	70/1
Hermansverk	—	27	1.25
Hestmannen	6*	54	3.3/1.115
Hol	—	28	1.296
Hommelfjell	—	39	14.6
Hovdefjell	7	—	62.5
Hvitingen	—	47	4.8
Ibestad	—	52	1.5
Jetta	8	—	95
Kappfjell	8*	—	2
Karasjok	5*	—	10
Kautokeino	3*	—	6
Kistefjell	8*	23	87.5/3.87
Kongsberg	4	43/28	100/175/1.1
Kongsvinger	9	28	60/250
Kopparen	11	23	45/60
Kristiansund	—	40	2.35
Kveøy	—	31	1.8
Kvisvik	—	32	2.15
Lifjell	52	56	1/1.3
Lifjell, Stavanger	—	37	24
Lillehammer	—	53	4.8
Lyngen	5*	—	3
Lyngdal	9	33	80/1.3
Lønahorgi	11**	—	25
Melhus	2	30	100/10.5
Mistberget	44	54	1.15/5.752
Mofjellet	—	37	1.4
Mosvik	7	37	34/360
Narvik	10*	24	130/100
Nittedal	—	25	1.3
Nordbykollen	—	22	3.5
Nordfjordeid	10**	27	3/1.45
Norhue	5	27	40/28
Nordkapp	6*	—	70
Norefjell	—	28	3.3
Oppstad	—	42	2.38
Oslo	6	12	100/90
Raufoss	—	33	1.519
Reinsfjell	6	29	85/28.4
Ringerike	41	23	2/17.447
Rubbestadfjell	—	29	6.667
Salten	7*	30	55/44.55
Skien	10	24	110/270
Skotterud	—	33	1.288
Snertingdal	37	46	1.05/1.1
Sogndal	7**	—	20
Sokna	—	26	1.512
Sollihøgda	—	39	1.676
Steigen	2*	—	80
Stord	5**	47	60/188.17
Spåkenes	52*	45	3.3/2.4
Stadlandet	—	47	2.6
Storberget	—	25	1.488
Store Jækkir	11*	22	4/1.2
Storhaugen, Målselv	—	45	11.756
Stryn	—	23	5.356
Sundalsøra	—	47	1.65
Tana	8*	—	60
Torsvarde	—	25	1.32
Toåsen	—	52	1.07
Tresfjord	—	40	3.736
Trolltind	11*	—	110
Tromsø	—	33	2.566
Tron	9	23	67.5/5.77
Trysilfjell	40	54	6.9/13.9
Ullandhaug	35+	—	1.1
Varanger	2*	—	90
Vardheia	—	53	1.088
Vega	9*	22	50/105
Veggen	—	41	1.75
Vestre Slidre	—	25	1.25
Viktjernåsen	—	48	1.2
Vinstra	—	24	1.2
Volda	—	46	2.8
Øksnes	—	25	2.9
Øvre Eiker	—	37	1.95
Åkersten	—	40	3.2
Ål	—	52	1.819
Ålesund	—	45	4.7
Ålmenberget	—	24	1.3
Åndalsnes	—	22	1.45
Ånsmarka	—	49	1.396

*) carriers regional prgrs from NRK TV Nord
**) carries NRK Vestlandsrevyen
+) carries NRK Rogeland

EUROPE

Norway — Portugal

D.Prgr: NRK: basically 1600-2300, school programming 0800-1600. **Regional TV:** Vestlandsrevyen: Mon, Tue, Thu, Fri 1745-1800, Wed. 1730-1800. TV Rogeland: Wed 1730-1800. TV Nord: (see Vestlandsrevyen). **TV-2:** 1300-2300

PHILIPS PETROLEUM 66
TV1: Ekofisk ch55. V. relays NRK.
TV2: Ekofisk ch52. V. relays Film Net.

POLAND

TV-sets: 12m — **Colour:** PAL — **Systems:** D & K.

TELEWIZJA POLSKA S.A. (Gov.)
Ul. Woronicza 17, 00-999 Warszawa. ☎ +48 (22) 433361/445432. +48 (22) 447419/435779. **Cable:** Telewar. ① 825331. **WWW:** http://www.tvp.com.pl
L.P: Dir TVP1:vacant. Dir TVP2: Maciej DomanskTech. Dir: Rajmund Gruszka. Dir. Int. Rel: Jerzy Romanski.
Power = ERP Pol H exc. where indicated.

Prgr I	ch	R kW/Pol		ch	R kW/Pol
Bydgoszcz	1	120	Szczecin	12	100
Pila	2	50	Wroclaw	12	100
Warszawa	2	60	Konin	22	1
Kielce	3	100	Przemyśl	24	100
Zielona Góra	3	200	Warszawa	27	20
Suwalki	5	100	Kalisz	28	1
Lódź	7	80	Plock	29	1000
Bialystok	8	100	Jelenia Góra	30	200
Katowice	8	200	Rabka	31	2
Koszalin	8	60V	?	32	?
Lublin	9	80	Poznan	33	1
Poznań	9	150	Bydgoszcz	36	650
Olsztyn	9	100V	Skierniewice	37	1
Gdańsk	10	100	Opole	40	700
Kraków	10	200	Bydgoszcz	41	650
Zamość	10	50V	Siedlce	52	300
Luban	11	1	Klodzko	52	100
Rzeszów	12	100V	Lezajsk	58	1
+124 low power st's					

Prgr II	ch	R kW/Pol		ch	R kW/Pol
Wroclaw	2	1	Zakopane	34	30
Lublin	2	0.4V	Lobez	35	40
Gdynia	3	2	Zamosc	36	300
Kielce	5	2	Klodzko	38	300
Rzeszów	7	0.5V	Lomza	38	7
Lódź	10	1V	Czluchów	39	13
Warszawa	11	50	Plock	39	1000
Bialystok	11	1V	Wroclaw	25	1000
Gizycko	11	1	Olsztyn	26	270
Koszalin	11	0.5V	Poznań	27	250
Bydgoscz	12	0.6V	Kielce	28	1000
Katowice	21	450	Zielona Góra	29	1000
Elblag	21	20	Rzeszów	29	700
Tarnów	22	20	Szczecin	30	600
Lublin	23	500	Kalisz	31	1
Opole	23	600	Krakow	33	300
Pila	24	150	Swinoujscie	33	10
Lebork	25	40	Konin	34	2
Wroclaw	25	1000	Jelenia Góra	35	200
Lezajsk	26	1	Lobez	35	25
Olsztyn	26	130	Bydgoszcz	36	650
Poznan	27	300	Suwalki	36	1000
Kielce	28	1000	Rabka	36	10
Zielona Gora	29	1000	Zamość	36	250

Prgr II	ch	R kW/Pol		ch	R kW/Pol
Gdansk	37	600	Szczawnica	39	10
Siedlce	37	300	Slupsk	40	15
Plock	39	1000	Przemyśl	41	100
Czluchow	39	30	Wisla	41	30
+87 low power st's.					

Prgr. III (local on Progr. II transmitters): Katowice ch 21. **D. Prgr:** W 0800-1000, 1500-1530.
Satellite TV Relay: Katowice ch R6 20kW — **Echo (priv):** Wro-claw ch 28 1kW — **RAI Uno Relay:** Krakow ch 50 50kW.

TELE-9
Kielce ch R5, Tarnow ch R22, Kielce ch R28, Rzeszow ch R29, Kraków ch R33, Przemyʹsl ch R41.

TV Polsat
Marchakovska 83, P-00517, Warsaw 84. ☎ +48-22-295684.
Stations:

Loc	ch/Pol	Loc	ch/Pol
Biala Potlaska	32H	Olsztyn	60H
Chelm	21H	Ostroleka	21H
Choragewica	53H	Palac Kultury	35H
Chrzelice	57H	Pila	57H
Chiechanow	52H	Piotrków Trybun.	34H
Czestochawa	34H	Plock	21H
Elblag	23H	Poznan-Piatkowo	50H
Gorzów Wielk.	26H	Przemysl	56H
Góra Siéza	59H	Radom	53H
Góra Skrzyczne	58H	Rzeszów	48H
Jezów Sudecki	57H	Siedlice	57H
Kalisz	56H	Sieradz	57H
Katowice-Bytków	47H	Skierniewice	24H
Kielce	22H	Suwalki	41H
Klodzko	R8H	Tarnobrzeg	41H
Kolowo	48H	Tarnów	60H
Konin	58H	Trebice	53H
Koszalin	60H	Urzud Miasta	60H
Krosno	60H	Walbrzych	49H
Lebork	57H	Waly Pistwskie	30H
Leszno	57H	Wloclawek	22H
Lomza	57H	Zakopane	51?
Lódz	49H	Zamosc	53H
Lublin	35H	Zielona Góra	R10H
Okskywie	57H		

PORTUGAL

TV-sets: 3m — **Colour:** PAL — **System:** B&G.

RADIOTELEVISÃO PORTUGUESA (RTP) (Comm.)
Head Office: Av. 5 de Outubro 197, 1094 Lisboa.
☎ +351 (1) 7931 774. **Cable:** Televisão. ① 14527-RTP RE P. +351 (1) 7931 758. **Mailing** Apartado 2934, Lisboa.
L.P: SM: José Eduardo Moniz. TD: Ismael Augusto.
Film Buyer: José Eduardo Moniz.
Pol H unless stated otherwise.

RTP Channel I

Stations	Ch	kW		Ch	kW
Muro	2	10/67V	S. Miguel	6	2/20
Lousa	3	12/60	Marao	6	5/40
Marofa	5	2/16	Bornes	7	5/12
Mendro	5	5/30	Lisboa	7	10/100
Montejunto	6	5/22	Valença	7	1/7

Portugal — Russia

Stations	Ch	kW		Ch	kW
Foia	8	2/20	Palmela	22	2/115
Leiranco	8	2/4	Mosteiro	24	1/9
Gardunha	8	1/3	S. Macario	47	2/67V
Porto	9	10/100			

+42 repeaters.
D.Prgr: Mon-Fri 0900-0030, Sat 0900-0230, Sun 0900-0030.

RTP Channel II

Stations	Ch	kW		Ch	kW
Mosteiro	21	1/9	Gardunha	34	2/18
Lisboa	25	10/405	Marao	35	10/300
Bornes	25	10/200	Porto	41	10/100
Lousa	26	20/540	Montejunto	46	10/200
Mendro	27	20/560	Valença	46	2/70
Muro	27	10/500	Foia	47	10/550
S. Miguel	31	10/250	Marofa	48	10/300
Palmela	32	2/128	Macario	50	2/67V
Leiranco	34	5/40			

S+20 repeaters.
D.Prgr: Mon-Fri 1200-0050/0130, Sat 0900-0130, Sun 0900-0100.

PRIVATE STATIONS

SIC-Sociedade Independente de Comunicacao, SA
⌂ Estrada da Outorela 119, Carnaxide 2795 Linda-a-vehla, Lisboa. ☎ +351 (1) 417 3138. 📠 +351 (1) 418 7156.
L.P: Chmn: Francisco Pinto Balsemao. Dir. of Prgrs: Dr. Emilio Rangel. Dir. of Mktg: Dr. Hugo Correia Pires. Tech. Dir: Eng. triga de Sousa.

TVI, Televisao Independente, SA
⌂ Rua 3, Matinha, Edifício Alteio, 6°, 1900 Lisboa.
☎ +351 (1) 4347500. 📠 +351 (1) 4355075
E-mail: tvi-sa@individual.eunet.pt **WWW:** http://www.tvi.pt
L.P: Chmn: Roberto Carmeiro. PD: Diogo Gaspar Ferreira. Tech. Dir: Joao Penha Lopes.

ROMANIA

TV-sets: 7.277.000 — **Colour:** PAL — **System:** D & G

RADIOTELEVIZIUNEA ROMÂNÂ (Gov.)
⌂ 191 Calea Dorobanti, Bucharest. ☎ +40 (1) 212 0290. 📠 +40 (1) 312 0381. ① 10182 TVR. **Cable:** Radioteleviziunea.
L.P: DG: Dumitru Popa. PD: Mamase Radnev.
Stations: Pol. H (except where indicated)
1st. National Program

Location	ch.	kW	Location	ch.	kW
Balota-Turnu S.	2	15	Piatra Neamt	6	5
Baneasa-Dobrogea	3	15	Vacareni-Galati	7	15V
Bistrita-Heniu	3	15	Varatec	7	50
Oradea	3	120	Sibiu-Paltinis	7	50
Semenic-Resita	3	15	Vaslui	7	5.0
Bucaresti	4	50	Litoral-Constanta	8	10V
Suceava	4	100	Bihor-Vascau	8	2.5V
Harghita-Gheorgh.	5	50	Rarau-Cimpulung	8	15H
Birlad	5	50	Cerbu-Novaci	8	150
Bucegi-Ploiesti	6	150	Iasi	9	25
Mahmudia-Delta	6	50V	Magura-Odobesti	9	10V
Zalau	6	15	Timisoara	9	50

Location	ch.kW		Location	ch.kW	
Brasov	10	10	Botosani-Hirlau	11	15
Topolog-Constanta	10	50	Cozia-Rimnicu V.	12	10
Mogosa-Bia Mare	10	15	Tulcea	12	2
Bacau	10	10	Tirgu Mures	12	10
Paring-Petrosani	10	10	Comanesti	12	20
Mangalia	11	1	Magura Boiului-D.	12	20
Feleac-Cluj	11	25	Siria-Arad	12	20

+ 400 lps.

2nd National Prgram

Location	ch.	kW	Location	ch.	kW
Bucaresti	2	150	Timisoara	21	1000
Nucet II	5	0.05	Craiova	21	1
Iasi	6	1.3	Vaslui	28	0.1
Cluj	8	1.3	Brasov	34	50
Pitesti	9	0.5	Resita	36	1.5
Piatra Neamt	11	0.1	Constanta	42	0.05

LD.Prgr: Prgr 1: Mon-Fri 0800-2200, Sat/Sun 0700-2300; Prgr 2: 1500-2200

PRIVATE STATIONS
More than 80 local st's are operating incl. 8 st's in Bucharest are as follows:
Antena 1, Poesti 25-27, 70000 Bucharest. ☎ +40 (1) 212 0619. 📠 +40 (1) 212 0188 — DG: F. Bratescu.
Canal 31, 155 Piata Victorei Di, 7th Floor, 70411 Bucharest. ☎ +40 (1) 210 6628 — DG: Adrian Sirlon.
Canalul de Stiri, Calea Victorei 133-5, s 1, Bucharest. ☎ +40 (1) 312 4348. 📠 +40 (1) 312 0349.
Pro TV, Bd Carol 1 109, etaj s2, Bucharest. ☎ +40 (1) 312 4218. 📠 +40 (1) 312 4228.
Tele 7 ABC, Christo Botev 8, s3, Bucharest. ☎ +40 (1) 312 1695. 📠 +40 (1) 611 6576 — DG: Mihai Tatulia.
Tele America, Blvd. Armata Poporului 1-3, s6, Bucharest. ☎ +40 (1) 311 0419. 📠 +40 (1) 311 0417.
Tele Europa Nova, Dr. Lister 6, s5, Bucharest. ☎ +40 (1) 623 6661. 📠 40 (1) 312 1324.
TV Sigma, Armata Poporului Blvd. 1, complex Leu-Facultatea de Electronic Crp A, Et 8, Bucharest. ☎ +40 (1) 631 4734 — MD: Constantin Crbu.

RUSSIA

TV-sets: 50m — **Colour:** SECAM — **Systems:** D & K.

NB: Due to incomplete information, details of channels used by the various broadcasters cannot be given. These details will be added in future editions as information become available.

ORT (formerly Ostankino) (Public sce.)
⌂ 12 Academika Korolyova Str, Moscow 127000. ☎ +7 (095) 217 7898. 📠 +7 (095) 288 9542 — **L.P:** DG: Sergei Blagovolin.

MOSCOW TV
⌂ 12 Akademika Korolyova Str, Moscow 127000. ☎ +7 (095) 217 5158. 📠 +7 (095) 216 5401.
L.P: Gen. Dir: Algar Misan. MD: Natalya Smirnova.

TV CHANNEL 2x2
⌂ 12 Alademika Korolyova Str, Moscow 127000. ☎ +7 (095) 217 7094. 📠 +7 (095) 215 2063.
L.P: GM: Vladimir Troepolski. PD: Victor K. Litenko.

EUROPE   Russia — Slovenia

TELEXPRESS - CHANNEL 31
✎ 15 Akademika Koroleva Str, Moscow 127000 ☎ +7 (095) 282 4260. 🗎 +7 (095) 276 8892.

NTV
✎ Novy Arbat Str 36, Moscow 121102. ☎ +7 (095) 290 7077. 🗎 +7 (095) 290 9757.

MOSCOW TV 6
✎ Ilyinka 15, Moscow. ☎ +7 (095) 206 8423. 🗎 +7 (095) 206 0886 — **L.P:** Exec. Mgr: Tatiana Voronovich.

INDEPENDENT BROADCASTING SYSTEM (MVS)
✎ Suite 303, 8A, Suvorovsky Blvd, Moscow 121019. ☎ +7 (095) 291 1787. 🗎 +7 (095) 291 2174.

RTR NETWORK
✎ 5 Yamskogo Polya 19/21, Moscow 125124. ☎ +7 (095) 250 0511. 🗎 +7 095 250 0105.
L.P: Chmn: Oleg Poptsov. Dep. Chmn: Segei Skvortsov. DG: Anatoly Lysenko. Int. Dir: Sergei Erofeev. Tech. Dir: Stanislav Bunevic.

RUSSIAN UNIVERSITY CHANNEL
✎ 5 Yamskogo Polya 19/21, Moscow 125124. ☎ +7 (095) 213 1754. 🗎 +7 (095) 213 1436.

REGIONAL TELEVISION
✎ Akademika Pavlova Str 3, St Petersburg 197022. ☎ +7 (812) 238 6073. 🗎 +7 (812) 238 5807.

6TR PETERSBURG - Channel 5
✎ 6 Chapygina Str, St Petersburg 197022. ☎ +7 (812) 234 3763. 🗎 +7 (812) 234 1416 — **L.P:** Chmn: Oleg Rudny.

RUSSKOYE VIDEO - CHANNEL 11
✎ Malaya Nevka 4, St Petersburg 197022. ☎ +7 (812) 234 4207. 🗎 +7 (812) 234 0088.

AS BAIKAL TV
✎ 6 Shelehov, Shelehov 666020. ☎ +7 (395) 109 3303. 🗎 +7 (395) 103 1917.

KALININGRAD STATE TV
✎ 19 Klinicheskaja Str, Kaliningrad 236016. ☎ +7 (0112) 452 700. 🗎 +7 (0112) 452 233

STATE NATIONAL TV AND RADIO COMPANY OF SAKHA REPUBLIC
✎ 48 Ordjonikidze Str, Yakutsk 677892. ☎ +7 (411) 225 3169. 🗎 +7 (411) 225 3176.

TV COMP FOURTH CHANNEL
✎ P.O. Box 751, Yekaterinburg 620069. ☎ +7 (3432) 232 041. 🗎 +7 (3432) 236 033.

SAN MARINO

Colour: PAL — **Systems:** B&G

SAN MARINO RTV spa
✎ V. le J. F. Kennedy, 13 - Città.
☎ +39 549 882000.
L.P: DG: Raviele Gianni.

SLOVAKIA

TV-sets: 1.800.000 — **Colour:** PAL System B; SECAM — **Systems** D&K

SLOVAK TELEVISION
✎ Mlynska Dolina 28, Bratislava 845 45. ☎ +42 (7) 727 448. 🗎 +42 (7) 729 440.
L.P: DG: Jozef Darmo. PD: Milan Polak. Int. Rel: Mikulas Gavala.

ST1: (PAL)

Stations	ch	kW	Pol.		ch	kW	Pol.
Bratislava	2	150	H	Stúrovo	9	1.5	V
Námestovo	4	0.8	H	Trencin	10	1.6	V
Bardejov	4	1.2	H	Zilina	11	100	V
Poprad	5	80	V	M. Kamen	12	15/6	V
Kosice	6	100	V	Bratislava	31	2	H
B. Bystrica	7	100	H	B. Stiavnica	40	9	H
Ruzemberok	9	1	H	Zilina	41	10	H
Sucise	9	1.6	H				

D.Prgr: Mon-Fri 0900-1200, 1600-2330, Sat 0830-0045, Sun 0825-2400

ST2 (SECAM)

Stations	ch	kW	Pol.		ch	kW	Pol.
Modrý Kamen	21	100	H	Poprad	30	600	H
Kosice (city)	21	2	H	Sturovo	31	100	H
Zilina	22	100	H	B. Brystica	32	600	H
Roznava	22	100	H	S. N. Sazavou	32	100	H
Trencin	23	300	H	Lucenec	33	100	H
Snina	23	15	H	Olomouc	33	100	H
Kosice	25	600	H	Novi Jisin	34	100	H
Mikulov	26	300	H	Zilina	35	1000	H
B. Stiavnica	26	300	H	Susice	35	100	H
Ruzemberok	27	18	H	Frýdec-Mistec	37	300	H
Bratislava	27	1000	H	B. Mikulás	37	100	H
Stara Lubovna	27	100	H	C. Budejovice	39	600	H
Námestovo	29	100	H	N. Mesto	39	600	H

D.Prgr: Mon-Fri 1530-2330, Sat/Sun 1000-1230, 1530-2330
F.PI: ST2 will be privatized during 1997. 9 bids have been received.

MARKYZA (Comm.)
✎ Palisady 39, Bratislava 81106. ☎ +42 (7) 531 6610. 🗎 +42 (7) 531 4061 — **L.P:** CEO: Pavol Rusko.

SLOVENIA

TV-sets: 600.000 — **Colour:** PAL — **Systems:** B & H

TELEVIZIJA SLOVENIJA
✎ Kolodvorska 24, 61000 Ljubljana. ☎ +38 (61) 131 1533. 🗎 +38 (61) 131 9171. ℗ 32283. **WWW:** http://www.rtvs.si
L.P: DG: Zarko Petan. PD (TV): Janez Lombergar. Int. Rel: Boris Bergant.
Stations: System B=ch2-12, System H=ch21-68.
ChE. audio powers 1/10 of vision powers indicated.
Main stations

	Ch. 1	kW	Ch.2	kW	Pol
Beli Kriz	33	20	46	20	H
Krim	31	10	44	10	H
Krvavec	5	50	21	27	H
Kuk	40	12.5	26	12.5	V

Slovenia — Spain

	Ch.1	kW	Ch.2	kW	Pol
Kum	3	50			H
	38	20	32	20	H
Nanos	6	35	400	41	H
Plesivec	6	35	49	400	H
Pohorje	11	30	56	1000	H
Trdinov Vrh	35	25	48	25	H
Tinjan	38	10	49	100	H
Tsrtelj			60	10	H

+194 low power sts (ch. 1), 180 low power st's (ch. 2)
Prgr. 1: Mon-Fri 0750-2215 (Thu 2300, Fri 2330), Sat 0700-1115, 1330-2345, Sun 0725-2230.
Prgr. 2: Mon-Fri 1430-2300, Sat 1530-0030, Sun 0900-2300.

TV Koper-Capodistria

Location	ch	kW	Pol
Nanos	27	400	H
Beli Kriz	58	20	H

+30 low power sts.
D.Prgr: Mon-Fri 1245-2400, Sat 0900-2400, Sun 0930-2400.

PRIVATE STATIONS

KANAL A
✉ Tivolska 50 pp 44, 61101 Ljubljana. ☎ +386 (61) 133 4133. 📠 +386 (61) 133 4222 — **L.P:** GM: Frank Dovecar. Mktg. Dir: Vlasta Bostjancic. Tech. Dir: Mladen Uhlik.

MMTV
✉ Zorgova 70, 61231 Ljubljana. ☎ +386 (61) 161 2525. 📠 +386 (61) 374 554.
L.P: Pres: Marjan Meglic. PD: Andrej Meglic. CE: Tomislav Kalan.

PRO PLUS/POP TV
✉ Kranjceva 26, 61113 Ljubljana. ☎ +386 (61) 189 3200. 📠 +386 (61) 189 3204 — **L.P:** Gen. Dir: Marjan Jurenec.

SPAIN

TV-sets: 17.000.000 — **Colour:** PAL — **Systems:** B&G.

RADIOTELEVISION ESPAÑOLA (RTVE)
✉ Prado del Rey, 28023 Madrid. Torrespaña C/ O'Donnell 77, 28007 Madrid. ☎ +34 (1) 346 8754. 📠 +34 (1) 581 7125. ⓓ 22053. **WWW:** http://www.rtve.es/rtve/
L.P: Dir. Gen: Jordi Garcia Candau. Ex.Dir.TVE, S.A: Ramón Colom Esmatges. Dir. of Int. Coop: Alfonso Callego. TD: Francisco Baquedano. Film Buyer: Fernando Moreno.
Stations: (System B&G): Powers = ERP. Audio powers 10% of vision powers indicated.
N.B: TVE's 1st. and 2nd. prgr. are also relayed via satellite. The 1st. prgr is relayed under the name TVE Internacional. All Pol H. unless otherwise indicated.

TVE 1st Prgr.

Station	ch (kW)	Station	ch (kW)
Pechina	41 100	Sierra de Lujar	7 150
Jerez de la Fron.	26 32	Huelva	39 250
Cordoba	11 10	Sierra Almaden	5 40
Parapanda	9 24	Mijas	65 200

Station	ch (kW)	Station	ch (kW)
Guadalcanal	4 120	Izaña	3 200
Valencina	63 79	Tibidabo	4 150
Arguis	64 36	Gerona	49 126
Camarena de la Sierra	9 53V	Alpicat	5 15
		Soriguera	39 40
Inoges	5 24	La Musara	8 22
La Muela	3 35	Monte Caro	10 60
Gamoniteiro	3 50	Aitana	3 60
Alfabia	6 50	Benicasim	7 10V
Alfabia	54 119	Torrente	5 20
Lierganes	40 100	Montanchez	11 45
Villadiego	8 13	Ares	5 50
Villadiego	47 40	Santiago de Compostela	4 112
Castropodame	7 22V	Paramo	9 9
Matadeon	8 12	Parada del Sil	5 40V
El Cabaco	9 63	Domayo	10 10
Soria	39 16	Madrid	4 15
Zamora	31 160	Navacerrada	2 250
Chinchilla	11 40	Navacerrada	34 604
La Mancha	9 60	Torrespaña	49 117
Isleta	6 4	Murcia	59 160
Montaña Mina	32 6	Monreal	29 158
Pozo de las Nieves	10 10	Sollube	4 60
		Jaizquibel	54 158
Fuencaliente	7 10	Zaldiaran	45 15

TVE 2nd Prgr.

Station	ch (kW)		
Pechina	47 100	Pozo de las Nieves	59 48
Jerez de la Frontera	23 32	Temejereque	52 25
		Fuencaliente	27 148
Cordoba	21 200	Izaña	45 87
Parapanda	23 160	Tibidabo	41 28
Sierra de Lujar	57 250	Tibidabo	31 144
Huelva	45 250	Gerona	55 126
Sierra Almaden	39 260	Alpicat	49 160
Mijas	59 200	Soriguera	45 40
Guadalcanal	40 32	La Musara	57 300
Valencina	52 79	Monte Caro	26 225
Alpicat	55 112	Aitana	32 200
Arguis	58 36	Benicasim	59 200
Camarena de la Sierra	41 200	Monte Caro	23 232
Monte Caro	43 60	Torrente	22 160
Inoges	57 200	Montanchez	23 158
La Muela	46 158	Ares	22 200
La Muela	33 158	Santiago de Compostela	45 316
Gamoniteiro	39 158	Santiago de Compostela	2 40
Alfabia	48 119	Paramo	57 158
Lierganes	46 100	Parada del Sil	47 400
Villadiego	41 200	Domayo	39 141
Castropodame	21 200	Madrid	21 100
Matadeon	33 28	Navacerrada	24 870
El Cabaco	39 315	Navacerrada	27 604
Soria	45 16	Torrespaña	55 117
Zamora	37 160	Murcia	65 160
Chinchilla	43 158	Monreal	23 158
La Mancha	58 200	Archanda	22 50
Isleta	28 20	Jaizquibel	48 158
Montana Mina	35 6	Zaldiaran	39 15

D.Prgr: TVE 1 0800-0230 (24 H. at weekends), TVE 2 0800-0200.

TV Networks in autonomous areas
TELEVISIO DE CATALUNYA (Aut)
Televisió de Catalunya, S.A. (TV3 and Canal 33)

EUROPE — Spain

Jacint Verdaguer, s/n 08970- Sant Joan Despí, Catalunya. ☎ +34 (9) 3 499 9333. 🖷 +34 (9) 3 473 1964 ① 53280 TVDC E. **L.P**: Dir: Jaume Ferrús i Estopa. T.D: Pere Vila. Film Buyer: Jaume Santacana.

Location	TV3	C33	Power
Alt Camp			
Pont d'Armentera	34	42	0.6W
Alt Empordà			
Cadaqués	42	40	5W
Maçenet de Cabrenys	62	43	200W
Portbou	41	47	20W
Alt Urgell			
Arsequel	22	26	20W
Coll de Nargó	38	40	1W
Oliana	34	27	5W
Organyà	47	51	200W
Os de Civis	30	34	1W
Valls de Valira	22	40	20W
Vilanova de Banat	21	23	0.6W
Alte Ribagorça			
Campament de Tor	23	27	1W
El Pont de Suert	31	51	100W
Senet	52	49	1W
Taüll	45	37	1W
Vall de Boí	52	55	20W
Vilaller	22	26	5W
Anoia			
Calaf	59	63	20W
Castellfollit de Riubregós	33	43	1W
Igualada	62	60	20W
Pobla de Clar/Capellades	37	33	5W
Bages			
Boixadors	36	38	20W
Cardona	62	64	5W
Ministrol	65	61	5W
Ministrol de Calders	50	56	1W
Montserrat	28	40	200W
Mura	43	35	0.6W
Sallent	57	38	5W
Baix Camp			
Escornalbou	33	30	5W
Pratdip	52	31	0.6
Riudecols	22	26	1W
Vandellós	33	22	5W
Centre Emissor la Mussara	63	25	10/5kW
Baix Erbre			
El Perelló	45	47	5W
Centre Emissor Montcare	29	36	5/1kW
Baix Empordà			
Calonge	61	63	100W
L'Estartit	60	26	20W
Palafrugell	23	27	20W
Sant Feliu de Guixols	24	28	20W
Baix Llobregat			
Baix Llobregat	49	21	20W
Begues	22	25	1W
Molins de Rei	64	60	20W
La Palma	24V	36V	1W
Vallirana	51	54	5W
Baix Penedès			
El Vendrell	17	21	100W
Barcelonès			
Centre Emissor Collserola	44	23	20/10kW
Berguedà			
Guardiola I	60	62	100W
Guardiola II	24	28	1W
Puig-Reig	36	57	20W
Saldes	24	28	1W

Location	TV3	C33	Power
La Cerdanya			
Aràncer	43	47	1W
Alp	51	—	0.6W
La Cerdanya	62	41	200W
La Molina	24	28	1W
Martinet	44	50	1W
Prullans	51	54	1W
Conca del Barberà			
Montblanc	43	23	100W
S. Coloma Queralt	36V	38V	5W
Garraf			
Garraf	52	56	5W
S. Pere de Ribes	42	32	100W
Garrotxa			
La Vall de Bianya	61	65	—
Les Lloses	52	45	—
Olot	21	40	40W
Montagut	36	43	20W
Gironès			
Sarrià de Ter	40	57	20W
Centre Emissor Rocacorba	52	45	10/5kW
Maresme			
Tiana	57	61	—
Argentona	25	37	1W
Cabrils/Mataró	54	52	100W
Calella	53	59	100W
Calella/Blanes	29	37	10W
Canet	52	45	1W
Llavaneres	29	37	1W
Teià/Alella	29	37	5W
Montsià			
Alcanar	27	25	5W
S. Carles de la Ràpita	24	30	20W
Ulldecona	42	32	100W
La Noguera			
Ager	35	54	20W
Artesa de Segre	46	34	20W
Camarasa	33	25	1W
Ponts	35	25	5W
Vila Nova de Meià	51	33	5W
Osona			
Bellmunt d'Osona	46	43	100W
Collsuspina	26	60	100W
Olost	52	45	1W
Viladrau	40	44	1W
Pollars Jusà			
Boixols	60	63	1W
Central de Cabdella	55	59	0.6W
Comiols	61	54	20W
Erinyà	65	62	1W
Llimiana	21	25	5W
Mur	28	32	100W
Pobla de Segur	23	25	5W
Senterada	53	50	5W
Torre de Cabdella	47	37	5W
Xeralló	22	24	1W
Pollars Sobirà			
Alins	36	34	5W
Alt Aneu	22	26	10W
Aneu	51	53	1W
Boldis	30	32	0.5W
Escaló	23	21	2W
Esterri de cardos	48	61	1W
Gerri de la Sal	47	49	5W
Llavorsí	31	35	1W
Centre Pic de l'Orri	42	56	5/2.5kW
Rialp	48	51	1W

Spain

Location	TV3	C33	Power
Tavascan	23	27	0.6W
Unarre	60	62	5W
Tornafort-Sort	24	26	5W
El Priorat			
El Priorat	34	38	100W
Ulldemolins	25	31	5W
Ribera d'Ebre			
Flix I	61	54	20W
Torre de lEspanyol	53V	60V	5W
Riba-Roja	44	46	1W
Flix II	34	38	0.6W
Ripollès			
Camprodon	22	26	20W
Espinabell	21	25	1W
Gombrèn	33	29	1W
Núria	22	—	0.6W
Ribes de Freser	54	44	20W
Ripoll I	39	63	100W
Ripoll II	21	23	1W
Setcases	57	61	1W
S. Joan de les Abadesses	29	—	0.6W
Taga	25	23	5W
Segarra			
Segrià			
Torà	37	64	1W
Almenar	35	25	20w
Mangraners	25	33	20W
Centre Emissor Alpicat	52	38	10/8kW
La Selva			
Lloret	50	56	5W
Arbúcies	42	36	20W
Osor	28	24	1W
S. Hilari Sacalm	26	22	5W
Tossa	25	37	5W
Solsonès			
La Coma	35	29	5W
S. Llorenç de Morunys	26	32	5W
Solsona	36	29	5W
Tarragonès			
Tarragona	21	27	0.5W
Terra Alta			
Gandesa	61	54	—
Val d'Aran			
Bagergue	41	43	5W
Bausen	54	52	1W
Bossost	43	37	20W
Cap de Vaqueira	30	28	20W
Sa Nela	61	53	5W
Vallès Occidental			
Gallifa	39	37	1W
Montcada	65	21	100W
S. Llorenç Savall	37	39	1W
Vallès Oriental			
Aiguafreda	42	38	5W
Bigues	59	61	5W
El Figaró	37	39	1W
Montseny	22V	36V	2W
Sant Celoni	61V	65V	5W
S. Fost de Campsentelles	37	39	1W
Sant Quirze Safaja	21	25	0.6W
Vallromanes-Monternés	50	52	5W
Andorra			
Andorra la Vella			
Les Escaldes/ Sant Julià/			
S. Coloma/El Serrat	33-26	63-47	—
Pas de ka Casa/Grau Roig	40	37	—

Location	TV3	C33	Power
Encamp i Conillo	65	58	—
Soldeu i Vall d'Incles	49	52	—
La Massana i Ordino	40	46	—
Pal	38	44	—
Arinsal	59	53	—
Arans	56	50	—
Sant Julià i Fontaneda	40	46	—
Mallorca			
Palma/Inca/Manacor/			
Pla de Palma i Soller	46	51	—
Alcúdia	29	35	—
Cap de Pera	21	33	—
Pollensa	27	21	—
Calvià i Palma Nova	24	—	—
Son Cervera	28	—	—
Sant Salvador	33	—	—
Andraitx	30	—	—
Cala Sant Vicenç	27	—	—
Sa Racó	49	—	—
Sant Telm	27	—	—
Bunyola	21	—	—
Llunch	24	21	—
Lloseta	27	—	—
Mancor de la Vall	26	23	—
Menorca			
Maó i sud de l'illa	27	31	—
Ciutadella i nord de l'illa	46	51	—
Pitiüses			
Eivissa/S. Antoni/S. Eulàlia/			
S. Gertrudis i Formentera	25	38	—
S. Josep i part de S. Antoni	56	52	—
Eivissa Ciutat	43	50	—
Cala Sant Vicenç	56	63	—
Castelló			
Castelló General	34	—	—
Vilafranca de Maestrat	50	—	—
Benicarló/Baix Maestrat	29	—	—
Vall d'Uxo-Plana Baixa	45	—	—
Morella Els Ports	22	—	—
Cinctorres Ports	38	—	—
València			
València, General	37	—	—
Sueca/Sollana/Carlet/Benifaió/			
part Ribera Alta i Baixa	37	34	—
Vilamarxant (Camp de Túria)	41	—	—
Algemesí (Ribera Alta)	30	—	—
Montroi, Montserrat d'Alcalà/			
Reial de Montroi	24	—	—
Cullera (Ribera Baixa)	56	—	—
Tavernes de Valldigna/Simat/			
Benefaió (Safor)	31	—	—
Gandia/Font d'en Carrós/			
Oliva Bellreguard (Safor)	48	—	—
Ontinyent (Val d'Albaida)	39	—	—
Sagunt (Camp de Morvedre)	34	—	—
Alacant			
Alacant/Xixona/S. Vicent del Raspeig/			
S.Joan d'Alacant i Mutxamel	41	—	—
Alcoi	45	—	—
Callosa/Toravella/Almoradi	41	—	—
Crevillent (Baix Vilanopó)			
Santa Pola (Baix Vilanopó)	39	—	—
Dènia i La Marina Alta	44	—	—
Benidorm/			
Alfas del Pi, VilaJoiosa	47	—	—
D.Prgr: TV3 1115-0030, Canal 33 1900-2400.			

EUROPE — Spain — Sweden

TV DE GALICIA (TVG)
✉ Apt. 707, San Marcos (Santiago de Compostela).
☎ +34 (81) 565141. 🖷 +34 (81) 562886. ℑ 97012.
L.P: Dir: Xerardo R. Rodríguez.
Station: ch 42. In Galician: 1200 (Sun 1030)-2300 (approx).

TELEVISIO VALENCIANA (Gov.) CANAL 9
✉ 46100 Burjassot, Valencia. ☎ +34 (6) 364 1100. 🖷 +34 (6) 363 9516.
L.P: GM: Amaden Fabregat-Manes. MD: Rafael Cano-Baron.

TV VASCA-EUSKAL TELEBISTA (Gov.)
✉ Barria Lurreta, 48200 Durango (Vizcaya). ☎ +34 (94) 6816600. 🖷 +34 (94) 6816526. ℑ 34441.
L.P: Dir: Koldo Anasagasti.
Stations: ch 42 (ETB-1) & ch 49 (ETB-2) +160 low power st's.
D.Prgr: ETB-1 1400-2000 — ETB-2 1430-2100.

COMMERCIAL STATIONS

ANTENA TELEVISION (Antena 3)
✉ (See Satellite Section)
L.P: Pres: Javier de Godó; G.M: Manuel Martín Ferrand; Film Buyer: Condorcet da Silva Costa.

Projected Autonomous Networks:
Radiotelevision Madrid, Radiotelevision Navarra, TV Andalucia, TV Cantabria & TV Valenciana.

Stations	Tele5	Antena3	C. Plus	kW
Albacete-Chinchilla	50	53	56	50
Alicante-Aitana	50	53	60	100
Almería-Pechina	58	61	64	100
Avila-San Mateo	59	62	65	1
Barcelona-Tibidado	27	34	47	100
Bilbao-Archanda	59	62	65	50
Burgos	30	33	36	1
Cádiz-San Cristobal	49	53	55	30
Castellón-Desierto	49	52	55	40
Ciudad Real-Atalaya	23	26	29	1
Córdoba-Lagar	58	61	64	100
Cuenca-San Cristóbal	37	60	63	1
Gerona-Rocacorba	32	35	38	50
Granada-Parapanda	50	53	56	100
Guadalajara	31	37	51	1
Huelva-Punta Umbría	32	35	56	100
Huesca-Arguis	21	24	27	30
Jaen-Sierra Almadén	32	35	49	100
La Coruña-Ares	35	62	65	100
Las Palmas-La Isleta	32	35	38	20
León-El Portillo	49	52	55	1
Lérida-Alpicat	59	62	65	50
Logroño-Moncalvillo	40	46	48	20
Lugo-Páramo	23	41	44	100
Madrid-Torrespaña	59	62	65	100
Málaga-Mijas	39	42	45	100
Mallorca-Alfabia	58	61	64	100
Montánchez	59	62	65	100
Murcia-Carrascoy	38	42	44	50
Orense-Barbadanes	26	40	43	1
Oviedo-Gamoniteiro	28	32	35	100
Palencia-Villamuriel	48	51	54	1
Pamplona-S. Cristóbal	49	52	55	1
Pontevedra-Tomba	31	35	41	1
Salamanca-Teso	29	60	63	1
San Roque-Carbonaras	21	24	27	20
San Sebastián-Ulia	31	41	44	1
Santander-Peña Cabarga	29	60	63	50
Santiago-Pedroso	38	56	59	100
Segovia	48	51	54	1
Sevilla-Valencia	38	41	44	50
Soria-Santa Ana	21	24	27	1
Tarragana-La Musara	37	50	53	20
Tenerife-Izaña	23	26	29	100
Teruel-Santa Bárbara	26	30	33	1
Toledo-Los Palos	30	53	56	1
Valencia-Torrente	40	43	46	100
Valladolid-Contienda	50	53	56	1
Vigo-Domayo	54	61	64	100
Vitoria-Zaldiarán	29	32	35	15
Zamora-El Viso	58	61	64	100
Zaragoza-La Muela	22	30	54	100

TELEVISION MURCIANA
✉ Plateraa 44, 230001 Murcia
☎ +34 (68) 212 224. 🖷 +34 (68) 214 673.

AFRTS (US Air Force Europe)
Addr. Det. 1 AFEBS, Zaragoza AB, c/o APO New York 09286-5000, USA.
Station: Zaragoza chA4 (Power: cable distribution), Torrejon chA4 2kW.
D.Prgrs: W 12h; Sat/Sun 15h.
NBS TV: Rota chA2 2kW.

SWEDEN

TV-sets: 3.750.000 — **Colour:** PAL — **Systems:** B & G.

TERACOM SVENSK RUNDRADIO AB
This company has the responsibility for the distribution of the prgrs produced by Sveriges Television AB and TV4 Nordisk TV.
HQ: Medborgarplatsen 3, Stockholm (✉ Box 17666, S-118 92 Stockholm. ☎ +46 (8) 671200. 🖷 +46 (8) 6712001.
L.P: Pres. & CEO: Valdemar Persson.

SVERIGES TELEVISION AB (Non-comm.)
✉ Oxenstiernsgatan 26-34, Stockholm.
Postal ✉ S-105 10 Stockholm. ☎ +46 (8) 7840000 & 7847400. 🖷 +46 (8) 7841500. ℑ 10000. **Cable:** Broadcast. **WWW:** http://www.svt.se
L.P: Chmn. Board of Governors: Anna-Greta Leijon. Man. Dir: S. Nilsson. Dir. Prgr. (SVT1): I. Bengtsson. Dir. Prgr: (SVT2): Hans Bonnevier. Dir. Eng: S.O. Ekholm. Dir. Staff: Peter Fogelmarck. Contr. News and Curr. Aff. (SVT1). Stig Fredrikson. Head News. and Curr. Aff. (SVT2): Jan Axelsson. Contr. SVT International: Åke Källqvist. Hd. Legal Dpt/Adm. Britt-Marie Blanck. Hd International Rel: Cecilia Hallgren Järeborg. Press and Information:Jan-Olof Burnhier.

TV4 NORDISK TELEVISION Co. (Comm.)
✉ S-115 79 Stockholm. ☎ +46 (8) 644 4400 🖷 +46 (8) 644 4440 ℑ 14124. **WWW:** http://www.tv4.se
L.P: Chmn: Erik Belfrage; Man Dir: Lars Weiss; Hd. of Inf. & Pub. Rel: Helga Baagøe; Dept Dir of Prgrs: Thomas Nilsson; Hd. of News: Kerstin Persdotter; Hd. of Eng: Olle Mossberg.

Sweden — Switzerland

Location	stations: Kanal 1 TV 2 (Sverige)				Nordisk tv 4	
	Ch	kW	Ch	kW	Ch	kW
Arvidsjaur	5	60	21	1000	24	1000
Älvsbyn	4	60	36	1000	52	1000
Ånge	8	15	42	250	52	250
Bäckefors	8	60	29	1000	49	1000
Bollnäss	6	60	29	1000	49	1000
Borås	6	3	42	1000	55	1000
Borlänge	10	60	47	1000	60	1000
Emmaboda	8	60	31	1000	47	1000
Filipstad	9	5	33	1000	23	1000
Finnveden	41	1000	48	1000	58	1000
Gällivare	9	60	33	1000	43	1000
Gävle	9	60	27	1000	30	1000
Göteborg	9	60	30	1000	46	1000
Hagfors	—	—	—	—	38	?
Halmstad	7	60	24	1000	45	1000
Helsingborg	9	1	30	10	41	10
Hörby	43	1000	33	1000	50	1000
Hudiksvall	11	4	31	1000	44	1000
Jönköping	8	1	28	15	31	15
Kalix	8	60	35	1000	29	1000
Karlshamn	11	30	26	1000	44	1000
Karlskrona	9V	3	34	20	41	20
Karlstad	5	1	43	20	46	20
Kiruna	6	60	29	1000	42	1000
Kisa	11	30	49	1000	56	1000
Köpmannholmen	—	—	—	52	?	
Loffstrand	—	—	—	—	44	?
Lycksele	8	15	45	1000	48	1000
Malmö	10	3	27	150	47	150
Mora	8	20	22	1000	25	1000
Motala	7	10	52	1000	39	1000
Nässjö	10	60	22	1000	25	1000
Norrköping	5	60	31	1000	54	1000
Örebro	2	60	48	1000	58	1000
Örnsköldsvik	6	7	39	400	42	400
Östersund	4	100	27	1000	45	1000
Östhammar	11	30	26	1000	48	1000
Överkalix	10	5	45	1000	48	1000
Pajala	7	60	34	1000	47	1000
Skellefteå	8	35	46	1000	49	1000
Skövde	3	60	37	1000	47	1000
Sollefteå	8	60	46	1000	49	1000
Stockholm	4	60	23	1000	42	1000
Storuman	10	60	33	1000	43	1000
Sundsvall	5	60	47	1000	59	1000
Sunne	7	60	50	1000	53	1000
Sveg	2	60	21	1000	24	1000
Tåsjö	9	60	37	1000	40	1000
Trollhättan	7	1	51	20	41	20
Uddevalla	33	1000	23	1000	43	1000
Uppsala	6	10	49	200	52	200
Varberg	10	1	49	200	43	200
Väddö	8	200	—	—	—	—
Vännäs	2	60	47	1000	50	1000
Västerås	8	10	31	1000	51	1000
Västervik	6	60	26	1000	43	1000
Virserum	—	—	—	—	42	?
Visby	9	60	41	1000	44	1000
Vislanda	39	1000	32	1000	56	1000

N.B: All sts. H unless indicated otherwise.

FINNISH TELEVISION RELAY
Stockholm (Nacka) ch39 1000kW(ERP). Pol H.
Relay of Finnish Prgr II: Mon-Fri 1530-2115, Sat 1200-2200, Sun 1200-2115.
+ 2 lp st.

SWITZERLAND

TV-sets: 2.602.023 — **Colour:** PAL — **Systems:** B&G.

SBC—SWISS BROADCASTING CORPORATION
The SBC's Television programme services are an integral part of the Swiss Broadcasting Corporation.

Addr: SBC, Giacomettistrasse 3, CH-3000 Berne 15.
Cable: Ra-dif. ☎ +41 (31) 3509111. ③ 911590 SSR ch. 🖹 +41 (31) 3509256.
L.P: Pres. SRG: Eric Lehmann. DG: Antonio Riva. Secr. Gen & Dir. Legal Dept: Beat Durrer. Dir. Finance: François Landgraf. Dir. Eng: Daniel Kramer. Dir. Human Resources: Raymond Zumsteg. Television Affairs: Tiziana Mona. Radio Affairs: Félix Bollmann. Dir. Communication & Marketing: Roy Oppenheim. Press Officer: Dr. Oswald Sigg.
Prgr. Sce. in German: TV Dir: Peter Schellenberg, Schweizer Fernsehen DRS, Fernsehstrasse 1-4, 8052 Zürich. ☎ +41 (1) 305 6611. ③ 823823 TVZ. 🖹 +41 (1) 305 5660.
WWW: http://www.srg-ssr.ch/srg/
Prgr. Sce. in French: TV Dir: Guillaume Chenevière, Télévision suisse romande, TSR, 20 Quai Ernest Ansermet. B.P. 234, 1211 Geneva 8. ☎ +41 (22) 708 9911. 🖹 +41 (22) 7811908. TV Prgr. Dir:Raimond Vouillamoz. **WWW:** www.tsr.ch
Prgr. Sce. in Italian: RTSI, Radiotelevisione Svizzera di lingua italiana, Via Canevascini, P.O. Box 6903, Lugano.Reg. Dir: M. Blaser, TV Prgr Dir: D. Balestra. ☎ +41 (91) 58 5111. 🖹 +41 (91) 589150. Studio Televisione, Casella postale, CH-6949 Comano. ☎ +41 (91) 585111. 🖹 +41 (91) 585355.
WWW: http://www.rtsi.ch/

SCHWEIZ 4/SUISSE 4/SVIZZERA 4 (Pub.)
✉ Giacomettistrasse 1. CH-3000 Bern 15, Switzerland. ☎ +41 (31) 350 9444. 🖹 +41 (31) 350 9725. ③ 911 590 ssr ch.
L.P: Dir: Dario Rabbiani
D.Prgr: National multi-lingual TV channel in cooperation with private program producers.

Prgrs as follows:
Schweizer Fernsehen DRS: DRS1, TSR2, TSI2
Télévision Suisse romande: TSR1, DRS2, TSI1
Televisione Svizzera italiana: TSI1, DRS3, TSR3
Schweiz 4: DRS4, TSR4, TSI4
Stations (Systems B&G)

Loc.	Progr.	ch	Power
Ausserberg	DRS2	37	1kW
Ausserberg	DRS3	40	1kW
Bantiger	DRS1	2	47kW
Bantiger	DRS2	50	120kW
Bantiger	DRS3	40	123kW
Bantiger	DRS4	43	92kW
Bantiger	TSR1	10	26kW
Cardada	TSI2	21	3.5kW
Cardada	TSI3	31	3.6kW
Cardada	TSI4	68	4.5kW
Castel San Pietro	TSI1	56	7.7kW
Castel San Pietro	TSI2	39	5.4kW
Castel San Pietro	TSI3	42	6.1kW
Celerina	DRS1	9	3.7kW

EUROPE — Switzerland — Ukraine

Loc.	Progr.	ch	Power
Celerina	DRS2	49	5.2kW
Celerina	DRS3	33	4.3kW
Celerina	TSI1	7	3.6kW
Chamossaire	TSR2	57	2kW
Chamossaire	TSR3	60	2kW
Chasseral	DRS1	62	17.5kW
Chasseral	TSR1	22	12.1kW
Chasseral	TSR2	59	15.1kW
Chasseral	TSR3	56	15.8kW
Chasseral	TSR4	25	12.4kW
Chaux-de-Fonds, La	TSR1	9	8.5kW
Chaux-de-Fonds, La	TSR2	32	9.1kW
Chaux-de-Fonds, La	TSR3	35	9.8kW
Dôle, La	TSR1	4	107.1kW
Dôle, La	TSR2	31	192.7kW
Dôle, La	TSR3	34	186.2kW
Dôle, La	TSR4	69	251.1kW
Feldis	DRS2	24	1.7kW
Feldis	DRS3	21	1.7kW
Gebidem	DRS1	11	5.1kW
Gebidem	DRS2	52	15.4kW
Gebidem	DRS3	55	15.1kW
Gorduno	TSI4	63	1.0kW
Haute-Nendaz	TRS1	7	2.3kW
Haute-Nendaz	TSR2	43	10.7kW
Haute-Nendaz	TSR3	46	11.2kW
Klewenalp	DRS3	55	2kW
Klewenalp	DRS3	58	2kW
Mont-Pèlerin	TSR1	11	2.6kW
Mont-Pèlerin	TSR2	44	2.4kW
Mont-Pèlerin	TSR3	47	2.4kW
Mont-Pèlerin	TSR4	52	2.7kW
Monte Ceneri-Passo 1	TSI1	5	4.9kW
Monte Ceneri-Passo 1	TSI2	46	34kW
Monte Ceneri-Passo 1	TSI3	49	34kW
Monte Ceneri-Passo 1	TSI4	55	37.9kW
Monte Morello	TSI1	68	6.7kW
Monte Morello	TSI2	62	4.2kW
Monte Morello	TSI3	66	6.7kW
Monte San Salvatore	TSI1	10	16.1kW
Monte San Salvatore	TSI2	54	106.7kW
Monte San Salvatore	TSI3	57	113.5kW
Monte San Salvatore	TSI4	60	114.8kW
Niederhorn	DRS1	12	0.9kW
Niederhorn	DRS1	53	19.4kW
Niederhorn	DRS2	27	8.4kW
Niederhorn	DRS3	30	9.2kW
Niederhorn	DRS4	65	16kW
Olten	DRS1	42	13.8kW
Olten	DRS2	45	13.8kW
Olten	DRS3	63	13.8kW
Ordens, Les	TSR1	7	3.6kW
Ordens, Les	TSR2	31	6 3kW
Ordens, Les	TSR3	34	6.3kW
Pfänder	DRS2	49	13.4kW
Pfänder	DRS3	62	14.2kW
Piz Lagalb	TSI1	55	1.1kW
Piz Lagalb	TSI2	37	1.1kW
Piz Lagalb	TSI3	40	1.1kW
Pizzo Matro	TSI2	29	7kW
Pizzo Matro	TSI3	32	7.3kW
Pizzo Matro	TSI4	22	5.9kW
Ravoire	TSR1	9	3.9kW
Ravoire	TSR2	51	5.6kW
Ravoire	TSR3	54	5.6kW
Rigi	DRS1	6	38.5kW
Rigi	DRS2	32	70.8kW
Rigi	DRS3	29	70.8kW
Säntis	DRS1	7	36.6kW
Säntis	DRS2	31	59.5kW
Säntis	DRS3	34	62.3kW
Schaffhausen	DRS1	47	5.3kW
Schaffhausen	DRS2	50	5.3kW
Schaffhausen	DRS3	60	3.6kW
Schüpfheim	DRS2	51	1kW
Schüpfheim	DRS3	54	1kW
Sedrun	DRS2	55	1kW
Sedrun	DRS3	52	1kW
St. Chrischona	DRS1	11	37.6kW
St. Chrischona	DRS2	46	104.7kW
St. Chrischona	DRS3	49	104.7kW
St. Niclaus/VS	DRS1	64	1.2kW
Uetliberg	DRS1	3	74.1kW
Uetliberg	DRS2	23	105.9kW
Uetliberg	DRS3	26	105.9kW
Vallée de Joux	TSR2	46	1kW
Vallée de Jouz	TSR3	49	1.1kW
Valzeina	DRS1	10	0.7kW
Valzeina	DRS2	56	11.6kW
Valzeina	DRS3	53	11.6kW
Wattenwil	DRS2	57	6kW
Wattenwil	DRS3	60	6kW
Ziegelbrücke	DRS2	51	1.3
Ziegelbrücke	DRS2	54	1.3

UKRAINE

TV-Sets: 1000 — **Colour:** SECAM — **Systems:** D&K.

UKRAJINSKA TELEBAČENNJA (Gov.)
+ vul. Chreščatik 26, 252001 Kyjiv. (+380 (44) 2290638.
📠 +380 (44) 2296945.

UT-1
(*Not ERP)

Stations	ch	kW*		ch	kW*
L'viv	1	5	Kirovograd	6	5
Sovetskij	2	5	Simferopol'	6	5
Izmajil	3	0.1	Ivano-Frankivs'k	7	5
Komis Zorja	3	5	Kam'jans'ke	7	5
Nikopol'	3	2.5	L'viv	8	25
Kyjiv	4	50	Kryvyj Rig	9	5
Kovel'	5	5	Krasnogorivka	10	25
Sevastopol'	5	5	Mykolajiv	10	5
Ternopil'	5	25			

+various low power translators.

NB: The above list of main transmitters is not complete.
D.Prgr: 0600-2200.

UT-2
(*Not ERP)

Stations	ch	kW*		ch	kW*
Kramators'k	1	0.1	Krasnoperekops'k	10	0.1
Ivano-Frankivs'k	2	5	Pervomejs'k	12	0.1
Simferpol'	3	5	Pryluky	12	0.1
Kirovogard	5	0.1	Kirovograd	21	5
L'viv	6	5	Mariupol'	25	5
Sevastopol'	8	0.1	Krasnogorivka	28	—
Kyjiv	9	50	Komiš Zorja	28	0.1
Odesa	9	5	Kam'jans'ke	29	—

*various low power translators.

NB: The above list of main transmitters is not complete.
D.Prgr: 0500-2200. Includes relays of Rossijskoje televidenije (RTV) from Moscow.

Relays of ORT (Russia)
(*Not ERP)

Stations	ch	kW*		ch	kW*
L'viv	1	40	Pervomejs'k	7	5
Simferopol'	1	25	Sovetskij	7	5
Kryvyj Rig	1	5	Komš Zorja	8	5
Kyjiv	2	50	Krasnogorivka	8	25
Mykolajiv	2	5	Starobil's'k	8	5
Rivne	3	5	Ivano-Frankivs'k	9	5
Kam'jans'ke	4	5	Izjum	11	25
Mariupol'	5	5	Kirovograd	11	5
Odesa	5	5	Melitopol'	11	5
Kovel'	6	25	Sevastopol'	11	5
Kotovs'k	6	5			

+various low power translators.

NB: The above list of main transmitters is not complete.
D.Prgr: Relays of ORT from Moscow. Further details see under Russia.

UNITED KINGDOM

TV-sets: approx 20m — **Colour:** PAL — **System:** I.

BRITISH BROADCASTING CORPORATION
Television Centre, London W12. ☎ +44 (181) 743 8000. ① 265781. 🗎 +44 (181) 749 7520.
L.P: MD: Will Wyatt. Asst. MD: Jane Drabble. Contr. BBC-1: J. Powell. Contr. BBC-2: Alan Yentob. CE: Peter Marchant. Publ: Keith Samuel.
Stations: See below.
D.Prgr: BBC1: Mon-Fri 0600-2400; Sat/Sun 0700-2400 (closing times vary). **BBC2:** 1500-2400 (Sat/Sun 0100) approx.

S4C, Welsh Fourth Channel Authority
Clos Sophia, Cardiff, CF1 9XY. ☎ +44 (1222) 747444. 🗎 +44 (1222) 754444. **WWW:** http://www.s4c.co.uk

INDEPENDENT TELEVISION COMMISSION
33 Foley Street, London W1P 7LB. ☎ 0171 255 3000. 🗎 0171 306 7800. **E-mail:** 100731.3515@compuserve.com
L.P: Chief Exec: Peter Rogers. Dep. Chief Exec: Clare Mulholland. Dir. of Finance: Sheila Cassells. Dir. of Public Affairs: Paul Smee. Dir. of Prgrs and Cable: Sarah Thane.Dir. of Eng: Gary Tonge. Dir. of Advertising and Sponsorship: Frank Willis.Contr. of Admin: Don Horn. Secr. to the Commission: Michael Redley. Dep. Dir. of Prgrs: Robin Duval. Dep. Dir. of Cable:Anthony Hewitt.
Function: The ITC is the public body responsible for licensing and regulating commercially funded television services provided in and from the UK. These include Channel 3 (ITV), Channel 4, Channel 5, public teletext and a range of cable, local delivery and satellite services. They do not include services provided by the BBC or by S4C, the fourth channel in Wales.

Program contractors licensed by the ITC:

Anglia Television (1), Anglia House, Norwich NR1 3JG. ☎ +44 (1603) 615151. 🗎 +44 (1603) 631032. **WWW:** http://www.anglia.co.uk/
Border Television (3), The Television Centre, Carlisle, CA1 3NT. ☎ +44 (1228) 25101. 🗎 +44 (1228) 41384. **E-mail:** ian@border-tv.com **WWW:** http://www.border-tv.com
Carlton Television Ltd. (8), 101 St. Martin's Lane, London WC2N 4AZ. ☎ +44 (171) 240 4000. 🗎 +44 (171) 240 4171. **WWW:** http://www.carltontv.co.uk
Central Independent Television (2), Central House, Broad Street, Birmingham, B1 2JP. ☎ +44 (121) 643 9898. 🗎 +44 (121) 616 4766.
Channel Four Television Corporation, 124 Horseferry Road, London SW1P 2TX. ☎ +44 (171) 306 8333. **WWW:** http://www.channel4.com
Channel Television (4), The Television Centre, St. Helier, Jersey, Channel Islands. ☎ +44 (1534) 68999. 🗎 +44 (1534) 59446/24770.
Data Broadcasting International Ltd, Allen House, Station Road, Egham, Surrey, TW20 9NT. ☎ +44 (1784) 471515.
GMTV Ltd, The London Television Centre, London SE1 9LT. ☎ +44 (171) 827 7000. 🗎 +44 (171) 827 7001.
Grampian Television (5), Queen's Cross, Aberdeen, AB9 2XJ. ☎ +44 (1224) 646464. 🗎 +44 (1224) 635127.
Granada Television (6), Granada Television Centre, Manchester, M60 9EA. ☎ +44 (161) 832 7211. 🗎 +44 (161) 839 0454. **WWW:** http://www.granadatv.co.uk
HTV Wales (7), The Television Centre, Culverhouse Cross, Car-diff, CF5 6XJ. ☎ +44 (1222) 590590. 🗎 +44 (1222) 597183.
HTV West (7a), The Television Centre, Bath Road, Bristol, BS4 3HG. ☎ +44 (1272) 778366. 🗎 +44 (1272) 722400.
London Weekend Television (LWT) (8), South Bank Television Centre, London, SE1 9LT. ☎ +44 (171) 620 1620. 🗎 +44 (171) 928 6948.
Meridian Broadcasting Ltd (10), Television Centre, Southampton, SO14 0PZ. ☎ +44 (1703) 222 555. 🗎 +44 (1703) 335050. **WWW:** http://www.meridian.tv.co.uk
S4C, Welsh Fourth Channel Authority, Clos Sophia, Cardiff, CF1 9XY. ☎ +44 (1222) 747 444. 🗎 +44 (1222) 75 4444. **WWW:** http://www.s4c.co.uk
Scottish Television (9), Cowcaddens, Glasgow, G2 3PR. ☎ +44 (141) 300 3000. 🗎 +44 (141) 300 3030. **WWW:** http://www.scotnet.co.uk/stv/
Teletext UK Ltd, 101 Farm Lane, London SW6 1QJ. ☎ +44 (171) 386 5000. 🗎 +44 (171) 386 5002.
Tyne Tees Television (11), The Television Centre, City Road, Newcastle upon Tyne, NE1 2AL. ☎ +44 (191) 261 0181. 🗎 +44 (191) 261 2302.
Ulster Television (12), Havelock House, Ormeau Road, Belfast, BT7 1EB. ☎ +44 (1232) 328122. 🗎 +44 (1232) 246695.
Westcountry Television Ltd (13), Western Wood Way, Lan-guage Science Park, Plymouth, PL7 5BG. ☎ +44 (1752) 333333. 🗎 +44 (1752) 333 444.
Yorkshire Television (14), The Television Centre, Leeds, LS3 1JS. ☎ +44 (532) 438283. 🗎 +44 (532) 445107.

EUROPE United Kingdom — Yugoslavia

INDEPENDENT TELEVISION NEWS (ITN)
200 Gray's Inn Road, London WC1X 8XZ. ☎ +44 (171) 833 3000.

INDEPENDENT TELEVISION ASSOCIATION (ITVA)
Knighton House, 56 Mortimer Street, London W1N 8AN. ☎ +44 (171) 612 8000. 📠 +44 (171) 580 7892.

STATIONS:

(Pol=H)	BBC1	BBC2	ITV	Ch4	kW (ERP)
So. Ea. England (+ 66 relays)					
Bluebell Hill (10)	40	46	43	65	30
Crystal Palace (8)	26	33	23	30	1000
Dover (10)	50	56	66	53	100
Heathfield (10)	49	52	64	67	100
Oxford (2)	57	63	60	53	500
So. We. England & Channel Isl. (+ 81 relays)					
Beacon Hill (13)	57	63	60	53	100
Caradon Hill (13)	22	28	25	32	500
Huntshaw Cross (13)	55	62	59	65	100
Redruth (13)	51	44	41	47	100
Stockland Hill (13)	33	26	23	29	250
Fremont Point (4)	51	44	41	47	20
So. England (+ 38 relays)					
Hannington (10)	39	45	42	66	250
Midhurst (10)	61	55	58	68	100
Rowridge (10)	31	24	27	21	500
We. England (+ 51 relays)					
Mendip (7a)	58	64	61	54	500
Ea. England (+ 18 relays)					
Sandy Heath (1)	31	27	24	21	1000
Sudbury (1)	51	44	41	47	250
Tacolneston (1)	62	55	59	65	250
Ce. England (+ 60 relays)					
Ridge Hill (2)	22	28	25	32	100
Sutton Coldfield (2)	46	40	43	50	1000
The Wrekin (2)	26	33	23	29	100
Waltham (2)	58	64	61	54	250
No. England (+ 52 relays)					
Belmont (14)	22	28	25	32	500
Emley Moor (14)	44	51	47	41	870
No. We. England (+ 66 relays)					
Winter Hill (6)	55	62	59	65	500
No. Ea. England (+ 64 relays)					
Bilsdale (11)	33	26	29	23	500
Caldbeck (3)	30	34	28	32	500
Chatton (11)	39	45	49	42	100
Pontop Pike (11)	58	64	61	54	500
Scotland (+ 188 relays)					
Angus (5)	57	63	60	53	100
Black Hill (9)	40	46	43	50	500
Sandale	22	—	—	—	500
Craigkelly (9)	31	27	24	21	100
Darvel (9)	33	26	23	29	100
Durris (5)	22	28	25	32	500
Eitshal (5)	33	26	23	29	100
Keelylang Hill (5)	40	46	43	50	100
Knock More (5)	33	26	23	29	100
Rosemarkie (5)	39	45	49	42	100
Rumster Forest (5)	31	27	24	21	100
Selkirk (5)	55	62	59	65	50
No. Ireland (+ 41 relays)					
Brougher Mt. (12)	22	28	25	32	100
Divis (12)	31	27	24	21	500
Limavady (12)	55	62	59	65	100

Wales (+ 162 relays)

	BBC1	BBC2	ITV	S4C	kW (ERP)
Blaen-plwyf (7)	31	27	24	21	100
Carmel (7)	57	63	60	53	100
Llanddona (7)	57	63	60	53	100
Moel-y-Parc (7)	52	45	49	42	100
Presely (7)	46	40	43	50	100
Wenvoe (7)	44	51	41	47	500

NB: The number in brackets indicates the ITC company responsible for prgrs on ITV (3rd column).

CHANNEL 5 BROADCASTING LIMITED
22 Long Acre, London WC2E 9LY. ☎ +44 (171) 550 5555. 📠 +44 (171) 550 5554. **WWW:** http://www.channel5.co.uk
L.P: Chief Exec. Officer: David Elstein. Chief Operating Officer: Ian Ritchie. Dir. of Programming: Dawn Airey. Dir. of Sales: Nick Milligan. Dir. of Marketing & Communications: David Brook. Dir. of Finance: Damian Harte.

Location	ch	kW (ERP)	Pol.
Plympton	30	2	V
Londonderry	31	10	V
Tay Bridge	34	4	V
Nottingham	34	2	V
Fawley	34	1	H
Fenton	35	F.Pl.	V
Hannington	35	F.Pl.	H
Ridge Hill	35	F.Pl.	H
Sudbury	35	F.Pl.	H
The Wrekin	35	F.Pl.	H
Waltham	35	F.Pl.	H
Emley Moor	37	870	H
Black Hill	37	500	H
Croydon	37	250	H
Mendip	37	126	H
Lichfield	37	100	H
Presely	37	100	H
Black Mountain	37	50	H
Cambret Hill	37	20	H
Redruth	37	3	H
Sandy Heath	39	10	H
Storeton	39	2.8	V
Craigkelly	48	4	H
Winter Hill	48	12.5	H
Churchdown Hill	48	1	H
Oxford	49	40	H
Selkirk	52	50	H
Tacolneston	52	4	H
Perth	55	2	V
Belmont	56	50	H
Caldbeck	56	10	H
Blaen Plwyf	56	4	H
Fenham	56	2	V
Chelmsford	63	1	H
Durris	67	100	H
Mounteagle	67	100	H
Sheffield	67	2.5	H
Huntshaw Cross	67	2	H
Burnhope	68	50	H

YUGOSLAVIA

TV-sets: 1.642.522 — **Colour:** PAL—**System:** B&G.

UDRUŽENJE JUGOSLOVENSKIH RADIOTELEVIZIJA d.o.o. (JRT)

☐ JRT-Permanent Services — Television Department: Hartvi-gova 70/I, 11000 Beograd. ☎ +381 (11) 434-910. ① 11469 yu jurate — **Televizija Srbije-Televizija Beograd:** Takovska 10, 11000 Beograd. ☎ +381 (11) 342-001. ① 11884 — **Televizija Srbije-Televizija Novi Sad:** Kamenicki put 45, 21000 Novi Sad. ☎ +381 (21) 56-855. **Televizija Srbije-Televizija Priština:** Beogradska 66, 38000 Pristina. ☎ +381 (38) 31 211 — **Televizija Crne Gore:** Cetinski put bb, 81000 Podgorica. ☎ +381 (81) 41 529. **Stations** (System B) (Ch 2-12), System G (Ch 21-69). **Ch E.** Audio powers 1/10 of vision powers indicated. St's below 3kW not mentioned.

MONTENEGRO
TV Crne Gore	I	II	III	IV	kW
Bjelasica	6	12	37	62/43	100/100/1/500/1
Sudjina Glava	12	6	24	39	5/5/15/15
Sjenica	6	33	23	25/29	1.5/15/15/15
Luštica	4	26	33	39/42	1/10/10/10
Lov´cen	8	31	10	35/67	100/1000/20/1000/500
Volujica	6	12	24	38	1/1/10/10
Mužura	12	23	33	43/53	1/10/10/10/10

+321 low power st's.
Local/private TV sts: Montena TV, 81000 Podgorica (ch 53); Montenegro TV, 81000 Podgorica (ch 59)

SERBIA
TV Srbije-TV Beograd	I	II	III	kW
Kopaonik	3	41		50/1000
Jastrebac	5	27	33	100/1000/1000
Avala	6	22	28	100/1000/30
Deli Jovan	6	23	43	10/50/100
Tornik	7	53	59	1/50/50
Besna Kobila	8	49	59	10/300/300
Ovčar	8	42	56	10/400/10
Maljen	9	26	32	5/2.5/250
Crveni Cot	10	24	30	100/1000/1000
Tupižnica	10	25	31	35/500/500
Crni Vrh (Svetozarevo)	11	35	38	35/500/500
Cer	7	37	34	3/300/300

+288 low power st's.
Local/private TV sts in Beograd:
TV Pink, Bul. Lenjina 2, 11070 Beograd (ch 59+5 relays). BK Telecom, Omlandinskih Brigada 1, 11070 Beograd (ch 12 + 4 relays). TV Politika, Makedonska 29, 11000 Beograd (ch 43 + 4 relays). NTV Studio "B", Masarikova 5, 11000 Beograd (ch 53 + 4 relays). TV Art, VI. Kovacevica 6, 11000 Beograd (ch 38 + 2 relays) TV Palma. Banijska 2, 11080 Beograd (ch 34 + 1 relay)

TV Srbije-TV Novi Sad	I	II	III	kW
Subotica	5	43		35/1000
Venac	41		48	1000/1000
Vršac	39	56		1000/1000

+13 low power st's.

TV Srbije-TV Pristina	I	II	III	kW
Goleš	7	44		35/600
Cviljen	9	21		10/400

+63 low power st's.
Other stations: there are numerous local/private TV st's in Serbia and Vojvodina.

AFRICA

ALGERIA

TV-sets: 2m — **Colour:** PAL — **System:** B.

ENTREPRISE NATIONALE DE TÉLÉVISION (E.N.T.V.) (Gov.)
☐ 21 Blvd. des Martyrs, Algiers 16000. ☎ +213 (2) 780310. ① 66101 or 65282. ▤ +213 (2) 601922.
L.P: DG: Zemzoum Zoubir. Dir. of Inf: M. Ibrahim. Dir. Tec: M. El Ksouri. Dir. Ext. Rel: M. Bey.
Stations: Pol: H.

	ch	kW (ERP)		ch	kW (ERP)
M. Cid	5	150/30	Adrar	7	17/1.7
Ain-n-Sour	5	100/20	Mecheria	8	30/6
Reggane	5	25/2.5	Chrea	9	120/24
Ghardaia	5	11/1.1	Touggourt	9	12/1.2
Akfadou	6	100/20	Metlili	10	150/30
Nador	6	150/30	Tessala	10	150/30
Ouargla	6	5/0.5	Algiers	11	20/4
Constantine	7	100/20	Aflou	11	150/30
Aouilef	7	19/1.9	In Amenas	11	10/1

Low power st's not mentioned.
D.Prgr: 1500-2300 (Thurs/Sat/Sun 1300-2400).

ANGOLA

TV-sets: 50.500 — **Colour:** PAL — **System:** I.

TELEVISÃO POPULAR DE ANGOLA (Gov.)
☐ Avenida Ho-Chi-Min, P.O. Box 2604, Luanda. ☎ +244 (2) 320025. ▤ +244 (2) 391091. ① 3238, 4153, 4157 TPA-AN. **Cable:** TPA.
L.P: MD: Carlos Chunha. Head of Prgrs: António Pedreira.
Station: ch9 13kW (ERP) + st's at Benguela, Huambo, Lubango, Namibe, Cabinda, Bié
D.Prgrs: 1730-2300 (Mon-Fri), 1400-2300 (Sat), 0900-2300 (Sun).

BENIN

TV-sets: 30.000 — **Colour:** SECAM. — **System:** K.

OFFICE DE RADIODIF. ET TV DE BENIN (ORTB) (Gov.)
☐ P.O. Box 366, Cotonou. ☎ +229 3010628.
L.P: Dir. Gen: Nicolas Benon. Dir. TV: Michèle Badarou. Chief of Sce. (TV): Marcellin Illougbade. Head of Prgrs: Didier Falde.
Station: ch4 20kW.
D.Prgr: Mon-Fri 1800-2100, Sat/Sun 1700-2200.

BOTSWANA

TV-sets: 13.800 — **Colour:** SECAM — **System:** K.

GABORONE TELEVISION CORP.
☐ Private Bag 0060, Gaborone. ☎ +267 352541. ▤ +267 357 138.

AFRICA Botswana — Cote d'Ivoire

L.P: Dep. Dir. of Prgrs: Mrs. B. Tafa; Dep. Dir. of Prod: Mr. S. Moribame.
Station: ch. not known.
D.Prgr. in English: 1700-2000.

BURKINA FASO

TV-sets: 45.500 — **Colour:** SECAM — **System:** K.

TÉLÉVISION NATIONALE BURKINA (Gov.)
✉ B.P. 2530, Ouagadougou. ☎ + 226354773/306621. ⓓ 5317.
L.P: DG: Mahamoudou Ouedraogo.
Station: Ouagadougou & Bobodioulasso ch6 50/10W V.
D.Prgr: Tues-Fri 1833-2220; Sat 1508-2000; Sun 1205-1300, 1833-2200 — **Projected:** Ouagadougou 10kW H.

BURUNDI (Rep.)

TV sets: 4500 — **Colour:** SECAM — **System:** K.

TÉLÉVISION NATIONALE DU BURUNDI (Gov.)
✉ B.P. 1900, Bujumbura. ☎ +257 2247 60. ⓓ 5119.
L.P: DG: Louis-Marie Nindorera; Hd. of Prgr: Leonidas; Director TV: Clément Kirahagazwi.
Station: ch 25 0.5kW — **D.Prgr:** 1600 (Sat/Sun 1400)-2200.

CAMEROON

TV-sets: 15.000 — **Colour:** PAL — **System:** B

CAMEROON RADIO AND TELEVISION (CRTV) (Gov.)
✉ P.O. Box 1634, Yaoundé. ☎ +237 42 6060/7211/9440. 🖷 +237 204340. ⓓ 8888 KN.
L.P: Dir. Gen: Pr. Gervais Mendoze.
Stations: ch 8 10kW

CANARY ISLANDS

TV-sets: 240.000 — **Colour:** PAL — **Systems:** B & G.

TELEVISION ESPAÑOLA EN CANARIAS
✉ 69 Calle Buenos Aires, Santa Cruz de Tenerife. ☎ +34 (22) 216200.
Stations: TV1: Santa Cruz de Tenerife (Izana) chE3 350/35kW H, Fuencaliente chE7 4.7/0.47 H, Poso de las Nieves ch10 10/1H, Arrecife ch 32 6/0.6 kW H + 38 relays. **TV2:** Fuencaliente ch27 170/17kW H, La Isleta ch28 18.6/1.86kW H, Arrecife ch38 6/0.6kW H, Izana ch45 87/8.7kW H, Puerto Rosario ch52 23.8/2.38kW H, Los Christianos ch 57 6/0.6kW H, Pozo de las Nieves ch59 48/4.8kW H + 13 relays.
D.Prgr. TV1: W 0615-0020 (Sun 0745-0020); **TV2:** W 1745-2300; Sun 0845-2325 (closing time may vary).

COMMERCIAL STATIONS

CANAL BUENAS NUEVAS
✉ Calle Sao Paulo 45, Santa Cruz de Tenerife. ☎ +34 (22) 279442 — **Station:** Cebadal ch21.

CANARYVISION
✉ Calle Arequipa 10, Santa Cruz de Tenerife. ☎ +34 (22) 470366 — **Station:** Cebadal ch25.

TELE GRAN CANARIA
✉ Calle Sao Paulo 46, Santa Cruz de Tenerife. ☎ +34 (22) 464722 — **Station:** ch40.

ONDA TELEVISION MASPALOMAS (OTM 6)
✉ Calle Galdar 48, San Agustin, Playa del Ingles. ☎ +34 (22) 772445, 773737 — **Stations:** Guia ch42, Cumbre ch46.

LIBERTAD TELEVISION
✉ Avda. Escaleritas 112, Escaleritas. ☎ +34 (22) 251440.
Station: Escaleritas ch50.

ANTENA 3 TELEVISION
✉ Eduardo Benot 3, Santa Cruz de Tenerife. ☎ +34 (22) 275242 — **Stations:** Cumbre ch36, Isleta ch38,

CENTRAL AFRICAN REP.

TV-sets: 7.500 — **Colour:** SECAM — **System:** K.

RADIODIFFUSION-TÉLÉVISION CENTRAFRIQUE
✉ P.O. Box 940, Bangui. ☎ +236 613242.
L.P: MD: Paul Service; Hd. of Prgrs: Henri-Gustav Hytayu.
Stations: N/A.

CHAD

TV-sets: 50.000 — **Colour:** SECAM — **System:** D.

TÉLÉTCHAD (Gov.)
✉ B.P. 74, N'Djamena. ☎ +235 51 29 23. ⓓ 5307.
L.P: Dir: Hourmadji Houssa Doumgor. Adj. Dir: Houssa Dago.
Station: N'Djamena ch7 (offset) 100W.
D.Prgr. in French/Arabic: 1800-2100 4 days per week.

CONGO (People's Rep.)

TV-sets: 8.500 — **Colour:** SECAM — **System:** K.

RADIODIFFUSION TÉLÉVISION CONGOLAISE
✉ 2241, Brazzaville. ☎ +242814574/814273/814030. **L.P:** D.G: J.F. Sylvestre SOUKA. **Station:** ch7 10/20kW H. **D.Prgr:** 1730-2300

COTE D'IVOIRE

TV-sets: 810.000 — **Colour:** SECAM — **System:** K.

TÉLÉVISION IVOIRIENNE (Gov.)
✉ 08 B.P. 883, Abidjan 08. ☎ +225 439039. 🖷 +225 222297. ⓓ 26110 ditele.
L.P: Dir. Gen: Mamadou Berté.

Cote d'Ivoire — Gambia

Station	ch	kW	Pol
Koun	4	10/2.5	V
Tiémé	4	10/2.5	V
Bouake	4	0.1	V
Abobo*	4	0.1	H
Séguéla	5	10/2.5	H
Digo	5	2/1	H
Dimbroko	6	2/1	H

Station	ch	kW	Pol
Touba	6	1/1	H
Man	7	10/2.5	H
Dabakala	7	2	H
Abidjan	8	10/2.5	H
Niangbo	8	10/2.5	H
Niangué	8	10/2.5	H
Bouaflé	9	10/2.5	V

*) 2nd prgr for Abidjan only.
D.Prgrs: 1st Prgr: Mon-Wed 1200-1330 & 1900-2300, Thurs 1200-1300 & 1600-2300, Sat 1200-0130, Sun 1030-2330. **2nd Prgr:** Mon-Fri 2030-2300, Sat 1600-2030 (Sun no transmissions). **Bouaké Regional Prgrs:** Thurs 1200-1300, Fri 1700-1830.

DJIBOUTI (Rep.)

TV-sets: 17.000 — **Colour:** SECAM — **System:** K.

RADIO TÉLÉVISION DE DJIBOUTI (Gov.)
P.O. Box 97, Djibouti. ☎ +253 352294. ③ 5863 DJ.
L.P: DG: Mohamed Tara Moussa.
Station: Djibouti chK'6 0.05kW H.
D.Prgr: 1500-2000. **F.Pl:** 5 relay st's.

EGYPT

TV-sets: 5m — **Colour:** PAL — **System:** B&G.

EGYPTIAN RADIO AND TV UNION (Gov.)
TV Bldg, Cornish El-Nil, Maspero Cairo. ☎ +20 (2) 757155. **Cable:** Cibrotev. ③ 92152 Karadun.
L.P: Chmn: Amin Bassounia. Head of TV: Sohair El Atreby.

Stations	I	II	kW	Pol	Stations	I	II	kW	Pol
Asswan	5	9	67	H	Idfu	8	11	165	H
Cairo	5	9	200	H	Mahalla	8	10	1600	H
Hurghada	5	7	89	V	Abu Simbil	9	11	0.05	H
Nag Hamadi	5	8	17	H	Ras Gharib	9	11	66	H
Port Said	5	7	10	V	Salum	9	11	6.9	H
Sidi Barani	5	7	11.2	H	Qena	9	6	80	H
El Farafra	5	7	10	V	Assiut	10	6	67.2	H
Alexandria	6	11	110	H	El Kharga	10	8	40	V
Beni Ali	6	9	5	V	Kom Ombo	10	7	40	H
El Arish	6	10	182	V	Matruh	10	8	39.2	H
Isna	6	9	17	H	Bawiti	10	8	21.9	H
Siwa	6	8	10	V	Beni Suef	11	7	110	V
Dahab	6	8	9.33	V	Ismailia	11	9	260	V
El Daba	7	5	2.95	V	Luxor	11	7	19	H
Quseir	7	5	50	H	Safaga	11	9	50	V
Sohag	7	11	52	H	Negala	22	25	5	H
Suez	7	5	20	H	Nuweiba	26	35	25.7	V
Managem	7	5	2.5	H	Sh. El-Sheikh	27	33	8.91	H
Baharia Baris	7	5	10	V	Taba	32	37	25.7	H
El Dakhla	8	6	22.4	H	Barnis	24	29	830	H
El Minya	8	5	165	V	El Hamam	48	51	69.2	H

D.Prgr: Prgr. I: 0400-0100.
Prgr. II: 0500-2400.
Prgr. III: Cairo ch7 100/10kW H, ch21 100/10kW H.
D.Prgr: 1100-2330.
Prgr. IV: Suez ch54 20kW H, Ismailia ch31 260kW V, Port Said ch42 10kW V, Cairo ch40 8.9kW.

D.Prgr: 1400-2330 (Mon: 1200-2330).
Prgr. V: Alexandria ch36 670kW H, Cairo ch46 9.5kW
D.Prgr: 1400-2400 (Thu 1400-0500), (Fri 0800-2400)
Prgr VI: Cairo ch43 9kW H, Mahalla ch49 32kW H
D.Prgr: 1500-2400 (Fri 1400-2400)
Prgr VII: Beni Ali ch22 1.32kW V, El Minya ch39 5.6kW V, Cairo ch42 6.5kW H, Assiut ch48 11.7kW H, Beni Suef ch51 4.3kW V, Fayoum ch55 10.7kW V
D.Prgr: 1400-2200
Esat Prgr: Shalatin ch96 10kW H, Cathrine ch12 1kW H, Marsa Alam ch21, Halayeb ch25 1kW, Natron ch12 1kW, Hasana ch 34, Nekhel ch34, Ras Sedr ch27 0.4kW, Abu Zonema ch33 o.4kW
D.Prgr: 24h.

EQUATORIAL GUINEA

TV-sets: 2.500 — **Colour:** SECAM — **System:** B.

TELEVISION NACIONAL (Gov.)
Malabo Bioko Norte.
L.P: Dir: Antonio Nkulu Oye.
Station: chE2.
No further information available.

ETHIOPIA

TV Sets: 150.000 — **Colour:** PAL — **System:** B (Pol=H).

ETHIOPIAN TELEVISION (Gov.)
P.O. Box 5544, Addis Ababa. ☎ +251 (1) 116701.
③ 0980 21429 ETV ET.
L.P: St. Mgr: Wole Gurmu. TD: Bermanu Sintayehu. Film Buyer: Mrs. Almaz Degene.

City	ch	kW	City	ch	kW
Asmara	5	5	Mekele	7	4
Bahir Dar	5	1.5	Gondar	7	2.5
Dire Dawa	5	1	Dessie	9	5
Addis Ababa	7	5	Deber	9	1
Harar	7	5	Jima	11	2.5

D.Prgr: Mon-Fri 1900-2230, Sat/Sun 1800-2400 (Sun 2300).

GABON

TV-sets: 40.000 — **Colour:** SECAM — **System:** K.

RADIODIFFUSION-TÉLÉVISION GABONAISE (Gov.)
B.P. 10150, Libreville. ☎ +241 732152. 🖷 +241 732153. **Cable:** Radiotelegabon.
L.P: DG (TV): Jules César Lekogho.
Stations: Libreville chK4 & K8 2kW H, Port Gentil chK10 0.1kW H, Moanda 1kW (relay) + 5 low power relay sts.
D.Prgr: 1800-2200; ch10: 1800-2100 (relay Libreville).

GAMBIA

TV-Sets: 6.000 — **Colour:** PAL — **System:** B.

RADIO GAMBIA
Mile 7, Banjul. ☎ +220 495101/495921. 🖷 +220

AFRICA Gambia — Madeira

495102/495923.
L.P: Senior Op's Mgr (Broadcasting): Momodou Cham.
Contr. TV Prgrs: Lasana Jobarteh.

GHANA

TV-sets: 800.000 — **Colour:** PAL — **System:** B.

GHANA BROADCASTING CORPORATION
✉ P.O. Box 1633, Accra. ☎ +233 (21) 221161. 🖷 +233 (21) 773240 . ℗ 2114 GBC GH.
L.P: DG: Dr Kofi Frimpong. Dir. of TV: Prof. Mark Duodu. Dep. Dir. of TV (News): Berfi Apenteng. Dep. Dir. of TV (Prgrs): H. Torto-Gilbertson.
Stations: *) colour; others monochrome.
Accra ch4 (5kW) & 9* (10kW), Kisi ch2 (5kW) & 11* (10kW), Jamasi ch3 (5kW) & 5* (10kW), Bolgatanga ch6 (5kW) + 8 translators.
D.Prgr: 1630 (SS 0930)-0100. Incl. relays of CNN International.

GUINEA (Republic)

TV sets: 65.000 — **Colour:** PAL — **System:** K.

RADIODIFFUSION TÉLÉVISION GUINÉENNE (Gov.)
✉ B.P. 391, Conakry. ☎ +224 442205. ℗ 22341 rtg ge conakry - Guinée.
L.P: Dir. Gen: B. Camara. Dir. of Prgrs: B. Kaba.
Stations: Conakry ch 5 1kW + Kindia ch4 0.2 kW, Faranah ch5 0.5kW, Labé ch7 8kW, Mamou/Mali ch9 0.2kW, Kankan ch9 1kW.
D.Prgr: 1700-2000 (approx.)

KENYA

TV-sets: 500.000 — **Colour:** PAL — **System:** B.

KBC TV (Gov, Comm.)
✉ Box 30456, Nairobi. ☎ +254 (2) 334567. 🖷 +254 (2) 220675 ℗ 25361 KBC KE.
L.P: MD: Philip. Okundi. Contr. TV Prgrs: Joseph Murema. Asst. Mgr. Tech. Sces (TV): Ben Muinde.
Stations: ch4 (Limuru) 10/1kW, Timboroa ch2 10/1 kW, Mombasa ch6 10/1kW, Mazeras ch6 5/1kW, Nyeri ch10 10/1kW, Nyambene ch11 10/1kW
D.Prgrs: 1400 (Sat/Sun 1100)-2100.

COMMERCIAL STATIONS

KENYA TELEVISION NETWORK (KTN-TV)
✉ P.O. Box 56205, Nairobi. ☎ +254 (2) 227122. 🖷 +254 (2) 214467. **WWW:** http://www.kenyaweb.com/ktn/ktn.html
L.P: Chmn: Mwakio Sio. MD: Sam Shollei. CEO: Steve Crozier. Mktg. Mgr: Patrick Ndeda. Tech. Mgr: Francis Kimore.
Station: UHF ch 62 (Nairobi), ch59 (Ngong).
D.Prgr: 24hrs with own programming, also incl. relays of CNN.
F.Pl: extend coverage to coast and we. Kenya by the end of 1997.

LESOTHO

TV-sets: 250.000 — **Colour:** PAL — **System:** I.

LESOTHO NATIONAL BROADCASTING SERVICES
✉ P.O. Box 552, Maseru 100. ☎ +266 323561. 🖷 +266 310003. ℗ 4340 LENA 10
L.P: Ag. Cont. of TV: Litebo Tshola.
D.Prgr: No information available.

LIBERIA

TV-sets: 45.000 — **Colour:** PAL — **System:** B.

LIBERIAN BROADCASTING CORPORATION (Gov. Comm.)
✉ P.O. Box 10-594, Monrovia. ☎ +231 271250. ℗ 44249 BROADCAST.
L.P: DG: Jesse B. Karnley.
Station: ELTV chE6 1/0.1kW + 4 low power repeaters.
D.Prgr: W 1815 (Sat 1615)-2300; Sun 1415-2230.

LIBYA

TV-sets: 550.000 — **Colour:** PAL — **System:** B.

PEOPLES REVOLUTION BROADCASTING TV (Gov.)
✉ P.O. Box 333, Tripoli — **L.P:** Dir: Youssif Debri.

Stations:	kW chE	(ERP)Pol		kW chE	(ERP)Pol	
Benghazi	5	10/2	V	Khoms	8	5/1 V
Tripoli	6	20/4	H	Tobruk	8	5/1 H
Derna	6	5/1	V	Yefren	9	20/4 H
Elmarj	7	5/1	V	El Beida	9	5/1 H
Houn	7	1/0.2	H	Misurata	10	5/1 H
Sirte	7	1/0.2	H	Egdabia	11	15/3 H

+ 1 low power repeater.
D.Prgr's: 1700-2230.

MADAGASCAR

TV-sets: 130.000 — **Colour:** SECAM — **System:** K.

RADIO TELEVISION MALAGASY (Gov.)
✉ P.O. Box 442, Antananarivo. ☎ +261 (2) 21784. ℗ 22506.
L.P: Dir: M. Rabesahala.
Stations: Antananarivo ch5 1kW H + 36 low power st's.
D.Prgr: 1600-1930.

MADEIRA

TV-sets: 80.700 — **Colour:** PAL — **System:** B.

RADIO TELEVISÃO PORTUGUESA, E.P. (Comm.)
✉ Rua das Maravilhas 42, 9000 Funchal. ☎ +351 45197/8. 🖷 +351 48859. ℗ 72478.
L.P: Gen. Mgr: A. Abreu.

Stations: Funchal (Pico do Silva) chE5 20/4kW H, with repeaters on ch5 100kW H, 6 20kW H, 8 100kW H (+4 low power st's not mentioned.)
N.B: only RTP's 1st. prgm. is broadcast.
D.Prgr: 1800 (Sat/Sun 1400)-2300.

MALI

TV-sets: 10.000 — **Colour:** SECAM — **System:** K.

RADIODIFFUSION TÉLÉVISION DU MALI
🖃 B.P. 171, Bamako. ☎ +223 (22) 2019 2243 08.
L.P: DG: Abdoulaye Sidibe. DG Adj: Sidki Konate. CE: Mahamadou Sow.
Stations: Bamako ch5 10kW + 2 repeaters.
D.Prgr: 1854 (Sat/Sun 1454)-2300.

MAURITANIA (Islamic Rep.)

TV-sets: 1100 — **Colour:** SECAM — **System:** B.

TÉLÉVISION NATIONALE (Gov.)
TVM (Television du Mauritanie)
🖃 B.P. 5522, Nouakchott. ☎ +222 53303. ⓓ 5817.
Station: Nouakchott chE5 (2x1kW).
D.Prgr: 2000-2245.

MAURITIUS

TV-sets: 156.850 — **Colour:** SECAM — **System:** B.

MAURITIUS BROADCASTING CORP. (Comm.)
🖃 1 Louis Pasteur Str, Forest Side. ☎ +230 675 5001/2. 📠 +230 675 7332. ⓓ 4320 MAUBROAD
L.P: Chairm: Mr Denis Rivet; DG: Mr Bijaye Madhou; Dep Chief Eng: Mr. Pather Amoordalingum
Stations:
TV 1: Malherbes ch. E4 15 kW (H) ERP + 7 relay st's 0.4kW ERP. **TV 2:** Malherbes ch. E5 15 kW (H) ERP + 4 relay st's 0.4kW ERP
D.Prgr: 1200 (Sat 0530/Sun 0600)-1900.

MAYOTTE

TV-sets: 3.500 — **Colour:** SECAM — **System:** K.

R.F.O.—MAYOTTE
🖃 B.P. 103, F-97610 Pamandzi, Ile de Mayotte. ☎ +269 601017. 📠 +269 601852.
L.P: St. Dir: Robert Xavier. Dir. Tec: Serge Sulpice-Timothee.
Stations: Lavigie ch 9H 100W, Mamadzou ch 7H 50W, Lima Combanich 4H 200W.
D.Prgr: Mon-Fri 1100-2100, Sat/Sun 0800-2100.

MOROCCO

TV-sets: 1.210.000 — **Colour:** SECAM — **System:** B.

RADIODIFFUSION TÉLÉVISION MAROCAINE (Gov.)
🖃 1, rue Al Brihi, Rabat. ☎ +212 (7) 704963. 📠 +212 (7) 722047. ⓓ 36577.
L.P: DG: Mohamed Tricha. Dir. of TV: Mohamed Issari. Head of Ext. Rel: A. Bekkali Abdellatif.
Stations: Pol: H.

	ch	kW		ch	kW
Zerhoun	4	120/12	Tan Tan	8	11/1.1
Zaio	4	9/1.8	Safi	8	20/2
Laayoune	E4	316/31.6	Tazerkount	8	90/18
Oujda	5	267/26.7	Tazekka	8	4/0.8
Boukhouali	5	150/15	Touzarine	9	9/1.8
Tanger	5	20/4	Tiguelmamine	9	9/1.8
S. Bounouara	5	11/1.1	S. Bounouara	9	11/1
Oukaimeden	6	18/3.6	Biougra	9	4/0.8
Azougar	6	9/1.8	Casablanca	10	180/18
Dakhla	6	11/1.1	Hafa Safa	10	9/1.8
Figuig	6	9/1.8	Ourzazate	10	267/26.7
Rabat	7	180/18	Bouarfa	10	267/26.7
Izeft	7	14/2.8	Essaouira	11	20/2

+ 35 low power relay sts.
N.B: All ch's M except where otherwise indicated.
D.Prgr. in Arabic/French: Mon-Thurs 1215-1415 & 1700-0100v, Fri/Sat/Sun 1215-0100v.

2M INTERNATIONAL (Comm.)
🖃 KM7, 300 route de Rabat, Casablanca. ☎ +212 (2) 354444.
📠 +212 (2) 354071.
L.P: MD: Tawfik Bennani-Smires. Tech. Dir: Driss Anouar. Comm. Dir: Mouhaddab Khadija.
D.Prgr: 15h daily, partly subscription.
F.Pl: Pan-African satellite sce.

MOZAMBIQUE

TV sets: 35.000 — **Colour:** PAL — **System:** B.

TELEVISÃO DE MOÇAMBIQUE TVM (Gov.)
🖃 C.P. 2675, Maputo. ☎ +258 744788 or 741395.
ⓓ 6-346 TEVEMP MO.
L.P: Dir: Botelho Moniz. Dt.Tec: Jaime Ferreira. Film Buyer: Arlando Tembe.
Station: Maputo ch33 1kW.
D.Prgr: 24h weekly.

NAMIBIA

TV-sets: 38.500 — **Colour:** PAL — **System:** I.

NAMIBIAN BROADCASTING CORPORATION
🖃 P.O. Box 321, Windhoek 9000. ☎ +264 (61) 215811. 📠 +264 (61) 2912291. ⓓ 50908 - 622/708.
Cable: Broadcast.
L.P: DG: Dan Tjongarero. Sen. Contr. Prgrs: Gabriel Haindaka. CE: Martin Venter. Head of PR: Cyril Lowe.
Stations

	ch		ch
Rundu	4	Signalberg	9
Keetmanshoop	4	Rösing	10
Paresis	5	Klein Windhoek	10
Windhoek	6	Mariental	10
Erongo	7	Waterberg	10

+ 3 low power st's and approx. 20 private low power relay st's.
D.Prgrs: Mon-Sat: 1400-2100, Sun: 0600-1100, 1400-2100 in English.

AFRICA — Niger — Seychelles

NIGER

TV-sets: 25.000 — **Colour:** SECAM — **System:** K.

TÉLÉ-SAHEL (Gov.)
✉ B.P. 309, Niamey. ☎ +227 723155. ☯ 5229 NI.
L.P: Dir. Gen: DG: Abdou Souley. TD: Zoudi Issouf.

Stations	ch	kW	Pol		ch	kW	Pol
Agadez	4	10	H	Dogondoutchi	7	1	H
Dosso	4	10	H	Gaya	8	1	V
Zinder	5	10	H	Niamey	9	10	H
Arlit	6	1	H	Konni	9	10	H
Maradi	7	10	V	Diffa	9	10	H

+ 7 low power relay st's.
D.Prgr: 0630-1130, 1430-1730.

NIGERIA

TV-sets: 6.100.000 — **Colour:** PAL — **Systems:** B.

NIGERIAN TELEVISION AUTHORITY (Gov.)
✉ Television House, PMB 120005, Victoria Island, Lagos. ☎ +234 (1) 614966/615154/612529. 🖷 +234 (1) 2610289. ☯ 21245 NTA HQ.
L.P: D G: Mohammed Ibrahim. Dir. Tec: Isaac Wakombo. Dir. of Prgrs: Prince Bayo Sanda.
National TV Production Centre.
TV House, Victoria Island, PMB 12005, Lagos.
L.P: Man. Dir: A. Micika.

REUNION

TV-sets: 90.500 — **Colour:** SECAM — **System:** K.

SOCIÉTÉ NATIONALE DE RADIO-TÉLÉVISION FRANÇAISE D'OUTRE-MER (RFO)
✉ 1, rue Jean-Chatel, F-97716 Saint-Denis. ☎ +262 406767. 🖷 +262 406771.
L.P: Regional Dir: Jean-Philippe Roussy.
Stations: P. Textor chK'9 0.5kW + 18 low power repeaters.
D.Prgr: Prgr.I: 1300 (Sat/Sun 1100)-1930; Prgr.II: no times available.

private stations

ANTENNE REUNION (Priv., Comm.)
✉ 33 Chemin Vavangues, 97400 Saint Denis. ☎ +262 48 2828. 🖷 +26248 2829
Stations: Saint Leu ch36 1kW, Sainte Suzanne ch42 2kW, Le Port ch57 9kW, Saint Pierre ch61 2kW, Saint Joseph ch55 1,7kW, Saint Benoit ch26 2,4kW, Saint Denis ch33 2kW, Piton Textor ch56 2kW.
D.Prgr: 0830-1830

CANAL REUNION (Priv., Comm.)
✉ 35 Chemin Vavangues, 97400 Saint Denis. ☎ +262 29 0202. 🖷 +262 29 1709
Stations: Saint Joseph ch52 1.7kW, Sainte Suzanne ch39 2kW, Le Port ch54 9kW, Saint Denis ch25 2kW, Piton Textor ch53 2kW, Saint Pierre ch26 2kW.
D. Prgr: 0245-2200, Sun 24h

TV SUD (Priv., Comm.)
✉ 10 rue Aristide Briand, 97430 Le Tampon. ☎ +262 57 4242

Stations: Saint Pierre ch58 2kW, Les Avirons ch60 0.72kW
D.Prgr: 1400-1800

TV-4 (Priv., Comm.)
✉ 8 chemin Fontbrune, 97400 Saint Denis. ☎ +262 52 7373
Stations: Saint Leu ch49 1kW, Saint Denis ch52 2kW, Sainte Suzanne ch31 2kW, Le Port ch65 9kW, Piton Textor ch63 2kW
D.Prgr: 0230-2130

SÃO TOMÉ E PRINCÍPE

TV-Sets: 21.000 — **Colour:** PAL — **System:** B & G.

TELEVISÃO DE SÃO TOMÉ E PRINCÍPE
✉ P.O. Box 393, S. Tomé, Republica de S. Tomé e Príncipe, Africa. ☎ +239 (12) 21041/22970 🖷 +239 (12) 21942.
L.P: Dir: Carlos Teixeira d'Alva
Stations: One 2kW transmitter in S. Tomé and one 10Watts transmitter in Principe covering 80% of the area. Channels: 11, 7, 5.
D.Prgr: RTP Internaçional is also relayed.

SENEGAL

TV-sets: 61.000 — **Colour:** SECAM — **System:** K

RADIODIFFUSION TÉLÉVISION SÉNÉGALAISE (RTS) (Gov.)
✉ B.P. 1765, Dakar. ☎ +221 21780. 🖷 +221 223490.
L.P: DG: Guila Thiam. Dir. of TV Sces: Babacar Diagne. Dir. of Tech. Sces: Seydou Diallo. Head of Ext. Affairs: Ka Aissatou.
Station: Dakar ch7 10kW
D.Prgr: 1900-2300 (Sat-Sun 1200-2330)

CANAL HORIZONS SÉNÉGAL (Pay TV Sce.)
✉ 31 ave. Albert Sarraut (B.P. 1390), Dakar. ☎ +221 232525. 🖷 221 233030.
L.P: Pres: Fara N'Diaye. DG: Jacques Barbier de Crozes. Dep. DG: Anne Marie Senghor. Tech. Coordinator: Issa Laye Diop.

SEYCHELLES

TV-sets: 14.000 — **Colour:** PAL — **System:** B.

SEYCHELLES BROADCASTING CORPORATION
✉ P.O. Box 321, Hermitage, Mahe. ☎ +248 22416. 🖷 +248 22564. ☯ 2315 INFO TV SEZ.
L.P: MD: Ibrahim Afif. Dir. Admin. and Personnel: Fauzia Rose. Sen. Eng: (TV): Joyvani Chetty. Chief Editor (TV): Ms. Marie-Claire Elizabeth. Prgr. Mgr (TV): Jean-Claude Matombe.
Stations: La Misère chE2 1kW, St. Louis chE7 6kW + 9 low power repeaters.
D.Prgr: Mon-Fri 1345-1830, Sat 1200-1830, Sun 1000-1830

SIERRA LEONE

TV-sets: 25.000 — **Colour:** PAL — **System:** B.

SIERRA LEONE TELEVISION (Gov. Com.)
✉ Private Mailbag, Freetown. ☎ + 232 (22) 40403/40906. ① RADTEX 3334 SL.
L.P: Ag. Head of Prgrs: G. Felix-George. Sen Eng. (TV): W.A.I. George.
Station: chE2 1kW H, chE7 126kW H.
D.Prgr: 1745-2330.

SOMALIA (Rep.)

TV sets: N/A — **Colour:** PAL — **System:** B.

MINISTRY OF INFORMATION
✉ P.O. Box 1748. ① 999621 Mogadishu.
L.P: Dir. of TV project: A. Ali Askar. Dir. Tec: A. Hassan.
Station: ch6 1kW.
D.Prgrs: 1700-2000 (Fri 2100).

SOUTH AFRICA

TV-sets: 3.485.000 — **Colour:** PAL — **System:** I.

SENTECH (PTY) LTD.
✉ Private Bag X06, Honeydew 2040. ☎ +27 (11) 475 1596. 🖷 +27 (11) 475 5112. **WWW:** http://www.sentech.co.za
Sentech is responsible for the signal distribution of all radio and TV services in South Africa.

SOUTH AFRICAN BROADCASTING CORPORATION (SABC)
✉ (Head Office): Broadcasting Centre, Auckland Park, Johannesburg 2092/Private Bag XI, Auckland Park 2006. ☎ +27 (11) 714 9111. 🖷 +27 (11) 714 3106.
Cable: Broadcast.
WWW: www.sabc.co.za
L.P: Chmn (Board): Dr. Ivy Matsepe-Casaburri. Group Chief Exec: Zwelakhe Sisulu. Chief Exec. (Signal Distribution): Neël Smuts. Chief Exec. (Operations): Gert Claassen. Chief Exec. (Human Resources): Ntombi Langa. Ag. Group GM (Finance): Talib Sadik. Sen. GM Strategic Planning: Solly Mokoetle. Group Sces. Co-ordinator: Leslie Xinwa.

CCIR System 1 (PAL colour) used on bands III/IV/V using ch4-13 (174-254MHz) and 21-68 (470-845MHz), chs 12/38 not used. Sound/Vision spacing is + 6MHz.

SABC-1:
Stations: 100kW (ERP) exc. *) 1-10kW

Location	ch	Pol	Location	ch	Pol
Durban	4	H	Pomfret*	6	H
Kimberley	4	H	Prieska*	6	V
Potgietersrus	4	H	Springbok*	6	V
De Aar	5	H	Volkrust*	6	V
George*	5	V	Walvis Bay*	6	V
Pretoria	5	H	Port Elizabeth	7	H
Theunissen	5	H	Queenstown	7	H
Donnybrook*	6	H	Welverdiend	7	H
Graaff-Reinet*	6	V	Villiersdorp	7	H
Cape Town	8	V	Eshowe	36	H
Garies*	8	H	Mooi River*	37	H
Grahamstown	8	H	Springfontein*	37	H
Kuruman	8	H	Hoedspruit*	39	H
Louis Trichardt*	8	V	Matjiesfontein*	39	H
Port Shepstone	8	V	Vryheid*	39	H
Bloemfontein	9	H	Carnarvon*	40	H
East London	9	H	Matatiele*	40	H
Johannesburg	9	H	Cradock*	40	H
Napier*	9	V	Middelburg (Tvl)	41	H
Piketberg	9	H	Senekal*	42	H
Oudshoorn	9	H	Williston*	42	V
Thabazimbi	9	V	Klerksdorp	45	H
Victoria West	9	H	Blouberg*	45	V
Vanrhynsdorp	10	H	Ubombo*	45	H
Beaufort West*	10	H	Carolina*	50	H
Pofadder	10	V	Piet Plessis*	50	H
Upington	10	H	Zeerust	52	H
Piet Retief*	11	H	Dullstroom*	53	H
Riversdale	13	H	Greytown*	53	H
Davel	22	H	Noupoort*	54	H
Calvinia*	22	H	Bethlehem	55	H
Enzelsberg	22	H	K. William'sTown*	56	H
Ladismith (C.P.)*	22	H	Ladybrand*	56	H
Bedford*	23	H	Rustenburg*	56	H
Boesmanskop*	23	H	Douglas*	57	H
Nelspruit	24	H	Kroonstad	57	H
Petrus Steyn*	24	H	Willowmore*	57	H
Kareedouw*	25	H	Suurberg*	59	H
Glencoe	27	H	Aliwal North	61	H
Schweizer Reneke	33	H	Christiana*	62	H
			Tzaneen	64	H

+ 70 gap fillers and estimated 400 privately owned low power tr's.
D.Prgr: 0400-0615 (Mon-Fri), 1300 (Sat 0400, Sun 1130)-2200 (Sat 2230).

SABC-2

Location	ch	Pol	Location	ch	Pol
Hoedspruit	43	H	Kimberley	24	H
Nelspruit	36	H	Klerksdorp	37	H
Potgietersrus	13	H	Durban	13	H
Pretoria	11	V	Glencoe	23	H
Tzaneen	56	H	Cape Town	34	V
Johannesburg	13	H	East London	4	H
Welverdiend	10	H	Port Elizabeth	13	H
Bloemfontein	44	H			

+27 gap fillers

SABC-3

Location	ch	Pol	Location	ch	Pol
Aliwal North	53	H	Grahamstown	5	H
Bethlehem	63	H	Greytown	61	H
Bloemfontein	13	H	Hoedspruit	47	H
Boesmanskop	27	H	Johannesburg	6	H
Cape Town	5	V	Kimberley	7	H
Christiana	58	H	K. William's Town	60	H
Cradock	48	H	Klerksdorp	41	H
Davel	30	H	Kroonstad	61	H
Donnybrook	9	H	Kuruman	11	H
Durban	7	H	Ladybrand	60	H
Dullstroom	61	H	Louis Trichardt	11	V
East London	13	H	Middelburg	45	H
Eshowe	28	H	Nelspruit	32	H
George	11	V	Nongoma	58	H
Glencoe	31	H	Oudtshoorn	6	H

AFRICA — South Africa — Uganda

Location	ch	Pol	Location	ch	Pol
Piet Retief	5	H	Thabazimbi	9	H
Piketberg	6	H	Theunissen	8	H
Port Elizabeth	4	H	Tzaneen	60	H
Port Shepstone	5	V	Ubombo	37	H
Potgietersrus	7	H	Upington	25	H
Pretoria	8	V	Villiersdorp	10	H
Queenstown	4	H	Volksrust	9	V
Riversdale	8	H	Vrÿheid	47	H
Rustenburg	64	H	Welverdiend	4	H
Schweizer-Reneke	25	H	Zeerust	44	H
Senekal	38	H			

+ 47 Gap fillers.
D.Prgrs: broadcasts in various languages.

M-NET TELEVISION (Pay channel, Comm, Priv.)
✉ P.O. Box 4950, Randburg 2125. ☎ +27 (11)329 5156. 📠 +27 (11) 329 5166 — **L.P:** PD: Sheryl Raine. Mktg. Dir: Etienne Heyns. Head of PR: John Badenhorst.
Stations: Bloemfontein ch6 (H), Alverstone & Pt. Elizabeth ch10 (H), Constantiaberg ch11 (V), Pretoria ch21 (H), Johannesburg ch39 (H) + 16 repeaters, Durban ch10, Cape Town ch11, George ch7, Newcastle ch62.
D.Prgr: Mon-Fri 0830-1030 & 1300-2300; Sat/Sun 0500-2300. **Indian Prgr:** Sun 0530-0830. **Portuguese Prgr:** Sun 0830-1130.

SUDAN

TV-sets: 250.000 — **Colour:** PAL — **System:** B.

SUDAN TELEVISION (Gov, Comm.)
✉ P.O. Box 1094, Omdurman. ☎ +249 (11) 55022. ③ 28002, 28053. **Cable:** Sudan TV.
L.P: Head of Directorate: Hadid al-Sira.
Stations: Omdurman chE5 5kW H, Gezira chE7 10kW, Atbara chE9 0.5kW.
D.Prgr: 1500-2200.

SWAZILAND

TV sets: 12.500 — **Colour:** PAL — **Systems:** B & G.

SWAZILAND TELEVISION AUTHORITY
✉ Swazi TV, P.O. Box A146, Swazi Plaza, Mbabane. ☎ +268 43036/7. 📠 +268 42093. ③ 2138 WD.
L.P: MD: Dan S. Lamini.
Stations: Bulembu ch5 1.5kW H, Ntondozi ch15 15kW H + 7 relay st's.
D.Prgr. in English: 1600-2100.

TANZANIA

TV-sets: 80.000 — **Colour:** PAL — **System:** B.

TELEVISION ZANZIBAR TVZ (Gov.)
✉ P.O. Box 314, Zanzibar. ☎ +255 (54) 32816/7. ③ 57200 TVZ TZ.
L.P: MD: Jama A. Simba. TD: George H. Majaliwa. Film Buyer: Jaffar S. Kassingo; Prod. Mgr: Abdulhamid H. Dau.
Stations: Unguja chE21 40kW, Pemba chE9 40kW (ERP).

D.Prgr: 1645-1900 (Sat/Sun/National Holidays 1645-2000).

INDEPENDENT TELEVISION (ITV)
Stations: Dar es Salaam UHF ch24
D. Prgr: 1400-1505/1700-1800 (weekdays), local prgr. Sundays and Tuesdays from 1600-1630

TOGO

TV sets: 150.000 — **Colour:** SECAM — **System:** K.

TÉLÉVISION TOGOLAISE (Gov.)
✉ B.P. 3286, Lomé. ☎ +228 215357. 📠 +228 215786. ③ 5320 ATOP.
L.P: Dir. of TV: Yao Martin Ahiavee. Tech. Dir: Vokou Raphaël Soumsa. Film Buyer: Ayi Léopold Mamavi.
Stations: Mt. Agou ch6 10kW V, Lomé ch8 1kW H, Aldjo-Kadara ch8 10kW H + 2 relay st's.
D.Prgr: Mon-Fri 1830-2230, Sat/Sun 1230-2400.

TUNISIA

TV-sets: 650.000 — **Colour:** SECAM — **System:** B&G.

ENTREPRISE DE LA RADIODIFFUSION-TÉLÉVISION TUNISIENNE E.R.T.T. (Gov.)
✉ 71, Ave de la Liberté, 1002 Tunis Belvedere. ☎ +216 (1) 287300, 782700 . 📠 +216 (1) 781058. ③ 14960.
L.P: Dir: Abdeh Afidh Hardudm.
Stations: Network I (Arabic)

	ch	kW	Pol		ch	kW	Pol
Remada	4	32/6.4	H	Gafsa	10	400/80	V
Kasserine	6	340/68	H	Zaghouan	11	280/56	H
Boukornine	7	12.5/1.25V		A. Draham	12	32/6.4	H
Sfax	8	365/73	H	Zarzis	12	180/36	H
Goraa	9	32/6.4	H				

Network II: (French) Pol=H.

	ch	kW		ch	kW
Zarzis	22	1500	Ain Drahan	35	87
Biadha	23	780	Chambi	40	1410
Boukornine	26	44	Ghraba	51	1800
Zaghouan	33	1425	Bizerte	53	40
Remada	33	89	Goroa	55	89

low power st's not mentioned.
D.Prgr: Netw. I & II: 1600-2300.

UGANDA

TV-sets: 115.000 — **Colour:** PAL — **System:** B (Pol=H).

UGANDA TELEVISION (Gov.)
✉ P.O. Box 7142, Kampala. ☎ +256 (41) 254461. ③ KNOLLEDGE 61084.
L.P: Ag. Dir: J.M.A. Obo.
Stations: Kampala chE5, Lira chE7, Masaka chE8, Mbale chE8, Mbarara chE10, Soroti chE10. Gulu chE9, Jinja chE11 + low power repeater at Kabale.
D.Prgr: Mon-Fri 1500-2100; Sat/Sun 1200-2100.

319

ZAIRE (Rep. of)

TV-sets: 22.000 — **Colour:** SECAM — **System:** K.

OZRT (Gov.)
✉ B.P. 3171, Kinshasa, Gombe 3164. ☎ +243 23171.
① 21583.Vozakin ZR.
L.P: Dir. Gen: B. Dongo. Dir. Tec: S. Lepamabla
Stations: a) regular; b) regular but not fulltime; c) irregular.

Location	ch	kW	Location	ch	kW
Kinshasa	5a	27	Uvira	6c	24
Kananga	4b	2	Kambove	7a	2
Kamina	4b	2	Goma	7a	2
Gbadolite*	4a	2	Gemena	8c	24
Kolwezi	5a	1	Kisangani*	8a	2
Mbuji Mayi	6b	2	Bandundu	8b	2
Mbandaka	6b	2	Lubumbashi*	9a	24
Kalemie	6a	2	Bukavu	9a	2
Kindu	6b	2	Isiro	9c	2

*) own studio facilities.
D.Prgr: Mon/Tues/Thurs/Fri 1130-1330 & 1630-2300, Wed 1130-2300, Sat/Sun 0900-2300. Relayed on Intelsat 66°E, C-band. Tr's from inland towns are dependent on power supplies (i.e. fuel availability).

private/commercial stations

ANTENNE A (Private/Comm.)
✉ Av. du Port 4, Building Forescom 2e floor, Kinshasa/Gombe. P.O. Box 2681 Kinshasa 1. ☎ +243 21736/24818/25308
L.P: P.D.G: A Pinhas; Dir Gen: Igal Avivi Neirson; Tech Dir: Ranny Ranny Shoket.
D.Prgr: Mon-Fri: 1430-0030, Sat1430-0130, Sun 1000-2330.
Station: ?

CANAL Z (Comm.)
✉ 6, av. du Port, Kinshasa/Gombe, P.O. Box 614 Kinshasa I. ☎ +243 20239.
L.P: Dir Gen: Frederic Flasse
D.Prgr: 1500-2400
Station: Kinshasa UHF ch23

ZAMBIA

TV-sets: 200.000 — **Colour:** PAL — **System:** B.

TELEVISION-ZAMBIA (Gov.)
✉ Broadc. House, P.O. 50015, Lusaka 229648. ☎ +260 (1) 220864-74; P.O. Box 20748, Kitwe. ☎ +260 (2) 223555. ① 41221 ZA. **Cable:** Broadcasts TV.
L.P: Act. Controller: Miss Emelda Yumbe.

Stations	chE	kW	Pol		chE	kW	Pol
Solweri	3	0.075H		Kitwe	9	200	H
Kapiri Mposhi	6	200	H	Lusaka	10	200	H
Pemba	8	200	V	Senkobo	10	200	V
Kasama	8	16	H	Chipata	11	16	H
Mubwa	8	0.1	H				

D.Prgrs: Mon-Thur. 1700-2230; Fri-Sat 1500-2400; Sun 1500-2230.

ZIMBABWE

TV-sets: 375.000 — **Colour:** PAL — **System:** B.

ZIMBABWE BROADCASTING CORPORATION (Independent Statutory Body, Comm.)
✉ P.O. Box HG 444, Highlands, Harare. ☎ +263 (4) 498659, 498670, 498630, 498620. 🖷 +263 (4) 498613. ① 24175 ZBCHOVZW.
L.P: DG: Edward Moyo. Dir. Eng. and Tech. Sces: Elliot Muchimbiri.
Contr. TV Prgrs: Mr Nyasha Masiwa.

Stations	ch	EkW	Pol	Stations	ch	EkW	Pol
Gweru	2	17.6	H	Kadoma	6	4	H
Bulawayo	3	3	H	Bulawayo	6	6	V
Harare	4	20	H	Mutare	7	12	V
Harare	5	30	H	Masvingo	7	6	H
Gwendingwe	5	7	H	V. Falls	7	1	H
Rukotso	5	6	V	Gwanda	9	6	H
Kariba	5	1	H	Gweru	11	25	H
Kamativi	5	16	V	Mutoroshanga	12	8	H

D.Prgr: English/Shona/Ndebele 0700-2200.

NEAR & MIDDLE EAST

AFGHANISTAN

TV-sets: 100.000 — **Colour:** PAL — **System:** B.

RADIO-TELEVISION OF AFGHANISTAN (RTA) (Gov.)
✉ P.O. Box 544, Kabul. ☎ +93 25460/25373. ① 24288 (AFGRTV AF).
L.P: Gen Pres: Shamsul Haq Arayanfar. Tech Adv: Eng. Faizuddin Ferogh
Station: Kabul ch5 10/1kW H.
D.Prgr: 1330-1830.
Reg. st's are operating in 9 different provinces at 1330-1530 or 1430-1630 (approx.).

ARMENIA

TV-sets: N/A — **Colour:** SECAM — **Systems:** D&K.

ARMENIAN TELEVISION (Gov.)
✉ Alek Manukyan 5, 375025 Yerevan. ☎ +374 (2) 552502. 🖷 +374 (2) 551513.

Stations (Prgr. 1)	ch	Stations (Prgr. 2)	ch
Yerevan	1	Yerevan	11

D.Prgr: 1730-2100. **D.Prgr:** 1730-2030.
NB: Prgr. 1 is carried by various main and relay transmitters throughout the country. Further details not known.

ORT

Stations	ch
Yerevan	8

D.Prgr: Relays of ORT from Moscow. Further details see under Russia.

NEAR & MIDDLE EAST — Armenia — Cyprus

ROSSIJSKOJE TELEVIDENIJE (RTV)
Stations   ch
Yerevan   5
D.Prgr: Relays of RTV from Moscow. Further details see under Russia.
NB: ORT and RTV are carried by various main and relay transmitters throughout the country. Further details not known.

AZERBAIJAN

TV-sets: N/A — **Colour:** SECAM — **Systems:** D&K

AZERBAYCAN RESPUBLIKASI RABITA NAZIRLIYI (Gov.)
Addr: Azerbaycan pr. 33, 370139, Baki. ☎ +994 (12) 93 0004. 🖷 +994 (12) 98 3325. ☉ 142105 INAM SU
L.P: Minister: Sirus Abasbeyli

TELERADIO (Gov.)
✉ Azerbaycan pr. 33, 370139, Baki. ☎ +994 (12) 98 8066. 🖷 +994 (12) 98 3325. ☉ 142472 DALGA

AZERBAYCAN RADIOTELIVIZIYA SIRKETL (Gov.)
✉ Mehdi Huseyin küçäsi 1, 370011 Baki. ☎ +994 (12) 39 8585. 🖷 +994 (12) 39 5452. ☉ 142214 TEMBR

AZERBAYCAN MILLI TELEVIZIYASI (Nat. TV, Gov.)
D.Prgr: Progamming in Azeri and Russian. 14 hrs. a day.
Stations: 17 powerful (5kW) tx, 48 100 W tx, 83 (1-30 W) tx.

BM-TV
Private TV program, 500 W. transmitter relay in Baki
D.Prgr: In Azeri and Russian 5 hrs.a day

TRT-1
Turkish Gov. TV program, 100 W transmitter relay prgr in Baki
D.Prgr: In Turkish

Stations	National TV	Moscow TV1	Russian TV
Baku	3 (5kW)	7 (5kW)	12 (5kW)
Gyandzha	11 (5kW)	8 (5kW)	31 (5kW)
Shusha	2 (5kW)	12 (5kW)	30 (1kW)
Duzdag	3 (5kW)	9 (5kW)	5 (5kW)
Ali-Bairamly	4 (0.1kW)	10 (5kW)	28 (1kW)
Lerik	6 (5kW)	2 (5kW)	? (5kW)
Ordubad	11 (5)	8 (5)	2 (0.1)
Dzhalilabad	1 (5kW)	—	7 (0.1kW)
Danachi	3 (5kW)	7 (5kW)	5 (0.1kW)
Lenkorañ	8 (5kW)	3 (0.1kW)	10 (0.1kW)
Dzhebrail	6 (5kW)	3 (0.1kW)	12 (0.1kW)
Imishly	11 (5kW)	7 (0.1kW)	21 (1kW)
Kuba	6 (5kW)	12 (0.1kW)	8 (0.1kW)
Shakuk	2 (5kW)	7 (0.1kW)	—
Astaka	9 (20kW)	5 (5kW)	2 (0.1kW)
Akstafa	7 (5kW)	—	—
Yvanovka	5 (5kW)	33 (2kW)	—
Kelbadzhari	3 (5kW)	6 (0.1kW)	—
Geokchay	3 (0.1kW)	8 (5kW)	25 (0.1kW)
Dashkesan	7 (0.1kW)	9 (0.1kW)	2 (0.1kW)
Sabirabad	3 (0.1kW)	2 (0.1kW)	12 (0.1kW)
Sheki	5 (0.1kW)	12 (0.1kW)	10 (0.1kW)
Mingechaur	1 (0.1kW)	9 (0.1kW)	4 (0.1kW)
Agdam	7 (0.1kW)	3 (0.1kW)	10 (0.1kW)
Kedabek	1 (0.1kW)	8 (0.1kW)	4 (0.1kW)
Zangelan	11 (0.1kW)	3 (0.1kW)	1 (0.1kW)
Shemakha	9 (0.1kW)	11 (0.1kW)	6 (0.1kW)
Divichi	12 (0.1kW)	5 (0.1kW)	26 (0.1kW)
Yardymly	5 (0.1kW)	8 (0.1kW)	26 (0.1kW)
Oguz	3 (0.1kW)	9 (0.1kW)	7 (0.1kW)
Kubatly	11 (0.1kW)	3 (0.1kW)	—
Lachin	9 (0.1kW)	12 (0.1kW)	—
Kürdamir	3 (0.1kW)	6 (0.1kW)	—
Gabala	1 (0.1kW)	4 (0.1kW)	—
Bakda	1 (0.1kW)	—	—
Salyany	—	8 (0.1kW)	3 (0.1kW)
Almaly	2 (0.1kW)	—	6 (0.1kW)
Dzhafarabad	6 (0.1kW)	—	1 (0.1kW)
Ger-Ger	12 (0.1kW)	10 (0.1kW)	—
Tauz	12 (0.1kW)	—	—
Kemekly	3 (0.1kW)	—	—
Shikhly	12 (0.1kW)	—	—
Sadhrak	—	5 (0.1kW)	—
Shakur	—	7 (0.1kW)	—
Neftechala	—	4 (0.1kW)	—
Jergündzh	—	10 (0.1kW)	—

+ 81 low power st's not mentioned.

BAHRAIN

TV-sets: 270.000 — **Colour:** PAL — **Systems:** B&G.

BAHRAIN TELEVISION (Gov, Comm.)
✉ P.O. Box 1075, Bahrain. ☎ +973 781888/ 686000. 🖷 +973 681544. ☉ 8311.
L.P: Dir: Dr. H. Al-Umran. Head of Prgrs: Fowzia Zainal. Head of Mktg: Maria Khoury. Dir. Pub. Rel: Ahmed Al Sherooqi.
Station: chE4, chE44, chE55, ch57.
D.Prgr: ch4: Main Arabic ch. loc. & int. news, cultural and variety prgrs, Arabic & English feature films.
ch44: Satellite broadcasts, loc. & int. sports. Also rel. Egypt satellite ch.
ch55: Main English ch. Also rel. CNN.
ch57: Transmits BBC World Sce. TV, 24h.

CYPRUS

TV-sets: 103.000 — **Colour:** PAL — **Systems:** B/G.

CYPRUS BROADCASTING CORPORATION
✉ P.O. Box 4824, 1397 Nicosia. **Cable:** Broadcasts. ☎ +357 (2) 422231. 🖷 +357 (2) 335010.
L.P: DG: George Potamites. Dir. of Tech. Sces: Andreas Michaelides. Dir. of Prgrs: Panos Ioannides. Head of Public and Int. Rel: Nayia Roussou. Head of News & Current Affairs: Evangelos Louca. Head of TV prgrs: A. Papayrinnis. Head of TV and Radio Studios: Chrysanthos Hadjioannou.
Stations: Mt. Olympus chE6 40/4kW V + 35 low power st's.
Ch.1 in Greek: 1300 (Sun 0800)-2300.
Ch. 2 in Greek, Turkish, English: 1400 (Sun 1000)-2300.

Cyprus — Israel

LOGOS TV (Comm.)
20 St. Avgoustinou, Archangelos, Nicosia. ☎ +357 (2) 355595. 🖹 +357 (2) 355737.
L.P: Chmn: Michalis Colocasides. DG: Frixos Klenatous.

LUMIERE TELEVISION (Comm. Pay-TV sce.)
Papabisiliou Bldg, 70 Kennedy Ave, Nicosia. ☎ +357 (2) 311080. 🖹 +357 (2) 415767.
L.P: Pres: Chris Economides. Dir: Joe Avraamides.

BFBS Akrotiri (SSVC)
BFPO 57, Dhekelia Mil 381. ☎ +357 (474) 8518
Stations: ch 69 and 66 in ESBA, ch 60 and 68 in the WSBA.
D.Prgrs: relays of English prgrs. + live prgrs. from satellite.

NORTHERN CYPRUS
TV-sets: 75.000 — **Colour:** PAL — **Systems:** B & G.

BAYRAK RADIO & TELEVISION CORP.
Yeni Organize Sanayi Bolgesi, Lefkosa via Mersin 10, Turkey. ☎ +90 392 228 5555. 🖹 57264 brtk. **E-mail:** brt@cc.emu.edu.tr
WWW: http://www.cc.emu.edu.tr/press/brt/brt.htm
L.P: DG: Ismet Kotak. Head of Tr's: A.Ziya Dïncer. Head of Admin: Süleyman Türem. Head of Prgrs: Hüseyin Cobanoglu. Head of Sales: Mehmet Kircailiar. Head of Int. N: Huriye Dimililer.
Stations: Sinan dagi ch8 100kW (Prgr. 2) — Selvilitepe ch40 450kW (Prgr. 3), ch44 450kW (Prgr. 1) + 4 relay st's.
D.Prgrs: no details received.

IRAN

TV-sets: 7.000.000 — **Colour:** SECAM — **System:** B.

ISLAMIC REPUBLIC OF IRAN TELEVISION (Gov.)
P.O. Box 19395 3333, Tehran; P.O. Box 15875-4344, Tehran ☎ +98 (21) 298053, 290079, 96715150. 🖹 +98 (21) 295056/ 291051. **Cable:** "IRIB IR". ① 213910/212431/212797/213253.
L.P: Pres: H.E. Dr. Ali Larijani. Gen Dir. Int. Affairs: Dr. A. Ghasemzadeh.
D.Prgr:
Network I: 0545-1200 & 1600-2400 (on holidays 1400-2400).
Network II: 0800-1300 & 1400-2400.
Network III: 0830-1100 & 1730-2300
Local st's (28) Abadan, Ahwaz, Alamdeh, Ardebil, Baharlo, Bakhtaran, Bandar-Abbas, Booshehr, Esfahan, Hamedan, Kerman, Khoramabad, Kohe Genon, Kohe Noor, Kosangar, Mashhad, Oromieh, Rasht, Sanandaj, Sari, Shahrkord, Shiraz, Tabriz, Tehran, Yazd, Zahedan, Zanjan, Zibakenar + 450 low power repeaters.

IRAQ

TV-sets: 1 m — **Colour:** SECAM — **System:** B.

IRAQI BROADCASTING AND TELEVISION ESTABLISHMENT (IBTE)
Salhiya Baghdad. ☎ +964 (1) 884 4412, 884 4413. 🖹 +964 (1) 541 0480. ① 212246 IDAH.

L.P: DG: Dr. Sabah Yaseen; Dir PR: Mr. Abdul-Ilah Al-Musfir.
Stations: 1=1st Prgr; 2=2nd Prgr; R=Regional Prgr.

Location	chE	kW/Pol	Location	chE	kW/Pol
Misan (1)	5	144H	Baghdad (1)	9	360H
Muthanna (1)	6	144V	Basrah (1)	9	360V
Al-Taamin (R)	6	144H	Nenavah (1)	9	180V
Baghdad (2)	7	288H	Al-Taamin (2)	10	144H
Nenavah (2)	7	144V	Misan (2)	11	180H
Al-Taamin (1)	8	180H	Basrah (2)	12	144V
Muthanna (2)	8	180V			

D.Prgrs: Prgr 1: 1400 (Fri 0700)-1930 (Fri 2100); Prgr 2: 1400-2100.

ISRAEL

TV-sets: 1.500.000 — **Colour:** PAL — **Systems:** B&G.

ISRAEL TELEVISION (operated by the IBA)
P.O. Box 7139, Jerusalem 91071. ☎ +972 (2) 291888. 🖹 +972 (2) 292944. ① 25301 — **L.P:** DG: Ayre Mekel.

Stations	chE	kW	Stations	chE	kW
M. Ramon	5	15/1.5	Acco	24	50/5
Acco	7	20/2	Jerusalem	27	1/0.2
Eilat	7	0.15/0.015	Tel Aviv	28	60/6
			Zefat	34	22/2.2
Eitanim	8	200/20	Haifa	36	40
Manara	8	1.5/0.15	Karmiel	39	1.5/0.15
Arad	9	10/1	Jerusalem	40	3/0.3
Grofit	10	15/1.5	Manara	44	20/2
Haifa	10	100/10	Haifa	46	200/20
Bar Jehuda	10	1.5/0.15	Jerusalem	56	1.5/0.15
Beersheba	11	100/10	K. Hayarden	60	20/2

+31 low power repeaters.
Ch. 1 in Hebrew/Arabic: 0400-0600, 1430-2300v
Ch. 3: broadcasts only via satellite.

ISRAEL EDUCATIONAL TELEVISION
14 Klausner Str, Tel Aviv. ☎ +972 (3) 5434343. ① 342325 ITCIL.
L.P: Gen. Mgr: Y. Lorberbaum. Dir. of Eng: A. Kaplan. Dir. of Prgrs: Mrs. Y. Prener.
Stations: uses same tr's as the IBA (see above).
D.Prgr. Educational: 0600-1530 (Fri 1430). (Sat: no trs).

CHANNEL 2
97 Jaffa St, Jerusalem 94340. ☎ +972 (2) 242776/242750. ① 25678 ITRA. 🖹 +972 (2) 242720.
L.P: Man. Dir: Oren Tokatly.
Stations: Eitanim chE22 50kW (ERP), Acco ch E27 2kW, Beersheba chE35 6kW — **D.Prgr:** 1100 (Sat 0800)-2200v.

PALESTINIAN BROADCASTING CORP. TV
Ramallah- Um Al Sharyet. ☎ +972 (2) 656 4017/9. 🖹 +972 (2) 6564029.
L.P: Dir: Radwan Abu Hayash. Tech. Coordinator: Hisham Makki.
Station: UHF ch23, VHF ch4.
D.Prgr: 1600-2000 incl two news bulletins

NEAR & MIDDLE EAST

JORDAN

TV-sets: 250.000 — **Colour:** PAL — **Systems:** B.

JORDAN RADIO & TELEVISION CORP.
◨ P.O. Box 1041 or 2333, Amman. ☎ +962 (6) 773 111/9, 777 151/5, 779 111, 638 760, 638 766/7. ▤ +962 (6) 751 503, 788 115. **Cable:** Television. ⓓ 24213 RTVENG JO.
L.P: DG: Mr. Ihsam Ramzi Shikim; Dir Tel: Mr. Nasser Judeh; Dir Int. Rel: Mrs Fatima Masri
Stations: 1st Prgr (in stereo): Suweilih chE3 104kW H, Aqaba ch9 5kW, Ras Munif-Ajlun ch9 500kW H, Deir Alla ch26 6kW H + low power repeaters. **2nd Prgr:** Aqaba ch5 5kW H, Suweilih ch6 108kW H, Ras Munif-Ajlun ch11 500kW H, Deir Alla ch29 6kW H + low power repeaters. **1st Prgr in Arabic (in stereo):** 1330 (Fri 0800)-2200. **2nd Prgr:** 1600-2200. **French:** 1600-1730. **N:** 1700. **Hebrew:** 1730-1800. **Arabic:** 1800-1830 (rel. 1st Prgr). **English:** 1830-2200. **N:** 2000.+ 43 low power sts.

KUWAIT

TV-sets: 800,000 — **Colour:** PAL — **System:** B & G

KUWAIT TELEVISION (Gov.)
◨ (Administration) Ministry of Information, P.O. Box 193 Safat, 13002 Safat, Kuwait. ☎ +965 (24) 150301. ▤ +965 (24) 34511. **Cable:** ALIRSHAD. (Television) **Kuwait Television,** P.O. Box 621 Safat, 13007 Safat, Kuwait. ☎ +965 242 3774. ▤ +965 245 6660/243 9667. **Tlx:** (496) MI 46285 KT, (496) KTV 22169 KT
L.P: Minister: Sheikh Saud Nasir Al-Saud Al-Sabah; Asst. Under Secr. for TV Affairs: Rida Yousef Al-Feeli; Dir. of Eng. (TV): Abdulazeez Al-Baghli.
Stations: 1st Prgr. in Arabic: ch8, ch9, ch24, ch26, ch38, ch45. 0500-2100.
2nd Prgr. in English: ch10, ch11, ch39, ch47: 1100-2100.
3rd Prgr (sports): ch11 & ch47 (occasional)
4th Prgr (Arabic and English entertainment): 2100-0500 on tr's of 1st and 2nd Prgr.

OTHER STATIONS

MBC (Middle East Broadcasting Centre), Arabic relay ch12, Sat-Thu. 1100-0030, Fri 0900-0030, Sun-Wed. 1200-2400
ESC (Egyptian Satellite Channel), Arabic relay ch5, Sat-Fri. 24h.
Kuwait Space Channel: see satellite section ARABSAT 1C 31° East.

LEBANON

TV-sets: 1.100.000 — **Colour:** SECAM — **Systems:** B & G.

TÉLÉ-LIBAN (Gov.)
◨ B.P. 115054, Hazmieh, Beirut. ☎ +961 (1) 405100. ▤ +961 (1) 457253 — **L.P:** DG: Jean-Claude Boulos.

Jordan — Oman

Stations	ch	kW)	Prgr
Fih	2	1/0.2	1
Fih	8	10/1	2
Jounieh	2	1/0.2	2
Zahle	8	10/1	2
Beit Mery	2	1/0.2	3
Beyrouth	9	5/0.5	3
M. el Chouf	4	60/6	2
Fih	10	10/1	1
Beyrouth	5	50/5	1
Zahle	10	10/1	1
Fih	6	10/1	3
Beit Mery	11	10/1	1
Beyrouth	7	5/0.5	2

+ 4 low power repeaters
D.Prgrs: 1st and 2nd Prgr (Arabic/French/English): 1000-2200; **N. in English:** 1600 (Prgr. 1). **N. in French:** 1615 (Prgr. 2). **3rd Prgr.** (French/English) 1630-2200.

MIDDLE EAST TELEVISION (Comm.)
◨ P.O. Box 5689, Nicosia, Cyprus (Studios in Marjayoun). **Addr.** in **USA:** 977 Centreville Turnpike, Virginia Beach, VA 23463-0001, USA. ☎ +1 (804) 579 3419. ▤ +1 (804) 579 3417.
L.P: GM: Wes Hylton. Prgr. Mgr: Tom Foley.
Station: chE12 100kW (Pol:V) — chE5 10W.
D.Prgrs: 1130-2200 (approx) in English & Arabic.
N. in English: 1700.

LEBANESE BROADC. CORP. INT. (Comm.)
◨ P.O. Box 16-5853 Beirut. ☎ +961 (9) 938938. ▤ +961 (9) 937916. **E-mail:** lbci@lbci.com.lb **WWW:** http://www.lbci.com.lb
L.P: SM: Pierre Al Daher. TD: Nasim Boustany. Film Buyer: Selim El-Azar.
Stations: ch12H 60kW; ch33H 325kW; ch10H 35kW; ch9H 35kW; ch5H 35kW.
D.Prgr: 0445-2200.

FUTURE TELEVISION (Comm.)
◨ White House, Rue Spears, Sanayeh, Beirut, Lebanon. ☎ +961 (1) 347121/4/5/8, 340303, 341303. ▤ +961 (1) 602310.
E-mail: future@future.com.lb **WWW:** http://www.future.com.lb
Stations: ch28, ch37, ch46, ch52.
D.Prgr: 1100-2200.

MURR TELEVISION (MTV) (Comm.)
◨ Fouad Chehab Ave(P.O. Box 166000) - Fassouh - MTV Bldg. ☎ +961 (1) 217000. ▤ +961 (1) 423121. **E-mail:** mtv@dm.net.lb
WWW: http://www.dm.net.lb/mtv/
L.P: Pres: Michel El Murr.
Stations: ch28, ch38, ch48, ch68.
D.Prgr: 0700-2200.

OMAN (Sultanate of)

TV-sets: 1.500.000 — **Colour:** PAL — **Systems:** B & G.

SULTANATE OF OMAN TELEVISION (Gov.)
◨ P.O. Box 600, 113 Muscat, Oman. ☎ +968 603222. ▤ +968 602381. ⓓ 5454/5151.

L.P: DG: Ali Bin Abdallah Al Mujeni. DG (Tech. Affairs): H.Y.Al-Kindy.

Location	ch	kW	Pol	Location	ch	kW	Pol
Qurum	6	120/24	H	Shinas	5	7	
Sur	7	6/1.2		Salalah	10	300/60	V
Buraimi	8	160/32		Saham	10	25/5	
Thumrait	8	140/28	V	Nizwa	10	200/20	
Barka	51	70		Bahla	5	4	
Bilad Bani				Fine Peak	56	10	
Bu Ali	9	5		Ibri	55	20	

+ 25 low power relay st's.
D.Prgr: 0400-2100.

QATAR

TV-sets: 250.500 — **Colour:** PAL — **System:** B

QATAR TELEVISION SERVICE (Gov.)
▣ Min. of Information and Culture, QTV. P.O. Box 1944, Doha. ☎ +974 89 4444. 🖷 +974 86 4511.
Cable: TELEVISION DOHA. **Tlx:** 4040 TEEVEE DH
L.P. Asst. Under Secretary for Radio & TV: Mr. Abdul Rahman Saif Al-Madhadi. Dir. of TV: Mr. Saad Al-Rumehi. Dir of Eng: Mr. Hussain A. Jaffar. Dir of News: Mr. Abdullah Al Haj
Stations: Doha ch9 200kW, Jamiliyah ch11 (repeater) 600kW, Doha ch37 695kW, ch49 (repeater), Jamiliyah ch 52 (repeater)
D.Prgr: 1200-2200 UTC (Fri. 0600-2200), 0600-2200 UTC (during Summer)

SAUDI ARABIA

TV-sets: 4.700.000 — **Colour:** SECAM & PAL — **System:** B & G.

SAUDI ARABIAN TELEVISION (Gov.)
▣ P.O. Box 57137, Riyadh 11574. ☎ +966 (1) 4014440. 🖷 +966 (1) 4044192/4054176. ① 401030 SJ.
L.P: Asst. Dep. Minister for TV Affairs: Dr. Ali M. Al Najai.

1st Prgr:

Location	ch	kW	Pol	Location	ch	kW	Pol
Abu Qumais	5	200	H	Makkah	7	52	H
Riyadh	5	45	H	Al Baha	7	94	H
Madinah	5	170	H	Jizan	7	160	H
Abha	5	170	V	K. Khaled City	7	2	H
Hail	5	42	H	Al Qurayyat	7	1	H
Zilfi	5	9	V	Al Ula	7	5	H
Arar	5	14	H	Buraydah	8	98	H
Tayma	5	5	H	Tabarjal	8	12	V
Bijadiah	5	9	V	Hafr Al Batin	9	16	V
Harradh	5	5	H	Al Hassi	9	8	H
Halat Ammar	5	8	H	Uqlat Assuqr	9	9	H
Al Hariq	5	2	H	Al Muwayh	9	9	V
Linah	5	5	H	N. Abu Qasr	9	7	H
Dammam	6	800	H	Skaka	9	5	H
Taif	6	47	H	Badr Hunayn	9	5	H
Yanbu	6	46	H	Sajir	9	13	V
Baqa	6	5	V	Jebel Khasheb	9	10	H
Sharourah	6	42	V	Al Bad	9	1	H
Al Quwaiyah	6	5	V	Jeddah	10	27	H
Hanakiyah	6	5	H	Tabuk	10	15	V
Athnen	6	6	H	Ranyah	10	7	H
M. Ad Dahab	6	5	H	Al Ardiyah	10	7	H
Khamasin	6	6	H	Bani Malik	10	2	H
Ar Rass	10	5	H	Turaif	24	3	H
Wadi Al Freah	10	9	H	Hufuf	25	25	H
Jebel Garah	10	3	H	Al Musayjid	25	11	H
Al Summan	10	29	H	Al Khurma	26	7	H
Tathlith	10	9.5	H	Dammam	27	44	H
Al Howaidh	10	4.7	H	Layla	27	7	H
Al Kharj	11	5	V	Haql	27	7	H
Al Majmaah	11	160	H	Rafha	28	7	H
Al Sulaiyal	11	5	H	Al Nimas	30	6	H
Bani Saad	12	6	H	Al Alaya	31	10	H
Owegliyah	12	26	H	Bishah	39	6	H
Al Harjah	21	2	H	Turabah	40	7	H
Al Ahmer	22	16	H	Umm Lajj	41	7	H
Shaqra	22	100	H	Rabigh	42	8	H
Durma	22	8	H	Al Hawtah	45	7	H
Khybar	22	7	H	Qaryat Al Ulya	46	17	H
Al Qunfudah	22	7	H	Tuwal	50	100	H
Al Nuariyah	22	8	H	Al Khafji	51	40	H
Al Sudah	22	19	H	Al Wajh	53	7	H
Artawiyah	23	9	H				

+ 25 Stations not mentioned.

2nd Prgr:

Location	ch	kW	Pol	Location	ch	kW	Pol
Makkah	5	52	H	Sharourah	8	42	H
Skaka	5	5	H	Khamasin	8	6	H
K.Khaled City	5	2	H	Abha	9	170	V
Buraydah	6	98	H	Al Baha	11	106	H
Riyadh	7	45	H	Hafr Al Batin	11	17	H
Madinah	7	170	H	Jeddah	12	27	H
Hail	7	42	H	Al Howaidh	12	4.7	H
Zilfi	7	9	V	Shaqra	25	100	H
Arar	7	14	H	Hufuf	28	25	H
Taif	8	47	H	Dammam	29	890	H
Tabuk	8	15	V	Tuwal	53	100	H
Yanbu	8	51	H	Al Khafji	53	40	H

+ 6 Stations not mentioned.
D.Prgr: 1st Prgr: (Arabic): Sat-Wed 0700-0930, 1400-2130; Thurs/Fri: 0700-2130. **2nd Prgr:** (English): 1400-2130 (Thurs/Fri also 0600-0900).

CHANNEL 3 TV (Non-Comm, Private Co.)
▣ Bldg. 3030 LIP, Dhahran.
☎ 875-4634. **Cable:** Aramco, Dhamedia.
Station: chE3 Dhahran 5kW.
D.Prgr: 0600-2030 (or 2100). **Colour:** PAL.

SYRIAN ARAB REP.

TV-sets: 700.000 — **Colour:** PAL — **System:** B.

SYRIAN ARAB TELEVISION (Gov.)
▣ Ommayyad Square, Damascus. ☎ +963 (11) 720700. 🖷 +963 (11) 720700. ① 411223.
L.P: DG: Khudr Omran. Dir. Eng: M. Bara. Dir. PR: Mrs. Awafet Haffar.
Stations: Pol = H. +) Prgr 2, others Prgr 1.

Location	chE	kW	Location	chE	kW
Holms	2	?+	Deir-Al-Zoor	6	100/10
Abou-Kmal	3	200/20	Soueida	7	350
Nabi-Saleh	3	100/10	Homs	7	200/20
Hassakeh	4	200/20	Aein-Al Arab	7	100/10
Aleppo	5	200/20	Tabqua	8	100/10
Damascus	5	100/10	Kaldoun	8	10/1

Location	ch	E	kW
Slenfeh	9		200/20
Salhieh	11		30/3
Afrien	11		10/1
Palmyra	11		10/1
Al-Malkieh+	12		200/20
Lattakia	26		60

+ 36 low power & repeater st's.
Prgr 1: 1100 (Fri 0700)-2130.
Prgr 2 in Arabic/English/French: 1700-2130.

TURKEY

TV-sets: 10.530.000 — **Colour:** PAL — **System:** B.

TURKISH RADIO TELEVISION CORPORATION
TRT-TV Department, TRT Sitesi Katib Ablok Oran 06450, Ankara. ☎ +90 (312) 4904 983/986. 📠 +90 (313) 4904 985.
L.P: Head of TV Dept: Çetin Izbul; Dir Ankara TV: Sabahattin Alpdogan; Tech Dir: Sükrü Sipka.

TV-1	chE	kW		chE	kW
Ankara	5	100	Hatay	8	10
Istanbul	5	100	Canakkale	8	30
Antalya	5	30	Usak	8	30
Agri	5	30	Ersincan	8	30
Giresun	5	30	Nigde	8	30
Cizre	5	30	Afyon	9	30
Adana	6	30	Amasya	9	100
Erzurum	6	100	Edirne	9	30
Bursa	6	100	Diyarbakir	9	30
Kastamonu	6	30	Trabzon	9	30
Aksehir	6	30	Kars	9	100
Malatya	7	30	Zonguldak	9	30
Eskisehir	7	100	Izmir	10	100
Kirsehir	7	30	Gaziantep	10	30
Samsun	7	30	Konya	10	100
Silifke	7	30	Adapazari	10	30
Van	7	30	Bingöl	10	30
Aydin	7	30	Tokat	10	30

+ 476 low power repeaters not mentioned.

TV-2*	chE		chE
Elbistan	5	Trabzon	28
Adapazari	21	Agri	32
Cizre	21	Elbistan	34
Kars	21	Ankara	37
Adana	22	Isparta	40
Canakkale	22	Ersincan	42
Eskisemir	22	Kastamonu	48
Van	22	Mus	48
Zonguldak	23	Istanbul	51
Erzurum	23	Samsun	52
Izmir	23	Bursa	55
Kayseri	26	Diyarbakir	56
Gaziantep	27	Giresun	57
Denizli	28	Mugla	55
Edirne	28		

*) all tr's 450kW (ERP).
D.Prgr: TV-1: 24 hrs. **TV-2:** Mon-Fri 1700-0100, Sat/Sun 0930-0300(Sun 0100)
TV-3, TV-4, TV-5
Stations: (Istanbul area) TV-3 ch48, TV-4 ch54, TV-5 ch24.
D.Prgr: TV-3 Wed/Thu/Sat/Sun1340-2400, Mon/Tue/Fri 1700-0100; TV-4 1330-2400.

UNIVERSITE TECHNIQUE
Stations: Istanbul chE4, 0.5kW.

UNITED ARAB EMIRATES

TV-sets: 170.000 — **Colour:** PAL — **Systems:** B & G.

UNITED ARAB EMIRATES TELEVISION SERVICE (Gov.)
P.O. Box 637, Abu Dhabi. ☎ +971 (2) 452000. 📠 +971 (2) 461823. ⓓ 22557 teevee em.
WWW: http://www.ecssr.ac.ae/05uae.6television.html
L.P: Hd. of Eng: Mustafa Hamouda Ishag. Dir. Gen: Ali Obaid.

Stations	Prgr. 1	Prgr. 2	kW
Mohawi	7	48	40/5
Abu Dhabi	11	—	2
Al-Ain	5	35	40/5
Umm Al Quwain	6	43	40
Ras Al Khaimah	11	—	10
Fujairah	9	36	10
Habshan	3	—	40
Dibba	21	23	0.5
Jabel Al Dhanna	31	—	40
Khorfakkan	28	—	0.02

F.Pl:	Prgr. 1	Prgr. 2	kW
Al Khan	62	—	?
Liwa	25	28	?
Al-Wagen	29	32	?
Habshan	—	59	?
Jabel Dhanna	—	61	?

D.Prgr: Prgr. 1 (Arabic main prgr), Prgr.2I (Arabic/English/French): 0600-1005, 1200-2030 (Fri & holidays 0600-2030).

U.A.E. RADIO AND TELEVISION-DUBAI (Gov.)
P.O. Box 1695, Dubai. ☎ +971 (4) 470255. ⓓ 45605 DRCTV EM. **WWW:** http://www.ecssr.ac.ae/05uae.6television.html
L.P: Dir. Gen: Abdul G. Al Sayeed Ibrahim. Contr. of Prgrs: Nasib Bitar. Contr. of Eng: Ahmed Najeeb.
Stations (Systems B & G): Audio powers are 1/10 of the vision powers indicated.

Location	ch	kW (ERP)	Pol	Prgr
Trade Centre	2	150	H	1
Zabeel	10	455	H	1
Zabeel	33	1700	H	2
Jebel Hatta	41	1600	H	2

D.Prgrs: Arabic (prgr. 1) on chs 2, 10 & 41: 1200 (Fri 1000)-2130 (or 2030). **English** (prgr. 2) on ch33: 1300 (Fri 1200)-2000 (approx.).

SHARJAH TV (Gov.)
P.O. Box 111, Sharjah. ☎ +971 (6) 547755.
Stations: ch22, 28, 54, 57.
D.Prgr: 1300-2000 (in Arabic).

YEMEN (Rep. of)

TV-sets: 100.000 — **Colour:** PAL/NTSC — **System:** B.

YEMEN RADIO & TV CORP. (Gov.)
P.O. Box 2182, Sana'a. ☎ +967 (1) 230654. 📠 +967 (1) 230761. ⓓ 2645 YARTV.
L.P: Dir's: Ali Caleh Algamrah, Mohammed Abdul Gawi.
Stations: chE5 (0.2kW), chE6 (2 st's, 4 & 1kW), chE7 (2kW), chE8 (4kW), chE10 (4kW), chE11 (1kW), chE13 (2kW) + low power repeaters.
D.Prgr: 1300-2100.

ASIA

BANGLADESH

TV-sets: 600.000 — **Colour:** PAL — **System:** B/G.

NATIONAL BROADCASTING AUTHORITY BANGLADESH TELEVISION (Gov, Comm.)
⌐ Television Bhaban, P.O. Box 456, Dhaka-1219. ☎ +880 (2) 400131/9. 🗎 +880 (2) 832927. ① 675624 BTV BJ.
L.P: DG: Ahmad Rislat. Dep. DG (Prgrs): Mustafa Kamal Sayed. Dep. DG (News): Farooq Alamgir. Chief Eng: S.D. Khan.

Stations	chE	kW		chE	kW
Chittagong	5	15	Dhaka I	9	60
Rangpur	6	360	Cox's Bazar	10	7
Sylhet	7	136	Khulna	11	91
Satkhira	7	250	Mymensingh	12	390
Rangamati	8	55	Noakhali	12	390
Natore	8	90			

D.Prgr: 1100 (Fri 0900)-1740. Also carries prgrs of BBC, CNN and the Open University — **F.Pl:** 2nd channel.

Private TV: The govt. has decided to allow private TV st's to operate in the future.

BRITISH INDIAN OCEAN TERR.

TV-sets: 650 — **Colour:** NTSC — **System:** M.

AF DIEGO GARCIA TELEVISION (AFRTS)
⌐ US Navy Broadcasting Service, Detachment, Box 14, US Navy Support Facility, F.P.O. San Francisco, Calif. 96685, USA.
L.P: St. Mgr: J.A. Riccio. Dir. Tec: R.L. Newton.
Station: chA8/chA10 0.2kW H.
D.Prgr: ch 8: 0600 (Sat/Sun 0300)-2200; ch 10 24h.

BRUNEI DARUSSALAM

TV-sets: 95.000 — **Color:** PAL — **System:** B

RADIO TELEVISION BRUNEI (RTB) (Gov.)
⌐ Jabatan Perdana Menteri, Bandar Seri Begawan 2042, Negara Brunei Darussalam. ☎ +673 (2) 243 111. 🗎 +673 (2) 241882. **Cable:** Broadcast
L.P: Dir: Pg Dato Haji Badaruddin. Bin Pg Hj Ghani. Dep. Dir: Haji Md Yusof Hj Abd Rahman. Head of Eng: Kim Sam Lee. Hd of Prgrs: Mrs. Pg Hjh Normah Pg Hj Daud. Prgr. Prod. Mgr (TV): Haji Mohd Hussain Abdul Rahman.
Stations: Bt. Subok ch5 10kW H; Bt. Andulau ch8 20kW H.
D.Prgr: 0800 (Fri/Sun 0030) -1600.

CAMBODIA

TV-sets: 70.000 — **Colour:** PAL — **System:** B/G.

CAMBODIAN TELEVISION (Gov.)
⌐ 19, Street 242, Chaktomuk, Daun Penh. Phnom Penh.
☎ +855 (2983) 22349/(4449) 24149

L.P: St. Mgr: Tan Yan. TD: Uy Thuon.
Stations: ch7 1kW.
D.Prgr: 0715 PM-0745 PM.

INTERNATIONAL BROADCASTING CORP. Ltd. (IBC)
⌐ Borei Keila Street No. 169, Sangkat Vealvong, Phnom Penh City. ☎ +855 (23) 66061, 66064. 🗎 +855 (23) 66063

CHINA (People's Republic of)

TV sets: 300m — **Colour:** PAL — **System:** D.

CHINA CENTRAL TELEVISION (CCTV Gov.)
⌐ 11 Fuxing Lu, Haidian, Beijing 100859, China. ☎ +86 (10) 850 0000. 🗎 +86 (10) 851 3025. ① 222299 CCTVE CN.
WWW: http://www.wtdb.com/CCTV/about.htm
L.P: Pres: Yang Weigwang. VP: Yu Guanghgua. GM TV Prgr Agency: Xu Xiongxiong. Dir. Int. Rel: Zhao Yuhui.
1st Prgr (news & information network): 2250-1700 on ch2 (Beijing) /Teletext Prgr: 0200-1600.
2nd Prgr (economic and social education prgrs) : 0030-1700 on ch8 (Beijing).
3rd Prgr (entertainment prgrs): 0030-1600 on ch15 (Beijing).
4th Prgr : news and entertainment prgrs. Channel 4 is also broadcast outside of China to dozens of television stations in Asia, Australia, Africa, and in the former Soviet Union republics. 2020-2220 on ch29 (Beijing/cable).

Location	1st Prgr	2nd Prgr	3rd Prgr
Beijing	2	8	15
Tianjin	5	36	—
Shijiazhuang	1	13	—
Taiyuan	7	14	—
Hohhot	2	26	—
Shenyang	5	33	—
Changchun	2	13	—
Harbin	8	14	—
Shanghai	5	—	—
Nanjing	38	15	—
Hangzhou	6	—	—
Hefei	11	2	—
Fuzhou	10	—	—
Nanchang	9	—	—
Jinan	8	25	—
Zhengzhou	2	7	—
Wuhan	8	25	—
Changsha	21	—	—
Guangzhou	8	—	—
Nanning	12	20	—
Haikou	7	—	—
Chengdu	8	10	—
Guiyang	2	—	—
Kunming	4	15	23
Lhasa	4	12	—
Xi'an	8	16	—
Lanzhou	4	10	—
Xining	11	15	—
Yinchuan	4	24	—
Urumqi	12	15	—

Other TV Stations: There are more than 31 Provincial TV Stations and nearly 3,000 City-TV Stations.

China — Georgia

Location	Provincial TV chs			City TV chs	
Beijing	6	21		27	
Tianjin	12	29	17	23	
Shijiazhuang	10	32		4	
Taiyuan	9	14		12	
Hohhot	10	20		4	
Shenyang	10	12	21	2	27
Changchun	7	19		9	39
Harbin	1			6	20
Shanghai	8	14	26		
(Shanghai Eastern TV) 20					
Nanjing	1+10	4	32	12	28
Hangzhou	4	22		11	41
Hefei	4			9	
Fuzhou	2	7		4	
Nanchang	7	13		1	
Jinan	2	15		6	
Zhengzhou	9			12	
Wuhan	4	2		19	
Changsha	9	27		4	40
Guangzhou	2	14	21	34	
Nanning	4	14		10	
Haikou	2	12		32	
Chengdu	2	27	21	15	
Guiyang	4			9	
Kunming	2+9	21	27	11	30
Lhasa	8	6		10	
Xi'an	4			10	
Lanzhou	2	8		18	
Xining	4	8	27	21	
Yinchuan	8			18	
Urumqi	4	6	8	21	27	33
Zhengzhou	9			12	
Wuhan	4	2		19	
Changsha	9	27		4	40
Guangzhou	2	14	21	34	
Nanning	4	14		10	
Haikou	2	12		32	
Chengdu	2	27	21	15	
Guiyang	4			9	
Kunming	2+9	21	27	11	30
Lhasa	8	6		10	
Xi'an	4			10	
Lanzhou	2	8		18	
Xining	4	8	27	21	
Yinchuan	8			18	
Urumqi	4	6	8	21	27	33

BEIJING TELEVISION (BTV)
📧 Bejing — **Station:** ch33 — **D.Prgr:** 0030-1600.

ORIENTAL TV
📧 Shanghai. ☎ +86 (21) 322 3007. 📠 +86 (21) 320 7368.

CHINA (Republic of) TAIWAN

TV-sets: 7m — **Colour:** NTSC — **System:** M.

CHINA TELEVISION COMPANY (Comm.)
📧 No. 120 Chung-Yang Road, Nankang District, Taipei. ☎ +886 (2) 783 8308. 📠 +886 (2) 782 6007. ① 25080 CHINA TV.
L.P: Pres: Hu Ping Chung. CE: Mr. Chen.
Stations: BEF21 chA9 180/40kW (No. Taiwan), BEF22 chA10 150/17kW (Ce. Taiwan), BEF23 chA9 180/80/20kW (So. Taiwan).
D.Prgr: W 0400-0500, 0930-1515; Sat 0450-1615; Sun 0250-1515.

CHINESE TELEVISION SERVICE (Comm.)
📧 100 Kuang Fu South Road, Taipei. ☎ +886 (2) 751 0321. 📠 +886 (2) 751 6019.
L.P: Chmn: Chien-Chiu Yee. Pres: Shih-shung Wu. CE: Shiao Ho Whu.
Stations: BET31 chA11 200kW (No.We.), BET32 chA8 190kW (Ce.), BET33 chA11 131kW (So.We.), BET34 chA8 24.5kW (No.Ea.), BET35 chA11 18kW (Ce.Ea. Taiwan), BET36 chA8 9.5kW (So.Ea. Taiwan), BET37 ch35 1702kW (No.We.), BET39 ch36 1383kW (Ce.), BET40 ch34 1383kW (So.We.), BET41 ch33 316kW (No.Ea.), BET42 ch35 316kW (Ce.Ea.), BET43 ch33 316 kW (So.Ea.), BET38 ch33 283kW (We.) + 2 relay st's.
D.Prgr: W 2320-0120, 0340-1600; Sun 2320-1600.

EDUCATIONAL TV
BET37 chA35 1717kW (No.We. Taiwan), BET38 chA33 285kW (Ce.No. Taiwan), BET39 chA36 1500kW (Ce. Taiwan), BET40 chA34 1500kW (So.We. Taiwan), BET41 chA33 354kW (No.Ea. Taiwan), BET42 chA35 343kW (Ce.Ea. Taiwan), BET43 chA33 334kW (So.Ea. Taiwan).
D.Prgr: W 1030-1435, Sun 1130-1505.

TAIWAN TELEVISION ENTERPRISE Ltd. (Comm.)
📧 No. 10, Pa Te Rd, Section 3, Taipei 10560. ☎ +886 (2) 7711515. 📠 +886 (2) 7413626. **Cable:** Television. ① 25714 TV TAIWAN.
L.P: Chmn: Ching-Teh Hsu. Pres: Walter C.H. Wang. Executive vice president: Wei-Yung Lee. Vice president: Hsiang-chuan Hsiung. Vice president: Shen-Wen Lee. Manager, Engineering Department: To-Hui Yang. Manager, Business Department: Wen-Lung Liu. Film buyer: Nancy Hu. Manager, News Department: Victor S.T. Chang. Manager, Sports Department: Jason Liao. Manager, Program Department: Sheng Chu-yu. Dir. Inf. Center: Ko-Jan Hwang.
Stations: BET21 chA7 282/95kW (No. Taiwan), BET22 chA12 162/136/6.5kW (Ce. Taiwan), BET23 chA7 136/162/5.5kW (So. Taiwan), BET24 chA10 4.0/0.4kW (I-Lan), BET25 chA7 12/36kW (Hua Lien), BET 26 chA10 1.9kW (Taitung) + 7 Transposer stations.
D.Prgr: Mon-Fri 2159-0030, 0311-0810, 0839-1610; Sat 0244-1740; Sun 2339-1610.

GEORGIA

TV-sets: N/A — **Colour:** SECAM — **Systems:** D & K.

GEORGIAN TELEVISION (Gov.)
📧 M. Kostavas 68, 380071 Tbilisi. ☎ +7 (8832) 362294.
📠 +7 (8832) 362319 (manual).
L.P: Chief-editor, Information Sector: Areshidze Mamuka.
Stations (Prgr. 1)

	ch		ch
Tbilisj	4	Kutaisi	10
Suchumi	5	Gori	12
Batumi	8		

+ various translators.

D.Prgr: Mon-Fri 1000-1100, 1200-2100. Sat/Sun 0900-2100.
Stations (Prgr. 2) ch
Tbilisi 28
D.Prgr: Current schedule not available.

ORT Relays
Stations	ch		ch
Batumi	3	Tbilisi	9
Kutaisi	7	Gori	31
Suchumi	9		

+ various translators.
D.Prgr: Relays of ORT-1 from Moscow. Further details see under Russia.

ROSSIJSKOJE TELEVIDENIJE (RTV)
Stations	ch
Tbilisi	6

+ various translators.
D.Prgr: Relays of RTV from Moscow. Further details see under Russia.

HONG KONG

TV-sets: 1.749.000 — **Colour:** PAL — **System:** I.

TELEVISION BROADCASTS LIMITED
TV City, Clear Water Bay Road, Kowloon. ☎ +852 2719 4828. +852 2358 1337.
L.P: MD: Louise Page. GM (Corporate Affairs): Alex Ying Ki Luen. GM (International Operations): Ken Lam Kon Leung. Film Buyer: Sophia Chan (Jade Network), Musetta Wu (Pearl Channel).
Stations: chE21 & chE25, 10kW.
D.Prgr: Jade network on ch21: 2245-2030 (approx.). Pearl network on ch25: 0000 (Sat 0100, Sun 0030)-1730 (approx.).

ASIA TELEVISION LIMITED
81 Broadcast Drive, Kowloon. ☎ +852 2992 8888. +852 2338 0438. ② HX44680. **Cable:** ASIATV. **E-mail:** atv@hkatv.com **WWW:** http://www.hkatv.com
L.P: CEO: Mr. Mark Lee; Dep CEO: Mr. Clarence Chang; Controller: Mr. Jermyn Lynn
Stations: chE23 & chE27, 10kW.
D.Prgr: Home Channel in Chinese on ch23: 2230-1930. World Channel in English on ch27: 24h.

INDIA

TV sets: 50m (est.) — **Colour:** PAL — **System:** B

DOORDARSHAN INDIA (Gov. semi-comm.)
Directorate General of Doordarshan, Mandi House, Copernicus Marg, New Delhi-110 001. ☎ +91 (11) 382094-99. +91 (11) 3386507. ② 81-31-65290/66413 dgdd in. **Cable:** tvgeneral.
L.P: DG: K.S. Sarma. CE: K.C.C. Raja.
Programming/News:
Doordarshan Kendra, Akashvani Bhavan, 1 Sansad Marg, New Delhi-110 001. ☎ +91 (11) 3382021/3715411(pabx). Add. DG (News): D.C. Bhaumik. PD: M.B. Pahari.
Programming:
Central Production Centre, Doordarshan, Asiad Village Complex, Siri Fort, New Delhi-110 016. ☎ +91 (11) 6462539/6462446.

Sations: Only high power tr's (1 kW and above) are listed below.

location	ch	kW	Relay
Mumbai	4	10	
Calcutta	4	10	
Delhi	4	10	
Hyderabad	4	10	
Lucknow	4	10	
Chenai	4	10	
Srinagar	4	10	
Pathankot	4	10	Delhi
Ahmedabad	5	10	
Bangalore	5	10	
Bhopal	5	10	
Jaipur	5	10	
Kanpur	5	10	Lucknow
Pune	5	6	Mumbai
Raipur	5	1	Delhi
Silchar	5	10	Delhi
Mumbai 2	6	1	Delhi 2
Calcutta 2	6	10	Delhi 2
Imphal	6	1	Delhi
Kasauli	6	10	Delhi
Chennai 2	6	10	Delhi 2
Muzaffarpur	6	1	Delhi
Rajkot	6	10	Ahmedabad
Allahabad	7	10	Lucknow
Amritsar	7	10	Jaladhar
Asansol	7	10	Calcutta
Delhi 2	7	10	
Gulbarga	7	1	Bangalore
Kodaikanal	7	10	Chennai
Nagpur	7	10	Mumbai
Panaji	7	10	Mumbai
Sambalpur	7	1	Cuttak
Pij	7	1	Ahmedabad
Vijayawada	7	10	Hyderabad
Visakapatnam	7	10	Hyderabad
Agartala	8	10	Delhi
Cuttack	8	10	
Kohima	8	1	Delhi
Kurseong	8	10	Calcutta
Varanasi	8	10	Lucknow
Aizawl	9	1	Delhi
Agra	9	10	Lucknow
Gorakhpur	9	10	Lucknow
Guwahati	9	10	
Indore	9	10	Bhopal
Jalandhar	9	10	
Thiru'puram	9	10	
Murshidabad	10	10	Calcutta
Mussoorie	10	10	Lucknow
Ranchi	10	10	Patna
Tura	10	10	Delhi
Dibrugarh	11	10	Guwahati
Patna	11	10	
Poonch	11	10	Delhi
Bathinda	12	10	Jalandhar
Kochi	12	10	Thiru'puram
Dwarka	12	10	Ahmedabad
Jammu	12	10	Delhi
Shillong	12	1	

+ 23 additional tr's 10kW or 1kW (channels unknown). Total transmitters = 562.
D Prgr: National: 1500-1800v (extended telecasts during live coverage). **Network:** 0130-0315 (Sun 0830), 0830-1043, 1420-1430. **General/Regional:** W 1130-1510, Sun 0315-0330/1400-1415. **Country-**

ASIA — India — Japan

wide **Classroom:** 0730-0830, 1030-1130 (W). Open University: 0100-0130 (Mon/Wed/Fri) Educational Television: 0500-0530 (W), summer vacations till 0730. The **Metro** (Entertainment) channels are networked from Delhi to Ch. 2 of Mumbai, Calcutta & Chennai, and are also available on Insat-2B. Regional Channels: available on Insat-2C.

INDONESIA

TV sets: 11.000.000 — **Colour:** PAL — **System:** B.

TELEVISI REPUBLIK INDONESIA TVRI (Gov.)
☞ Jalan Jerbang Pemuda, Senayan, Jakarta. ☎ +62 (21) 5733135/2279. 🖷 +62 (21) 5732408. **Cable:** telviri Jakarta. ① 073-46154 tvri jkt.
L.P: MD: Azis Husein. TD: Djoko Widayat. Film Buyer: Adi Kasno.
Key Stations

Loc.	chE	kW	Loc.	chE	kW
Ujung Padang	4	1	Pontianak	7	10
Medan	5	10	Jakarta	8	10
Banda Aceh	5	10	Yogyakarta	8	10
Padang	5	10	Denpasar	8	5
Banjarmasin	5	10	Kupang	8	5
Jakarta	6	5	Palembang	9	5
Bandung	6	10	Balikpapan	9	1
Semarang	6	5	Surabaya	9	10
Ambon	7	5	Manado	9	1

+ 10 relay st's.
D.Prgr: W 0930-1630 (Sat 1700), Sun & Holidays 0100-0630, 0930-1630. **English:** 1130-1200. **N:** 1130.

commercial stations

RCTI (PT Rajawali Citra Televisi Indonesia)
☞ Jl. Raya Perjuangan No. 3, kb. Jeruk, Jakarta 11000.
☎ +62 (21) 530 3540/3550/3564. 🖷 +62 (21) 549 3852/3846.
L.P: Pres Dir: Mr. Muchamad Ralie Siregar; TM: Mr Doopy Irwan
Stations: Ambon ch11, Balikpapan ch4, Batam ch43, Denpasar ch35, Jakarta ch43, Malang ch40, Manado ch30, Semarang ch30, Ujung Padang ch33
D.Prgr: 90 hrs per week.

SCTV (PT Surya Citra Televisi)
☞ JL Raya Darmo Permai III, Surabaya 60189. ☎ +62 (31) 714 567/714 033. 🖷 +62 (31) 717 273
Stations: Balikpapan ch11, Bandung ch52, Banjarmasin ch11, Batam ch47, Dili ch11, Malang ch46, Manado ch34, Mataram ch11, Medan ch35, Solo ch44, Surabaya ch43, Ujung Padang ch35, Yogjakarta ch34

TPI (PT Cipta Televisi Pendidikan Indonesia)
☞ Jalan Pintu II - Taman Mini Indonesia Indah, Pondok Gede, Jakarta Timur 13810. ☎ +62 (21) 841 2473 to 83 (HQ). 🖷 +62 (21) 841 2470/1
L.P: Sa'Dullah Sulchan; GM: Syamsudin C. Haesy
Stations: Jakarta ch34 & 37
D.Prgr: Mon-Fri: 2230-1800, Sat: 2230-1800

ANTEVE (PT Cakrawala Andalas Televisi)
☞ Mulia Center Building, 19th Floor, Jl. HR Rasuna Said Kav. X-6 No.8, Jakarta 12940. ☎ +62 (21) 522 2084 to 86, 522 9175. 🖷 +62 (21) 522 2087, 522 9174.
L.P: GM: Mr. Dennis M. Cabalfin
Stations: unknown

IVM (PT. Indosiar Visual Mandiri)
☞ Jl. Damai No 11, Daan Mogot, Jakarta 11510. ☎ +62 (21) 567 2222, 568 8888. 🖷 +62 (21) 565 2221
Stations: unknown

JAPAN

TV-sets: 100m — **Colour:** NTSC — **System:** M.

NIPPON HOSO KYOKAI (Japan Broadc. Corp.) (non Gov., non Comm.)
☞ 2-2-1, Jinnan, Shibuya-ku, Tokyo 150-01. ☎ +81 (3) 3485-6517, 3481-1362. 🖷 +81 (3) 3481-1576.
WWW: http://www.nhk.or.jp
L.P: Pres: Mikio Kawaguchi; Exec. Dir: Michio Futami; Eng: Hiroaki Ohtsuka.
1)= General Sce; 2)= Educational Sce. Call: JO(call) TV.

Main st's	Call	ch	kW	Prgr
Tokyo	AK	1	50	1
Kofu	KG	1	3	1
Fukuyama	DP	1	1	1
Yamaguchi	UC	1	1	1
Nagasaki	AC	1	1	2
Asahikawa	CC	2	1	1
Kushiro	PC	2	1	1
Muroran	IZ	2	1	2
Akita	UB	2	5	2
Fukushima	FD	2	3	2
Nagano	NK	2	1	1
Shizuoka	PB	2	1	2
Osaka	BK	2	10	1
Matsuyama	ZB	2	5	2
Kumamoto	GB	2	1	2
Okinawa	AP	2	5	1
Sapporo	IK	3	10	1
Kitami	KP	3	1	1
Sendai	HK	3	10	1
Tsuruoka	-	3	1	1
Aomori	TG	3	5	1
Tokyo	AB	3	50	2
Kofu	KC	3	3	2
Nagoya	CK	3	10	1
Fukui	FC	3	3	1
Toyama	IG	3	3	1
Hiroshima	FK	3	10	1
Okayama	KB	3	10	1
Tottori	LG	3	1	1
Tokushima	XK	3	1	1
Fukuoka	LK	3	10	1
Nagasaki	AG	3	1	1
Kagoshima	HG	3	5	1
Oita	IP	3	3	1
Hakodate	VK	4	1	1
Obihiro	OG	4	1	1
Yamagata	JC	4	2	1
Morioka	QG	4	3	1
Kanazawa	JK	4	3	1

Hamamatsu	DG	4	1	1
Tottori	LC	4	1	2
Kochi	RK	4	1	1
Sendai	HB	5	10	2
Aomori	TC	5	5	2
Okayama	KK	5	10	1
Kagoshima	HC	5	5	2
Tsuruoka	-	6	1	2
Matsue	TK	6	1	1
Matsuyama	ZK	6	5	1
Kochi	RB	6	1	2
Fukuoka	LB	6	10	2
Kitakyushu	SK	6	1	1
Hiroshima	FB	7	10	2
Fukuyama	DO	7	1	2
Yamagata	JG	8	3	1
Morioka	QC	8	3	2
Niigata	QK	8	5	1
Kanazawa	JB	8	3	2
Hamamatsu	DC	8	1	2
Miyazaki	MG	8	1	1
Asahikawa	CG	9	1	1
Kushiro	PG	9	1	1
Muroran	IQ	9	1	1
Akita	UK	9	5	1
Fukushima	FP	9	3	1
Nagano	NB	9	1	2
Nagoya	CB	9	10	2
Shizuoka	PK	9	1	1
Fukui	FG	9	3	1
Yamaguchi	UG	9	1	1
Kumamoto	GK	9	1	1
Hakodate	VB	10	1	2
Toyama	IC	10	3	2
Sapporo	IB	12	10	2
Obihiro	OC	12	1	2
Kitami	KD	12	1	2
Niigata	QB	12	5	2
Osaka	BB	12	10	2
Matsue	TB	12	1	2
Kitakyushu	SB	12	1	2
Miyazaki	MC	12	1	2
Oita	ID	12	3	2
Okinawa	AD	12	5	2
Kobe	PP	28	10	1
Otsu	QP	28	1	1
Tsu	NP	31	5	1
Kyoto	OK	32	10	1
Wakayama	RP	32	1	1
Takamatsu	HP	37	10	1
Tokushima	XB	38	10	2
Saga	SP	38	5	1
Gifu	CP	39	5	1
Takamatsu	HD	39	10	2
Saga	SD	40	5	2
Nara	UP	51	1	1

+ approx. 6900 st's.
D.Prgr: 1)Gen. TV: 2055-1507. 2)Educ. TV: 2100-1502

THE NATIONAL ASSOCIATION OF COMMERCIAL BROADCASTERS IN JAPAN
Addr: 3-23, Kioi-cho, Chiyodaku, Tokyo 102.
☎ +81 (3) 5213-7700. 🖹 +81 (3) 5213-7701. **Cable**: Mimporen Tokyo. ③ 2325163 NABTYO J.

TV Networks & Key Stations

Japan News Network (28 st's)
Tokyo Broadc. System, Inc. (TBS)
✉ 3-6, Akasaka 5-chome, Minato-ku, Tokyo 107-06.
☎ +81 (3) 3746 1111. 🖹 +81 (3) 3588-6378. **L.P**: Pres: Hirozono Isozaki.
Station: JOKR-TV ch6 50kW(+ 89 relay st's). **D.Prgr**: 24h

Nippon News Network (30 st's)
Nippon Television Network Corp. (NTV)
✉ 14, Niban-cho, Chiyoda-ku, Tokyo 102-40.
☎ +81 (3) 5275-1111. 🖹 +81 (3) 5275-4008. **L.P**: Pres: Seiichiro Ujiie.
Station: JOAX-TV ch4 50kW (+ 89 relay st's). **D.Prgr**: Approx 2000-1800

Fuji News Network (26 st's)
Fuji Television Network, Inc. (CX)
✉ 3-1, Kawada-cho, Shinjuku-ku, Tokyo 162
☎ +81 (3) 3353-1111. 🖹 +81 (3) 3358-1747. **L.P**: Pres: Hisashi Hieda.
Station: JOCX-TV ch8 50kW(+ 89 relay st's). **D.Prgr**: Approx 2010-1900.

All-Nippon News Network (23 st's)
Asahi National Broadc. Co., Ltd. (TV Asahi) (ANB)
✉ 1-1, Roppongi 1-chome, Minato-ku, Tokyo 106.
☎ +81 (3) 3587-5111. 🖹 +81 (3) 3505-3539. **L.P**: Pres: Kunio Ito.
Station: JOEX-TV ch10 50kW (+ 88 relay st's). **D.Prgr**: Approx 2030-1800.

TXN Network (6 st's)
Television Tokyo Channel 12 Ltd. (TX)
✉ 3-12, Toranomon 4-chome, Minato-ku, Tokyo 105-12.
☎ +81 (3) 3432-1212. 🖹 +81 (3) 5473-3447. **L.P**: Pres: Naomichi Sugino.
Station: JOTX-TV ch12 50kW (+ 86 relay st's). **D.Prgr**: Approx 2055-1800.

ARMED FORCES RADIO & TV SERVICE (U.S. Mil.)
✉ (Misawa) OLAA, AFPBS, APO San Francisco 96519 - (Okinawa) Det 2, AFPBS, APO San Francisco 96239
Stations: Misawa ch66 1kW, Iwakuni ch66 0.4kW — Okinawa ch8 40kW + additional st's at Iwakuni & Sasebo on cable only.

KAZAKHSTAN

TV-sets: N/A — **Colour**: SECAM — **Systems**: D&K.

KAZAKH TELEVISION (Gov.)
✉ Jeltoksan kösesi 175, 480013 Almati. ☎ +7 (3272) 695188. 🖹 +7 (3272) 631207.

Stations	ch		ch
Karağandi	1	Akmola	10
Semey	4	Öskemen	10
Aktöbe	8	Oral	11
Almati	10	Petropavlovsk	12

+ further main transmitters and translators. Complete details not known.
D.Prgr: Mon-Fri 0255-0800, 1200-1830. Sat/Sun 0300-1830.

ASIA — Kazakhstan — Korea

PRIVATE STATIONS (details Incomplete)

Rakhat: ch26
TAN: channel unknown
TRK 32: ch31
TRK Shakhar [Kazakhcity] Ltd: ch46.

ORT-1 RELAYS

Stations	ch		ch
Aktöbe	1	Akmola	7
Öskemen	1	Semey	8
Oral	2	Karağandi	12
Almati	3		

+ further main transmitters and translators. Complete details not known.
D.Prgr: Relays of ORT-1 from Moscow.

OTHER STATIONS

Location	ch	Location	ch
Almati	7	Almati	40

D.Prgr: The above transmitters carry relays of private programme producers, Kyrgyz and Turkish TV.

KOREA (Dem. People's Rep.)

TV-sets: 2.000.000 — **Colour:** PAL/NTSC — **System:** D & K/M

THE RADIO AND TELEVISION BROADCASTING COMMITTEE OF THE DEMOCRATIC PEOPLES REPUBLIC OF KOREA (KRT) (Gov.)

✉ Chonsung-dong, Moranbong District, Pyongyang.
☎ +850 (2) 816 035. 🖷 +850 (2) 812 100
L.P: Chairm: Chong, Ha-Chol; Dir: Chun, Li-Ji; Head of Tech: Chol, Li-Yong.

Location	ch	kW	Location	ch	kW
Sangmasan	1	10	Kanggye	8	70
Chayubong	2	30	Jaedoksan	9	30
Suryongsan	2	30	Unjubong	9	70
Pegebong	3	30	Wangjesan	9	30
Hamhung	3	70	Sepo	9	70
Wonsan	4	10	Sinyang	9	30
Songjinsan	4	20	Wonsan	10	70
Jajiryong	5	30	Haeju	11	70
Paekam	5	10	Sambongsan	11	10
Sambongsan	5	10	Jonchon	11	10
Kangryong	5	30	Songsan	12	10
Kumgangsan	5	30	Jajiryong	12	30
Chongjin	6	70	Chongjin	12	70
Hyangsan	6	10	Haksongsang	12	20
Sepo	6	70	Misan	12	70
Sinuiju	6	70	Pyongyang	12	700
Sariwon	7	30	Rimbong	12	10
Chayubong	8	30	Sobaeksan	12	30
Haksongsan	8	20	Tokusan	12	20

Stations(ERP over 10kW):
D.Prgr: W 0900-1400; Sun: 0000-0300, 0600-1400
NB: In early 1997, transmissions were irr. due to power shortages.

MANSUDAE TELEVISION

✉ Mansudae, Pyongyang.
Station: ch5 350kW(ERP).
D.Prgr: Sun: 0100-0400, 0700-1300.

KAESONG TELEVISION

✉ Kaesong.
Stations: Kaesong ch8 30kW(ERP), Pyongyang ch9 140kW(ERP).
D.Prgr: W 0900-1400; Sun: 0000-0300, 0600-1400

KOREA (Rep.)

TV-sets 10.430.000 — **Colour:** NTSC — **System:** M

KOREAN BROADCASTING SYSTEM (KBS) (Public Corporation)

Addr. 18 Yoido-dong Youngdungpo-gu, Seoul 150-790. ☎ +82 (2) 781 2001/2, 781 1460/1, 781 5108. 🖷 +82 (2) 781 2099, 781 1496-7, 781 5199. **WWW:** http://www.kbs.co.kr
L.P: Pres: Hong, Too-Pyo; Dir Int. Rel: Ms Cha, Myonh-hee; Tech Coord: Ahn, Dong-Su
Stations: 1=Prgr. 1; 2=Prgr. 2. Call: HL—TV.

Location	ch	kW	Prgr.	Location	ch	kW	Prgr
Chinju	3	1	1	Ch'unch'on	12	1	1
Yosu	4	10	1	Chonju	13	2	2
Ch'ongju	4	1	1	Cheju	13	1	2
Seoul	5	10	1	P'ohang	13	1	1
Ulsan	5	2	1	P'ohang	20	10	2
Namwon	5	1	1	Ch'unch'on	22	30	2
Ch'angwon	6	10	1	Andong	23	10	2
Taejon	6	10	1	Yosu	24	30	2
Kangnung	6	10	2	Ch'ongju	24	5	2
Ch'unch'on	6	1	2	Kwangju	25	30	2
Seoul	7	50	2	Seoul	26	10	2
Chonju	7	10	1	Ulsan	27	10	2
Pusan	7	10	2	Chinju	27	5	2
T'aebaek	7	1	1	Mokp'o	27	0.5	1
Taegu	8	10	1	Kangnung	28	5	2
Ch'unch'on	8	10	1	Ch'ungju	30	1	2
Seoul	9	50	1	Wonju	31	10	2
Pusan	9	10	1	Seoul	32	10	1
Kangnung	9	10	1	Taejon	35	5	2
Cheju	9	1	1	Seoul	37	10	2
Cheju	10	5	2	T'aebaek	37	5	2
Ch'ongju	10	1	1	Teagu	38	30	2
Wonju	10	1	1	Taejon	42	10	2
Chinju	10	1	1	Chinju	42	5	2
Kwangju	11	10	1	Ch'angwon	45	30	2
Wonju	11	1	1	Taejon	47	10	1
Andong	11	1	1	Ch'ongju	48	0.5	2
Cheju	12	5	1	Taejon	51	10	2
Kangnung	12	1	1	Wonju	52	5	2
Taejon	12	1	1	Andong	53	5	2
Ch'ungju	12	1	1	Namwon	57	10	2
Andong	12	1	1				

+ approx 500 relay st's less than 1kW.
D.Prgr.: 1st prgr: W 2100-0100, 0830-1500; Sat/Sun: 2100-1500; 2nd prgr: W 2100-0100, 0830-1500; Sat/Sun: 2100-1530.

EDUCATION BROADCASTING SYSTEM

✉ 92-6., Umyeon-dong, Seocho-gu. Seoul 137-791.
☎ +82 (2) 521 1586/1988/1989/1357. 🖷 +82 (2) 521 0241/522 8043
L.P: Pres: Dr. Chung, Yun Choon; Dir: Ms Chung, Hyo-soon; GD: Park, Myung-ha

Korea — Malaysia

Stations (operated by KBS): Call: HLQK-TV.

Location	ch	kW	Location	ch	kW
Seoul	13	10	Taejon	29	5
Kwangju	19	30	Yosu	30	30
Kangnung	19	10	Kangnung	34	5
Cheju	20	10	Chinju	36	5
Ch'ungju	21	30	Ch'angwon	39	30
Ulsan	21	10	Taejon	39	30
Namwon	22	10	Wonju	40	10
Pusan	23	30	Seoul	43	1
T'aebaek	25	10	Taegu	44	30
P'ohang	26	10	Chonju	45	30
Cheju	26	1	Andong	47	5
Ch'unch'on	28	30	Taejon	53	10
Andong	29	10	Ch'ongju	54	5
Seoul	13	10	Taejon	29	5
Kwangju	19	30	Yosu	30	30
Kangnung	19	10	Kangnung	34	5
Cheju	20	10	Chinju	36	5
Ch'ungju	21	30	Ch'angwon	39	30
Ulsan	21	10	Taejon	39	30
Namwon	22	10	Wonju	40	10
Pusan	23	30	Seoul	43	1
T'aebaek	25	10	Taegu	44	30
P'ohang	26	10	Chonju	45	30
Cheju	26	1	Andong	47	5
Ch'unch'on	28	30	Taejon	53	10
Andong	29	10	Ch'ongju	54	5

+ approx 200 relay st's less than 1kW.
D.Prgr: W 0730-0010; Sun: 2200-1500.

MUNHWA BROADC. CORP. (Comm.)
✉ 31 Yoido-dong Youngdungpo-gu, Seoul 150-728.
☎ +82 (2) 789 2851/3521. 📠 +82 (2) 782 3094/0294.
WWW: http://www.mbc.co.kr
L.P: Pres: Mr. Kang, Sung-Koo; Dir: Mr. Song, Iljun;
Dir TV Eng: Mr Jung, Jai-Soon
Call: HL—TV.

Location	ch	kW	Location	ch	kW
Andong	5	1	Yosu	28	2
P'ohang	6	1	Kangnung	31	10
Cheju	7	5	Ulsan	33	10
Mokp'o	7	0.5	Ch'ongju	33	5
Taejon	8	5	Wonju	34	10
Chinju	8	1	Ch'ungju	36	10
Kwangju	9	10	Seoul	38	10
Taegu	10	5	Taejon	40	5
Chonju	10	5	Seoul	41	10
Ch'unch'on	10	1	Samch'ok	43	5
Seoul	11	50	Taejon	45	10
Pusan	11	10	Masan	48	5
Cheju	11	1	Wonju	49	5
Masan	13	5	Kwangju	51	10
Samch'ok	22	5	Andong	59	5
Ch'unch'on	24	30			

+ approx 150 relay st's less than 1kW.
D.Prgr: W 2100-0100, 0830-1500; Sat/Sun: 2100-1500.

SEOUL BROADCASTING SYSTEM (Comm.)
Addr. 10-2 Yoido-dong Youngdungpo-gu, Seoul 150-010. ☎ +82 (2) 786 0792, 780 0006. 📠 +82 (2) 785 6171
L.P: Pres: Mr Yoon, Hyuck-ki; Mr. Park, Jin
Station: HLSQ-TV ch6. **D.Prgr:** W 2100-0100, 0830-1500; Sat/Sun: 2100-1500.

SATELLITE AND TV HANDBOOK

AMERICAN FORCES KOREA NETWORK (US Mil.)
✉ Unit #15324, APO AP 96205-0097, USA. ☎ +82 (2) 7914 6495 — **L.P:** Commanding Officer: LTC Cad C. Starr.
Stations: chA2: P'algonsan, Tongduch'on, Pusan, Chinhae, Hoi-dok, Wonju, Kunsan — chA6: Munsan — chA12: Ch'unch'on, Taegu, Taejon — chA13: Kwangju — chA34: Seoul — chA49: Osan, Susong, Taegu — chA70: P'yongt'aek, Tongduch'on — chA75: Tongduch'on. **D.Prgr:** 24h.

KYRGYZSTAN

TV-sets: N/A — **Colour:** SECAM — **Systems:** D & K.

KYRGYZ TELEVISION (Gov.)
✉ pr. Molodoj Gvardii 63, 720885 Biskek. ☎ +7 (3312) 253404. 📠 +7 (3312) 257930 — **L.P:** Gen. Dir: Tugelbay Kazakov.
Stations: current details not known.
D.Prgr: 1300-1600.

TV RELAYS
Various main and relay transmitters throughout the country relay prgrs from Moscow, as well as Uzbek and Kazakh TV, and prgrs of TRT from Turkey.

LAOS (Peoples Dem. Rep.)

TV-sets: 80.000 — **Colour:** PAL — **System:** B

LAO NATIONAL TELEVISION (TVNL) (Gov)
✉ P.O. Box 310 Vientiane Lao PDR. ☎ +856 4475/4523/4425
L.P: DG: Dr. Khekkeo Soisaya; Dir. Tech: Mr. D. Sisombath
Stations: Vientiane stations: ch9 (5kW), ch23 (0.1kW), ch12 (1kW); Savannakhet station: ch12 (1kW)
D.Prgr: ch9, 2330-0030 (UTC) (Sat-Sun 2330-0230 UTC), evening prgr.:1130-1600 (UTC); ch23, 2330-0400 (UTC); ch12, 2330-0030 (UTC), 1000-1600 (UTC); Savannakhet station: 1130-1600 (UTC

MACAU

TV-sets: 70.300 — **Colour:** PAL — **System:** I.

TELEDIFUSÃO DE MACAU (TDM SARL)
Addr. P.O. Box 446, Macau. ☎ +853 520204/6. 📠 +853 520 208. ☉ 88309 RADIO OM
L.P: Chairman: Stanley Ho; Exec. Vice Chairman: Maria do Carmo Figueiredo.
Stations: Portuguese Ch. — Ch30 0.2kW+1 repeater of 10W.
Chinese Channel — Ch32 0.2kW+1 repeater of 10W.
D.Prgr: Portuguese Channel: appr. 1858-2400; Chinese Channel 0730-0900, 1815-2400.

MALAYSIA (Federation of)

TV-sets: 9.4m* — **Colour:** PAL — **System:** B.

Peninsular Malaysia

RADIO TELEVISION MALAYSIA (Gov.)
✉ Dept. of Broadc, Angkasapuri, Kuala Lumpur 50614.

ASIA   Malaysia — Pakistan

☎ +60 (3) 282 5333/3140. 🗎 +60 (3) 282 4735. **Cable:** Tivimalsia, Kuala Lumpur. ③ MA 31383 Kuala Lumpur.
WWW: http://www.asiaconnect.com.my
L.P: DG: Dato' Jaafar Kamin; Int Rel: Ms Nawiyah Che'Lah; Dir Eng: Mr. Lal Singh

Location	TV1 ch	TV1 kW(ERP)	TV2 ch	TV2 kW(ERP)
Genting Sempah	2	112	10	112
Johor Baru/Gunong Pulai	3	38	10	142.5
Tangkak	4	154	7	77.4
Kuala Lumpur	5	100	8	100
Alor Setar	5	100	8	100
Kuantan	5	127	8	127
Kuala Pilah	5	7.95	8	7.95
Kuala Lipis	5	20	8	20
Mersing	6	100	9	100
Ipoh	6	190	9	190
Melaka	6	185	9	185
Kota Bharu	6	250	9	250

+ low power relay st's.
D.Prgr: TV1: Mon-Wed 0900-1600, Thurs/Sat 0700-1600; Fri/Sun 0200-1600. **TV2:** 0900 (Fri/Sun 0700)-1600.

SYSTEM TV MALAYSIA BERHAD (TV3)
🖃 Sri Pentas (Ground Floor, South Wing) No. 3, Persiaran Banjar Utama, 47800 Petaling Jaya Selangor Darul Ehsan. ☎ +60 (3) 716 6333. 🗎 +60 (3) 716 133. ③ 33014 STMB MA.
L.P: MD: Khalid Hj Ahmad. Eng. Mgr: Rahmad A Kadir.
Stations: Kuala Lumpur ch12 124/25kW H.
Repeater st's of high power in Kuantan (ch 11), Ipoh (ch 11), Gunung Ledang (ch 11), Johore Bahru (ch 26), Kedah (ch 29), Kota Bharu (ch 27), Kuala Terengganu (ch 11), Dungun (ch 27), Tampin (ch 23), Taiping (ch 41), Kuching (ch 12), Sibu (ch 11), Kota Kinabalu (ch 29), Ulu Kali (ch 29) & Machang (ch 11).
D. Prgrs: Mon-Wed 0830-1615, Thurs-Fri 0630-1615, Sat/Sun 0100-1615.

METROVISION
🖃 33 Jln Delima 1/3 Subang, Hi-Tec Industrial Park 40000, Shahalam. ☎ +60 (3) 732 8000. 🗎 +60 (3) 732 8932.
L.P: GM (Prgrs): Tunku Yahaya. GM (Mktg.): Lim Eng Kien.

East Malaysia
TV MALAYSIA SABAH AND SARAWAK (Gov.)
Addr. P.O. Box 1016, 88614 Kota Kinabalu. ☎ +60 (88)52711. **Cable:** Broadcasts, Sabah. ③ MA 80061 Kota Kinabalu.
L.P: Prgr. Contr: M.A. Mahmood.
Stations: Pol = H.

Location	TV1 chE	TV2 chE	kW (ERP)
Sipitang	11	3	13.6
Lahad Datu	7	10	100
Sandakan	8	5	100
Tawau	9	6	33
Tambunan/Keningau	9	12	36
Kudat	9	6	100

K.K. (Lawa Mandau)	10	7	100
Limbang	2	4	13
Sibu	8	5	100
Bintulu	9	6	25
Simangang (Sri Aman)	9	6	120
Miri	10	12	55
Kuching	10	7	220
Kapit	4	10	18

D.Prgrs: as for Peninsular Malaysia.

MALDIVES (Rep. of)

TV sets: 4.750 — **Colour:** PAL — **System:** B.

TELEVISION MALDIVES (Gov.)
🖃 Buruzu Magu, 20-04 Male'. ☎ +960 323 105, 324 105. 🗎 +960 325 083. ③ 66183 TVM MF.
L.P: DG: Mr. Hussain Mohamed
Station: ch7 1kW H.
D.Prgr: 0300-0500 (Fri), 1200-1620.

PAKISTAN

TV-sets: 2.080.000 — **Colour:** PAL — **System:** B.

PAKISTAN TELEVISION CORPORATION LTD.
🖃 Federal Television Complex, P.O. Box 1221, Islamabad. ☎ +92 (51) 828723. 🗎 +92 (51) 823406/812202.
Cable: Pakteevee. ③ 5833 PPV RP. (Prgr. Centres at Lahore, Karachi, Peshawar, Quetta and Islamabad).
L.P: MD: Mrs Raana Shaikh. TD: Mohammad Kamil.

Stations	ch	kW	Stations	ch	kW
Karachi	4	60	Mangora	7	10
Sakessar	4	114	Muree	8	180
Lak Pass	4	8	Shikarpur	8	213
Kala Shah Kaku	5	400	Quetta	8	1.25
Nurpur	5	170	Shujabad	8	178
Ghazaband	5	10	Thana Bola Khan	9	205
Quetta	6	30	Cherat	10	170
Thandiani	6	5	Sahiwal	10	277
Faisalabad	6	20	J. din Wali	10	200
Sibi	6	6	Pasrur	10	6.8
Sakessar	7	168	Tando Allahyar	11	400
Karachi	4	60	Mangora	7	10
Sakessar	4	114	Muree	8	180
Lak Pass	4	8	Shikarpur	8	213
Kala Shah Kaku	5	400	Quetta	8	1.25
Nurpur	5	170	Shujabad	8	178
Ghazaband	5	10	Thana Bola Khan	9	205
Quetta	6	30	Cherat	10	170
Thandiani	6	5	Sahiwal	10	277
Faisalabad	6	20	J. din Wali	10	200
Sibi	6	6	Pasrur	10	6.8
Sakessar	7	168	Tando Allahyar	11	40

+7 low power st's.
D.Prgr: 1130-1930. **N:** 1400, 1600.

SHALIMAR TELEVISION NETWORK (STN)
🖃 P.O. Box 1246, Islamabad. ☎ +92 (51) 856 171. 🗎 +92 (51) 261 225. **Cable:** SUPERSOUND Islamabad
L.P: MD: M Arshad Choudhry. CE: Agha Nasir.

PHILIPPINES (Rep. of the)

TV-sets: 7m — **Colour:** NTSC — **System:** M.

NATIONAL TELECOMMUNICATIONS COMMISSION (Department of Transportation and Communications)
✉ 855 Vibal Bldg, Esda Corner Times Str, Quezon City.
L.P: Commissioner: Josefina Lichauco. Dep. Commissioners: Aloysius R. Santos, Florentino L. Ampil. Chief, Broadcast Sce. Dept: Carlos D. Saliuan Jr.
Stations C=City.

	Call	chA	kW	Location
1)	DWWX-TV	2	35	Quezon C.
2)	DXRV-TV	2	1	Ihgan C.
3)	D-3-Z0-TV	3	100	Baguio, C.
12)	—	3	1	Metro Manila
1)	DYCB-TV	3	10	Cebu C.
11)	DYLL-TV	3	500	Zamboanga C.
3)	DWGT-TV	4	25	Metro Manila
1)	DYXL-TV	4	2	Bacolod C.
1)	DXAS-TV	4	10	Davao C.
2)	DYXX-TV	6	1	Iloilo C.
2)	DZBB-TV	7	25	Quezon C.
2)	DYSS-TV	7	5	Cebu C.
5)	DXSS-TV	7	1	Davao C.
2)	DWAI-TV	7	1	Naga C.
4)	DYKB	8	5	Bacolod C.
4)	DZKB-TV	9	25	Quezon C.
4)	DYKC-TV	9	5	Cebu C.
6)	DXLA-TV	9	2.5	Zamboanga C.
2)	DYAF-TV	10	5	Bacolod C.
2)	DWMJ-TV	10	1	Baguio C.
2)	DWMT-TV	10	25	Benguet
2)	DWLA-TV	12	10	Legaspi C.
2)	D-12-ZB	12	1	Batangas
4)	DZBS-TV	12	5	Baguio C.
8)	PR-TV	12	?	Tacloban C.
9)	DXNS-TV	12	1	Cotabato C.
3)	—	13	20	Metro Manila
10)	DZTV-TV	13	6.25	Quezon C.
7)	DXGL-TV	13	2	Butuan C.
10)	DXTV-TV	13	5	Davao C.
10)	DYTV-TV	13	12.5	Cebu C.

low power st's not mentioned.
Addresses & other information:
1) ABS-CBN Broadc. Corp, Mother Ignacia Ave, 1100 Quezon C. — Mon-Fri 2300-1530 (Sat/Sun 0000-1730).
2) Republic Broadc. System Inc. EDSA, Diliman, Quezon City, Metro Manila. Mon-Fri 0800-0130, Sat 0730-0130, Sun 0730-0200.
3) PTV Channel 4, Media Center, Bohol Ave, Quezon City — 0425 (Sat/Sun 0300)-1730.
4) RPN Channel 9, Broadcast City, Capitol Hills, Quezon City — 0230-1600.
5) Southern Broadc. Netw, 3881 E. Vallejo Str, Santol Sta. Mesa, Metro-Manila.
6) First United Broadc. Corpo, Lozenzo Bldg, 787 Vito Cruz Str, Metro-Manila. ☎ +63 583082. 🖹 +63 583082.
7) Philippine Electronic & Communications Institute, Montilla Blvd, Butuan C — Mon-Fri 1000-1500, 0500-2200.
8) East Visayas Broadc, 2647 Donada Str, Malate, Metro-Manila.
9) Cotabato TV Corpo, Regional Complex, Catabato City.
10) Intercontinental Broadc. Corp., Quezon C.
11) Zamboanga TV Corp., Zomboanga C.
12) RT Broadcasting Spec. Rm.9, RC Poblete Bld., 17 Sen. Gil Y. Puyat Av. Makati, Metro Manila.

PEOPLE'S TELEVISION NETWORK, INC (Comm.)
✉ Broadcast Complex, Visayas, Avenue, Quezon City 1100. ☎ +63 (2) 921 2344/2451. 🖹 +63 (2) 921 1777/7310/886, 922 9112/6064
L.P: Chairm: Ms Lourdes I Ilustre; GM: Mr Ramon S. Diez; CE: Mr Antonio M Leduna

ARMED FORCES RADIO & TV SERVICE (U.S. Mil.)
✉ Det 1, AFPBS, APO San Francisco 96274, USA.
Stations: Olongapo ch14 0.25kW, Angeles City ch17 1kW, San Miguel ch40 0.03kW, Baguio ch14 0.3kW, San Fernando ch17 0.25kW, Capas ch17 25W.

SINGAPORE

TV-sets: 1.060.000 — **Colour:** PAL — **System:** B.

TELEVISION CORPORATION OF SINGAPORE (TCS)
✉ Caldecott Broadcast Centre, Andrew Rd. Singapore, 299939. ☎ +65 2560401. 🖹 +65 2538808. ① RS 39265 SBCGEN.
WWW: http://rock.tcs.com.sg
L.P: Chief Exec. Officer: Leo Cheok Yew. PR Mgr: Julie Lee. VP (Prgrs): Daniel Yun. VP Eng: Tay Joo Thong.
Stations: Singapore chE5/chE8 120kW.
D.Prgr: chE5: Sun-Thurs 0600-0300, Fri & Sat 0600-0400; chE8: Mon-Fri 0800-1230, Sat 0800-0230, Sun 0800-1230

TELEVISION TWELVE (TV-12)
✉ 12 Prince Edward Road, #05-00 Bestway Building, Singapore 0207. ☎ +65 225 8133. 🖹 +65 220 3881, 225 0966.
L.P Chief Exec. Officer: Sandra Buenaventura. VP Programming: Amy Chua. Transmission Mgr: Mr. H.T. Lau
Stations: chE12 120kW & E24
D.Prgr: Mon-Fri 0600-1230, Sat 0200-1230, Sun 0900-1230

SRI LANKA

TV-sets: 1.500.000 — **Colour:** PAL — **System:** B.

INDEPENDENT TELEVISION NETWORK (ITN)
✉ Wickramasinghepura, Battaramulla. ☎ +94 (1) 864591. 🖹 +94 (1) 864591. ① 22445.
L.P: Chairm: Mr G.B Rajapakse; GM: Mr Bertie Galahitiyawa; DE: Mr W.S.E. Fermando
Station: Wickramasinghepura ch12 10kW (ERP), Yatiyantota ch9 10kW, Deniyaya ch5 1kW.
D.Prgr: 1800-2300 (approx).

SRI LANKA RUPAVAHINI CORPORATION (SLRC)
✉ P.O. Box 2204, Colombo 7. ☎ +94 (1) 580136. 🖹

ASIA — Sri Lanka — Uzbekistan

+94 (1) 580929. **Cable:** Rupavahini. ① 22148 SLTV CE.
L.P: DG: Mr. W.D Jayasinghe. Dep. DG: A. Senadheera. Dep. DG (Prgrs): Lucien Bulathsinghala. Dep. DG (Eng): Upali Arambewela.
Stations: Kokavil ch8 20 kW, ch11 1 kW Sooriyakanda +2 low power repeaters.
D.Prgrs: 1130-1700. Sun 1000-1700.

TELSHAN NETWORK (PVT) Ltd. (TNL TV)
☞ Innagale Estate Dampe-Piliyandala. ☎ +94 (1) 575436 430 859 ≣ +94 (1) 575436,574 962
L.P: Chairm MD: Mr. Shantilal Nilkant Wickremesinghe
Stations: Colombo ch 21 22kW; Piliyandala ch 26 22kW, ch 3 20kW; Nuweraeliya ch 4 40kW; Polgahawela ch 3 1kW; Ratnapura ch 26 1kW; Hantana (Kandy) ch21 22kW.

EAP NETWORK (PVT) LTD.
☞ 676 Galle Rd, Colombo 3. ☎ +94 (1) 503819 (9 lines). ≣ +94 (1) 503788. **E-mail:** eapnet@slt.lk
L.P: Chairperson: Mrs. Soma Edirisinghe. MD: Jeevaka Edirisinghe. Dir/GM: Rosmand Senaratne.

Location	ETV1	ETV2	kW
Colombo	37	35	1
Deniyaya	31	35	1
Hantana	37	33	1
Matale	31	35	1
Nuwara Eliya	37	33	1

D.Prgr: 24h. ETV1 rel. Sky News. ETV2 rel. Star TV.

MTV CHANNEL (PVT) LTD.
☞ 109 Collets Bldg., Rt. Hon. D. S. Senanayake Mawatha, Colombo 8. ☎ +94 (1) 689324-6. ≣ +94 (1) 689328.
Stations: Depanama ch23 1kW, Nuwaraeliya ch25 1kW.
D.Prgr: 24h. Rel. BBC World + local prgrs 1130-1700.

TAJIKISTAN

TV-sets: N/A — **Colour:** SECAM — **Systems:** D & K.

TAJIK TELEVISION (Gov.)
☞ Behzod küça 7, 734013 Dushanbe. ☎ +7 (3772) 224357.
L.P: Head of TV: Mirbobo Mirrahimov.
Stations and **D.Prgr:** current details not available.

TV RELAYS
Various main and relay transmitters throughout the country relay programmes of ORT-1 from Moscow, as well as TRT from Turkey and IRIB from Iran. Further details not known.

THAILAND

TV-sets: 3.300.140 — **Colour:** PAL — **Systems:** B&M.

TELEVISION OF THAILAND (Gov.)
☞ Public Rel. Dept., 26th Floor Fortune Town Bldg, Ratchadapisek Road, Huay Khwang, Bangkok 10310. ☎ +66 (2) 248 1601/8088 to 94. ≣ +66 (2) 248 1601/1655/2155. ① 72243 PRDTHAI TH.
Stations: Bangkok ch11 200kW (ERP) + 27 relay st's.
D.Prgrs: 0930-1400.

BANGKOK ENTERTAINMENT CO. Ltd.
(Licensed through Mass Communications Organisation of Thailand).
☞ 1126/1 New Petchbury Road, Bangkok 10400.
☎ +66 (2) 2539970-3. ≣ +66 (2) 2539978. ① 82616 BECOM TH. **Cable:** BANGERTAIN.
L.P: Prgr. Dir: Pravit Maleenont. TD: Manoontham Thachai.
Stations: Bangkok ch3 650 kW + 32 relay st's not mentioned.
D.Prgr: Mon-Fri 1400-2400, SS 0800-2400.

MASS COMMUNICATIONS ORG. OF THAILAND
☞ 222 Thanon Ysok Asok-Dindaeng, Bangkok 10210.
☎ +66 (2) 2450700. ① 84577 MOT BKK TH.
Stations: Bangkok ch9 20/4kW + 32 relay st's not mentioned.
D.Prgrs: 0850 (Sat/Sun 0025)-1700.

THE ARMY TELEVISION HSA-TV (Gov, Comm.)
☞ Phaholyothin Rd, Sanampao. Bangkok 10400. ☎ +66 (2) 2710069. ≣ +66 (2) 2712510. ① 81080 ATV TH.
L.P: DG: Maj. Gen. Vijit Junapart.
Stations: Bangkok ch5 20/4kW + 18 relay st's not mentioned.
D.Prgr: 0900 (Sat/Sun 0100)-1700.

BANGKOK BROADCASTING & TV CO. LTD. (Comm.)
☞ P.O. Box 4-56, Bangkok 10900. ☎ +66 (2) 2781255. ≣ +66 (2) 2701976. ① 82730 BBTV TH.
L.P: St. Man: Chatchur Karnasuta. TD: Supoch Sangsayan.
Stations: Bangkok ch7 20kW + 22 relay st's not mentioned.
D.Prgr: 0900 (Sat/Sun 0100)-1700.

TURKMENISTAN

TV-sets: N/A — **Colour:** SECAM — **Systems:** D&K.

TURKMEN TELEVISION (Gov.)
☞ Machtumkuli 89, 744000 Asgabat. ☎ +7 (3632) 251515. ≣ +7 (3632) 251421 (manual).
Stations: Chardzhev chR10 (TMT-1), R8 (TMT-2), R3 (ORT), all 5kW.
D.Prgr: TMT-1: MF 0150-0540, 655-1020, SS 0150-2030; TMT-2: MF 0200-0300, 1300-1930, SS 0200-0830, 1200-1930. rel: ORT, Moscow (24hrs) & TRT-1 (Turkey): 1200-1600.
TMT-1/2, ORT on nationwide networks, TRT-1 in Asgabat.

UZBEKISTAN

TV-sets: N/A — **Colour:** SECAM — **Systems:** D&K.

UZBEK TELEVISION (Gov.)
☞ Navoii küçasi 69, 700011 Toskent. ☎ +7 (3712) 495214.
L.P: Chmn: Eerkin K. Haitboev.

Stations: (Prgr. 1)	ch	**Stations:** (Prgr. 2)	ch
Toskent	5	Toskent	1

D.Prgr: Current schedule not available.

ORT
Stations	ch
Toskent	3

D.Prgr: Relays of ORT from Moscow. Further details see under Russia.

ROSSIJSKOJE TELEVIDENIJE (RTV)
Stations	ch
Toskent	9

D.Prgr: Relays of RTV from Moscow. Further details see under Russia.

TV RELAYS
Stations	ch
Toskent	40

D.Prgr: Relays of Kazakh, Kyrgyz and Tadzhik TV, as well as TRT from Turkey.

VIETNAM

TV-sets: 2.500.000 — **System:** M — **Colour:** NTSC/SECAM

TELEVISION VIETNAM (Gov.)
✉ 59 Giang Vo Street, Hanoi. ☎ +84 (43) 43188/55933. 🖷 +84 (43) 55332. ① 412279 THVN/VT.
L.P: DG: Ho Anh Dung. Dir of loc. TV: Nguyen Van Nhuong.
Stations: Hanoi ch2, ch6 (SECAM); Ho Chi Minh City ch9 240kW; Hue, Cantho & Quinhon ch7 and DaNang ch13 (NTSC). Details of st's in Vinh & Nha Trang not available. 13 additional relay tr's.
D.Prgr: Central Prgr. from Hanoi: Mon-Fri 90 min's a day, Sat/Sun 3h (prgrs in black & white only).
Ch 6: Hanoi City Sce. **Ch. 2:** Provincial Sce. for Red River delta. **Other st's:** 25% in colour (no further details available).

PACIFIC

AUSTRALIA

TV-sets: 8m — **Colour:** PAL. **System:** B.

AUSTRALIAN BROADCASTING CORPORATION (ABC)
The 'National Television Service' is controlled by the Australian Broadcasting Corporation which is responsible to Parliament through the Minister for Transport and Communications.

In addition to the terrestrial transmitters listed below, ABC also broadcasts to remote areas using five transponders of the Australian domestic satellite system to cover the five main geographic regions and time zones. This service is known as the 'Homestead and Community Broadcasting Satellite Service' (HACBSS) and is carried in the frequency band 12 250 to 12 750 MHz on satellites which are owned and operated by Optus Communicatons Pty Ltd. Except in the States of Victoria and Tasmania, all transmitters outside the State capital cities are fed by satellite. The satellite signal is 'B-MAC' encoded and carries three sound-only programs, one of which is stereophonic, in addition to the television signal and its associated sound channels.

✉ GPO Box 9994, Sydney NSW 2001. ☎ +61 (2) 437 8000. 🖷 +61 (2) 9950 3055. **WWW:** http://www.abc.net.au
L.P: Ag. Dir. TV: Penny Chapman.
Stations: Audio powers are 1/10 of vision powers when a single sound channel is used. Stations below 10 kW ERP are not listed. Pattern: 'O': omnidirectional, 'D': directional. Channels 3, 4, 5 and 5A are progressively being phased out to make way for FM radio and other services — the expected new channel numbers and other information covering changes arising from the first stages of the clearance of stations from 'Band II' are given in parentheses in the table.

Call	ch	ERP(kW)	POL	Pattern	Location
Australian Capital Territory					
ABC	3	100	V	O	Canberra
ABN	2	100	H	O	Sydney
New South Wales					
ABMN	0	100	H	O	Wagga Wagga
ABCN	1	100	V	O	Orange
ABTN	6	100	V	D	Taree
ABDN	2	100	H	D	Grafton/Kempsey
ABQN	11	200	V	O(D)	Dubbo
ABHN	5A	100	H	O	Newcastle
ABWN	56	600	H	D	Illawarra
ABRN	6	100	V	D	Lismore
ABGN	7	100	H	O	Griffith
ABUN	7	100	H	D	Tamworth
ABSN	8	100	V	D	Bega/Cooma
Victoria					
ABAV	1	100	H	D	Albury
ABEV	1	100	V	D	Bendigo
ABV	2	100	H	O	Melbourne
ABSV	2	100	V	D	Swan Hill
ABGV	40	100	V	D	Shepparton
ABRV	11	100	H	D	Ballarat
ABLV	40	100(1000)	H	D	Traralgon
ABMV	6	100	H	D	Mildura
ABWV	5A	66	H	D	Western Victoria
ABWV	45	100	H	D	Horsham/Dimboola
Queensland					
ABSQ	1	100	H	D	Warwick
ABQ	2	100	H	O	Brisbane
ABDQ	332	100	H	O(D)	Toowoomba
ABRQ	9	100	H	D	Rockhampton
ABTQ	3	100	H	D	Townsville
ABMQ	8	180	H	D	MacKay
ABWQ	6	150	V	D	Maryborough
ABNQ	9	100	H	D	Cairns
ABQ	49	50	H	D	Gold Coast
South Australia					
ABGS	1	100	H	D	Mount Gambier
ABNS	1	100	V	D	Port Pirie
ABS	2	100	H	D	Adelaide
ABRS	3	150	V	D	Loxton
ABS	30	10	H	D	Wudinna
Western Australia					
ABW	2	100	H	O	Perth
ABAW	2	100	V	O	Albany
ABSW	5	100	H	O	Bunbury
ABCW	5A	100	H	O	C. Agricultural
ABGW	6	10	H	D	Geraldton
ABW	8	60	H	O	Wagin
ABCMW	8	10	H	D	Morawa
ABW	46	10	H	D	Dalwallinu
ABW	60	60	H	D	Moora

PACIFIC — Australia

Call   ch ERP(kW) POL Pattern Location

Tasmania
ABT   2   100   H   O   Hobart
ABNT 32 100(300) H   D   Launceston

Northern Territory
ABD   6   30   H   D   Darwin

+ 343 low power services with ERPs below 10kW, made up of 110 translators and 233 satellite-fed transmitters.
+ 46 (approx.) low power (up to 150 watts ERP), privately owned transmitters/translators which are licensed under a 'self-help' scheme to carry ABC television programs.

Daily Prgrs: Mon-Fri 2100-1400; Sat/Sun 24h.

SPECIAL BROADCASTING SERVICE (SBS)

Locked Bag 028, Crows Nest, NSW 2065. ☎ +61 (2) 9430 2828. +61 (2) 9430 3700 — P.O. Box 294, So. Melbourne, VIC 3205. ☎ +61 (3) 9685 2828. +61(3) 9686 7501.
WWW: http://www.sbs.com.au
L.P: MD: Malcolm Long.
Stations: All stations horizontal polarity unless specified.

	Ch.	kW		Ch.	kW
Sydney	28	300	Hobart	28	225
Melbourne	28	300	Gold Coast	61	50
Brisbane	28	300	Newcastle	45	600
Adelaide	28	300	Illawarra	53	600
Perth	28	300	Ulladulla	30(V)	400
Bendigo	29	100	Ballarat	30	300
Canberra	28	200	Traralgon	34	800

+ estimated 35 low power transmitters with kw(ERP) below 10 kw. not mentioned.

IMPARJA TELEVISION PTY LTD.

PO Box 2924, Alice Springs, NT 0871. ☎ +61 (89) 523744, +61 (89) 531014. ① AA81166.
This st. is owned by the Central Australian Aboriginal Media Association (CAAMA).
Station: ch9 250kW.

Commercial TV Stations

FEDERATION OF AUSTRALIAN COMMERCIAL TELEVISION STATIONS

44A Avenue Rd, Mossman, NSW 2088. ☎ +61 (2) 960 2622. + 61 (2) 969 3520. ① 121542.

Main Networks

THE SEVEN NETWORK

Television Centre, Mobbs Lane, Epping, NSW 2121. ☎ +61 (2) 858 7777. +61 (2) 858 7888. ① AA20250.
WWW: http://www.seven.com.au
(5 owned and 9 affiliated st's).

THE NINE NETWORK

P.O. Box 27, Willoughby, NSW 2068. ☎ +61 (2) 430 0444. +61 (2) 436 2193.

NETWORK 10 AUSTRALIA

P.O. Box 10, Lane Cove, NSW 2066. ☎ +61 (2) 887 0222.

Stations: The first two letters are an abbreviation of the name of the licence, the third indicates the state and the numeral signifies the channel

Call	on air ID	affiliation	tx's
1) RTQ	WIN TV N.Queensland	9 Network	70
2) STQ	Sunshine Television	7 Network	67
3) TNQ	TEN North Queensland	10 Network	64
4) BTQ	Seven Network	7 Network	4
5) QTQ	Nine Network	9 Network	5
6) TVQ	Ten Network	10 Network	3
7) NBN	NBN Television	9 Network	38
8) NEN	Prime Television	7 Network	36
9) NRN	TEN Northern NSW	10 Network	37
10) CBN	Prime Television	7 Network	39
11) CTC	TEN Capital	10 Network	44
12) WIN	WIN Television	9 Network	46
13) ATN	Seven Network	7 Network	6
14) TCN	Nine Network	9 Network	6
15) TEN	Ten Network	10 Network	6
16) AMV	Prime Television	7 Network	80
17) BCV	TEN Victoria	10 Network	25
17) GLV	TEN Victoria	10 Network	41
18) VTV	WIN Television Victoria	9 Network	71
19) ATV	Ten Network	10 Network	9
20) GTV	Nine Network	9 Network	9
21) HSV	Seven Network	7 Network	9
22) TNT	Southern Cross Netw.	7/10 Netw.	30
23) TVT	WIN TV Tasmania	9 Network	28
24) ADS	Ten Network	10 Network	9
25) NWS	Nine Network	9 Network	8
26) SAS	Seven Network	7 Network	9
27) NEW	Ten Network	10 Network	4
28) STW	Nine Network	9 Network	4
29) TVW	Seven Network	7 Network	4
30) BKN7	Broken Hill TV	—	1
31) MTM	—	—	3
32) STV8	Sunrayia TV	—	1
33) GTS	—	—	8
34) RTS 5a	—	—	1
35) SES	—	—	5
36) GTW	—	—	4
37) SSW	Golden West Network	—	10
37) VEW	Golden West Network	—	8
38) NTD	—	—	2
39) ITQ 8	—	—	1
—) QQQ	—	—	75
40) IMP	Imparja Television	—	68
—) WAW	—	—	100

addresses:
1) WIN Television, P.O. Box 568, Rockhampton QLD 4700
2) Sunshine Television, P.O. Box 30, Maryborough QLD 4650
3) TEN North Queensland, P.O. Box 1016, Townsville QLD 4810
4) Seven Network Limited, P.O. Box 604, Brisbane QLD 4001
5) Queensland Television Limited, P.O. Box 72, Brisbane QLD 4001
6) Ten Network, P.O. Box 751, Bribane QLD 4001
7) NBN Television, P.O. Box 750L, Newcastle NSW 2300
8) Prime Television (Northern) Pty Ltd, P.O. Box 317, Tamworth NSW 2340.
9) TEN Northern NSW, P.O. Box 920, Coffs Harbour NSW 2450
10) Prime Television (Southern) Pty Ltd, P.O. Box 465, Orange NSW 2800
11) Capital Television, Private Mailbag 10, Dickson ACT 2602

12) WIN Television, Television Ave, Mt St Thomas NSW 2500
13) Seven Network Ltd, Mobbs Lane, Epping NSW 2121
14) Nine Network, P.O. Box 27, Willoughby NSW 2068
15) Ten Network, 44 Bay St, Ultimo NSW 2007
16) Prime Television (Victoria) Pty Ltd, Union Road, Lavington NSW 2641
17) TEN Victoria, P.O. Box 888, Bendigo VIC 3550
18) WIN Television, P.O. Box 464, Ballarat VIC 3350
19) Ten Network, Level 3-6, Como Centre, TRB West, 620 Chapel Street, South Varra VIC 3141
20) Nine Network, P.O. Box 100, Richmond VIC 3121
21) Seven Network Ltd, P.O. Box 215D, Melbourne VIC 3001
22) Southern Cross Television, 37 Watchorn Street, Launceston TAS 7250
23) WIN Television, P.O. Box 1209M, Hobart TAS 7001
24) Ten Network, 125 Strangeways Tce, North Adelaide SA 5006
25) Southern Television Corp. Pty Ltd, P.O. Box 9, North Adelaide SA 5006
26) Seven Network Ltd, 45-49 Park Tce, Gilberton SA 5081
27) Ten Network, P.O. Box 1010, Mirrabooka WA 6061
28) Sunraysia Television Ltd, P.O. Box 99, Tuart Hill WA 6060
29) Seven Network, P.O. Box 77, Tuart Hill WA 6060
30) Broken Hill Television Ltd, P.O. Box 472, Broken Hill NSW 2880
31) Lochfield Consultants Pty Ltd, P.O. Box 493, Griffith NSW 2680
32) Sunrayia Television Ltd, P.O Box 1157, Mildura VIC 3500
33) Spencer Gulf Telecasters Pty Ltd, P.O. Box 305, Port Pirie SA 5540
34) Riverland Television Ltd, P.O. Box 471, Loxton SA 5333
35) South East Telecasters Pty Ltd, P.O. Box 821, Mt Gambier SA 5290
36) Geraldton Telecasters Pty Ltd, Cnr Fifth & Howard street, Geraldton WA 6530
37) Golden West Network Pty, Ltd, P.O. Box 112, Bunbury WA 6230
38) Territory Television Pty Ltd, P.O. Box 1764, Darwin NT 0800
39) Mt Isa Television Pty Ltd, P.O. Box 1557, Mt Isa QLD 4825
40) Imparja Television Pty Ltd, 14 Leichhardt Tce, Alice Springs NT 0871

COOK ISLANDS

TV-sets: 3500 — **Colour:** PAL — **System:** B.

COOK ISLANDS BROADCASTING CORPORATION (Gov.)
P.O. Box 126, Avarua, Rarotonga. ☎ +682 29460. +682 21907 Cook Islands Television (CITV).
Stations: VHF ch1 & 6 — **D.Prgr:** 6h a day, 5 days a week.

EASTER ISLAND

TV-sets: N/A — **Colour:** PAL — **System:** B.

TV RAPANUI
Hanga Roa, Isla de Pascua. ☎ +5639 223291
L.P: Dir. Gen: J. Edmund Paoa. Head Tec. Sce's: J. Pont Chavez.
Station: ch (unknown). **D.Prgr:** 0000-0600.

FIJI

TV-sets: N/A — **Colour:** NTSC — **System:** M.

FIJI TELEVISION
GPO Box 2442, Suva. ☎ +679 305100. +679 305077. **E-mail:** fijitv@is.com.fj — **L.P:** Chief Exec: Peter Wilson.

Pay-TV: the govt. is considering the award of a franchise to operate a Pay-TV sce.

GALAPAGOS ISLANDS

TV-sets: 4.000 — **Colour:** NTSC — **System:** M.

TELEGALAPAGOS (Cult)
ADDR: Misión Franciscana, Puerto Baquerizo Moreno, Isla San Cristobal, Galapagos, Ecuador.
LP: Dir. Gen: Mons. Manuel Valarezo. Dir. Tec: Germán Chiriboga. Film Buyer: Remigio Andrade.
Station: chA13 — **D.Prgr:** 2000-0400.

GUAM (US Terr.)

TV-sets: 75.000 — **Colour:** NTSC — **System:** M.

KUAM TELEVISION (Comm.)
Pacific Telestations, P.O. Box 368, Agana 96910. ☎ +671 6375826, 6376397. +671 6379865, 6379870.. **Cable:** Kuam.
L.P: Pres: L.S. Berger. Gen. Mgr: Greg Perez. Dir. Tec: K. Tydingco.
Station: chA8 21.9/3.8kW
D.Prgr: Mon-Fri 2000-1400 (also rel. CBS/NBC prgrs).

KGTF TELEVISION (Educ.)
Guam Educational Telecommunications Corporation, P.O. Box 21449, Guam, Marianas Is. 96921. ☎ +671 7342207. +671 7345483. ③ 6467 KGTF GM.
L.P: GM: Joseph E. Tighe. TD: Edmond Cheung. Film Buyer: Doris Gallo.
Station: chA12 27.6/5.5kW. **F.Pl.** ch 14 (Saipan & Tinian); ch 16 (Rota).
D.Prgr: 2000-1300.

KTGM-TV
692 Marine Dr, Tamuning 96911. ☎ +671 6498814. +671 6490371 — Station: chA14.

HAWAII (US State)

TV-sets: 552.500 — **Colour:** NTSC — **System:** M.

All st's comm. exc. 7) and 9) Educational, 18) religious.

Call	ch	kW	Call	ch	kW
1) KHON-TV	2	100/20	7) KMEB-TV	10	5/0.5
1a) KHBC-TV	2	2.29/0.458	9) KHET	11	150/30
2) KGMV	3	14.1/2.82	9a) KHAW-TV	11	2.09/0.27
3) KITV	4	100/20	10) KMAU-TV	12	30/4.36
16) KFVE	5	100/20	11) KHNL-TV	13	316/63.2
14) KVHF	6	52.6/6.7	12) KHVO	13	4.68/0.59
4) KAII-TV	7	29.8/5.9	13) KHAI-TV	20	537/63
5) KGMB-TV	9	209/41.8	14) KMGT	26	100/10
6) KGMD	9	9.68/1.71	15) KBFD	32	146/14.

PACIFIC
Hawaii — New Zealand

1) Burnham Broadc. Co., 1170 Auahi Str, Honolulu 96814-4975.
1a) Hilo Broadc. Corp, Box 4250 Hilo, Big Island 96720-0520.
2) Wailuku (Relay of st. 5).
3) TAK Communications, 1290 Ala Moana Blvd, Honolulu 96814-4299.
4) Wailuku (Relay of st. 1).
5) Lee Enterprises, 1534 Kapiolani Blvd, Honolulu 96814-3799. Mon-Fri 1600-1130; Sat/Sun 24h.
6) Hilo (Relay of st. 5).
7) Wailuku (Relay of st. 9).
9) Hawaii Public Broadc. Authority, 2350 Dole Str, Honolulu 96822-2495.
9a) Hilo (Relay of st. 1).
10) Wailuku (Relay of st. 3).
11) King Broadc. Co, 150-B Puuhale Rd, Honolulu, HI 96819.
12) Hilo (Relay of st. 3).
13) Honolulu Family TV Ltd, 735 Sheridan Str, Honolulu 96814-3095.
14) Mauna Kea Broadc. Co, 970 N. Kalaheo Ave, Honolulu 96734-1892.
15) The Allen Broadc. Corp, 1188 Bishop Str, Honolulu 96813-3314.
16) 315 Sand Island, Access Rd, Honolulu 96819-2245.

KIRIBATI

TV-sets: 685 — **Colour:** PAL — **System:** B.

RADIO KIRIBATI
Broadcasting House, P.O. Box 78, Bairiki, Taiwara, Rep. of Kiribati, Central Pacific. ☎ (686) 21187. (686) 21096
Station & D.Prgr: details unknown

MARSHALL ISLANDS

TV-sets: N/A — **Colour:** NTSC — **System:** M.

MBC-TV
Marshall Islands Broadcasting Company, Majuro 96960. ☎ +692 6253413.

AFRTS TELEVISION (Department of Defense)
Box 23, APO San Francisco, CA 96555, USA.
L.P: Netw. Mgr: Larry Malinowski.
Stations: ch9, ch13 0.25kW (24 h to Kwajalein Island & Roi-Namur Island).

MICRONESIA

TV-sets: 7.000 — **Colour:** NTSC — **System:** M

TV STATION POHNPEI (Comm.)
KPON-TV, Central Micronesia Communications, P.O. Box 460, Kolonia, Pohnpei, FSM 96941.
L.P: Pres: Bernard Hegenberger. Dir. Tec: David Cliffe.
Station: Pohnpei chA7 1kW + cable TV on ch's 4,5,9.

TV-STATION TRUK (Comm.)
Truk State, FSM 96942.
Station: TTTK chA7 0.1kW.
D.Prgr: 0400-1200 (approx).

TV-STATION YAP (Gov.)
WAAB-TV, Yap State, FSM 96943.
Station: chA7 1kW.
D.Prgr: 0400-1200 (approx).

NAURU (Republic of)

TV-sets: N/A — **Color:** NTSC — **System:** M

NAURU TELEVISION
Rep. of Nauru, Ce. Pacific. ☎ +674 4443190

NEW CALEDONIA

TV-sets: 35.500 — **Colour:** SECAM — **System:** K.

RFO-TV (Gov.)
Radio Télévision Française d'Outre Mer (RFO), BP G3 Mont Coffin, F-98848 Nouméa cedex. ☎ +687 274327. ☎ +687 281252 — **L.P:** Dir: Alain Le Garrec.
Stations: Mont Do chK4 2 x 1kW H, Noumea chK8 0.4kW H, Lifou chK7 0.4kW H (+ 25 low power repeaters).
D.Prgr: 2 channels, 8h a day.

private stations
CANAL CALEDONIE (Priv, Comm.)
8 rue de Verneilh, Noumea
Stations: Noumea Mt Coffyn ch43 11kW, Noumea (Town) ch33 0.06kW, Noumea Mt Koghi ch25 9kW
N.B: Canal Caledonie is a subscription sce and the signal is encrypted except: 1945-2020, 0125-0225, 0715-0910.
D. Prgr: W1945-1400 Sun 24h.

NEW ZEALAND

TV-sets: 1.100.000 — **Colour:** PAL — **System:** B.

TELEVISION NEW ZEALAND Ltd.
P.O. Box 3819, Auckland. ☎ +64 (9) 377 0630. +64 (9) 375 0918. ① TVNZACQ 60056.
L.P: Dep Group Exec: Darryl Dorrington. Contr of Eng:Neville Lane.
Stations: Channel One: Wellington ch1 100/20 kW H, Auckland ch2 100/20 kW H, Christchurch ch3 100/20 kW H, Dunedin ch2 50/10 kW H + relay st's at Hamilton ch1 100/20 kW V, Palmerston North ch2 100/20 kW V, Invercargill ch1 1/0.1 kW H + 18 medium and 412 low powered relay st's.
D.Prgr: 2100 (Sat 1900, Sun 2000)-1200 (Fri 1300, Sat 1400).
Channel Two: Auckland ch4 300/30 kW H, Wellington ch5 300/30 kW H, Christchurch ch3 300/30 kW H, Dunedin ch4 300/25 kW H + repeaters at Hamilton ch3 100/20 kW H, Palmerston North ch4 300/30 kW V, Invercargill ch3 1/0.1 kW H + 17 medium and 240 low powered repeaters.
D.Prgr: 1830-1200. Weekends 24hrs.

ACTION TV (Trackside)
P.O. Box 388-99 Wellington. ☎ +64 (4) 576 6999.
+64 (4) 576 6942
Stations: Auckland ch55 150/15kW, ch53 100/20kW,

ch58 50/5kW; Waikato ch56 100/20kW, ch50 50/5kW, ch47 50/5kW; Palmerston North ch?; Wellington ch56 150/15kW, ch58 100/20kW; Christchurch ch48 65/7kW.
D.Prgr: Fri 2200-0930, Mon 2330-0600, Tues 2345-0610, Thur 2250-1000.

TV3 (Comm.)
⌨ P.O. Box 5185, Auckland. ☎ +64 (9) 779 730. 📠 +64 (9) 366 7029.
L.P: Head of Netw. Programming: Kel Geddes.
Stations: Auckland ch7 100/20kW; Wellington ch11 100/20kW; Christchurch ch6 100/20kW; Dunedin ch10 100/20kW; Hamilton ch9 100/20kW; Palmerston North ch7 100/20kW; Invercargill ch7 1/0.1kW. (+ 11 medium powered relay sts.)
D.Prgr: Auckland ch7 H, Christchurch ch6 H, Hamilton ch9 V, Invercargill ch10 H, Wellington ch11 H, Dunedin ch10 H, Palmerston North ch7 V.

CANTERBURY TELEVISION (CTV)
⌨ 196 Gloucester Street, Christchurch.
D.Prgr: 0500-1200 UTC, Fri 2300-Sat 1200 — **Station:** ch48 65kW.

SKY NETWORK TELEVISION (pay-tv)
⌨ P.O. Box 9059, Auckland. ☎ +64 (9) 525 5555. 📠 +64 (9) 525 5725
Stations: Auckland ch27/29/30V (movies), ch31/33/52V (sports), ch43/45/54 (news); Waikato ch28/30V/31V (movies), ch32/34V/27V (sports), ch44/46V/51V (news); Palmerston North ch30 (movies), ch34 (sports), ch46 (news); Wellington ch28/30V/30/49 (movies), ch32/34V/34/53 (sports), ch44/54V/54/47 (news); Christchurchch30 (movies), ch34 (sports), ch46 (news).

NIUE ISLAND

TV-sets: N/A — **Color:** PAL — **System:** B

NIUE TV
⌨ Broadcasting Corp. of Niue, P.O. Box 23, Alofi. ☎ +683 4026. 📠 +683 4217 — **L.P:** GM: Hima Douglas

NORFOLK ISLAND

TV-sets: 900 — **Colour:** PAL — **System:** B.

NORFOLK ISLAND TELEVISION SCE. (Gov.)
⌨ New Cascade Rd, Norfolk Island 2899, Australia. ☎ +672 (3) 22137. 📠 +672 (3) 23298. ⓘ NV 30003
Station: Mt. Pitt ch7 0.02kW V.
D.Prgr: rel. ABC Australia from Optus Satellites.

N. MARIANA ISLANDS

TV-sets: 4.100 — **Colour:** NTSC — **System:** M.

MICRONESIA BROADC. CORPO. (Comm.)
⌨ c/o KUAM, Box 368, Agana, Guam 96920.
L.P: Pres: H. Scott Killgore. Gen. Mgr: T. Dickey. Asst. Gen. Mgr: A. Ocambo. Technician: M. Madaing.
Stations: WSZE-TV ch10 0.5kW (Saipan)
D.Prgr: 0600-1400.

PALAU (Rep.)

TV-sets: 1.600 — **Colour:** NTSC — **System:** M.

STV-TV KOROR (Comm.)
⌨ Koror, Palau 96940.
L.P: Mgr: David Nolan. Technician: Ray Omelen.
Station: Ngermit, Koror ch7 0.1kW.
D.Prgr: 0400-1400.

PAPUA NEW GUINEA

TV-sets: 100.000 — **Colour:** PAL — **Systems:** B & G.

EMTV (Comm.)
⌨ Media Niugini Pty. Ltd, P.O. Box 443, Boroko NCD. ☎ +675 3257322. 📠 +675 3254450.
L.P: Chief Exec: John Taylor. CE: Geoff Kong.
Stations: Burns Peak ch9 1.1kW, Air Niugini Hill ch31 0.17kW, Garden City ch68 0.02kW (all Port Moresby area)
F.Pl: st's at Mt. Hagen, Goroka, Lae, Rabaul.
D.Prgr: 12 h daily in English and Tok Pisin, 7 days a week.

POLYNESIA (French)

TV-sets: 26.500 — **Colour:** SECAM — **System:** K.

TELE TAHITI
⌨ Radio Télévision Française d'OutreMer (RFO), B.P. 125, Papeete, F-98 713 Polynésie Française. ☎ +689 430551. 📠 +689 413155. **E-mail:** rfopolyfr@mail.pf
WWW: http://www.tahiti-explorer.com/rfo.html
L.P: Dir: Claude Ruben. Chief Editor: Patrick Durand Gaillard. Dir. of Prgrs: Jean-Raymond Bodin.
Stations: Papeete chK4 0.1kW H, Mont Marau chK8 0.5kW H, Vaitape chK7 0.2kW V, Taravao chK4 0.1kW H (+ 8 low power repeaters).
D.Prgr: 0400 (Sat/Sun 0200)-0830.

commercial stations

CANAL POLYNESIE (Priv., Comm.)
⌨ Colline de Putiaoro, Papeete
Stations: Taravao ch26 2kW, Mont Marau ch43 55 kW, Punaauia ch55 1.2kW + 9 low power repeaters.
N.B: Canal Polynesie is a subscription sce and the signal is encrypted except: 1630-1720, 2225-2325, 0415-0610.
D.Prgr: W 1630-1100 Sun 24h.

SAMOA (American)

TV-sets: 8.000 — **Colour:** NTSC — **System:** M.

KVZK-TV (Gov.)
⌨ Office of Public Information, PO Box 3511, Pago Pago 96799
☎ + 684 6334191. 📠 +684 6331044
L.P: Dir: Vaoita Sava. Dir. Tec: Robert Blauvelt.
Stations: chA2/A4/A5 (72kW).
D.Prgr: ch2 & 5: 1830-1000; ch4: 1830-1100.

PACIFIC / NORTH AMERICA — Samoa — Canada

SAMOA (Western)

TV-sets: 5.000 — **Colour:** PAL — **System:** B.

TELEVISE SAMOA
✉ P.O. Box 1868, Apia. ☎ + 685 26641. 🖷 +685 24789.
L.P: GM: Tupai Kuka Brown.
Stations: Apia ch11 10W; Mount Vaea ch8 50W; Faleasiu ch10 10W; Mount Aflau ch4 50W, Mount Fiamoe ch6 10W; Api Park ch5 5W
D.Prgr: 0400-1030.

TONGA

TV-sets: 2500 — **Colour:** NTSC — **System:** M.

ASTL-TV3 (Comm.)
✉ P.O. Box 66, Nuku'alofa. ☎ +676 22325. 🖷 +676 22811
L.P: Pres: Latu Tupouniua
Station: Nuku'alofa ch3 50W
D.Prgr: Mo-Sat 1800-2000 & 0300-1000; Sun 0400-0900.

OCEANIA BROADCASTING NETWORK LTD
✉ P.O. Box 91, Nuku'alofa. ☎ +676 23314. 🖷 +676 23658

VANUATU

TV-Sets: N/A — **Colour:** NTSC — **System:** M

TV BLONG VANUATU
✉ Vanuatu Broadcasting and Television Corp, PMB 927, Port Vila. ☎ +678 25412. 🖷 +678 22026 — **L.P:** Mgr: Gaile Dantec..

WALLIS & FUTUNA

TV-sets: N/A — **Colour:** SECAM — **System:** K.

RADIODIFFUSION FRANCAISE D'OUT-RE-MER (RFO)
✉ B.P. 102, Mata Utu, 98600 Uvea, Iles de Wallis-et-Futuna, Pacifique sud (par Nouméa, Nouvelle-Calédonie). ☎ +681 722020. 🖷 +681 722346 — B.P. 20, Sigave, F-98620 Futuna. ☎ +681 723531. 🖷 +681 723534 — **L.P:** SM: Joseph Blasco.
Stations: ch6 & ch9 — **D.Prgr:** 5-6h each evening.

NORTH AMERICA

ALASKA (US State)

TV-sets: 2000 — **Colour:** NTSC — **System:** M.

TV-STATIONS: All comm. exc. 2, 4, 5, 7 (Educational).

Call	ch	kW	Call	ch	kW
1) KTUU-TV	2	100/10	6) KJUD-TV	8	0.24/0.05
1a) KATN-TV	2	28.2/5.5	7) KUAC-TV	9	46.8/8.33
2) KTOO-TV	3	2.45/0.49	8) KTVA-TV	11	50.7/5.07
3) KJNP-TV	4	19.1/3.31	9) KTVF-TV	11	27/5.5
4) KYUK-TV	4	4.67/1.16	10) KIMO-TV	13	39/7.8
4a) KTBY-TV	4	42.5/8.5	11) KTNL-TV	13	0.2/0.03
5) KAKM-TV	7	163/16.3			

1) Box 102880, Anchorage, AK 99510 — Gen. Mgr: Al Bramstedt Jr. 2230 (Sat/Sun 0000)-0900.
1a) P.O. Box 74730, Fairbanks, AK 99701.
2) Capital Community Broadc. Inc, 224 Fourth Str, Juneau 99801 — Pres. & Gen. Mgr: D. Rinker. Dir. Tec: J.W. Foster. 1930 (Sat/Sun 1530)-0800.
3) Evangelistic Alaska Missionary Fellowship, Box "0", North Pole 99705 — Pres. & Dir: D.L. Nelson. Dir. Tec: E. Nichols.
4) Bethel Broadc. Inc, Box 468, Bethel 99559 — Dir. Gen: J. Brigham.
4a) KTBY Inc, 1840 S. Bragnaw Str, Anchorage 99508 — Gen. Mgr: R.V. Bradley. Dir. Tec: E. Gjernes.
5) Alaska Public Television Inc, 2677 Providence Dr, Anchorage 99508 — Gen. Mgr: E. Sackett. Dir. Tec: F. Mengel.
6) 1107 West Eighth St., Suite 2, Juneau, AK 99801.
7) University of Alaska, Fairbanks 99701 — St. Mgr: Kathryn Jensen. Dir. Tec: David L. Walstad. 1800-0900.
8) Northern TV Inc, Box 102200, Anchorage 99510 — Dir. Tec: D. Milsap. 1700-1100.
9) Northern TV Inc, Box 950, Fairbanks 99707 — 1700-1100.
10) Alaska 13 Corp, 2700 Tudor Rd., Anchorage 99507 — Pres: D.L. Triplett.
11) Sitka Broadc. Co. Inc, Box 2668, Sitka 99835 — Gen. Mgr: D. Etulain. Mon-Fri 1900-0900, Sat 1700-0900, Sun 1400-0800.

ARMED FORCES RADIO & TV SCE. (US Mil.)
Addr: Navsta Box 14, FPO Seattle, WA 98791, USA
Station: Adak ch 8 & 10 (cable only).

BERMUDA

TV-sets: 30.000 — **Colour:** NTSC — **System:** M.

BERMUDA BROADCASTING COMPANY Ltd.
✉ P.O. Box HM 452, Hamilton. ☎ +1 (809) 295-2828. 🖷 +1 (809) 295-4282. ☉ 3702 ZBMBA.

local tv stations
There are three local commercial television stations in Bermuda. Reception is island-wide and no special cabling or antennas are required.
Stations: **ZFB-TV**, ch7, operated by the Bermuda Broadcasting Company Ltd. (US ABC affiliate); **ZBM-TV**, ch9, operated by the Bermuda Broadcasting Company Ltd (US CBS affilate); **VSB-TV**, ch11, operated by DeFontes Broadcasting Television Ltd. (US NBC Affiliate).

CANADA

TV-sets: 19.400.000 — **Colour:** NTSC — **System:** M.

CANADIAN BROADCASTING CORPORATION (SOCIÉTÉ RADIO-CANADA)
✉ 1500 Bronson Avenue. P.O. Box 8478, Ottawa,

Ontario K1G 3J5. ☎ +1 (613) 724 1200. 📠 +1 (613) 738 6887. Cable: Broadcasts: ③ 053-4260. **WWW:** http://www.tv.cbc.ca
L.P: Pres. and CEO: Perrin Beatty. Chmn, Board of Directors: Guylaine Saucier. Sen. VP, Resources: Louise Tremblay. Sen. VP, Media (Vacant). VP/Sen. Advisor, Office of the Pres. and CEO: Michael McEwen. VP, Internal Audit: Robert Hertzog. VP, General Counsel and Corporate Secr: Gerald Flaherty, Q.C. VP, Human Resources: George C. B. Smith. Exec. Dir, Media Accountability: Donna Logan. Sen. Dir. of Corporate Communications and Public Affairs: Charlotte O'Dea.
English Networks: P.O. Box 500, Station "A", Toronto, ON M5W 1E6. **Cable:** Broadcast. ③ 062-17796. ☎ +1 (416) 975-3311.
L.P: VP English Television Networks: Jim Byrd. Exec. Dir, News, Current Affairs & Newsworld, Television: Bob Culbert. Head, CBC Newsworld: Slawko Klymkiw. Exec. Dir, Media Operations: Michael Harris. Sen. Dir, Media and Public Relations: Tom Curzon. Sen. Dir, Broadcast Communications: Diane Kenyon.
French Networks: P.O. Box 6000, Montreal, PQ H3C 3A8. **Cable:** Radcan. ③ 05-267417. ☎ +1 (514) 285-3211.
L.P: VP, French Television: Michèle Fortin. DG, Communications: Raymond Guay. Dir. of Public Rel: Micheline Savoie. Exec. Dir, RDI, Renaud Gilbert. GM TV5 (Consortium Québec-Canada): Guy Gougeon.
CBC Engineering: 7925 Côte St-Luc Road, Montreal, PQ H4W 1R5. ③ 055-66287. ☎ +1 (514) 485-1301.
L.P: Sen. Dir, Eng: Brian D. Baldry.

Affiliated Stations: CKVR Barrie, Ont; CKX Brandon, MB; CJDC Dawson Creek, BC; CFJC Kamloops, BC; CHBC Kelowna, BC; CKWS Kingston, Ont; CKSA Lloydminster, AB; CHAT Medicine Hat, AB; CHNB North Bay, Ont; CHEX Peterborough, Ont; CKBI Prince Albert SK; CKPG Prince George BC; CDMI Quebec, PQ; CKRD Red Deer, AB; CHSJ Saint John, NB; CJIC Sault St. Marie, Ont; CKNC Sudbury, Ont; CJFB Swift Current, SK; CFTK Terrace, BC:CKPR Thunder Bay, Ont; CFCL Timmins, Ont.
Stations: Pol: H.

English Network	Call	chA	kW
St. John's, Nfld.	CBNT	8	196
Charlottetown, P.E.Q.	CBCT	13	178
Halifax, N.S.	CBHT	3	56
Montreal, Que	CBMT	6	100
Ottawa, Ont.	CBOT	4	100
Toronto, Ont.	CBLT	5	84
Winnipeg, Man.	CBWT	6	100
Regina, Sask.	CBKT	9	140
Edmonton, Alta.	CBXT	5	318
Vancouver, B.C.	CBUT	2	50
Yellowknife, N.W.T.	CFYK	8	2.4

CBC English TV is also distributed through 200 rebroadcasters

Addresses:
P.O. Box 12010 Stn. "A", St. Johns, Nfld. A1B 3T8.
P.O. Box 2230, Charlottetown, P.E.I. C1A 8B9.
P.O. Box 3000, Halifax, N.S. B3J 3E9.
P.O. Box 6000, Montreal, Que. H3C 3A8.
P.O. Box 3220, Stn. "C", Ottawa, Ont. K1Y 1E4.
P.O. Box 500, Stn. "A", Toronto, Ont. M5W 1E6.
P.O. Box 160, Winnipeg, Man. R3C 2H1.
2440 Broad Str, Regina, Sask. S4P 4AI.

P.O. Box 555, Edmonton, Alta. T5J 2P4.
P.O. Box 4600, Vancouver, B.C. V6B 4A2.
P.O. Box 160, Yellowknife, N.W.T. X1A 2N2.

French Netw.	Call	chA	kW
Moncton, N.B.	CBAFT	11	163
Montreal, Que.	CBFT	2	100
Quebec, Que.	CBVT	11	252
Ottawa, Ont.	CBOFT	9	128
Winnipeg, Man.	CBWFT	3	59.0
Regina, Sask	CBKFT	13	140
Edmonton, Alta.	CBXFT	11	90
Vancouver, B.C.	CBUFT	26	105

CBC French TV is also distributed through 200 rebroadcasters

Addresses:
P.O. Box 950, Moncton, N.B. E1C 8N8.
P.O. Box 10400, St-Foy, Que G1V 2X2.
Others: See English Netw.

Private TV Networks

ATLANTIC TELEVISION LTD. (ATV)
📧 P.O. Box 1653, Halifax, Nova Scotia B3J 2Z4. ☎ +1 (902)453-4000. 📠 +1 (902)454-3302. **E-Mail:** ASN@asn.ca **WWW:** http://www.atv.ca

CITY-TV
📧 299 Queen Str. W. Toronto, Ontario, M5V 2Z5. ☎ +1 (416) 591-5757. 📠 +1 (416) 591-7791. ③ 06218283. **WWW:** http://www.bravo.ca/citytv.html
L.P: GM: Dennis Fitzgerald.
Stations: Toronto ch57 310kW (ERP), Woodstock ch31 929kW (ERP). Ottawa ch65.
D.Prgr: 24h.

CANWEST GLOBAL SYSTEM
📧 81 Barber Greene Rd, Don Mills, ON, M3C 2A2. ☎ +1 (416) 446-5311. 📠 +1 (416) 446-5490.
Stations: CKVU Vancouver, BC; CKKX Calgary, AB; CISA Lethbridge, AB; CFRE Regina, SK; CFSK Saskatoon, SK; CKND Winnipeg, MB; CIII Toronto, Ont; MITV Saint John, NB; MITV Halifax/Dartmouth, NS.

CTV TELEVISION NETWORK LIMITED
Head Office: 42 Charles Str. E., Toronto, Ontario M4Y 1T5.
☎ +1 (416) 928-6000. ③ 06-22080. 📠 +1 (416) 928-0907.
L.P: Pres. & C.E.O: Murray H. Chercover; Chief Financial Officer and Treasurer: Duncan Morrison; VP Network Relations: Marge Anthony; VP Gov't Relations & Corporate Planning: John T. Coleman; VP Operations: Joseph A. Colson; VP Sports: Johnny Esaw; VP News, Features & Information Programming: Tim Kotcheff; VP Finance: Peter O'Neill; VP Sales: Peter Sisam; VP Programming: Philip Wedge; VP Entertainment Programming: Arthur Weinthal.
Affiliated Stations: NTV St. John's, Nfld; CJCB Sydney, N.S; CJCH Halifax, N.S.; CKCW Moncton, N.B.; CKNY North Bay, Ont.; CITO Timmins, Ont.; CJOH Ottawa, Ont.; CFTO Toronto, Ont.; MCTV Sudbury, Ont.; CKCO Kitchener, Ont.; CFCF Montreal, Que.; CKY Winnipeg, Man.; CKTV Regina, Sask.; CFQC Saskatoon, Sask.; CFRN Edmonton; Alta.; CFCN Calgary, Alta.; BCTV Vancouver, B.C.; CHEK Victoria, B.C.; CIPA Prince Albert, Sask.; CHBX Sault Ste. Marie, Ont.;

NORTH AMERICA   Canada — USA

CHFD Thunder Bay, Ont.; CICC Yorkton, Sask.; CITL Lloydminster, Alta.; CJBN Kenora, Ont.; CHRO Pembroke, Ont.

LE RESEAU DE TELEVISION (TVA)
◳ 2600 boul. de Maisonneuve, Montreal, PQ, H2L 4P2. ☎ +1 (514) 526-9251.
Stations (ch no in brackets): CFCM Quebec (4), CJPM Chicoutimi (6), CHLT Sherbrooke (7), CHEM Trois-Rivières (8), CIMT Rivière d. Loup (9), CFTM Montreal (10), CFER Rimouski (11), CFEM Rouyn-Noranda (13), CHOT Hull/Ottawa (40), CHAU-TV Carleton (5).

TÉLÉVISION QUATRE SAISONS
◳ 405 Oglivy Ave., Montreal, PQH3N 1M4. ☎ +1 (514) 495-6884. 🗎 +1 (514) 495-6231.
Stations: CFJP Montreal (35); CFGS Hull (49); CFRS Jonquière (4); CFAP Quebec (2); CFPC Rimouski (18); CFTF Rivière-du-Loup (29); CFKS Sherbrooke (30); CFKM Trois Rivières (16); CFVS Val d'Or (20/25).

Public TV Networks

SOCIETE DE RADIO-TELEVISION DU QUEBEC
◳ 800 rue Fullum, Montreal, Quebec H2K 3L7. ☎ +1 (514) 521-2424. ⌚ 05-25808.
L.P: Dir. Gen: Ms. Françoise Bertrand. Head of Prgrs: Pierre Roy.
Affiliated Stations: CKRS Jonquière; CKRN Rouyn-Noranda; CKRT Rivière-du-Loup; CKSM Sherbrooke; CKTM Troit-Rivières.

TVO (English) & TFO (French)
Ontario Educational Communications Authority.
◳ Box 200, Station Q, Toronto, Ont. M4T 2T1.
☎ +1 (416) 484-2600. 🗎 +1 (416) 484-7771. ⌚ 06-23547. **E-mail:** online@tvo.org. **WWW:** http://www.tvo.org
L.P: Chmn. & Chief Exec. Officer: B. Ostry. Chief Operating Officer: P. Bowers. Man. Dir. French Prgr. Sces: Jacques Bensimon. Man. Dir. English Prgr. Sces: Don Duprey. MD. Ext. Rel: Bill Roberts/Judith Tobin.
Station: CICA-TV chA19 1080/108kW + low power relays.

GREENLAND

TV-sets: 21.000 — **Colour:** PAL — **System:** B.

KALAALLIT NUNAATA RADIOA (Gov.)
◳ Kalaallit Nunaata Radioa TV, P.O. Box 1007, DK-3900 Nuuk/Godthåb. ☎ +299 25333. 🗎 +299 25042.
WWW: http://www.knr.gl
L.P: Head of Prgr's: H.P. Møller Andersen. Head of Production: Niels Pavia Lynge. Head of News: Maliinannguaq M. Mølgaard.
Stations: The KNR TV prgr's are aired locally using band 3 channels with 1 to 20 watts tr. power.
D.Prgr: 2000-2100, 2225-0230. **N: Greenlandic:** 2230-2255 MF. **Danish:** 2300-2330 (Wed 2400): rebroadcasts from the Danish TV aired 3 days after the actual broadc. date in Denmark. Prgrs mainly originate from Radio Denmark TV, plus 10 hours a week from the Danish TV2 + local commercials.
Text-TV: 24 hrs.
N.B: Local TV organizations mainly distribute via closed networks, and include some satellite relays.

AFRTS (US Air Force)
◳ OLA Det 1, AFABS, APO New York, NY 09121-5000—
Thule Air Base, Det 1, AFABS, APO New York, NY 09023-5000.
Stations: Søndre Strømfjord chA8 0.1kW H; Thule chA8, chA13 0.1kW (using NTSC colour).
D.Prgr: 1330-0630 (Sat 1300-0700).

ST. PIERRE ET MIQUELON

TV-sets: 2.200 — **Colour:** SECAM — **System:** K.

SOCIETE NATIONALE DE RADIO TÉLÉVISION FRANÇAISE D'OUTRE MER (RFO)
◳ BP 4227, F-97500 St. Pierre et Miquelon. ☎ +508 411111. 🗎 +508 412219 — **L.P:** Dir: Joseph Eden.
Stations (Pol H)
1st Prgr: St. Pierre chK4 0.1kW, chK8 0.5kW, chK39 5W — Miquelon chK6 0.1kW.
2nd Prgr: St. Pierre chK31 0.5kW, chK55 0.05kW — Miquelon chK56 0.2kW.
D.Prgr: 1st Prgr: 1800 (SS 1500)-0400. **2nd Prgr:** 2115-0330.

USA

TV-sets: 215m — **Colour:** NTSC — **System:** M.

Networks providing programming to local st's nationwide:

ABC TELEVISION DIVISION (Comm.)
◳ 77 West 66th Street, New York NY 10023. ☎ +1 (212) 456 6400. 🗎 +1 (212) 456 2795. ⌚ 422003.
E-mail: abcaudr@ccabc.com (Audience Information Section)
WWW: http://www.abctelevision.com
Owned stations: WABC-TV/New York, WLS-TV/Chicago, KGO-TV/San Francisco, KABC-TV/Los Angeles, KTRK-TV/Houston, KFSN-TV/Fresno, CA, WPVI-TV/Philadelphia, WTVD-TV/Durham, North Carolina. **Affiliates:** approx. 220.

AMERICA ONE TELEVISION (Comm.)
◳ 100 E. Royal Ln, Irving, TX 75039. ☎ +1 (214) 868 1000. 🗎 +1 (214) 868 1662. **E-mail:** a1tv@airmail.net
WWW: http://www.americaone.com

CBS, Inc. (Comm.)
◳ 51 West 52nd Str, New York, NY 10019. ☎ +1 (212) 975-4321. 🗎 +1 (212) 975 7452. **E-mail:** marketing@cbs.com
WWW: http://www.cbs.com
Owned Stations: WCBS-TV, New York; WBBM-TV, Chicago; KCBS-TV, Los Angeles; WCIX-TV Miami; WCCO TV, Minneapolis; WBAY TV, Green Bay. **Affiliated Stations:** 200.

FOX TELEVISION NETWORK (Comm.)
◳ P.O. Box 900, Beverly Hills CA 90213. ☎ +1 (310) 277 2211. **E-mail:** foxnet@delphi.com
WWW: http://www.foxnetwork.com

NATIONAL BROADCASTING COMPANY (Comm.)
◳ 30 Rockefeller Plaza, New York, NY 10112. ☎ +1 (212) 664-2074. 🗎 +1 (212) 664 7541. ⌚ 662131.

USA — Barbados

WWW: http://www.nbc.com
NBC Television Stations: WNBC-TV, New York; WRC-TV, Washington; WMAQ-TV, Chicago; WKYC-TV, Cleveland; KNBC-TV, Los Angeles; WCAU-TV, Philadelphia; KCNC-TV, Denver; WTVJ-TV, Miami. **Affiliated stations:** 208.

PUBLIC BROADCASTING SERVICE (Non-comm, Educ.)
☞ 1320 Braddock Place, Alexandria, VA 22314-1698. ☏ +1 (703) 739-5000. 🖷 +1 (703) 739-0775. **E-mail:** viewer@pbs.org (viewer mail). **WWW:** http://www.pbs.org
Member Stations: 314.

UNITED PARAMOUNT NETWORK (UPN) (Comm.)
☞ 5555 Melrose Avenue, Marathon 1200, Los Angeles, CA 90038. **WWW:** http://www.upn.com

UNIVISION (Spanish Language Network) (Comm.)
☞ 605 Third Ave, New York, NY 10158-0180. ☏ +1 (212) 455 5200. **WWW:** http://www.univision.net

WARNER BROTHERS TELEVISION NETWORK (WB)
☞ 4000 Warner Blvd, Bldg. #34R, Burbank, CA 91522. ☏ +1 (818) 954 6479. **E-mail:** wbnetwork@aol.com **WWW:** http://www.tv.warnerbros.com

Local stations: There over 1,500 TV stations across the U.S. The major cities have st's affiliated to each of the above-mentioned networks, and there may be additional local st's which generally broadcast movies, and re-runs of older programs. For details of satellite-delivered services, refer to the satellite section.

ARMED FORCES RADIO & TV SERVICE BROADCAST CENTER (AFRTS-BC) (Mil.)
☞ 1363 Z Street, Bldg 2730, March ARB, CA 92518. 🖷 +1 (909) 413-2234.
WWW: http://www.dodmedia.osd.mil/afrts_bc/ahome.htm
L.P: Dir. Programming: Robert W. Matheson. Dir. Eng. & Op's: Bruce V. Ziemienski.
The AFRTS Broadcast Center delivers programming to the AFRTS overseas audience through land-based outlets or Navy ships at sea. AFRTS has more than 450 land-based outlets located in over 140 countries and U.S. territories (including remote areas of Alaska). TV stations with over-the-air transmissions may be found at the end of listings for the respective countries.

TV MARTI (Gov.)
TV Marti is the television broadcasting service of the United States Information Agency, Office of Cuban Broadcasting.
☞ Washington, D.C. 20547, USA. ☏ +1 (202) 501-7210. 🖷 +1 (202) 208-7808.
L.P: Ag. Dir. Office of Cuba Broadcasting: Dr. Rolando Bonachea. Dir. of Tech. Op's (OCB): Michael Pallone.
Stations: VHF ch13 (0830-1100).

CENTRAL AMERICA & THE CARIBBEAN

ANTIGUA & BARBUDA

TV-sets: 28.000 — **Colour:** NTSC — **System:** M.

ANTIGUA & BARBUDA BROADCASTING SERVICE (Gov.)
☞ P. O. Box 590, St. John's. ☏ +1 (268) 4620010. 🖷 +1 (268) 4621622.
L.P: Prgr. Mgr. (TV): James Tanny Rose. CE: Denis Leandro. Film Buyer: J.T. Rose.
Stations: ABS-TV chA10 50/20kW H.
Relay: Montserrat ch13 10kW.
D.Prgr: 1900-0700.

CTV ENTERTAINMENT SYSTEMS (Comm.)
☞ 25 Long Str, St. Johns. ☏ + 1 (809) 4620346. 🖷 + 1 (809) 4624211 — **L.P:** Prgr. Dir: J. Cox.

ARUBA

TV-sets: 19.000 — **Colour:** NTSC — **System:** M.

TELE ARUBA (Comm.)
☞ P.O. Box 392, Oranjestad. ☏ + 297 (8) 47302. 🖷 + 297 (8) 41683. ① 5195 TELAR.
L.P: Gen. Mgr: Mrs. Jane Lampkin. CE: Miguel Roga.
Station: chA13 3/0.6kW H.
D.Prgr: 2030-0400.

BAHAMAS

TV-sets: 50.000 — **Colour:** NTSC — **System:** M.

BAHAMAS TELEVISION (owned and operated by the Broadc. Corp. of the Bahamas)
☞ P.O. Box N-1347, Nassau. ☏ +1 (242) 322 4623. 🖷 +1 (242) 322 3924 — **L.P:** Ag. Prgr. Dir (TV): Ms. R. Simmons.
Station: ZNS ch13 50kW (ERP).
D.Prgr: 2200 (Sat 2100, Sun 2000)-0400.

BARBADOS

TV-sets: 65.000 — **Colour:** NTSC — **System:** M.

CARIBBEAN BROADCASTING CORP. (Gov, Comm.)
☞ P.O. Box 900, Bridgetown. ☏ +1 (246) 429 2041. 🖷 +1 (246) 429 4795.
L.P: Prgr. Mgr (TV): Mrs. O. Cumberbatch.
Station: CBC-TV ch8 60/30kW, ch 9, ch14, ch18,

CENTRAL AMERICA & THE CARIBBEAN — Barbados — Dominican Republic

ch22, ch26.
D.Prgr: Mon-Fri 0940-1400 and 2000-0330, Sat 1200-0430, Sun 1200-0330.
Cable TV: A Gov. Sce, STV, provides two additional subscription ch's.
N.B: Wednesday and Friday closing may be after 0400 as movies are shown on these days. During close times classified commercials and coming attractions are broadcast continuously.

BELIZE

TV-sets: 23.457 — **Colour:** NTSC — **System:** M.

TROPICAL VISION (Comm.)
48 Albert Cattouse Bldg, Regent Str, P.O. Box 89, Belize City. ☎ +501 77246/7/8 — 📠 +501 (2) 75040 — **Station:** ch 7.

BAYMEN BROADCASTING NETWORK (Comm.)
27 Baymen Ave., Belize City. ☎ + 501 (2) 44400. 📠 +501 (2) 31242 — **Station:** ch 9.

COSTA RICA

TV-sets: 340.000 — **Colour:** NTSC — **System:** M.

Stations	chA	kW	Stations	chA	kW
2)	4	10	2)	9	10
1)	6	300	4)	13	2.5
3)	7	3605)	15	0.1	

Addresses & other information:
1) Corporación Costaricense de Televisión (Comm.), P.O. Box 2860, 1000 San José. ☎ +506 312222 — Dir. Gen: M. Sotela B. — 1530-0700.
2) Multivision, Apt 4666, 1000 San José. ☎ +506 334444. ① 3043 Televi. 📠 +506 211734 — Dir: Arnaldo Vargas V. — 1600-0600.
3) Televisora de Costa Rica, Apt 3876, San José. ☎ +506 322222. ① 2220 Teletica — 1730-0600 (+ 9 repeaters).
4) Rede Nacional de Televísion, Apt 7-1980, San José. ☎ +506 200071 — Dir. Gen: Dr. Ch. Zelaya Goodman — 2200-0400 (+ 2 repeaters).
5) Universidad de Costa Rica, San Pedro, Montes de Doa, San José — Dir. Gen: Dr. Sergio Guevara Fallas — 1600-2200. ☎ +506 340463. 📠 +506 256950.

CUBA

TV-sets: 2.500.000 — **Colour:** NTSC — **System:** M.

INSTITUTO CUBANO DE RADIODIFUSION (Gov.)
Television Nacional, Calle M No. 313, Vedado, La Habana.
Estudios en Pinar del Rio, Ciudad de la Habana, Santa Clara, Nueva Gerona, Camagüey, Holgüí, Santiago de Cuba & Guantánamo.

TELE REBELDE
Mazón No. 52, Vedado, La Habana. ☎ +537 (32) 3369. ① 511661.
L.P: VP: Gary Gonzalez.
Studios en Santiago de Cuba, Holguín & La Habana.

Stations: Pol H (no calls used).

ch	Location	kW	ch	Location	kW
2	Babiney, P.Río	8	11	Cienfuegos	8
2	La Habana	132	11	Salón, P.Río	32
4	Camagüey	53	12	Prov. Granmá	54
5	La Palma	8	12	Chivirico	56
5	Stgo. de Cuba	77.5	12	Ciego de Avila	56
5	Santa Clara	60	13	Pinar del Río	63
8	Holgüín	27.5	13	Guantánamo	79
9	Baracoa	12	13	Prov. Matanzas	120
10	I. Juventud	100			

+ 12 st's below 4kW.
D. Prgr: 2257-0500.

CUBAVISION (Gov.)
Calle M No. 313, Vedado, La Habana.

ch	Location	kW	ch	Location	kW
2	Stgo. de Cuba	32	7	Salón, P. Río	32
3	La Palma	8	8	Ciego de Avila	51
3	Holguín	27.5	8	I. Juventud	100
3	Santa Clara	60	9	Pinar del Río	63
6	Babiney, P.Río	6	9	Prov. Matanzas	120
6	Camagüey	56	10	Prov. Granma	54
6	La Habana	129	10	Chivirico	56
7	Baracoa	12	11	Guantánamo	79
7	Cienfuegos	8			

+ 12 st's below 4kW.
D.Prgr: 2227-0500.

AFRTS (US Navy)
US Naval Base, P.O. Box 22, FPO New York, NY 09406.
Station: Guantanamo Bay chA8 0.35kW.

DOMINICA

TV-sets: 5.200 — **Colour:** NTSC — **System:** M.

MARPIN-TV (Comm.)
P.O. Box 382, Roseau. ☎ +1 (767) 4484107. 📠 +1 (767) 4482965 — **L.P:** Prgr. Mgr: Ron Abraham.

DOMINICAN REPUBLIC

TV-sets: 728.000 — **Colour:** NTSC — **System:** M.

VHF Stations	chA	VHF Stations	chA
1) Sto. Domingo	2	5) Cibao/Costa Norte	11
2) Cibao/Costa Norte	2	3) Barahona	12
3) Sto. Domingo	4	3) Dajabon	12
3) Cibao/Costa Norte	4	3) Descubierta	12
3) Alto Bandera	5	3) El Cercado	12
4) Sto. Domingo	6	3) Enriquillo	12
4) Cibao/Costa Norte	6	3) La Romana	12
5) Sto. Domingo	7	3) Pedernales	12
6) Santiago	7	3) Puerto Plata	12
2) Sto. Domingo	9	8) Sto. Domingo	13
2) La Romana	9	1) Cibao/Costa Norte	13
7) Cibao/Costa Norte	9	**UHF Stations**	**chA**
7) Sto. Domingo	11	5) La Naviza	70

N.B: Because of contradictory information, no transmitter power figures are given.

Addresses and other information
1) Teleantillas, Autopista Duarte Km.7, Sto. Domingo.

Dominican Republic — Haiti

☎ +1 (809) 567-7751.
2) Color Vision, Corporación Dominicana de Radio & TV, Emilio Morel esq. Lulu Perez, Ens. La Fé, Sto. Domingo. ☎ +1 (809) 556-5876.
3) RadioTelevisión Dominicana (Gov.), Dr. Tejeda Florentino 8, Sto. Domingo. ☎ +1 (809) 689-2120 — Dir: George Rodriguez.
4) Canal 6, Circuito Independencia C-A, Mariano Cestero esq. Enrique Henriquez, Sto. Domingo. ☎ +1 (809) 689-8151.
5) Rahintel, Av. Indepencia, Centro de los Heroes, Sto. Domingo. ☎ +1 (809) 532-2531.
6) Canal 7 Cibao, Edificio Banco Universal, Calle del Sol, Santiago. ☎ +1 (809) 583-0421.
7) Telesistema, 27 de Febrero, Sto. Domingo. ☎ +1 (809) 567-1251.
8) TV 13, Av. Pasteur esq., Santiago, Sto. Domingo. ☎ +1 (809) 687-9161.

EL SALVADOR

TV-sets: 500.700 — **Colour:** NTSC — **System:** M.

chA	kW)		chA	kW		chA	kW
1)	2	100/25	3)	6	150/30	4) 10	109/5
2)	4	75/37.5	4)	8	109/5		

Addresses & other information:
1) Canal Dos SA(Comm.), Ap. Postal 720, San Salvador. ☎ +503 236744 — 1730-0530.
 2) Canal Cuatro(Comm.), Ap. Postal 720, San Salvador. ☎ +503 244633 — 2100(Sat/Sun 1500)-0600.
3) Canal Seis(Comm.), Km. 6, Carretera Panamericana a Santa Tecla, San Salvador. ☎ +503 235122 — 2300 (Sat/Sun 1700)-0600.
4) Television Cultural Educativa, Ap. Postal 4, Santa Tecla. ☎ +503 280499 — Dir. Gen: Maura Echaverria — 1300-0500.

GRENADA

TV-sets: 15.000 — **Colour:** NTSC — **System:** M.

GRENADA BROADCASTING CORPORATION (GBC).
🖃 Morne Jaloux, P.O. Box 535, St. George's, Grenada. ☎ +1 (473) 444 5521/22 (PBX). 📠 +1 (473) 444 5054.
L.P: MD: Cecil Benjamin. Sen. Eng: John Phillip
Station: ch7 (4kW), ch11 (5kW).

GUADELOUPE

TV-sets: 150.000 — **Colour:** SECAM — **System:** K.

RFO-GUADELOUPE
🖃 B.P. 402, F-97163 Point-à-Pitre Cedex. ☎ +590 939696. 📠 +590 939682. ⓒ 919064
L.P: Dir: R.Surjus. Editor-in-Chief: Philippe Goudé. PD: L.Francil. Head Communications Dept: Sonia Gémieux.
Stations: Basse Terre chK5 2kW + low power repeaters.
D.Prgr: 1900-0230.

SAINT MARTIN
An RFO st. is operating on chK7 0.1kW relaying RFO Guadeloupe.

commercial stations

ARCHIPEL 4 (Priv., Comm.)
🖃 Résidence Les Palmiers, Gabarre 2, 97110 Pointe a Pitre. ☎ +590 8363 50
Stations: Morne a Louis ch53 1.3kW

CANAL ANTILLES (Priv., Comm.)
🖃 2 lot. Les Jardins de Houelbourg, 97122 Baie Mahault. ☎ +590 26 8179
Stations: Morne a Louis ch 58 1.3kW, Basse-Terre ch42 60kW + 5 low power repeaters under 1kW.
N.B: Canal Antilles is a subscription sce and the signal is encrypted except: 1120-1130, 1630-1730, 1820-0010.
D. Prgr: W 1100-0500 Sun 24h

TCI GUADELOUPE (Priv., Comm.)
🖃 Montauban, 97190 Gosier.
Station: Basse-Terre ch32 60kW

GUATEMALA

TV-sets: 475.000 — **Colour:** NTSC — **System:** M.

Stations: 2)Gov, others comm.
Call	chA kW/Pol	Call	chA kW/Pol
1)TGV-TV	3 240H	1) TGV-TV	10 25H
2)TGCE-TV	5 ?	4) TGMO-TV	11 316/63
3)TGVG-TV	7 108/22H	5) TGSS-TV	13 25/5

Addresses & other information:
1) Radio-Television Guatemala, Apt. 1367, Guatemala. ☎ +502 (2) 922491. **Cable:** Teletenango — Pres: Lic. M. Kestler F. Dir. Tec: E. Sandoval — 1200-0600 (+ 2 repeaters).
2) Television Cultural Educativa, 4a Calle 18-38, Zona 1, Guatemala. ☎ +502 (2) 531913 — 2200-0500.
3) Televisiete, Apt. Postal 1242, Guatemala. ☎ +502 (2) 62216. **Cable:** TV siete — 1800-0600(+ 3 repeaters).
4) Teleonce, Ca. 20, 5-02, Zona 10, Guatemala. ☎ +502 (2) 682165 — 1800-0600 (+ 2 repeaters).
5) Trecevision, 3a Calle 10-70, Zona 10, Guatemala. ☎ +502 (2) 63266. ⓒ 6070 Trece GU — 1800-0600 (+ 12 repeaters).

HAITI

TV-sets: 25.000 — **Colour:** NTSC — **System:** M.

TÉLÉVISION NATIONALE D'HAITI (Gov., Cult.)
🖃 P.O. Box 13400 Delmas 33, Port-au-Prince. ☎ +509 (1) 63324/64049/62202.
L.P: Dir: Mme. Jacqueline André.
Stations: Port-au-Prince chA8 0.3kW, chA10 5kW.
D.Prgr: Mon-Fri 2100-0400 in French & Creole.
F.Pl: chA12 Cap. Haïtien.

TÉLÉ HAITI S.A. (Comm.)
🖃 B.P. 1126, Port-au-Prince. ☎ +509 (1) 23000.
L.P: Dir: Walter Bussenius.
Operates cable TV 24h in Port-au-Prince area.

CENTRAL AMERICA & THE CARIBBEAN — Honduras — Mexico

HONDURAS (Rep.)

TV-sets: 160.000 — **Colour:** NTSC — **System:** M.

Stations (all comm.)

Call	chA	kW	Call	chA	kW
1) HRJS-TV	2	25	2) HRTS-TV	7	10/2
2) HRCV-TV	3	2/0.4	4) HRLP-TV	7	2/0.4
4) HRLP-TV	4	2/0.4	1) HRJS-TV	9	25
3) HRTG-TV	5	10/2	3) HRTG-TV	9	5/1
3) HRTG-TV	5	5/1	5) HRNQ-TV	13	2/0.4
6) HRGJ-TV	6	10/2			

Addresses & other information:
1) Corp. Centroamericana de Comunicaciones, S.A. de C.V, Apt. Postal 120, San Pedro Sula — Pres: J.J. Sikaffy. Exec. Vice Pres: F.J. Sikaffy. Dir. Tec: R. Beurket & A. Pinto — 1530-0400 (+ 2 relays).
2) Telesistema Hondureño, Apt. Postal 642, Tegucigalpa — 1530-0400 (+ 6 relays).
3) Compañia Televisora Hondureña, Apt. Postal 734, Tegucigalpa — 1230-0400 (+ 9 relays). ☎ +504 (32) 7835. 🗎 +504 (32) 0097.
4) Compañia Centroamericana de TV, Apt. Postal 68, Tegucigalpa — 1430-0400.
5) Cruceña de Televísíon, Casilla 3424, Tegucigalpa — Pres: Lic. Ivo Kuljis F. Dir. Gen: Lic. Walter Gasser Diaz C. — 1700-0800.
6) Compañia Broadcasting, Apt. Postal 882, Barrio Rio Piedras — 1130-0500.

JAMAICA

TV-sets: 500.000 — **Colour:** NTSC — **System:** M.

JAMAICA BROADCASTING CORPORATION (Comm.)
✉ Box 100, Kingston 10. ☎ +1 (876) 926 5620/9. 🗎 +1 (876) 929 1029. **Cable:** JARAD Jamaica. ⓓ 2218 BROADCORP JA.
L.P: DG: Claude Robinson. Dir. of Television: Keith Campbell. Mgr. Eng. Sces (TV): Norman Mighty.
Station: HWT ch 7, 8, 9, 10, 11, 12, 13.
D.Prgr: 2200(Sat 1930, Sun 1900)-0500.

MARTINIQUE

TV-sets: 65.000 — **Colour:** SECAM — **System:** K.

SOCIÉTÉ NATIONAL DE RADIO-TÉLÉVISION D'OUTRE MER (RFO)
✉ B.P. 662, F-97263 Fort de France Cedex. ☎ +595 595200.
L.P: Dir: Fred Jouhoud. CE: Jean Claude Arrivé.
Stations: Fort de France chK4 1kW (+ 9 relay st's). Pol: V.
D. Prgr: 2200 (Sun 2000)-0300.

commercial stations

ATV ANTILLES TELEVISION (Priv., Comm.)
✉ 28 rue Arawaks, 97200 Fort de France. ☎ +596 75 4444. 🗎 +596 75 5565
Stations: La Trinite ch39 7kW, Fort de France ch44 8 kW, La Morne Rouge ch52 1.4 kW, Riviere Pilote ch34 0.19kW

CANAL ANTILLES (Priv., Comm.)
✉ Centre Commercial La Galléria, 97232 Le Lamentin. ☎ +596 50 5787
Stations: La Trinite ch25 7 kW, Le Morne Rouge ch46 1.4kW, Port de France ch29 8kW, Saint Pierre ch34 0.6kW, Riviere Pilote ch50 0.19 kW.

TCI MARTINIQUE-TELE CARAIBES INTERNATIONAL (Priv., Comm.)
✉ Immeuble RCI/TCI-Zone industrielle-97232 Le Lamentin.
☎ +596 510606. 🗎 +596 518562.
D. Prgr: 11:00 a.m. - NOON-(rel. T.F.I. & Euronews)

MEXICO

TV-sets: 15m — **Colour:** NTSC. — **System:** M.

INSTITUTO MEXICANO DE TELEVISION (IMEVISION) (Gov. Agency)
✉ Ave. Periférico Sur 4121, Colonia Fuentes del Pedregal, 14141 México, DF. ☎ +52 (5) 5685684, 5681313.
L.P: Dir.Gral: Lic. Jose Antono Alvarez Lima.
STATIONS: México, D.F; IMT(7), DF(13), IMT(22) — Chihuahua: CH(2) — Monterrey: FN(8).
Red Nacional 7: Aguascalientes: LGA(10), Cd. Obregón: BK(10), Culiacán: BL(13), Guadalajara: SFJ(11), Hermosillo: TH(10), Irapuato: CCG(7), Jalapa: IC(13), Matamoros: OR(14), Mérida: DH(11), México DF: IMT(7), Nogales: FA(5), Oaxaca: PSO(10), Tampico: WT(12) + 51 others.
Red Nacional 13: Aguascalientes: JCM(4), Cd. Obregón: CSO (6), Cuernavaca: CUR(13), Guadalajara: JAL(13), Hermosillo: HSS(5), Irapuato: MAS(12), Mérida: MEY(2), Mexicali: AQ(5), México DF: DF(13), Monterrey: WX(4), Oaxaca: IG(12), Puebla: TEM(12), Saltillo: SAO(11), S.L. Potosi: CLP(6), Tuxtla Gutiérrez: AO(4) + 39 others. ch7 & 13 are national, 11 & 22 regional.
N.B: Early 1993, the Mexican Radio and Television Corporation ceased to exist and **Azteca Television** was established.

TELEVISA, S.A. (Priv, Comm.)
✉ Av. Chapultepec 28, 06724 México, D.F — ☎ +52 (5) 709-3333. ⓓ XEWTM 77-3154 — 🗎 +52 (5) 709-3021 — Chmn. of the Adm. Council: R. O'Farrill Jr. Pres. TV: Emilio Azcárraga.
STATIONS: XEW-TV(2), XHTV(4), XHGC(5), XEQ(9).
Cadena Canal 2: Acapulco: AP(2), Cd. Juárez: EPM(2), Cd. Obregón: BS(4), Córdoba/Orizaba: AH(8), Culiacán: BT(7), Chihuahua: FI(5), Guadalajara: EWO(2), Hermosillo: XEWH(6), Jalapa: AH(8), León: L(10), Los Mochis: BS(4), Mazatlán: OW(12), Matamoros: AB(7), Mérida: TP(9), México DF: XEW(2), Mexicali: BM(14), Monterrey: XHX(10), Morelia: KW(10), Nuevo Laredo: BR(11), Oaxaca: BN(7), Saltillo: AE(5), Tampico: XHD(2), Torreón: O(11), Tuxtla Gutiérrez: TX(8), Villahermosa: LL(13) + 28 others.
Cadena Canal 4: Córdoba/Orizaba: AI(10), Jalapa: AI(10), México DF: TV(4), Puebla: XEX-TV(7) + 2 others.
Cadena Canal 5: Aguascalientes: AG(13), Guadalajara: GA(9), Hermosillo: TH(10), Jalapa: AJ(5), México DF: GC(5), Monterrey: XET-TV(6), Saltillo: AD(7) + 3 others.
Cadena Canal 9: Mexico City. Ch2 & 5 are national, 4 & 9 regional. Part-time affiliates (after 2200) noted below.

347

Mexico — Puerto Rico

STATIONS originating local programming.
Call: XH— (unless otherwise stated).

No.	Call	ch	kW	No.	Call	ch	kW
1)	IA	2	30	**)	IMT	7	?
2)	CH	2	6	24)	GEM	7	1
*)	XEW-TV	2	100	25)	MZ	7	72
3)	AP	2	1.2	26)	GO	7	50
4)	KG	2	0.5	27)	FN	8	17.5
5)	FB	2	100	28)	US	8	1.6
6)	I	2	86.5	29)	TVL	9	250
7)	RIO	2	100	30)	FW	9	50
8)	FE	2	60	*)	Q	9	100
9)	FM	2	14.2	31)	K	10	8
10)	BC	3	100	32)	A	10	0.85
11)	PN	3	5.9	33)	L	10	150
12)	JMA	3	13	34)	IPN	11	300
13)	P	3	30	35)	WT	12	66
14)	Q	3	100	36)	MH	12	1
15)	Y	3	100	37)	ND	12	1
16)	LN	4	4	38)	AW	12	100
*)	TV	4	100	39)	CG	12	7
17)	AL	4	2	40)	AK	12	5
18)	G	4	88	41)	DE	13	50
19)	CC	5	3.5	42)	UT	13	2
20)	XEJ-TV	5	67.4	43)	ST	13	54
*)	GC	5	100	**)	IMT	22	?
21)	ETV	6	100	44)	S	23	18
22)	EDK	6	84	45)	IJ	44	330
23)	EWH	6	1.6				

*) see Televisa above — **)see Imevision above.

MONTSERRAT

TV-sets: 5.000 — **Colour:** NTSC — **System:** M.

ANTILLES TV LIMITED (Comm.)
✉ P.O. Box 342, Plymouth, Montserrat. ☎ +1 (664) 491 2226. 📠 +1 (664) 491 4511.
L.P: Gen. Mgr: K. Osborne. Dir. Tec: Z.A. Joseph.
Station: Chance Pic chA7 (48kW towards Dominica, 3kW towards Antigua & St. Kitts).
D.Prgrs: 1000-1110, 2000-0400 (Fri/Sat 0700).

NETHERLANDS ANTILLES

TV-sets: 35.000 — **Colour:** NTSC — **System:** M.

TELE CURAÇAO (Gov, Comm.)
✉ P.O. Box 415, Curaçao. ☎ +599 (9) 61288. 📠 +599 (9) 614138.
L.P: Gen. Mgr: Norbert Hendrikse. Prgr. Mgr: H. van der Beist. Dir. Tec: J. Rufina.
Station: chA8 20/5kW H/A6 — **D.Prgr:** 2000-0345.
Cable TV: Telecuraçao also provides a cable service relaying various U.S. satellite networks and two Venezuelan channels.

LEEWARD BROADCASTING CORPORATION — TELEVISION
✉ P.O. Box 375, Philipsburg, St. Maarten.
Station: chA7 5kW.
D.Prgr: 2030 (Sun 1800)-0300 (approx); Sat also 1200-1300.
NB: A Gov. st. on chA6 is reported operating from Saba.

NICARAGUA

TV-sets: 210.000 — **Colour:** NTSC — **System:** M.

SISTEMA SANDINISTA DE TELEVISION (Gov.)
✉ Km 31/2 Carretera Sur, Contig o Shell, Las Palmas, Managua. ☎ +505 (2) 660028/660879. 📠 +505 (2) 662411. ⓓ 1226 Sandino.
L.P: Hd. of Sales: Miguel Chivel.
Stations: chA2 25kW, chA6 25kW H, Chg 1 kW.
+ 7 low power repeaters.
D.Prgr: 2030-0600 (approx.)

PANAMA

TV-sets: 204.539 — **Colour:** NTSC — **System:** M.

Location	chA	kW	Location	chA	kW
1) Cerro Azul	2	650	6) Fort Davis	10	5
2) Cerro C. Chilibre	4	32	4) Cerro Oscuro	11	110
3) Cerro Azul	5	20	2) El Valle	12	5
6) Cerro Ancón	8	5	5) Cerro Oscuro	13	10
1) El Valle	9	130			

Addresses and other information
1) Televisora Nacional, Apt. 6-3092, El Dorado, Panamá — Dir. Gen: Lic. Alejandro Ayala V. — 1700-0500 (+6 relays).
2) RPC Television, Apt. 1795, Panamá 1 — 1330-0600 (+8 relays).
3) Panavision, Apt. 6-2605, El Dorado, Panamá (+ 1 relay).
4) Sistema de TV Educativa, Estafeta Universitaria, Universidad de Panamá — Dir. Gen: I. Velasquez de Cortes — Mon-Fri 2200-0400, Sat 1500-2200. **F.Pl:** 1 relay st at Ninguna in 1992.
5) Telemetro, P.O. Box 8-116, Panamá 8 — Dir. Gen: B. Marques — 1730-0530 (+ 1 relay).
6) Armed Forces Television (US Mil), Drawer 919, Bldg. 209, Fort Clayton — 24h.

PUERTO RICO

TV-sets: 830.000 — **Colour:** NTSC — **System:** M.

	Call	ch	kW
1)	WKAQ-TV	2	53.7/10.5
2)	WIPM-TV	3	72.4/7.24
3)	WAPA-TV	4	53.7/8.1
4)	WORA-TV	5	95.5/19.1
5)	WIPR-TV	6	53.7/5.37
6)	WLUZ-TV	7	166/22.4
7)	WSUR-TV	9	57.5/7.24
8)	WLII	11	200/39.8
9)	WOLE-TV	12	316/31.6
9a)	WPRV-TV	13	170/17
9b)	WSJU-TV	18	759
9c)	WKPV	20	100/10
8b)	WJNX	2	?
9d)	WSJN-TV	24	4384/490
12)	WMTY	40	—
10)	WVEO-TV	44	993/198
11)	WATX-TV	54	11.78/6.35
11a)	WUJA	58	55/5.5

1) Telemundo of Puerto Rico Inc., 383 Roosevelt Av., Hato Rey, PR 00918 — GM: Jose Ramos. CE: Jose Medina. 1210-0500 (+ 8 relay sts).

CENTRAL AMERICA & THE CARIBBEAN — Puerto Rico — Virgin Islands

2) Dept. of Education, Box 909, Hato Rey 00919 — 1800-0500 (Sat/Sun 1330-0400).
3) SFN Communications, Inc, GPO Box 2060, San Juan 00938.
4) Telecinco Inc, Box 43, Mayagüez 00708 — St. Mgr. & Film Buyer: E. Bado. Dir. Tec: G.A. Bonet. 1200-0400.
5) Avda. Hostos Hato Rey. Esquina Tous Urbanization Baldrich, 00919 Puerto Rico.
6) Ponce TV Corpo, Isabel Esq Montaner, Ponce 00731 — Pres: L.T. Muniz.
7) La Ramble, Ponce (belongs to 8).
8) American Colonial Broadc. Corp, Box S-4189, San Juan 00905. 1230-0430 — 8b) rel. 8.
9) Western Broadc. Corpo. of Puerto Rico, Box 1200, Mayagüez 00709 — 1245-0500 (partly rel. st. 3).
9a) No. 10 Simon Madeira Rio Piedras, 00929 — GM: Nacha Rivera.
9b) Three Star Telecast Inc, Box 18, Carolina 00628 — Dir. Gen.: Barakat Saleh. 1030-0400.
9c) Multi Media TV, Box 2556, San Juan 00936.
9d) belongs to 9c) 24h (mostly rel. st's 1 and 4). All News.
10) Siglares Iglesia Catolica Inc, Buzon C-339, Quebradillas 00742.
11) Arecibo Video Corp, Arecibo 00612. Pres: F. Velasquez.
11a) Community TV of Cagua, Box 6556, Caguas 00626.
12) Ana G. Mendez Foundation, Box 21345, Rio Piedras 00928 — GM: Gloria Hernandez. CE: Ariel Diaz.

AFRTS (U.S. Mil.)
U.S. Naval Station, Box 3029, FPO Miami 34051.
Stations: Roosevelt Roads chA40 1kW, Ft. Allen ch40 0.05kW, Viques Isl. ch56 0.05kW.

ST. KITTS & NEVIS

TV-sets: 9500 — **Colour:** NTSC — **System:** M.

ZIZ TELEVISION (Gov, Comm.)
P.O. Box 331, Basseterre, St. Kitts. ☎ +1 (809) 465 2621. 🖷 +1 (809) 465-5202.
L.P: GM: Mrs. Claudette Manchester. Ag. Producer TV: Barry Thomas.
Stations: Basseterre chA5 20/5kW H. + 3 repeaters.
D.Prgr: 2000-0430.

ST. LUCIA

TV-sets: 25.000 — **Colour:** NTSC — **System:** M.

HELEN TV (Comm.)
P.O. Box 621, Le Morne Castries. ☎ +1 (758) 4522 693. 🖷 +1 (758) 454 1737. ⓘ 6254 HTSTV.
L.P: MD: Linford Fevrier. CE: Stephenson Anius.
Station: ch's 4 & 11.

CABLEVISION
George Gordon Bld., Bridge Str. Castries, St. Lucia. P.O. Box 111, Castries. ☎ (758) 452 3301. 🖷 +1 (758) 453 2544. ⓘ 6362 — **Station:** ch's 16 on CATV.

ST. VINCENT

TV-sets: 20.000 — **Colour:** NTSC — **System:** M.

ST. VINCENT & THE GRENADINES BROADCASTING CORPORATION Ltd.
P.O. Box 617, Kingstown. ☎ +1 (809)-4561078. 🖷 +1 (809) 4561015.
L.P: Chief Eng: R.P. MacLeish.
Station: ch9. Relay sts. ch's 7, 11, 13 varying between 5-30 W. 1500-0900.
Coverage area: St. Vincent, Grenadines, St. Lucia & Grenada.

TRINIDAD & TOBAGO

TV-sets: 250.000 — **Colour:** NTSC — **System:** M.

TRINIDAD & TOBAGO TELEVISION CO. LTD.
Television House. 11A Maraval Rd., P.O. Box 665, Port of Spain. ☎ +1 (809) 622-4141-4. 🖷 +1 (809) 622-0344. **Cable:** Television Trinidad.
L.P: GM: Grenfell Kissoon. Chief Eng: D. Dhani.
Stations: chA2 12/6kW, chA13 2/0.2kW (ch's 9 & 14 since 1983).
D.Prgr: 0940-0400.

VIRGIN ISLANDS (American)

TV-sets: 31.500 — **Colour:** NTSC — **System:** M.

WBNB-TV
Box 1947, Charlotte Amalie, St. Thomas 00801.
☎ +1 (809) 774-0300. 🖷 +1 (809) 776-3511.
L.P: Sen. Vice Pres: J. Potter. St. Mgr: P. Stull.
Station: WBNB-TV chA10 113/76kW.

CARIBBEAN COMM. CORP.
1 Beltjen Place, St. Thomas, V1 00802. ☎ +809 77621 50. 🖷 +809 774 5029.
L.P: Hd. of Sales: Randolph H. Knight; MD: Andrea L. Martin.

VIRGIN ISLANDS PUBLIC TV-SYSTEM
Box 7879, Charlotte Amalie, St. Thomas 00801.
☎ +1 (809) 774-6255. 🖷 +1 (809) 774 7092.
L.P: TD: Leslie Hayes; Film Buyer: Lori Elskoe.
Station: WTJX-TV chA12 31.6/6.32kW.

VIRGIN ISLANDS (British)

TV-sets: 3.000 — **Colour:** NTSC — **System:** M.

TELEVISION WEST INDIES LTD. (Comm.)
P.O. Box 34, Broadcast Peak, Chawell, Tortola, BVI.
☎ +1 (809) 409 43332
Station: ZBTV (Tortola) chA5 30/3kW.

BVI CABLE TV
P.O. Box 694, Road Town, Tortola. ☎ (809) 495 3205.
L.P: MD: Todd Klindworth.

SOUTH AMERICA

ARGENTINA

TV-sets: 9.800.000 — **Colour:** PAL — **System:** N.

ASOCIACIÓN DE TELERADIODIFUSORAS ARGENTINAS (ATA)

✎ Av. Córdoba 323, 6to., 1054 Buenos Aires. ☎ +54 (1) 312-4208/4219/4533. 🖷 +54 (1) 312-4208. ① 17253 ATA AR. **Cable:** Teleradio Baires — **L.P:** Pres: Alejandro Enrique Massot. CE: Enrique Parodi.

Call	chA	kW (ERP)	City of location
1) LS86	2	100/10	La Plata
2) LT83	3	90/20	Rosario
3) LU89	3	30/3	Santa Rosa
4) LT84	5	36/6	Rosario
6) LV84	6	1.5/0.15	San Rafael
5) LU93	6	1.5/0.15	San Carlos de Bariloche
7) LS82	7	212/20	Buenos Aires
8) LU81	7	28/3	Bahía Blanca
9) LU84	7	24/3	Neuquén
10) LU90	7	30/3	Rawson
11) LV89	7	45/5	Mendoza
12) LW80	7	10/1	San Salvador de Jujuy
13) LW81	7	28/3	Santiago del Estero
14) LRI486	8	75/7.5	Mar del Plata
15) LV82	8	10/1	San Juan
16) LV85	8	18/1.8	Córdoba
17) LRK458	8	5/0.5	San Miguel de Tucumán
18) LS83	9	62/6	Buenos Aires
19) LU80	9	43/4	Bahía Blanca
20) LU83	9	6/0.6	Comodoro Rivadavia
21) LV83	9	52/5	Mendoza
22) LU85	9	1.5/0.15	Río Gallegos
23) LT81	9	1.5/0.15	Resistencia
24) LV91	9	5/0.5	La Rioja
25) LU82	10	295/30	Mar del Plata
26) LV80	10	80/8	Córdoba
27) LW83	10	6/0.6	San Miguel de Tucumán
28) LU92	10	50/5	General Roca
29) LRH450	10	22/2.2	Junín
30) LS84	11	180/20	Buenos Aires
31) LT88	11	230/20	Formosa
32) LU87	11	15/1.5	Ushuaia
33) LW82	11	11/1	Salta
34) LV81	12	170/20	Córdoba
35) LT85	12	20/2	Posadas
36) LU91	12	4/0.4	Trenque Lauquén
37) LS85	13	116/20	Buenos Aires
38) LT80	13	65/6	Corrientes
39) LT82	13	80/8	Santa Fé
40) LU88	13	15/1.5	Río Grande
41) LV86	13	5/0.5	Río Cuarto
42) LV90	13	12/1.2	San Luis

Associated to A.T.A.: 2), 4), 5), 8), 9), 12), 13), 14), 15), 16), 17), 19), 20), 21), 23), 25), 33), 36), 38), 39), 41).

Operated by State Administration: 3), 7), 10), 22), 24), 28), 29), 30), 31), 32), 34), 35), 37), 40), 42).

Addresses and other information:
1) Radiodifusora El Cármen S.A., Calle 27 e/530 y 531, 1900 La Plata. Pcia. Buenos Aires. (+5 relays).
2) Televisión Litoral S.A., Av. Godoy 8101., 2000 Rosario. Pcia. Santa Fé. (+ 5 relays).
3) Ruta 35 Km.322, Casilla Correo 139, 6300 Santa Rosa. Pcia. de La Pampa. (+7 relays).
4) Rader S.A., Av. Belgrano 1055, 2000 Rosario. Pcia. de Santa Fé. (+5 relays).
5) TV Río Diamante, Luzuriaga 360, 5600 San Rafael. Pcia. de Mendoza. (+3 relays).
6) Bariloche Televisión SRL., Elflein 251, 8400 San Carlos de Bariloche. Pcia. de Río Negro. (+5 relays).
7) Argentina Televisora Color S.A., Figueroa Alcorta 2977, 1425 Buenos Aires. Capital Federal. (+163 relays).
8) Telba S.A., Blandengues 225, 8000 Bahía Blanca. Pcia. de Buenos Aires. (+4 relays).
9) Neuquén TV S.A., Av. Argentina 1700, 8300 Neuquén. Pcia. de Neuquén. (+30 relays).
10) Av. Fontana 50, 9103 Rawson. Pcia. de Chubut. (+9 relays).
11) Garibaldi 7 Piso 5, 5500 Mendoza. Pcia. de Mendoza. (+8 relays).
12) R. Visión Jujuy S.A., Av. 19 de Abril 749, 4600 San Salvador de Jujuy. Pcia. de Jujuy. (+8 relays).
13) Pellegrini 345, 4200 Santiago del Estero. Pcia. de Santiago del Estero. (+7 relays).
14) Emisora Arenales de Radiodifusión S.A., Av. Luro 2907, 7600 Mar del Plata. Pcia. de Buenos Aires. (+ 8 relays).
15) Mitre 59 Oeste, 5400 San Juan. Pcia. de San Juan. (+5 relays)
16) DICOR S.A., Vélez Sarsfield 4300, 5000 Córdoba. Pcia. de Córdoba. (+16 relays)
17) Televisora Tucumana Color S.A., Av. Salta y Delfín Gallo. 4000, San Miguel de Tucumán. Pcia. de Tucumán.
18) Telearte S.A., México 990, 1097 Buenos Aires. Capital Federal.
19) Telenueva S.A., Sarmiento 64, 8000 Bahía Blanca. Pcia. de Buenos Aires. (+13 relays)
20) Comodoro Rivadavia TV, Rawson 1459, 9000 Comodoro Rivadavia. Pcia. de Chubut. (+7 relays).
21) Cuyo Televisión S.A., San Martín 1027, Galería Piazza, Local 10. 5500 Mendoza. Pcia. de Mendoza. (+5 relays)
22) H. Irigoyen 250, 9400 Río Gallegos. Pcia. de Santa Cruz. (+15 relays).
23) TV Resistencia S.A., Av. Alvear 50, 3500 Resistencia. Pcia. de Chaco. (+9 relays).
24) Av. Ortiz de Ocampo 1700, 5300 La Rioja. Pcia. de La Rioja. (+4 relays).
25) TV Mar del Plata S.A., Independencia 1163, 7600 Mar del Plata. Pcia. de Buenos Aires. (+4 relays).
26) SRTV S.A., Rivera Indarte 170, 5000 Córdoba. Pcia. de Córdoba. (+24 relays)
27) Televisora de Tucumán S.A., Av. Buenos Aires 296, 4000 San Miguel de Tucumán. Pcia. de Tucumán. (+2 relays)
28) Mitre y Sarmiento Piso 1, 8332 General Roca. Pcia. de Río Negro. (+10 relays).
29) Junín TV S.A., Belgrano 84, 6000 Junín. Pcia. de Buenos Aires. (+1 relay).
30) Telefé S.A., Pavón 2444, 1248 Buenos Aires. Capital Federal.
31) Tucumán 56, 3600 Formosa. Pcia. de Formosa. (+2 relays).

SOUTH AMERICA — Argentina — Brazil

32) Magallanes 1310, 9410 Ushuaia. Tierra del Fuego. (+1 relay).
33) CorTe S.A., España 475, 4400 Salta. Pcia. de Salta. (+4 relays).
34) Telecor S.A., Av. Fader 3469, Cerro de las Rosas. 5000 Córdoba. Pcia. de Córdoba. (+21 relays).
35) Rioja 161, 3300 Posadas. Pcia. de Misiones. (+11 relays).
36) Av. Leandro N. Alem 351, 6400 Trenque Lauquén. Pcia. de Buenos Aires.
37) ArTeAr S.A., Cochabamba 1153, 1147 Buenos Aires. Capital Federal.
38) Río Paraná TV SRL., Calle 13 s/n., 3400 Corrientes. Pcia. de Corrientes. (+3 relays).
39) Televisora SantaFesiona S.A., Blvd. Gálvez 840, 3000 Santa Fé. Pcia. de Santa Fé. (+6 relays).
40) Alberti 739, 9420 Río Grande. Tierra del Fuego.
41) Imperio Televisión S.A., Alberdi 823, 5800 Río Cuarto. Pcia. de Córdoba. (+6 relays).
42) Colón 925, 5700 San Luis. Pcia. de San Luis. (+6 relays). Federal.
38) Río Paraná TV SRL., Calle 13 s/n., 3400 Corrientes. Pcia. de Corrientes. (+3 relays).
39) Televisora SantaFesiona S.A., Blvd. Gálvez 840, 3000 Santa Fé. Pcia. de Santa Fé. (+6 relays).
40) Alberti 739, 9420 Río Grande. Tierra del Fuego.
41) Imperio Televisión S.A., Alberdi 823, 5800 Río Cuarto. Pcia. de Córdoba. (+6 relays).
42) Colón 925, 5700 San Luis. Pcia. de San Luis. (+6 relays).

3) Cristal de TV, Cas. 4399, Santa Cruz.
4) Benivision, Cas. 54, Trinidad.
4a) TV O Ltda, Cas 631, Oruro.
5) TV-Popular, Calle Juan de la Riva 1527, Casilla #8704, La Paz — 1200-0600 (+76 relays).
5a) Casilla 4573, Cochabamba.
6) Galavision, Cas. 495, Santa Cruz — 1000-0600.
7) TV-Universo, Av. Circunvalación, Santa Cruz — 24h.
8) Orureña TV, Cas. 14, Oruro.
10) America TV, Cas. 10076, La Paz. 24h.
11) TV Cochabamba, Cas. 1009, Cochabamba.
12) Empresa Nacional de TV, Cas. 900, La Paz. 24h.
13) Illimani de Comunicaciones, Oruro.
14) Trinivision, Cas. 333, Trinidad.
15) Rede ATB, Av. 6 de Agosto 2972, La Paz.
16) Cochabamba TV, Cas. 4545, Cochabamba.
17) Teleoriente, Cordillera 550, Santa Cruz.
18) Chuquisaqueña TV, Cas. 187, Sucre.
19) Teleandina, Cas. 1665, La Paz.
20) TV Universitaria, Cas. 21982, La Paz. 1900-0300.
22) Cruceña TV, Cas. 3424, Santa Cruz.
23) Sonomac/Tricolor, Av. 16 de Julio 1810, La Paz.
24) TV Integral, Cochabamba.
24a) SONOMAC, Cas. 21375, La Paz.
25) Grigota TV, Santa Cruz.
26) Tecnitron, Cas. 4410, La Paz.

BOLIVIA

TV-sets: 750.000 — **Colour:** NTSC — **System:** M&N.

Location	ch	A	kW	Location	ch	A	kW
1) La Paz	2			13) Cobija	9		
2) Cochabamba	2			16) Cochabamba	9		
3) Santa Cruz	2			17) Santa Cruz	9		
4) Trinidad	2			18) Sucre	9		
4a) Oruro	3			20) Potosi	9		0.1
5) La Paz	4		5	1) Oruro	10		
5a) Cochabamba	4			19) La Paz	11		
6) Santa Cruz	4		8	13) Tarija	11		
13) Trinidad	4			20) Cochabamba	11		1
7) Santa Cruz	5		2	20) Santa Cruz	11		10
8) Oruro	5			13) Potosi	11		
7) Trinidad	5		2	20) Trinidad	11		0.1
10) La Paz	6			20) Sucre	12		0.1
11) Cochabamba	6			1) Trinidad	12		
10) Oruro	6			20) La Paz	13		10
1) Cochabamba	7			7) Cochabamba	13		5
1) Sucre	7			22) Santa Cruz	13		2
12) La Paz	7			20) Oruro	13		2
1) Potosi	7			14) Trinidad	13		
1) Tarija	7			23) La Paz	15		
13) Oruro	8			24) Cochabamba	15		
14) Trinidad	8			24a) La Paz	15		
20) Tarija	8		1	25) Santa Cruz	15		1
15) La Paz	9			26) La Paz	20		

Addresses & other information
1) Television Boliviano, Cas. 4837, La Paz.
2) CCA-TV, Av. Heroinas 467, Cochabamba. 1500-0430.

BRAZIL

TV-sets: 32.600.000 — **Colour:** PAL — **System:** M. All st's comm. exc. where indicated.

ASSOCIACÃO BRASILEIRA DE EMISSORAS DE RADIO E TELEVISÃO (ABERT).
Hotel Nacional, s/5 a 8, C.P. 040-280, 70322-900 Brasilia, DF. ☎ +55 (61) 224 4600. 🖷 +55 (61) 321 7583.

SEARA — Serviços Associados de Rádio Ltda, Rua do Livramento 189, 20021 Rio de Janeiro, RJ. ☎ +55 (61) 243 2225.

STATIONS (relay st's omitted):

	ch	kW	Name and location
1)	2	13.17	TV Educativa, Manaus
2)	2	70/7	TV Guaiba, Porto Alegre
3)	2	100/10	TV Anhanguera, Goiânia
5)	2	78.25/8	TV Jornal do Comercio, Recife
6a)	2	6	TV Globo, Bauru
7)	2	57	TV Educativa, Rio de Janeiro.
8)	2	282/60	TV Cultura, São Paulo
8)	2	282	TV Geradora
9)	2	10	RBS TV, Erexim
9a)	2	120	TV Manchete, Fortaleza
10)	2	42	TV Educativa, São Luiz
11)	2	37.8	TV Educativa da Bahia, Salvador
12)	2	10.6	TV Educativa Espirito Santo, Vitória
63)	2	1	TV Liberal Marabé
12a)	2	8.5	TV Curitiba, Curitiba
13)	3	30.4	TV Nacional de Brasilia
6j)	3	10	TV Coroados, Londrina
16)	3	50	RBS TV, Blumenau
18)	3	1.6	TV Studios Rio de Janeiro, N. Friburgo
108)	3	10	RBS TV, Cruz Alta

Brazil — SATELLITE AND TV HANDBOOK

ch	kW	Name and location		ch	kW	Name and location
108)	4	5	Santa Angelo	71)	8	5	TV Campo Grande
19)	4	7.2	TV À Critica, Manaus	72)	8	31.6	RBS TV, Caxias do Sul
20)	4	100/10*	TV Aratu, Salvador	73)	8	20/4	TV Uberlândia, Uberlândia
21)	4	10.31	TV Gazeta, Vitória	74)	8	80 /8*	TV Record, São José do Rio Preto
22)	4	10/1*	TV Record, Franca	75)	8	9.2	TV Ajuricaba, Manaus
6)	4	100	TV Globo, Rio de Janeiro	76)	8	87.9	TV Atalaia, Aracaju
23)	4	14.77	TV Goyá, Goiânia	77)	8	8.91	TV de Fortaleza, Fortaleza
106)	4	3.16	TV Paranaiba-Prata	6k)	8	5	TV Cultura, Maringá
24)	4	1.8	TV Montes Claros, Montes Claros	108)	9	5	Santa Rosa
25)	4	12.3	TV Difusora do Maranhão, São Luis	79)	9	2/0.5	TV Borborema, Campina Grande
27)	4	120	TV Manchete, Belo Horizonte	80)	9	31.6	RBS TV, Rio Grande
29)	4	15.6	TV Guajará, Belém	81)	9	21.6	TV Educativa, Belo Horizonte
30)	4	10	TV Iguaçu, Curitiba	82)	9	76	TV Morado do Sol, Araraquara
31)	4	3.1	TV Tapajós, Santarém	83)	9	1.32	TV Eldorado, Criciuma
32)	4	6.7	TV Radio Clube Teresina, Teresina	106)	9	316	TV Paranaiba—Frutal
33)	4	15	RBS TV, Pelotas	83a)	9	320	TV Manchete, São Paulo, SP
34)	4	30/6	TV Rio Grandense, Porto Alegre	83b)	9	16.2	TV Record, Rio de Janeiro
38a)	4	116*	TVS, São Paulo	84)	10	138	TV Verde Mares, Fortaleza
39)	4	0.5	TV Roraima	85)	10	2	TV O Estado Chapeco
39)	4	4	TV Rondônia	85)	10	36	TV Planalto, Lages
39)	4	0.5	TV Acre	106)	10	3.16	TV Paranaiba—Araxa
92a)	4	100	TV Santa Cruz	106)	10	31.6	TV Paranaiba—Uberlândia
85)	4	1	TV O Estado Florianopolis	6c)	10	47	TV Globo Capital, Brasilia
88)	4	—	TV Bandeirantes, Brasilia	87)	10	30/6	TV Paranaiba, Uberlândia
4)	5	100	TV Alterosa, Belo Horizonte	87a)	10	?	TV Carima, Cascavel
39)	5	13	TV Amazonas, Manaus	88)	10	126	TV Bandeirantes, Porto Alegre
40)	5	2/0.5	TV Universitaria, Natal	88)	10	8.8	TV Bandeirantes, Pres. Prudente
40a)	5	5	TV Educativa, Teresina	88b)	11	70	TVS, Rio de Janeiro
41)	5	100/10*	TV Itapoan, Salvador	63)	11	2	TV Liberal—Castanhal
42)	5	10	RBS TV, Joinville	105)	11	10	TVU
43)	5	55	TV Educativa, Fortaleza	90)	11	5.6	TV Tibagi, Apucarana
6f)	5	2	TV Globo, Juiz de Fora	91)	11	5/1.5	TV Universitaria, Recife
46)	5	38/1	TV Alagoas, Maceio	92)	11	10	TV Gazeta, São Paulo
6b)	5	75	TV Globo, São Paulo	92a)	11	316	TV Bahia, Salvador
38b)	5	100*	TVS, Porto Alegre	108)	11	1	São Luiz Gonzaga
38c)	5	100*	Sistema B de TV, Belem	6d)	12	35	TV Globo, Belo Horizonte
92a)	5	100	TV Sudoeste	6i)	12	25	TV Paranaense, Curitiba
39)	6	1	TV Amapá	95)	12	5	TV Norte Fluminense, Campos
47)	6	24	TV Brasilia, Brasilia	96)	12	520	RBS TV, Porto Alegre
48)	6	8	TV Morena, Campo Grande	97)	12	31.6	RBS TV, Santa Maria
49)	6	12/2	TV Parana, Curitiba	6g)	12	1	TV Campinas, Campinas
17)	6	30/3*	TV Cruz Alta, Cruz Alta	99)	12	13	TV Independência, Cornelio Procopio
50)	6	70*	TV Tarobá, Cascavel	99)	12	?	TV Independência, Maringá
96)	6	20	RBS TV, Santa Cruz (F.Pl.)	100)	12	31.6	TV Cultura, Chapeco
51)	6	8.5	TV Cultura, Florianópolis	101)	12	316	RBS TV, Florianópolis
52)	6	400	TV Manchete, Rio de Janeiro	38e)	12	10	TVS, Belem
52a)	6	180	TV Manchete, Recife	105)	12	7.5	Televisão Imembui
53)	6	10	RBS TV, Bagé	102)	13	200	TV Bandeirantes, São Paulo
54)	6	14	TV Ribamar, São Luiz	6e)	13	330	TV Globo, Recife
54a)	6	27.6	TV Vitoria, Vitoria	103)	13	80	TV Brasil Central, Goiânia
55)	7	189/18.9*	TV Bandeirantes, Salvador	104)	13	31.6	RBS TV, Uruguaiana
56)	7	14/1.2	TV Tropical, Londrina	*) ERP.			
57)	7	20	TV Esplanada, Ponta Grossa				
58)	7	31.6*	TV Sudoeste do Parana, Pato Branco				
59)	7	316	TV Record, São Paulo				
88)	7	316/31.6	TV Bandeirantes, Belo Horizonte				
6h)	7	5	TV Ribeirão Preto, Ribeirão Preto				
62)	7	12	TV Gazeta de Alagoas, Maceió				
63)	7	60	TV Liberal, Belém				
64)	7	57.48	TV Bandeirantes, Rio de Janeiro				
65)	7	100	TV Educativa, Porto Alegre				
66)	7	40	RBS TV, Passo Fundo				
66a)	7	7.7	TV Uberaba, Uberaba				
99)	7	?	TV Independência, Curitiba				
67)	7	15.8	TV Capital, Brasilia				
68)	8	3.16/0.36*	TV Sul Fluminense, Barra Mansa				
70)	8	9.6	TV Brasil Oeste, Cuiabá				

Addresses and other information:

1) Rua Major Gabriel s/n, 69000 Manaus, AM — 2200-0400.

2) Rua Caldas Jr. 219, 90000 Porto Alegre, RS — 1200-0400.

3) Rua Thomaz Edson, Qd.7, 74000 Goiânia, GO— 0900-0600.

4) Av. A. Chateaubriand 499, 30150 Belo Horizonte, MG — 1100-0600.

5) Rua do Lima 250, Recife, PE — 1200-0430.

6) Rua Lopez Quintas 303 (Studio: Rua Von Martins), Jardim Botanico, 22463 Rio de Janeiro, RJ. Dir. Tec: A. Pontes Malta. 0900-0600.

6a) Rua Padre Anchieta 941, 17100 Bauru, SP.

6b) Alameda Santos 1893, 01419 São Paulo, SP —

SOUTH AMERICA — Brazil

0930-0300.
6c) SCS Q2, Bl. B. no 81, Edif. Bradesco, 70300 Brasilia, DF.
6d) Rua Rio de Janeiro 1279, 30000 Belo Horizonte. 1200-0300.
6e) Av. Dantes Barreto 1186, 50000 Recife, PE.
6f) Rua Ewbank da Camara 46, 36100 Juiz de Fora, MG — 0930-0530.
6g,h) Rua Javari 3099, 14110 Ribeirão Preto, SP.— Dir. Gen: A.C. Coutino Nogueira. Dir. Tec: A. João Filho. 0900-0600.
6i) Av. Batel 1393, 80000 Curitiba, PR — 1430-0500.
6j) Av. Tirandentes 1370, 86100 Londrina, PR — 2000-0300.
6k) Rua Sta. Joaquina de Vedruna 625, 87100 Maringá, PR.
7) Av. Gomes Freire 474-B, Rio de Janeiro, RJ.— 0930-0345.
8) Rua Cenno Sbrighi 378, Agua Branca, 05099 São Paulo, SP — St. Mgr: R. Muylaert. Dir. Tec: J. Munhoz — 1100-0300.
9) Rua Soledade 277, Erexim, RGS.
9a) Av. Antonio Sales Esq, 60000 Fortaleza — 1100-0500.
10) Av. Kennedy s/n, São Luis, MA — 1000-0300.
11) Rua Pedro Gama 413/E, 40000 Salvador, BA — 1330-0300.
12) Rua Pedro Palacio 99, 29000 Vitoria, ES.
12a) Rua Francisco Caron 29, Curitiba, PR.
13) Praça 31 de Marco s/n, 70000 Brasilia — 1400-0300.
16) Rua Getulio Vargas 32, 89100 Blumenau, SC.
17) Rua Jango Vidal 427, 98100 Cruz Alta, RS — 1000-0500.
18) Rua E. Brasilio 30, 28600 Nova Friburgo.
19) Estrada do Aleixo Km 3, 69060 Manaus, AM — 0900-0400.
20) Rua Pedro Gama 31, 40230 Salvador, BA — 0900-0500.
21) CP. 1070, 29050 Vitória, ES — 0900-0600.
22) Rua José Maria Medeiros 5120, 14400 Franca, SP — 1100-0500. ☎ +55 (16) 727 0400.
23) Av. Goiás 187, 74000 Goiânia, GO.
24) Praça dos Morrinhos S/N, Morrinhos. 0930-0530 (Sat/Sun 24h).
25) Av. Camboa do Maio 120, 65000 São Luis, Maranhão — 1000-0600.
29) Av. Governador José Malcher 1332, Belém, PR — 1400-0400.
30) Rua João Tscharnell 800, Jardim Merces, 80000 Curitiba, PR.
31) Av. Ismael Aranju 266, 68100 Santarem, PA.
32) CP. 209, 64000 Teresina, PI—Pres: S.F. de Alénear. Dir Tec: H.P. de Carvalho. 0900-0600.
33) Rua 15 de Novembro 612, Pelotas, RGS. 1200-0600.
34) Rua Orfanotrôfio 711, Alto Teresópolis, 90000 Porto Alegre, RS — 1000-0500.
35) Rua Alto do Morro, 49000 Aracaju, SE.
38a-e) Rua Dona Santa Velozo 535, 02050 São Paulo — 1000-0500 (on all st's).
39) Ave. Carvalho Leal 1270, Cachoeirinha, Manaus, AM — 0930-0500.
40) Rua Princesa Isabel 758, 59020 Natal, RN — Dir: C.A. Martins. Dir. Tec: R. de Andrade Martins — 1000-0300.
40a) Ave. Prof. V. Alencar s/n, 64065 Teresina, PI — Dir. Tec: F.J. de Paiva Ribeiro — 1000-0400.

41) Rua Ferreira Santos 5, 40000 Salvador, BA — Dir: A. Moraes. 0930-0430.
42) Rua Saguaçu s/n, Joinville, SC.
43) Rua Oswaldo Cruz, 1985, 60000 Fortaleza, CE — 1000-0200.
46) Rua Cel. Paranhos 305, Jacintinho, 57000 Maceió, AL.
47) Av. W-3, Setor de Radio e TV, Brasilia — 1300-0500.
48) Av. Eduardo E. Zahran s/n, Campo Grande, MS — 2030-0300.
49) C.P. 7061, 80000 Curitiba, PR — 1000-0500.
50) C.P. 1169, 85800 Cascavel, PR — 1030-0600.
51) Alto do Morro Antão, Florianópolis, SC.
52) Rua do Russel 766-804, 20000 Rio de Janeiro — Dir. Gen: R. Furtado. Dir. Tec: F. Cavalcanti — 1100-0500.
52a) 1100-0500.
53) Rua do Acampamento 2550, 96400 Bage, RGS.
54) Praça Bom Menino s/n, São Luiz, Maranhão.
54a) Av. Presidente Florentino Avidos, 350 7° anolar Vitoria ES.
55) Largo do Candomblé 19A, Salvador, B.A — Dir. Gen: F.H. Chagas. Dir. Tec: R. Blum. 0900-0500.
56) Rodovia Celso Garcia, Londrina, PR — 1300-0500.
57) Rua João França Silva 2885, Ponta Grossa, PR. 1030-0300.
58) C.P. 591, 85500 Pato Branco, PR — Dir: V. Hillesheim. Dir. Tec: L. Schmitz — 1000-0300.
59) Av. Miruna 713, Aeroporto, 01000 São Paulo, SP — 1100-0500. ☎ +55 (11) 542 9000.
59a) Campo de São Cristovão 105, Rio de Janeiro, RJ.
60) Rua Tomé de Souza 1251, 30000 Belo Horizonte, MG — Dir. Gen: M. Pereira Leite. Dir. Tec: O. Dominco Dalip. Film Buyer: R. Hachich Maluf. 1200-0600.
62) Av. Aristeu de Andrade s/n, Maceió, AL — 1830-0330.
63) Av. Nazaré, 350 Belem Pará, PA — 1200-0300.
64) Rua Alvaro Ramos 492, Bota fogo, Rio de Janeiro, RJ — 1230-0300.
65) Rua Correa Lima 2118, Morro Santa Tereza, 90640 Porto Alegre, RS — St. Mgr: A.C. Fedrizzi — 1145-0330.
66) Rua Princesa Isabel, 99100 Passo Fundo, RGS.
66a) Rua General Osorio 755, 38100 Uberaba, MG — 2000-0400.
67) Torre de Televisão de Brasília, Box 1, Brasilia, DF — 1440-0200.
68) C.P. 85919, 27400 Barra Mansa, RJ—Dir: O.M. Nora. Dir. Tec: C. Pina. 1000-0500.
70) Rua Giboia, s/n°, Bairro Concil, 78000 Cuiaba, MT—1030-0600.
71) Av. Calogeras 315, 79300 Campo Grande, MT — 0600-0200.
72) Av. Rio Grande do Sul, Caxias do Sul, RS — 1530-1730, 1900-0500.
73) Rua R.G. do Norte 1069, 38400 Uberlândia, MG.
74) Via Washington Luiz, km 436, 15100 São José do Rio Preto, SP — Dir: P.M. de Carvalho F. Dir. Tec: C. Victor Donato. 1100-0500.
75) Rua O.G. 18 Santo Antonio, 69000 Manaus.
76) Rua Claudio Batista 122 (ex Cláudio Batista S/N), 49045 Aracaju, SE.
77) Av. Desembargador Moreira 2565, Fortaleza, CE. 1300-0400.
79) Rua Venâncio Neiva 287, 2° andar, 58100 Campina Grande, PB — 1355-0400.
80) Rua Duque de Caxias 63, 1° e 2° andares, Rio Grande, RGS.

81) Av. Assis Chateaubriand 167, 30150 Belo Horizonte, MG — 0900-0500.
82) Praça José Palmores Lepre 99, 14800 Araraquara, SP — 24h.
83) Rua Silva Jardim 216, Cricuma, SC.
83a) Rua Bruxelas 193 Sumare, 01000 São Paulo, S.P.
83b) Rua G. Padilha 144, 20000 Rio de Janeiro.
84) Av. Desembargador Moreira 2430, Fortaleza, CE. 1405-0220.
85) Rua Carlos Jofre do Amaral 67, 88500 Lages, SC.
87) C.P. 210, 38400 Uberlândia, MG — St. Mgr: A. de Castro jun. Dir. Tec: M.J. Rodrigues dos Reis — 0900-0500.
87a) Av. Barao do Rio Branco, 1960, Sao Paulo.
88) Rua Delfino Riet 183, C.P. 1474, Porto Alegre, RS — 1300-0400.
88) Rua Radiantes 13, Sao Paulo, Brazil 05699.
90) Av. Santos Dumont 11, 86800 Apucarana, PR — 1200-0400.
91) Av. Norte 68, Santo Amaro, Recife, PE. 1900-0300.
92) Av. Paulista 900, São Paulo, SP. 1500 (Sat/Sun 1200)-0300.
92a) Estrada de São Lázaro 540, Federação, Salvador-BA, Brazil.
95) Av. 24 de Outubro 201, 28100 Campos, RJ — 0945-0530.
96) Rua Radio y TV Gaucha 189, 90650 Porto Alegre, RS — 0900-0500.
97) Av. 2 de Novembro s/n, 97100 Santa Maria, RS.
99) SSC — Sistema Sul de Comunicacão, R. Amaury Lange Silvério, 450 Pilarzinho, 82000 Curitiba, PR.
100) Estrada de Seara Km 3, 89800 Chapeco, SC.
101) Morro da Cruz, 88000 Florianópolis, SC.
102) Rua Radiantes 13, 05699 São Paulo, SP — 1300-0400.
103) Rua 201 no. 430, Vila Nova, Goiânia, GO.
104) Rua Domingos de Almeida 1722, 97500 Uruguaiana, RGS. 24h.
105) Avenida Norte, 68 Sante Amaro, Recife, Pernambuco.
106) Av. Prof. José Ignácio de Souza, 2710, 38400, Uberlandia.
107) Av. Mauricio Sirotsky Sobrinho, 25.
108) Caixa Postal 324, 98100, Cruz Alta.

CHILE

TV-sets: 2.371.520 — **Colour:** NTSC — **System:** M.

TELEVISION NACIONAL DE CHILE (Gov.)
✎ Bellavista 0990, Providencia, Santiago. ☎ +56 (2) 7077660.
📠 +56 (2) 7077761. ① 241375 TVNCH CL. **Cable:** TV Chile. **WWW:** http://www.tvn.cl
L.P.: DG: Jorge Navarrete Martinez. GM: Bartolome Dezerega Salgado. CE: Jaime Sancho Martinez.
Stations

Location	ch	kW	Location	ch	kW
El Roble	2	175	Cayumanqui	6	175
Valdivia	3	20	Santiago	7	340
Puerto Montt	4	20	Temuco	7	115
Coquimbo	4	20	El Roble	2	175
San Fernando	5	25	Valdivia	3	20
Antofagasta	6	20	Puerto Montt	4	20
Osorno	6	20	Coquimbo	4	20
Punta Arenas	6	20	San Fernando	5	25
Antofagasta	6	20	Vallenar	12	5
Osorno	6	20	Valparaíso	12	100
Punta Arenas	6	20	San Antonio	12	10
Cayumanqui	6	175	Copiapo	7	4
Santiago	7	340	Pto. Williams	8	1
Temuco	7	115	Ancud	8	2
Copiapo	7	4	Iquique	10	30
Pto. Williams	8	1	Castro	10	50
Ancud	8	2	Salvador	10	4
Iquique	10	30	Talca	10	20
Castro	10	50	Ovalle	10	3
Salvador	10	4	Vallenar	12	5
Talca	10	20	Valparaíso	12	100
Ovalle	10	3	San Antonio	12	10

+ 96 low power repeaters less than 1 kW.
D.Prgr: 1200 (Sat/Sun 1400)-0500.

CHILEVISION S.A. (Comm.)
✎ Ines Matte Urrejola, 0825, Santiago. ☎ +56 (2) 7372227. 📠 56 (2) 7377923. ① 340 492 TVCH. **E-mail:** chilevis.ionsa001@chilnet.cl **WWW:** http://www.cisneros.com/companies/broadcast/chilevision.htm
Stations: chA11 60/30kW (+ relay st. at Valparaíso ch10).
D.Prgr: 2145-0430.

TELETRECE
✎ Inés Matte Urrejola 0848, Santiago. ☎ +56 (2) 514000. ① 440182 TRECE CZ. 📠 +56 (2) 377044.
WWW: http://www.reuna.cl/teletrece/

Location	ch	kW	Location	ch	kW
Arica	8	21.7	San Fernando	5	37
Iquique	8	28.8	Talca	8	27
María Elena	11	20	Chillán	13	245
Chuquicamata	12	20	Concepción	5	56
Antofagasta	13	41.6	Angol	10	1.2
Copiapó	11	9.5	Victoria	10	1.2
La Serena	13	21.6	Temuco	4	45
Ovalle	5	2	Villarrica	9	2
San Felipe	7	21	Valdivia	12	45
Valparaíso	8	137	Osorno	9	47
Santiago	13	220	Puerto Montt	13	66.4
San Antonio	10	21.7	Ancud	5	7
San Fernando	5	37	Castro	12	73.8
Arica	8	21.7	Talca	8	27
Iquique	8	28.8	Chillán	13	245
María Elena	11	20	Concepción	5	56
Chuquicamata	12	20	Angol	10	1.2
Antofagasta	13	41.6	Victoria	10	1.2
Copiapó	11	9.5	Temuco	4	45
La Serena	13	21.6	Villarrica	9	2
Ovalle	5	2	Valdivia	12	45
San Felipe	7	21	Osorno	9	47
Valparaíso	8	137	Puerto Montt	13	66.4
Santiago	13	220	Ancud	5	7
San Antonio	10	21.7	Castro	12	73.8

+ 25 low power tr's less than 1 kW.

RED DE RADIOTELEVISION DE LA UNIVERSIDAD DEL NORTE (TELENORTE)
✎ Carrera 1625, Antofagasta. ☎ +56 222496 —
L.P.: GM: Juan Carlos Salas Floras.
Stations: ch3 Antofagasta, ch11 Arica, ch12 Iquique +7 low power sts.
D. Prgr: 1735-0435.

SOUTH AMERICA — Chile — Ecuador

CORPORACION DE TELEVISION DE LA UNIVERSIDAD CATOLICA DE VALPARAÍSO
✍ Agua Santo Alto 2455 (Casilla 4059), Viña del Mar.
☎ +56 (32) 610140. 🖷 +56 (32) 610505.
E-mail: ucvtelev.ision@chilnet.cl
Stations: chA4 (Valparaíso), chA5 (Santiago), ch7 (Puerto Montt), chA8 (La Serena).

RADIO COOPERATIVA TELEVISION S.A.
✍ Antonio Bellet 223, Santiago. ☎ +56 (2) 2360066.
🖷 +56 (2) 2352320. E-mail: canalroc.kpop002@chilnet.cl
Station: chE2.

MEGAVISION S.A.
✍ Av. Vicuña Mackenna 1348, Santiago. ☎ +56 (2) 5555400. 🖷 +56 (2) 5518916. E-mail: megavisi.onsa001@chilnet.cl
Station: chE9.

COLOMBIA

TV-sets: 7.029.000 — Colour: NTSC — System: M.

INSTITUTO NACIONAL DE RADIO Y TELEVISION (INRAVISION)
✍ Centro Administrativo Nacional, Via Eldorado, Bogotá.
☎ +57 (1) 2220700. 🖷 +57 (1) 222 0800. ① 43311
INRACO — L.P: Exec. Dir: Jose Jorgo Dangorich Castro.
Inravision leases airtime to 26 comm. companies. The three largest are:

Caracol
✍ AA 9291, Santafé de Bogot. ☎ +57 (1) 337 8866.
🖷 +57 (1) 337 7126. WWW: http://latina.latina.net.co/empresa/caracol/
Key st: Manjui ch7.

Punch
✍ Carrera 28, 49-98 Bogotá. ☎ +57 (1) 2174750.
Key st: Manjui ch9.

RTI
✍ Calle 19 N 4-56 Piso 2, Bogotá. ☎ +57 (1) 282 7700. 🖷 +57 (1) 284 9012. ① 43294.
L.P: Pres: Patricio Wills. Head of Prgrs: Patricio Wills.
Key st: Manjui ch11.

Station/Area	Netw I ch	kW	Netw II ch	kW (ERP)
Manjui (1)	7	668	9	668
La Rusia (2)	10	54	8	54
C. Oriente (3)	8	336	3	336
Tasajero (4)	5	150	11	100
Jurisdicciones (5)	6	468	4	372
Saboya (6)	13	27	11	20
Chigorodo (7)	2	15	11	15
El Ruiz (8)	12	54	10	54
Padre Amaya (9)	3	152	13	252
Galeras (10)	12	393	9	393
Planadas (11)	4	16	2	125
Munchique (12)	10	54	13	54
Paramo de Dominguez (13)	8	426	13	426
La Pita (15)	10	?	2	?

Station/Area	Netw I ch	kW	Netw II ch	kW (ERP)
La Popa (14)	7	10	9	10
Monteria (16)	7	27	9	25
Cerro Kennedy (17)	13	700	11	700
El Alguacil (18)	7	270	12	270
Arauca (19)	11	1.2		
Gabinete (20)	6	10	4	10
Buenavista (21)	9	5	7	5
Leticia (22)	8	54	13	54
Nieva (23)	12	54	10	54
Alto del Tigre (24)	12	16	2	10
San Andres (25)	11	27	9	15

+ 7 low power stations not mentioned.
Areas (Departments) served: 1) Bogotá (Sabana); Tolima, Huila; 2) Boyaca, Santander (South); 3) Santander (North), Arauca; 4) Santander (North & Central); 5) Santander, Magdalena (South); 6) Boyaco; 7) Uraba; 8) Manizales, Caldas; 9) Medellin, Antioquia (Central); 10) Nariño (Central); 11) Quindio; 12) Popayan; 13) Cauca & Cauca Valley; 14) Cartagena, Bolivar (North); 15) Bolivar, Sucre; 16) Cordoba; 17) Atlantico, Magdalena (North); 18) Guajira; 19) Arauca; 20) Caqueta, Huila; 21) Huila (South); 22) Leticia, Amazonas; 23) Huila (Central & North); 24) Meta; 25) San Andres.
Network III: Manjui (1) ch A 11 668kW.
D. Prgrs: 1630-1830 (Comm), 1830-2130 (Educ.), 2100-0500 (Comm).

ECUADOR

TV-sets: 900.000 — Colour: NTSC — System: M.

Stations	chA	kW	Stations	chA	kW
1) Guayaquil	2	6/0.6	9) Portoviejo	7	5
2) Quito	2	1/0.1	7) Quito	8	10/1
3) Cuenca	2	8	11) Quito	8	10
4) Quito	4	5/0.5	8) Portoviejo	9	3
5) Guayaquil	4	3/0.3	9) Guayaquil	10	10/1
) Quito	5	?	9) Quito	10	10
6) Esmeraldas	6	0.5/0.05	10) Quito	13	1/0.1

Addresses and other information:
1) Corporación Ecuatoriana de Televisión, Casilla 1239, Guayaquil. ☎ +593 (4) 300150. 🖷 +593 (4) 303677. ① 3409 TVDOSG. Cable: Teledos — 1200-0600.
2) Canal 2 Quito, Murgeon 732, Quito. ☎ +593 (2) 540877 — 1930-0730 (Sun 1330-0700).
3) Telecuenca, Canal Universitaria Catolica, Casilla 400, Cuenca. ☎ +593 827862. ① 4775 TELCUE ED — 1600-0400.
4) Teleamazonas, Av. Diguja 529 y Brazil, Quito. ☎ +593 (2) 430313.
5) Canal 4 Guayaquil S.A, 9 de Octubre 1200, Guayaquil. ☎ +593 (4) 308194.
6) Canal 6 Esmeraldas, Cas. 108, Esmeraldas. ☎ +593 (2) 710090. Cable: TECEM — 2300-0400.
7) Canal 8 Quito, Cas. 3888, Quito. ☎ +593 (2) 244888 — 1730-0530.
8) Manavision S.A, Apt. 50, Portoviejo — 1930-0530.
9) Canal 10, Guayaquil, Casilla 673, Guayaquil. ☎ +593 (4) 391555 — 1130-1430, 1630-0500 (+14 relays).
10) Canal 13 Quito, Rumipampa 1039, Quito. ☎ +593 (2) 242758.
11) Televisora Nacional, Bosmediano 447 y José Carbo, P.O. Box 6615, Quito.

FALKLAND ISLANDS

TV-sets: N/A — **Colour:** PAL — **System:** I.

BRITISH FORCES BROADCASTING SERVICE (BFBS)
✉ BFBS Falkland Islands, Mount Pleasant, BFPO 655.
☎ 32179. 🗎 32193.
L.P: St.Mgr: Steve Brearton. Gen. Eng: S. Brown. **Station:** Mount Pleasant ch 24 UHF/100 W. Port Stanley ch 30 UHF/15 W. Rebros ch40UHF.
D.Prgr: 4 h. of taped broadcasts from BBC and ITV London.
GUIANA (French)
TV-sets: 6.500 — **Colour:** SECAM — **System:** K.

RFO-GUYANE
✉ BP 7013, Cayenne Cedex. ☎ +594 299900. 🗎 +590 302649 — **L.P:** Dir: Henri Neron. T.D: Daniel Beugin.
Stations: Cayenne chK4 0.1kW, +8 low power repeaters.
D.Prgr: 2100-0130.

ANTENNE CREOLE (Priv., Comm.)
✉ 31 avenue Louis Pasteur, 97300 Cayenne. ☎ +594 31 2020
Stations: Kourou ch44 1kW, Cayenne ch39 3kW

GUYANA

TV-sets: 15.000 — **Colour:** NTSC — **System:** M.

GUYANA TELEVISION (Gov.)
✉ 68 Hadfield St., Lodge Georgetown. ☎ +592 (2) 69231-4/62691-4/58584.
L.P: GM: A. Brewster; CE: S. Goodman.
Station: Georgetown ch10 0.04kW.
D.Prgr: Sun 1500-1600, repeated 2100-2200.
Two private TV stations relay U.S. satellite sces.

PARAGUAY

TV-sets: 350.000 — **Colour:** PAL — **System:** N.

Stations	chA	kW	Stations	chA	kW
1) Encarnación	7	60/12	4) Asunción	9	60/12
5) Illar	7	10	4) P.J. Caballero	9	5/1
5) Encarnacion	7	10	5) Misiones	10	10
5) C. del Este	8	10	5) Villarrica	12	10
2) Pto. Stroessner	8	25/53)	4) Asunción	13	30/4

Addresses:
1) Television Itapua, Avda Irrazabal y 25 de Mayo, Encarnación — Gen. Mgr: J. Mateo G. — 1600-0400.
2) Televisora del Este, Area 5, Cd. Puerto Stroessner — Gen. Mgr: A. Villalba V. — 2000-0300 (Sat/Sun 1100-0330).
3) Canal 9 TV Cerro Cora SA. ✉ Avenida Carlos Antonio Lopez 572, Asuncion. ☎ +595 (84) 22226. 🗎 +595 (84) 498911. **L.P:** Hd. of Sales: Hugo Montgomery.
4) Teledifusora Paraguaya S.A., 8 Proy Lambare, Asunción — 1600-0400. ☎ +595 (21) 443093.
5) Sistema Nacional de Television

PERU

TV-sets: 2m — **Colour:** NTSC — **System:** M.

Stations: *) in stereo.

	chA	kW (ERP)		chA	kW (ERP)
1)	2	22.54/2.25	5)	8	5
5)	3	20	5)	9	315
2)	4	25/12.5	6)	11	30/6
5)	4	20	1)	13	?
3)	5	290/29	7)	27*	7
5)	6	2	8)	33*	10
4)	7	10/5			

Addresses & other information:
1) Compañia Latinoamericana de Radiodifusion S.A. Av. San Felipe 968, Jesús Mariá, Lima 11. ☎ +51 (14) 707272. 🗎 +51 (14) 712688.
2) Compañia Peruana de Radiodifusion, Cas. 1192, Lima. Dir. Gen: M. M. Arbulu B. Man. Dir: N. Gonzalez U. Dir. Tec: D. Capella. Operates st's in Piura(ch2 2 kW), Chiclayo(ch 4 5 kW), Trujillo(ch 4 2 kW), Tacna(ch 9 2 kW), Huancayo(ch 4 2 kW) + 59 repeaters — 1000-0600.
3) Panamericana de Television, Av. Arequipa 1110, Lima. 1500-0600 (+ 14 relay st's).
4) Empresa de Cine, Radio y Television Peruana, José Galvez 1040, Lima. Pres: Carlos Guillen B. 1900-0400 (+ 39 repeaters).
5) Andina de Television, Arequipa 3570, San Isidro, Apartado 270077, Lima. 1900-0400.
6) RBC Television, Juan de la Fuente 453, Miraflores, Lima.
7) Difusora Universal de Television, Paseo de la República 6099, San Antonio, Miraflores, Lima — Exec. Pres: J.L. Banchero H. 2200-0400.
8) Empresa Radiodifusora 1160 TV, Apt. Postal 2355, Lima. 800-0500.

SURINAME

TV sets: 43.000 — **Colour:** NTSC — **System:** M.

SURINAAMSE TELEVISIE STICHTING (STVS) (Gov./Comm.)
✉ P.O. Box 535, Paramaribo. ☎ +597 473100. 🗎 +597 477216. ➀ 271 STVS. **Cable:** Surteve.
Stations: Paramaribo chA8 6kW + 5 relay st's.
D.Prgr: 2130-0300 (+ Sat 1500-2130, Sun 1400-1700).

ALTERNATIEVE TELEVISIE VERZORGING (ATV Telesur)
✉ Adrianusstraat 1, Paramaribo. ☎ +597 410027, +597 470425. 🗎 +597 479260. ➀ 488 ATV TLS SN.
L.P: Mgr: Roy Doorson.
Stations: Paramaribo chA2 + chA12 + 2 relay st's.
D.Prgr: 1115-0320.

URUGUAY

TV-sets: 600.000 — **Colour:** PAL — **System:** N.

St's 1), 3), 8), 9), 13), 15), 16), 17), 18), 20), 23), 24) & 26) are affiliated to ANDEBU (Associación Nacional de Broadcasters Uruguayos) — All st's are comm.

SOUTH AMERICA

Uruguay

Location	ch	kW	Location	ch	kW
1) Artigas	3	0.5	14) P. de los Toros	9	0.1
2) Colonia	3	1/0.5	15) Maldonado	9	12
3) Paysandú	3	1/0.5	16) Bella Unión	10	0.1
4) Rio Branco	3	0.1	17) Montevideo	10	600
5) Montevideo	4	300	18) Rivera	10	1/0.1
6) Montevideo	5	96/9	19) Punta del Este	11	5/0.5
7) Rivera	5	7/0.7	20) Treinta y Tres	11	1/0.5
8) Rocha	7	16	21) Durazno	11	0.1
9) Tacuarembó	7	75	22) Fray Bentos	12	30/3
10) M. de Corrales	7	0.1	23) Melo	12	1/0.5
11) Melo	8	0.5/0.05	24) Montevideo	12	600/180
12) Rosario	8	75/7.5	25) Chuy	12	0.1
13) Salto	8	2/1	26) Minas	13	1/0.5

+10 st's 0.1kW (all rel. st 6).
Addresses and other information:
1) Tele-Artigas, Lecueder 291, 55000 Artigas — Dir. Gen: Carlos F. Falco. 1800-0300.
2) Televisora Colonia, W. Barbot 172, 70000 Colonia — Dir. Gen. G. de Gonzáles.
3) Río de los Pájaros TV, Av. España 1629, 60000 Paysandú — Dir Gen: A. Davison. Dir. Tec: H. Caporale.
4) 37100 Río Branco, Cerro Largo.
5) Monte Carlo TV Color, Paraguay 2253, 11800 Montevideo — Dir. Gen: H. Romay. Film Buyer: M. Fonticiella. Dir. Tec: J. Spinella — 1830/0400 (Sat/Sun 1300-0400).
6) S.O.D.R.E., Bul Artigas 2552, 11600 Montevideo.
7) Canal 5 S.O.D.R.E, Bulevar Artigas 2552, Montevideo. Gen. Man: Julio Frade. Dir. Tec: Pedro Narancio.
8) Tele-Rocha, Av. O. de los Santos 105, 27000 Rocha — Dir. Gen: M. Scherchener. Dir. Tec: J. Regalo. Film Buyer: L. Castillo.
9) Radiotelevisión "Zorilla de San Martín", 18 de Julio 302, 45000 Tucuarembó — Dir. Gen: D. Dini. S. St. Mgr: G. Valdés G. 2030 -0330. Dir. Tec: G. Acosta. Film Buyer: Jose Abbondanza.
10) 40002 Minas de Corrales, Rivera.
11) Canal 8 TV Melo, 18 de Julio 572, 37000 Melo, Cerro Largo — Dir: Raul Figueredo. Dir. Tec: Eduardo Baptista.
12) Canal 8 Rosario TV Color, Ruta 2 Km. 136.500, 70200 Rosario, Colonia — Dir: H. Fripp.
13) Televisora Salto Grande, Av. Viera 1280, 50000 Salto — Dir. Gen: Dr. Carlos A. Gelpi. Dir. Tec: K. Muguerza.
14) 45100 Paso de los Toros, Tacuarembó.
15) Canal 9 del Este TV Color, Av. Artigas 879, 20000 Maldonado — Dir. Gen: M. Scherschener. Dir. Tec: Fernando Bareño. Film Buyer: J. López — 2000-0400.
16) Telediez, General Rivera s/n, 55100 Bella Unión, Artigas — Dir. Gen: C. Gelpi.
17) SAETA TV Canal 10, Dr. Lorenzo Carnelli 1234, 11200 Montevideo — Dir. Gen: J. de Feo. Dir. Tec: Oscar Inchausti. Film Buyer: H. Villar — Mon-Fri 1930.
18) Tevediez, Sarandí 675, 40000 Rivera — Dir. Gen: A. Pereira. 2000-0300.
19) Canal 11 Punta del Este, Cantegril Country Club, 20100 Punta del Este, Maldonado — Dir. Gen: D. Romay — 1830-0400.
20) Televisora Treinta y Tres, Pablo Zufriategui 226, Treinta y Tres — Dir. Gen: A. Pinho. St. Mgr: A. Lagos. Dir. Tec: D. Ponce.
21) 97000 Durazno.
22) Canal 12 Río Uruguay, Cno. San Salvador s/n, 65000 Fray Bentos, Río Negro — Dir. Gen: D. Romay.
23) Melo TV, Castellanos 723, 37000 Melo, Cerro Largo — Dir. Gen: R. Lucas. Dir. Tec: C. Britos. Film Buyer: J. Lucas.
24) Teledoce Televisora Color, Enriqueta Compte y Riqué 1276, 11800 Montevideo — Dir. Gen: H. Scheck. Dir. Tec: M. Donnangelo. Film Buyer: C. Restano — 1810 (Sat 1600, Sun 1330)-0400.
25) 27100 Chuy, Rocha.
26) TV Cerro del Verdún, Treinta y Tres 632, 30000 Minas, Lavalleja — Dir. Gen: C. Falco. Dir. Tec: J. Rodriguez.

VENEZUELA

TV-sets: 3.855.480 — **Colour:** NTSC — **System:** M.

CAMARA VENEZOLANA DE LA TELEVISION
(Organization for private TV stations).
Ap. 60423, Chacao, Caracas 1050. ☏ +58 (2) 7814608. **Cable:** CAVETEL. ③ 21144 — **L.P:** Pres. H. Ponsdomenech.

TELEVISORA NACIONAL TVN (Gov.)
Ap. 3979, Caracas 1010-A. ☏ +58 (2) 239 9811
D.Prgr: 1800-0400 (actually relaying VTV).
Stations: Anzoategui ch13 64kW; Carobobo ch6 67kW; Bolivar ch5 57kW; D. Federal ch5 279kW; Falcon ch5 50kW; Lara ch13 64kW; Merida ch13 2kW; Tachira ch13 72kW; Tachira ch2 5kW; Zulia ch6 61kW.

VENEZOLANA DE TELEVISION "5" (Gov.)
Ap. 2739, Caracas 1010-A. ☏ +58 (2) 2399811. +58 (2) 35734. ③ 25401.
Stations: Caracas (Central St): chA5 210/105kW.

Repeaters	chA	kW		ch	kW
Margarita	5	99.6/49.8	El Tocuyo	9	2/1
La Grita	6	75.9/37.9	Terepalma	13	72/36
Los Olivos	5	40.9/20.9	Pto. La Cruz	13	72/36
Maracaibo	6	108/54	Táchira	13	72/36
Aleton	6	67.2/38.6	Litoral Central	13	60.25/30.12
Güigüe	6	67.2/38.6	Escuque	13	26.25/13.12
Cd. Bolivar	6	12.5/6.25	Caricuao	13	0.631

D.Prgr: 2200-0400 (approx.)

VENEZOLANA DE TELEVISION "8" (Gov.)
Ap. 2739, Caracas 1010-A. ☏ +58 (2) 349571.
Cable: VTV. ③ 25401-25412.
Stations: Caracas (Central st): chA8 190/95kW.

Repeaters	chA	kW		ch	kW
Mérida	5	5/2.5	Valencia	11	151/75.5
Barquisimeto	7	107.5/53.5	Litoral	11	79/49.5
Maracaibo	8	157/78.5	Pto. La Cruz	11	23/11.5
Boconó	8	2/1	Cd. Bolivar	11	23/11.5
Cd. Piar	8	0.15	Vidoño	11	23/11.5
Anaco	9	12.5/6.25	Pto. Cabello	12	20/10

D.Prgr: 1600-0600 (approx.)

commercial stations

AMAVISION (Cult & Rlgs)
Calle Selesiano, Colegio Pio XI, Puerto Ayacucho, Amazonas. ☏ +58 (2) 987 6190.

D.Prgr: 2200-0200
Stations: Puerto Ayacucho ch7 6kW

CANAL 10
✉ Av. Francisco de Miranda, con Principal de los Ruices, Centro Empresarial Miranda PHD, Caracas. ☎ +58 (2) 239 8679. 📠 +58 (2) 239 7757.
Station: Caracas ch10.

CANAL METROPOLITANO DE TELEVISION (Comm.)
✉ Av. Circumvalacion El Sol, Centro Professional Santa Paula, Torre B, Piso 4, Santa Paula, Caracas. ☎ +58 (2) 987 6190. 📠 +58 (2) 985 4856
D. Prgr: 2100-0400
Stations: Caracas ch51

NCTV (Comm.)
✉ Urv. La Paz, Avenida 57 y Maracaibo, Maracaibo. ☎ +58 (61) 512662. 📠 +58 (61) 512729.
L.P: Dir: Gustavo Ocando Yamarte.
Station: Maracaibo chA11 108/54kW (est).
D.Prgr: 1600-0400.

OMNIVISION (Comm.)
✉ Calle Milan, Edif. Omnivision, Los Ruices Sur, Caracas. ☎ +58 (2) 256 3586/256 5011. 📠 +58 (2) 256 4482
D. Prgr: 24 hrs.
Stations: Anzoategui ch25 10kW; Aragua ch26 10kW; Bolivar ch26 52kW; Carabobo ch24 10kW; D. Federal ch12 605kW, ch13 10kW; Lara ch21 10kW; Tachira ch24 10kW; Zulia ch23 10kW.

RADIO CARACAS TELEVISION RCTV(Comm.)
✉ Ap. 2057, Caracas. ☎ +58 (2) 256 3696. 📠 +58 (2) 256 1812. ① 21527.
D.Prgrs: 24 hrs.
Stations: Anzoategui ch3 210kW, ch2 20kW; Carabobo ch7 264kW; Bolivar ch2 80kW, ch3 2kW; Carabobo ch10 22kW; D. Federal ch10 22kW, ch2 132kW, ch7 22kW; Falcon ch10 330kW; Lara ch4 300kW; Monagas ch10 50kW; Portuguesa ch2 12kW; Tachira ch10 14kW, ch7 75kW; Trujillo ch7 5kW; Zulia ch2 400kW.

TELEVISORA ANDINA DE MERIDA (Cult & Rlgs)
✉ Av. Bolivar, Calle 23 entre Av. 4-5, Merida 5101. ☎ +58 (74) 525 785. 📠 +58 (74) 520 098
D.Prgr:1400-0200
Stations: Merida ch6 20kW; Tachira ch3 33kW.

TELE BOCONO (Cult.)
✉ Calle 3, Qta. Caleuche, El Saman. Bocono. ☎ +58 (72) 521 27. 📠 +58 (72) 524 85
D. Prgr: 2100-0300
Stations: Trujillo ch13 4kW.

TELECARIBE (Comm.)
✉ Centro Banaven (Cubo Negro), Torre C, Piso 1, of C-12, Chuao, Caracas. ☎ +58 (2) 911 964/913 089.
D. Prgr: 1000-0400
Stations: Anzoategui ch9 50kW; Nueva Esparta ch12 30kW.

TELECENTRO (Comm.)
✉ Avenide Pedro León Torres, esquina de la calle 47, Edificio Telecentro, Barquisimeto, (3001) Lara. ☎ +58 (51) 460 917/4525 27. 📠 same as telepone.

L.P: Dir: Jorge Felix
D.Prgr: 1030-0400
Station: Lara ch11 100kW.

TV GUAYANA (Comm.)
✉ Puerto Ordaz, Bolivar. ☎ +58 (86) 2299 08
D.Prgr: 2100-0300
Stations: Bolivar ch12 125kW, ch13 80kW

TELESOL (Comm.)
✉ Calle Sucre no 15, Cumana, Sucre. ☎ +58 (93) 6620 59. 📠 +58 (93) 6627 75
D.Prgr: 1800-0200
Station: Sucre ch7 12kW

TELEVEN (Comm.)
✉ C.C. Los Chaguaramos, Caracas. ☎ +58 (2) 6617 511. 📠 +58 (2) 6625 300
D.Prgr: 24 hrs.
Stations: Anzoategui ch6 94kW; Carabobo ch13 360kW; Bolivar ch10 125kW, ch9 125kW; D. Federal ch10 139kW, ch21 188kW, ch13 10kW, ch6 23kW; Falcon ch3 111kW; Lara ch9 420kW; Miranda ch13 61kW; Nueva Esparta ch10 375kW; Tachira ch3 120kW; Zulia ch13 182kW.

TELEVISORA REGIONAL DEL TACHIRA (Comm.)
✉ Av. Libertador, edif. Servicios Unidos, Piso 3, San Cristobal, Tachira. ☎ +58 (76) 4473 66. 📠 +58 (76) 4652 77
D. Prgr: 2000-0200
Stations: Tachira ch6 144kW.

TELEVISORA DE ORIENTE, TVO (Comm.)
✉ Puerto la Cruz, Anzoategui. ☎ +58 (82) 6621 63
D.Prgr: 1800-0300
Station: Anzoategui ch5 50kW.

VENEVISION (Comm.)
✉ Av. La Salle, Edif, Venevision,Colinas de Los Caobos, Caracas. ☎ +58 (2) 782 0111/4444/4356/4267.
Cable: Ventel.
WWW: http://www.venevision.com
L.P: Pres: Carlos Bardasano. GM: Manuel Fraiz Grijalba. Eng. Mgr: German Landaeta. PR: Mariela Castio.
Stations: Caracas (Central st): chA4 132/66kW.
Repeaters: Trujillo ch3(6kW), Valencia ch4(20), Puerto Cabello ch4(20), Maracaibo ch4(336), Puerto Ordaz ch4(20), El Tigre ch4(12), Barquisimeto ch6(198), Barcelona ch7(25), C. Bolivar ch7(10), Maturin ch7(25), La Guaira ch9(21), Maracay ch9(476), Mérida ch9(16), San Cristobal ch9(40), Valera ch9(12), Anyarito ch9(10), Caricuao ch9(11), Curimagua ch12(40), Guaramacal ch12(13kW).
D.Prgr: 24 hrs.

ANTARCTICA

TV-sets: N/A — **Colour:** NTSC — **System:** M

AMERICAN FORCES ANTARCTIC NETWORK (AFAN McMurdo)
✉ "Operation Deep Freeze", Fleet Post Office, San Francisco, California 96692, USA.
D.Prgr: The US Navy Antarctic support group operates six cable TV channels. incl. occ. local prgrs on ch13.

CHAPTER 16
WORLD TV BY COUNTRY

Afghanistan 320	Czech Republic 286
Alaska 341	Denmark 287
Albania 283	Djibouti 314
Algeria 312	Dominica 345
Angola 312	Dominican Republic 345
Antarctica 358	Easter Island 338
Antigua & Barbuda 344	Ecuador 355
Argentina 350	Egypt 314
Armenia 320	El Salvador 346
Aruba 344	Equatorial Guinea 314
Australia 336	Estonia 288
Austria 283	Ethiopia 314
Azerbaijan 321	Falkland Islands 356
Azores 283	Faroe Islands 288
Bahamas 344	Fiji 338
Bahrain 321	Finland 288
Bangladesh 326	France 289
Barbados 344	Gabon 314
Belarus 283	Galapagos Islands 338
Belgium 284	Gambia 314
Belize 345	Georgia 327
Benin 312	Germany 292
Bermuda 341	Ghana.......................315
Bolivia 351	Gibraltar 294
Bosnia-Hercegovina 284	Greece 294
Botswana 312	Greenland 343
Brazil 351	Grenada 346
British Indian Ocean Territory 326	Guadeloupe 346
Brunei Darussalam 326	Guam 338
Bulgaria 284	Guatemala 346
Burkina Faso 313	Guinea (Rep.) 315
Burundi 313	Guyana (Rep.) 356
Cambodia 326	Haiti 346
Cameroon 313	Hawaii 338
Canada 341	Honduras 347
Canary Islands 313	Hong Kong 328
Central African Republic 313	Hungary 295
Chad 313	Iceland 295
Chile 354	India 328
China (P.R.) 326	Indonesia 329
China (Republic of) 327	Iran 322
Colombia 355	Iraq 322
Congo 313	Ireland 296
Cook Islands 338	Israel........................322
Costa Rica 345	Italy 296
Côte d'Ivoire 313	Jamaica 347
Croatia 285	Japan 329
Cuba 345	Jordan 323
Cyprus 321	Kazakhstan 330

Kenya	315
Kiribati	339
Korea (D.P.R.)	331
Korea (Rep.)	331
Kuwait	323
Kyrgyzstan	332
Laos	332
Latvia	297
Lebanon	323
Lesotho	315
Liberia	315
Libya	315
Luxembourg	298
Macau	332
Macedonia	298
Madagascar	315
Madeira	315
Malaysia	332
Maldives	333
Mali	316
Malta	298
Marshall Islands	339
Martinique	347
Mauritania	316
Mauritius	316
Mayotte	316
Mexico	347
Micronesia	339
Moldova	299
Monaco	299
Montserrat	348
Morocco	316
Mozambique	316
Namibia	316
Nauru	339
Netherlands	299
Neth. Antilles	348
New Caledonia	339
New Zealand	339
Nicaragua	348
Niger	317
Nigeria	317
Niue Island	340
Norfolk Island	340
Northern Marianas	340
Norway	299
Oman	323
Pakistan	333
Palau	340
Panama	348
Papua New Guinea	340
Paraguay	356
Peru	356
Philippines	334
Poland	301
Polynesia (Fr.)	340
Portugal	301
Puerto Rico	348
Qatar	324
Réunion	317
Romania	302
Russia	302
Samoa (Am.)	340
Samoa (We.)	341
San Marino	303
São Tomé e Principe	317
Saudi Arabia	324
Senegal	317
Seychelles	317
Sierra Leone	318
Singapore	334
Slovakia	303
Slovenia	303
Somalia	318
South Africa	318
Spain	304
Sri Lanka	334
St. Kitts & Nevis	349
St. Lucia	349
St. Pierre & Miquelon	343
St. Vincent	349
Sudan	319
Suriname	356
Swaziland	319
Sweden	307
Switzerland	308
Syrian Arab Republic	324
Tajikistan	335
Tanzania	319
Thailand	335
Togo	319
Tonga	341
Trinidad & Tobago	349
Tunisia	319
Turkey	325
Turkmenistan	335
Uganda	319
Ukraine	309
United Arab Emirates	325
United Kingdom	310
Uruguay	356
USA	343
Uzbekistan	335
Vanuatu	341
Venezuela	357
Vietnam	336
Virgin Isl. (Am.)	349
Virgin Isl. (Br.)	349
Wallis & Futuna	341
Yemen	325
Yugoslavia	311
Zaire	320
Zambia	320
Zimbabwe	320

Address: ✉
Telephone: ☎
Facsimile: 📄
Telex: ⌕

CHAPTER 17
GLOSSARY

Actuator, Also referred to as black jack or jack. This is the outside device, powered and controlled by the indoor positioner, that actually moves the dish eastward and westward.
Aperture, The total reflective area of a satellite dish.
Aperture blockage, The blockage of the antenna's reflective area, often caused by the LNB casting a shadow.
Apex, Highest point.
Apogee, The orbital point at which the satellite is furthest away from the earth.
Apogee Kick Motor (AKM), A small jet engine, fired when the spacecraft has reached its apogee, that propels the satellite to its final geostationary position.
Arc, A bow or curve.
Atmospheric distortion, Certain weather related circumstances causing abnormal signal propagation characteristics.
Audio subcarrier, An audio signal transmitted within the bandwidth of a wider transponder signal.
Automatic Frequency Control (AFC), A circuit within the satellite receiver, preventing the receiver drifting of a chosen frequency.
Azimuth, A compass bearing in degrees, clockwise from true north.
Bandpass filter, A circuit within the satellite receiver, allowing only a specific range of frequencies to pass.
Bandwidth, The total range of frequencies over which the signal may vary in the course of its transmission.
Baseband, The crude, unmodulated audio and video signal, used by most decoders.
C-Band, The frequency range from 3700-4200 MHz (or 3.7-4.2 GHz.).
Carrier-to-Noise Ratio (C/N Ratio), The C/N Ratio is derived from a number of interrelated parameters and indicates how well the outside unit of a satellite system performs. A poor C/N Ratio (below the receiver's threshold) will cause sparklies.
Cassegrain, Dual reflector antenna using a convex hyperbolic subreflector and a parabolic main reflector.
Center Focus Antenna, Also referred to as prime focus antenna. This antenna has a perfectly round circumference, but the parabola can be deep or shallow.

Centrifugal force, The force which causes an object spinning round a center to tend to fly off.
Clarke Belt, The circular orbit at 35.800 Km. (22.247 Mls.) above the Equator, named after its discoverer Arthur C. Clarke, in which satellites appear to hover stationary.
Co-located, Also often mistakenly referred to as collocated. Two or more satellites put at the same orbital location (e.g. ASTRA 1A/B/C at 19.2° East).
Communications Payload, All receiving and transmitting equipment aboard a satellite.
Composite video, The complete video signal comprised of chrominance, luminance and sync information.
Declination angle, Also often referred to as declination offset. This is the angle in between the polar axis and the dish used to aim accurately at the Clarke Belt.
De-emphasis, A method to reduce noise.
Demodulator, A circuit within the satellite receiver, extracting the baseband signal from the carrier wave that was used for transmission.
Dielectric,
An insulating material used to keep the distance between the center conductor of a coaxial cable and the cable shield to a close tolerance.
Digital Satellite Radio (DSR), A method used by some radio broadcasters in Europe allowing the transmission of up to 16 digital radio stations on a single transponder.
Direct Broadcasting Satellites (DBS), Very powerful Ku-Band satellites transmitting only four or five channels directly to the end users. The DBS frequency range is 11.7-12.5 GHz. New American DBS (or BSS) satellites (e.g. Hughes' DirecTv) are able to broadcast up to 180 digital tv channels simultaneously. They broadcast in the 12.2-12.7 GHz frequency range.
Downconverter, Also referred to as front end. A downconverter is a circuit within a satellite receiver that converts a high frequency to a lower, intermediate frequency.
Downlink, The satellite-to-earth signal path.

Digital Television Broadcasting (DTB), Possible future global television standard.
Effective Isotropic Radiated Power (EIRP), A measure to indicate the transmitted signal strength of a satellite.
Elevation, The vertical angle of a satellite dish aiming up to a satellite.
Elliptical, Oval shaped.
Equatorial Orbit, Elliptical or circular orbit in the plane of the Equator.
EuroCrypt, D2-MAC encryption method, used in Europe.
F-connector, This type of connector is the industry standard used for attaching coaxial cables to the LNB, satellite tuner and in some countries (a.o. U.S.) the television set.
Focal length-to-Diameter Ratio (F/D Ratio), The relation between dish depth to the focal point. Shallow dishes have a high F/D Ratio, deep dishes a low F/D Ratio.
Feedhorn, The receptive part collecting signals reflected from an antenna's surface. It is located within the focus of an antenna.
Fish-bone pattern, FM distortion caused by two tv channels interfering with each other.
Flat Plate Antenna, Also often referred to as Squarial™. Special antenna configuration using a large number receptive elements rather than a dish reflector.
Focal point, Meeting point of rays.
Footprint, A geographical coverage area illuminated by the satellite's downlink antenna.
Front End, See downconverter.
Gain-to-(noise)Temperature Ratio (G/T Ratio), The measure used to indicate ability of the outside unit (antenna + LNB) to lift the reflected signal level (gain) above a certain noise level. The higher the G/T Ratio the better the performance.
Geostationary Orbit, See Clarke Belt.
Geostationary Transfer Orbit (GTO), Temporary elliptical orbit of a satellite directly after its launch, prior to its final geostationary orbit.
Geosynchronous Orbit, See Clarke Belt.
Gravity, Force of attraction between any two objects, especially the force which attracts objects towards the center of the earth.
Gregory, Dual reflector antenna using a concave hyperbolic subreflector and a parabolic main reflector.
Half-transponder, A broadcasting technique used to relay two tv programs via one transponder (usually with a 72 MHz bandwidth), by splitting this transponder into two (36 MHz wide) channels. As a result, the EIRP in half-transponder mode is reduced by half (i.e. minus 3 dB) as well.
HDTV, High Definition Television. A future video standard utilizing approx. 1200 scanning lines and a 16:9 screen format.
HEMT, High Electron Mobility Transistor. A transistor used in LNB's or other equipment requiring very low-noise components.
IC, Integrated Circuit. An electrical component used in modern equipment.
IF (Intermediate Frequency), A middle range frequency generated after downconversion, to be used for processing and filtering.
Inclined Orbit, An orbit offset in degrees from the plane of the Equator.
Inclinometer, An instrument used to measure the various angles necessary to accurately aim a satellite dish.
Interference, An undesired signal causing video and/or audio distortion.
Ionosphere, Also known as the Heaviside Layer. Set of layers of the earth's atmosphere, which reflect radio waves and cause them to follow the earth's contour.
Jack, See actuator.
Ku-Band, The frequency range from 10.7-12.75 GHz.
LNB, Also sometimes referred to as LNA or LNC. The outside device that receives ,amplifies and downconverts the weak satellite signal, reflected by the the dish and captured by the feedhorn.
Low Pass Filter (LPF), A filter that allows only one side of a sinus wave to pass.
MAC (Multiplexed Analogue Components), Broadcast transmission standard utilizing analog picture and digital audio components. Most commonly used variants are B-MAC, D-MAC and D2-MAC.
Main lobe, Signals reflected by the dish, coming directly from the center of the reflector.
Moire effect, The interference of chrominance and luminance, causing odd color patterning.
Mount, The structure that supports the satellite dish and allows accurate aiming towards a satellite. The most commonly used variants are Az/El mount and Polar mount.
MPEG-2, A digital video compression technique and possible global digital video standard.
Multi Focus Antenna, Spherical antenna configuration with multiple focal points, intended to receive various satellites simultaneously.
NTSC, US and Japanese tv broadcasting standard utilizing 525 scanning lines.
Offset Antenna, Widely used antenna configuration, usually oval shaped, in

which the feedhorn assembly appears offset from the reflector's center.
Orthogonal-Mode Transducer (OMT), Also referred to as Ortho-Coupler. An outside device which allows two opposite polarizations to be received simultaneously by two LNB's mounted in a 90° angle to eachother.
Orbit, The path followed by a heavenly body, e.g. a satellite, round another body, e.g. the earth.
PAL, European tv broadcasting standard utilizing 625 scanning lines.
Parabola, A plane curve formed by cutting a cone on a plane parallel to its side, in a way that the two arms get farther away from each other.
Phase Locked Loop (PLL), Electronic circuitry within a satellite receiver, used to eliminate frequency drifting.
Polar Orbit, An orbit perpendicular to the plane of the Equator.
Polarization, The plane of vibration of the electrical field of a signal radiated from a transmitting antenna. Signals can be horizontally, vertically, left-hand circular and right-hand circular polarized.
Polarizer, Outside devise in the wave guide between the feedhorn and the LNB, which allows receivers to pick up only one polarity.
Positioner, Indoor unit to control and power the ourdoor actuator. Nowadays often integrated in multi-satellite receivers.
Pre-emphasis, A method used to increase components of an FM signal in the higher frequencies before transmission. Used in conjunction with the proper amount de-emphasis at the receiving side, this method will reduce the higher noise level inherent in FM transmissions.
Prime Focus Antenna, See center focus antenna.
Radio Frequency (RF), Frequency range from approx. 10 kHz-100 GHz used for all kinds of communication. The satellite receiver's RF modulator generates a UHF frequency ranging from approx. 30-39 MHz.
Reflector, The dish-shaped reflective part of the outdoor unit.
SCART connector, Also known as Euro connector or Peritel connector. This connector was developed by French engineers and is currently the standard audio/video connector in Europe.
SCPC, Single Channel Per Carrier. A transmission system that uses a separate transponder for each channel. Widely used in the US for transmission of digital radio packages.
Scientific satellites, Non commercial satellites used for scientific purposes.
SECAM, French tv broadcasting standard, also used by some (former) communist countries.
Side lobe, Satellite signals, reflected by the dish, coming from an off-axis angle.
Signal-to-Noise Ratio (S/N Ratio), The ratio of signal power to noise level in a specific bandwidth, usually expressed in decibels.
Skew, A method used to fine adjust the feedhorn polarity detector (probe), by tilting it to lie in the plane of the incoming signal.
Slot, Place on the Clarke Belt with geographical coordinates, for one or more satellites.
SMATV, Satellite Master Antenna TV, or cable headend station.
Satellite News Gathering Unit (SNG Unit), Mobile uplink unit used to uplink tv (news) broadcasts to a satellite.
Sparklies, Also known as spikes or more technically "impulse noise". Black and white streaks in the picture caused by an insufficient C/N Ratio. Solution: bigger dish!!
Spherical Antenna, See multi focus antrenna.
Stationkeeping, The process of keeping a satellite in its correct position.
Threshold, The minimal C/N level required by a satellite receiver to deliver a sparkly free picture, measured in decibels.
Transponder, This is a compound word put together out of two words: trans(mitter) and (res)ponder. A satellite repeater, which receives, amplifies, downconverts and retransmits signals.
TVRO (TV receive only), A satellite system that can only receive, but not transmit signals.
Ultra High Frequency (UHF), The frequency range from 300-3000 MHz.
Uplink, The earth-to-satellite signal path.
Utility satellites, Group of commercial and non-commercial satellites used for telecommunications, radio, tv, weather monitoring or navigation purposes.
Very High Frequency (VHF), The frequency range from 54-216 MHz.
VideoCrypt, An encryption system widely used in Europe, e.g. the Sky Multi-Channels on the ASTRA satellites, which requires a viewing card (Smartcard).
Voltage Tuned Oscillator (VTO), An electronic circuit within the satellite receiver, which output oscillator frequency is controlled by voltage. Widely used in satellite receivers to tune in to the various transponders.
Wave Guide, A metal pipe of a specific diameter, used in the feed assembly to conduct microwave signals.`

CHAPTER 18
WORLD TIME TABLE

The differences marked + indicate the number of hours ahead of UTC. Differences marked – indicate the number of hours behind UTC. Variations from standard time during part of the year (in some countries referred to as Summer Time) are decided annually and may vary from year to year. N=Normal Time; S=Summer Time.

	N	S		N	S
Afghanistan	+4½	+4½	Burkina Faso	UTC	UTC
Alaska	–9	–8	Burundi	+2	+2
	–10	–9	Cameroon	+1	+1
Albania	+1	+2	Canada		
Algeria	+1	+1	a) NF, Labrador	–3½	–2½
Andorra	+1	+2	(So. Ea.)		
Angola	+1	+1	b) Labrador (rest),	–4	–3
Anguilla	–4	–4	NS, NB, PEI		
Antigua	–4	–4	c) ON, PQ	–5	–4
Argentina (Ea.)	–3	–2	d) MB	–6	–5
Argentina (rest)	–3	–3	e) AB, NWT	–7	–6
Armenia	+4	+4	f) BC, YT	–8	–7
Aruba	–4	–4	Cambodia	+7	+7
Ascension Isl.	UTC	UTC	Canary Isl.	UTC	+1
Australia			Cape Verde Isl.	–1	–1
Victoria & NSW	+10	+11	Cayman Isl.	–5	–4
Queensland	+10	+10	Ce. African Rep.	+1	+1
Tasmania	+10	+11	Chad	+1	+1
N. Territory	+9½	+9½	Chile	–4	–3
S. Australia	+9½	+10½	China (P.R.)		
(we. part)	+9	+10	Beijing	+8	+9
W. Australia	+8	+8	Urumqi	+6	+7
Austria	+1	+2	Christmas Isl.	+7	+7
Azerbaijan	+3	+4	Cocos Isl.	+6½	+6½
Azores	–1	UTC	Colombia	–5	–5
Bahamas	–5	–4	Comoro Rep.	+3	+3
Bahrain	+3	+3	Congo	+1	+1
Bangladesh	+6	+6	Cook Isl.	–10	–9½
Barbados	–4	–4	Costa Rica	–6	–5
Belarus	+2	+3	Côte d'Ivoire	UTC	UTC
Belgium	+1	+2	Croatia	+1	+2
Belize	–6	–6	Cuba	–5	–4
Benin	+1	+1	Cyprus	+2	+3
Bermuda	–4	–3	Czech Rep.	+1	+2
Bhutan	+6	+6	Denmark	+1	+2
Bolivia	–4	–4	Diego Garcia	+5	+5
Bosnia/Hercegovina	+1	+2	Djibouti	+3	+3
Botswana	+2	+2	Dominica	–4	–4
Brazil			Dom. Rep.	–4	–4
a) Oceanic Isl	–2	–2	Easter Isl.	–6	–5
b) Ea & Coastal	–3	–2	Ecuador	–5	–5
c) Manaos	–4	–3	Egypt	+2	+2
d) Acre	–5	–4	El Salvador	–6	–6
Brunei	+8	+8	Equatorial Guinea	+1	+1
Bulgaria	+2	+3	Estonia	+2	+3

	N	S		N	S
Ethiopia	+3	+3	Laos	+7	+7
Falkland Isl.	−4	−4	Latvia	+2	+3
(Port Stanley)	−4	−3	Lebanon	+2	+3
Faroe Isl.	UTC	+1	Lesotho	+2	+2
Fiji	+12	+12	Liberia	UTC	UTC
Finland	+2	+3	Libya	+1	+2
France	+1	+2	Lithuania	+2	+3
Gabon	+1	+1	Lord Howe Isl.	+10 1/2	+11
Galapagos Isl.	−6	−6	Luxembourg	+1	+2
Gambia	UTC	UTC	Macau	+8	+8
Georgia*	+4	+5	Macedonia	+1	+2
(* exc. Abkhasia	+3	+4)	Madagascar	+3	+3
Germany	+1	+2	Madeira	UTC	UTC
Ghana	UTC	UTC	Malawi	+2	+2
Gibraltar	+1	+2	Malaysia	+8	+8
Greece	+2	+3	Maldive Isl.	+5	+5
Greenland			Mali	UTC	UTC
Scoresbysund	−1	UTC	Malta	+1	+2
Thule area	−3	−3	Marshall Isl.	+12	+12
Other areas	−3	−2	Martinique	−4	−4
Grenada	−4	−4	Mauritania	UTC	UTC
Guadeloupe	−4	−4	Mauritius	+4	+4
Guam	+10	+10	Mayotte	+3	+3
Guatemala	−6	−5	Mexico	−6	−6
Guiana (French)	−3	−3	Micronesia		
Guinea (Rep.)	UTC	UTC	Truk, Yap	+10	+10
Guinea Bissau	UTC	UTC	Pohnpei	+11	+11
Guyana (Rep.)	−3	−3	Midway Isl.	−11	−11
Haiti	−5	−4	Moldova	+2	+3
Hawaii	−10	−10	Monaco	+1	+2
Honduras (Rep.)	−6	−6	Mongolia	+8	+9
Hong Kong	+8	+8	Montserrat	−4	−4
Hungary	+1	+2	Morocco	UTC	UTC
Iceland	UTC	UTC	Mozambique	+2	+2
India	+5 1/2	+5 1/2	Myanmar	+6 1/2	+6 1/2
Indonesia			Namibia	+1	+2
a) Java, Bali,			Nauru	+11	+11
Sumatra	+7	+7	Nepal	+5.45	+5.45
b) Kalimantan,			Netherlands	+1	+2
Sulawesi			Neth. Antilles	−4	−4
Timor	+8	+8	New Caledonia	+11	+11
c) Moluccas,			New Zealand	+12	+13
We. Irian	+9	+9	Nicaragua	−6	−6
Iran	+3 1/2	+4 1/2	Niger	+1	+1
Iraq	+3	+4	Nigeria	+1	+1
Ireland	UTC	+1	Niue	−11	−11
Israel	+2	+3	Norfolk Isl.	+11 1/2	+11 1/2
Italy	+1	+2	N. Marianas	+10	+10
Jamaica	−5	−4	Norway	+1	+2
Japan	+9	+9	Oman	+4	+4
Jordan	+2	+3	Pakistan	+5	+5
Kenya	+3	+3	Palau	+9	+9
Kazakhstan	+6	+7	Panama	−5	−5
Kiribati	+12	+12	Papua N. Guinea	+10	+10
Korea (Rep.)	+9	+10	Paraguay	−4	−3
Korea (D.P.R.)	+9	+9	Peru	−5	−4
	N	S	Philippines	+8	+8
Kuwait	+3	+3	Poland	+1	+2
Kyrgyzstan	+5	+6	Polynesia (Fr.)	−10	−10

	N	S		N	S
Portugal	UTC	+1	Tanzania	+3	+3
Puerto Rico	−4	−4	Thailand	+7	+7
Qatar	+3	+3	Togo	UTC	UTC
Reunion	+4	+4	Tonga	+13	+13
Romania	+2	+3	Transkei	+2	+2
Russia			Trinidad	−4	−4
Moscow	+3	+4	Tristan da Cunha	UTC	UTC
Novosibirsk	+7	+8	Tunisia	+1	+2
Khabarovsk	+10	+11	Turks & Caicos	−4	−4
Petropavlovsk	+12	+13	Turkey	+2	+3
	N	S	Turkmenistan	+5	+5
Rwanda	+2	+2	Tuvalu	+12	+12
Samoa Isl.	−11	−11	Uganda	+3	+3
S. Tomé	UTC	UTC	Ukraine	+2	+3
Saudi Arabia	+3	+3	United Arab Em.	+4	+4
Senegal	UTC	UTC	United Kingdom	UTC	+1
Seychelles	+4	+4	Uruguay	−3	−2
Sierra Leone	UTC	UTC	USA		
Singapore	+8	+8	a) Eastern*	−5	−4
Slovakia	+1	+2	*) Indiana	−5	−5
Slovenia	+1	+2	b) Central	−6	−5
Solomon Isl.	+11	+11	c) Mountain*	−7	−6
Somalia	+3	+3	*) Arizona	−7	−7
S. Africa	+2	+2	d) Pacific	−8	−7
Spain	+1	+2	Uzbekistan	+5	+5
Sri Lanka	+5½	+5½	Vanuatu	+11	+12
St. Helena	UTC	UTC	Venezuela	−4	−4
St. Kitts–Nevis	−4	−4	Vietnam	+7	+7
St. Lucia	−4	−4	Virgin Isl.	−4	−4
St. Pierre	−3	−2	Wake Isl.	+12	+12
St. Vincent	−4	−4	Wallis & Futuna	+12	+12
Sudan	+2	+2	Yemen	+3	+3
Suriname	−3	−3	Yugoslavia	+1	+2
Swaziland	+2	+2	Zaire		
Sweden	+1	+2	Kinshasa	+1	+1
Switzerland	+1	+2	Lubumbashi	+2	+2
Syria	+2	+3	Zambia	+2	+2
Tajikistan	+5	+5	Zimbabwe	+2	+2
Taiwan	+8	+8			

WHERE TO OBTAIN THE SATELLITE AND TV HANDBOOK

Australia
Dick Smith Electronics (Pty) Ltd., P.O. Box 321, North Ryde, NSW 2113
Bookwise, 54 Crickenden Rd, Findon 5023, South Australia

Austria
ERB Verlag, Eichenstr. 38, A-1120 Vienna

Finland
Tietoteos, Yläportti 1A, SF-02211 Espoo 21

France
Brentano's, 37 Avenue de l'Opera, F-75002 Paris

Germany
Walter Braun Verlag, Mercatorstr. 2, 47051 Duisburg 1

Greece
Librairie Cacoulides, Bld. Panepistimiou 25-29 GR 105 64 Athens

Hong Kong
Swindon Book Co. Ltd., 13-15 Lock Road, Tsim Sha Tsui, Kowloon

Israel
Steimatzky Ltd., P.O. Box 1444, Bnei Brak 51114

Italy
A.C. Distribuzione, Via Kramer 31, 20129 Milano

Netherlands
De Muiderkring B.V., P.O. Box 313, 1380 AH Weesp

New Zealand
Burnet Pollard Books, P.O. Box 149, Otaki 6471

Norway
BBV-Kjop, P.O. Box 88, N-1851 Mysen

South Africa
Technical Books (Pty) Ltd., P.O. 2866, Cape Town 8000
Verbatim Distributors, P.O. Box 190, Steenburg 7947

Spain
Mercambo S.A., Gran Vie de les Corts Catalanes 594, E-08007 Barcelona
Diaz de Santos S.A., Books Department, Legasca 95 E-28006 Madrid

Sweden
Radex, Box 726, S-251 07 Helsingborg

Switzerland
Thali A.G., CH-6285 Hitzkirch

Canada
General Publishing, 30 Lesmill Road, Don Mills, Ontario M3B 2T6

U.K. & Ireland
Windsor Books International, The Boundary, Wheatley Road
Garsington, Oxford OX44 9EJ

United States
Available at better Book Stores and Electronic Dealers everywhere

Far East, Southeast and Central Asia
Rotovision, SA, Route Suisse 9, CH-1295 Mies, Switzerland

Ask your local bookseller or electronic shop to order a copy for you.

CREDITS

The Satellite and TV Handbook, Fourth Edition 1997, was written and produced by Bart Kuperus (Managing Editor World Radio TV Handbook).

Special Thanks go to Andrew G. Sennitt (World Television), Richard R. Peterson (US DBS, Frequently Asked Questions List), Erik Nijhof (Netherlands), Jonathan Marks (Netherlands) and my wife, Liesbeth, for their help and support.